Industrial Oil Crops

Industrial Oil Crops

Editors

Thomas A. McKeon
United States Department of Agriculture, Agricultural Research Service
Western Regional Research Center
Albany, CA, USA

Douglas G. Hayes
Department of Biosystems Engineering and Soil Science
University of Tennessee
Knoxville, TN, USA

David F. Hildebrand
Department of Plant and Soil Sciences
University of Kentucky
Lexington, KY, USA

Randall J. Weselake
Department of Agricultural, Food and Nutritional Sciences
University of Alberta
Edmonton, AB, Canada

ELSEVIER

AMSTERDAM • BOSTON • HEIDELBERG • LONDON
NEW YORK • OXFORD • PARIS • SAN DIEGO
SAN FRANCISCO • SINGAPORE • SYDNEY • TOKYO

Academic Press is an imprint of Elsevier

AOCS PRESS

Academic Press and AOCS Press
Academic Press is an imprint of Elsevier
125 London Wall, London EC2Y 5AS, UK
525 B Street, Suite 1800, San Diego, CA 92101-4495, USA
50 Hampshire Street, 5th Floor, Cambridge, MA 02139, USA
The Boulevard, Langford Lane, Kidlington, Oxford OX5 1GB, UK

Published in cooperation with American Oil Chemists' Society www.aocs.org.
Director, Content Development: Janet Brown

Notices
Knowledge and best practice in this field are constantly changing. As new research and experience broaden our understanding, changes in research methods, professional practices, or medical treatment may become necessary.

Practitioners and researchers must always rely on their own experience and knowledge in evaluating and using any information, methods, compounds, or experiments described herein. In using such information or methods they should be mindful of their own safety and the safety of others, including parties for whom they have a professional responsibility.

To the fullest extent of the law, neither the Publisher nor the authors, contributors, or editors, assume any liability for any injury and/or damage to persons or property as a matter of products liability, negligence or otherwise, or from any use or operation of any methods, products, instructions, or ideas contained in the material herein.

British Library Cataloguing-in-Publication Data
A catalogue record for this book is available from the British Library

Library of Congress Cataloging-in-Publication Data
A catalog record for this book is available from the Library of Congress

ISBN: 978-1-893997-98-1

Publisher: Nikki Levy
Acquisition Editor: Nancy Maragioglio
Editorial Project Manager: Karen Miller
Production Project Manager: Caroline Johnson
Designer: Victoria Pearson

Typeset by TNQ Books and Journals
www.tnq.co.in

Contents

3. Polymeric Products Derived From Industrial Oils for Paints, Coatings, and Other Applications

Douglas G. Hayes and Marie-Josée Dumont

4. Castor (*Ricinus communis* L.)

Thomas A. McKeon

9. Coconut (*Cocos nucifera*)

Laura J. Pham

10. Tung (*Vernicia fordii* and *Vernicia montana*)

Jay Shockey, Timothy Rinehart, Yicun Chen, Wang Yangdong, Zhan Zhiyong and Hu Lisong

11. Emerging Industrial Oil Crops

Thomas A. McKeon

List of Contributors

Maythem AL–Amery Department of Plant and Soil Sciences, University of Kentucky, Lexington, KY, United States

Helen Booker Crop Development Centre, Department of Plant Sciences, University of Saskatchewan, Saskatoon, SK, Canada

Karen L. Caswell National Research Council, Saskatoon, SK, Canada

Guanqun Chen Department of Biological Sciences, University of Manitoba, Winnipeg, Manitoba, Canada

Yicun Chen Institute of Subtropical Forestry, Chinese Academy of Forestry, Fuyang, Zhejiang, China

Jonathan Curtis Agricultural Lipid Biotechnology Program, Department of Agricultural, Food and Nutritional Sciences, University of Alberta, Edmonton, AB, Canada

Marie-Josée Dumont Department of Bioresource Engineering, McGill University, Ste-Anne-de-Bellevue, QC, Canada

Alison M.R. Ferrie National Research Council, Saskatoon, SK, Canada

Hirotada Fukushige Department of Plant and Soil Sciences, University of Kentucky, Lexington, KY, United States

Linda M. Hall Department of Agricultural, Food and Nutritional Science, University of Alberta, Edmonton, AB, Canada

Douglas G. Hayes Department of Biosystems Engineering and Soil Science, University of Tennessee, Knoxville, TN, United States

Andrew Hebard Technology Crops International, Winston-Salem, NC, United States

David F. Hildebrand Department of Plant and Soil Sciences, University of Kentucky, Lexington, KY, United States

Amit J. Jhala Department of Agricultural, Food and Nutritional Science, University of Alberta, Edmonton, AB, Canada; Department of Agronomy and Horticulture, University of Nebraska-Lincoln, Lincoln, NE, United States

Gerhard Knothe U.S. Department of Agriculture, Agricultural Research Service, National Center for Agricultural Utilization Research, Peoria, IL, United States

Igor Kovalchuk University of Lethbridge, Lethbridge, AB, Canada

Hu Lisong Institute of Subtropical Forestry, Chinese Academy of Forestry, Fuyang, Zhejiang, China

Priti Maheshwari Lethbridge Research Station, Lethbridge, AB, Canada

Thomas A. McKeon United States Department of Agriculture, Agricultural Research Service, Western Regional Research Center, Albany, CA, United States

Peter B.E. McVetty Department of Plant Science, University of Manitoba, Winnipeg, MB, Canada

Elzbieta Mietkiewska Agricultural Lipid Biotechnology Program, Department of Agricultural, Food and Nutritional Sciences, University of Alberta, Edmonton, AB, Canada

Eric J. Murphy Department of Basic Sciences, School of Medicine and Health Sciences, University of North Dakota, Grand Forks, ND, United States; Agragen, LLC, Cincinnati, OH, United States; Agragen OY, Helsinki, Finland

Tolibjon Omonov Agricultural Lipid Biotechnology Program, Department of Agricultural, Food and Nutritional Sciences, University of Alberta, Edmonton, AB, Canada

Shreya Patel Department of Plant and Soil Sciences, University of Kentucky, Lexington, KY, United States

Laura J. Pham Oils and Fats Laboratory, National Institute of Molecular Biology and Biotechnology (BIOTECH), University of the Philippines at Los Baños, Laguna, Philippines

Tim Phillips Department of Plant and Soil Sciences, University of Kentucky, Lexington, KY, United States

Timothy Rinehart United States Department of Agriculture-Agricultural Research Service, Thad Cochran Southern Horticultural Laboratory, Poplarville, MS, United States

William Serson Department of Plant and Soil Sciences, University of Kentucky, Lexington, KY, United States

Jay Shockey United States Department of Agriculture-Agricultural Research Service, Southern Regional Research Center, New Orleans, LA, United States

Rodrigo M.P. Siloto Department of Agricultural, Food and Nutritional Science, University of Alberta, Edmonton, AB, Canada

David C. Taylor National Research Council of Canada, Saskatoon, SK, Canada

Randall J. Weselake Department of Agricultural, Food and Nutritional Sciences, University of Alberta, Edmonton, AB, Canada

Wang Yangdong Institute of Subtropical Forestry, Chinese Academy of Forestry, Fuyang, Zhejiang, China

Zhan Zhiyong Institute of Subtropical Forestry, Chinese Academy of Forestry, Fuyang, Zhejiang, China

Li-Hua Zhu Department of Plant Breeding, Swedish University of Agricultural Sciences, Alnarp, Sweden

Editor Biographies

Thomas A. McKeon is a research chemist at the Western Regional Research Center of the U.S. Department of Agriculture, Agricultural Research Service (USDA-ARS WRRC) in Albany, CA. He received his bachelor's of science in chemistry from Worcester Polytechnic Institute and PhD in biochemistry from the University of California at Berkeley. He was a postdoctoral associate in plant lipid enzymology in the Department of Biochemistry and Biophysics at UC Davis and visiting scientist in postharvest biochemistry and ethylene regulation in The Department of Vegetable Crops at UC Davis. He started his career at WRRC as a research chemist in 1981, initially in the area of postharvest biochemistry and starting research on castor oil biosynthesis and ricin detection in 1992. He has published 120 papers in peer-reviewed journals and book chapters. Tom has served as a division officer for the American Oil Chemists Society (AOCS) Biotechnology (BIO) Division from 2004 to 2010 and continues to serve the BIO Board on special assignment. He has served on the AOCS Books and Special Publications Committee and on the AOCS Board of Directors (2008–2010). He also served on the secretariat of the US-Japan Natural Resources (UJNR) Food and Agriculture Panel from 2006 to 2015. Tom is an editor for *Biocatalysis and Agricultural Biotechnology* and board member for the International Society for Biocatalysis and Agricultural Biotechnology (ISBAB). His current research interests include development of technologies to improve the processing of castor seed and elucidation of enzymology that regulates isoprenoid synthesis in plants.

Douglas G. Hayes is a professor of biosystems engineering at the University of Tennessee (UT). Doug received his bachelor's of science and PhD in chemical engineering at Iowa State University (1986) and the University of Michigan (1991), respectively, and served as a postdoctoral research chemist at the USDA/ARS/NCAUR (Peoria, IL) from 1991 to 1994. Doug served as a faculty member in the Department of Chemical and Materials Engineering at the University of Alabama in Huntsville before joining UT in 2004. Doug also holds an adjunct professorship in the UT Department of Chemical and Biomolecular Engineering and guest professorships at Wuhan Polytechnic University and Jinan University in China and is a UT–Oak Ridge National Laboratory Joint Faculty. Doug is a senior associate editor (SAE) for the *Journal of the American Oil Chemists' Society* and an AE for *Journal of Surfactants and Detergents*. Doug has served AOCS as an officer for the Biotechnology Division (Chair from 2014–2015), chair of the AOCS Professional Educators' Common Interest

Group, and a member of the AOCS Books and Special Publications Committee. Doug has published over 65 articles in peer-reviewed publications, 15 book chapters, and one edited book. Doug's research interests include applied biocatalysis, biobased surfactants and polymers, soft matter colloidal systems, and small-angle scattering methods.

David Floyd Hildebrand is a professor of plant & soil sciences at the University of Kentucky, Lexington, KY. David earned a BS in chemistry from the University of Maryland in 1977 and a second BS in agronomy, also from the University of Maryland. David earned an MS in plant breeding and genetics from the University of Illinois in 1980 and a PhD in plant breeding and genetics, also from the University of Illinois in 1982. David worked at the Kyoto University Research Institute for Food Science under a Japanese Society for the Promotion of Science Fellowship. David's research focuses on oil biosynthesis, breeding, and metabolic engineering of plants for improved edible and industrial uses. He has authored or co-authored more than 150 publications and 10 patents. David serves on several editorial boards including the *JAOCS* and *Biocatalysis and Agricultural Biotechnology* and served as the chairperson of the AOCS biotechnology division. David's recent and current teaching is a graduate-level plant biochemistry course, http://www.uky.edu/~dhild/biochem/welcome.html, and an introductory course on Food, Agriculture, Energy and the Environment.

Randall J. Weselake is a professor and Tier I Canada Research Chair in Agricultural Lipid Biotechnology in the Department of Agricultural, Food and Nutritional Science at the University of Alberta (Edmonton, Alberta, Canada). Since 1987, when he was a research associate at the National Research Council of Canada in Saskatoon, Randall's research has focused on the biochemistry and molecular biology of storage lipid metabolism in *Brassica napus* and other oil crops. Recently, he served as scientific director of the Alberta Innovates Phytola Center, which specializes in oilseed innovations, including research on the development of industrial oil crops. From 2007 to 2013, Randall was leader of the "Bioactive Oils Program" funded by AVAC Ltd., and from 2006 to 2011, he was co-leader of the Genome Canada/Genome Alberta project "Designing Oilseeds for Tomorrow's Markets". From 1989 to 2004, Randall was with the Department of Chemistry and Biochemistry at the University of Lethbridge (Alberta, Canada), serving as chair from 1996 to 1999. His doctoral research in plant biochemistry was conducted at the University of Manitoba and Grain Research Laboratory of the Canadian Grain Commission. Randall is joint editor-in-chief of the *American Oil Chemists' Society (AOCS) Lipid Library*, associate editor for *Lipids*, and editor for *Biocatalysis and Agricultural Biotechnology*. Randall has published more than 160 papers in refereed journals and books and is editor of the book *Teaching Innovations in Lipid Science*, which was published by the Taylor & Francis Group of the CRC Press and the AOCS Press. He is a fellow of both the *AOCS* and the *International Society of Biocatalysis and Agricultural Biotechnology.*

Preface

While seed oils and other vegetable oils are mostly thought of as food ingredients, many of these oils are also used for nonfood industrial applications. This book describes plants that produce oils that are chiefly used for industrial purposes and provides relevant information about the seeds and vegetative parts that produce oils and their uses. The initial chapters focus on key applications of industrial oils which include biodiesel polymers, paints, and coatings. Subsequent chapters deal with long-established crops, and entire chapters are devoted to these. Over the years, many plants that produce oils well suited to industrial use have been identified. Often these were wild or decorative species, referred to as "new crops," which are not well suited to large-scale production or cultivation without considerable agronomic improvement. We have identified and described the more prominent new crops in the subchapters collected in the chapter entitled "Emerging Oil Crops." This chapter is followed by a case study on the economic and marketing strategies underlying successful introduction of a new crop, in this case, the Ahiflower™.

In addition to describing the properties and uses of industrial oil crops, many chapters include information about attempts to genetically engineer the crop or to transfer genes from the industrial crop to a more manageable transgenic crop with more favorable agronomic characteristics. Some of the novel technologies developed for oil crop modification are more closely examined and explained in chapters specifically dedicated to these topics. The application of these and related technologies are brought to fruition in the final chapter, which is devoted to the development of crops that produce oil throughout the vegetative portion of the plant.

This volume went through several evolutionary steps. We would like to thank our current management team, Janet Brown and Lori Weidert of AOCS Press as well as Caroline Johnson of Elsevier, who has shepherded the final version for AOCS in conjunction with Elsevier Press. We would also like to thank the AOCS Press and the AOCS Books and Special Publications Committee, especially Magdi Mossoba, former Committee Chair for continuing patience and support from the early stages of the book's development. We hope this volume serves as a valued source of information in the area of industrial oil crops and believe that expanded cultivation of the crops described can play a significant role in replacing some petroleum-based products with products derived from these renewable resources.

Chapter 1

Introduction to Industrial Oil Crops

Thomas A. McKeon
United States Department of Agriculture, Agricultural Research Service, Western Regional Research Center, Albany, CA, United States

Douglas G. Hayes
Department of Biosystems Engineering and Soil Science, University of Tennessee, Knoxville, TN, United States

David F. Hildebrand
Department of Plant and Soil Sciences, University of Kentucky, Lexington, KY, United States

Randall J. Weselake
Department of Agricultural, Food and Nutritional Sciences, University of Alberta, Edmonton, AB, Canada

INTRODUCTION

Vegetable oils are derived from the seeds or fruit of certain crops and are most often used for food or animal feed. These oils are composed of triacylglycerols (TAG) and trace amounts of various organic compounds including sterols and antioxidants. For most of the commodity vegetable oils, the TAG are acylated with varying proportions of the same five fatty acids, the two saturates palmitate (16:0) and stearate (18:0), the monounsaturate oleate ($18{:}1\Delta^{9}$), and the polyunsaturates linoleate ($18{:}2\Delta^{9,12}$) and α-linolenate ($18{:}3\Delta^{9,12,15}$) (double bonds all *cis*) (Fig. 1.1). Even though they are primarily consumed as food, for thousands of years the oils containing these fatty acids have also served in certain non-food applications. The oils, hydrolyzates of the oils, or alkyl esters prepared from transesterification of the oils have been used in producing fuel for lighting, lubricants, soaps, cosmetics, and lacquers. These end uses have been supplied by seed oils including what are now major commodity oils derived from palm (eg, *Elaeis guineensis*), soybean (*Glycine max*), canola (*Brassica napus*), sunflower (*Helianthus annuus*), cottonseed (*Gossypium hirsutum*), olive (*Olea europaea*), corn (*Zea mays*), peanut (*Arachis hypogaea*), and other seeds, as well as animal fats.

Industrial Oil Crops. http://dx.doi.org/10.1016/B978-1-893997-98-1.00001-4

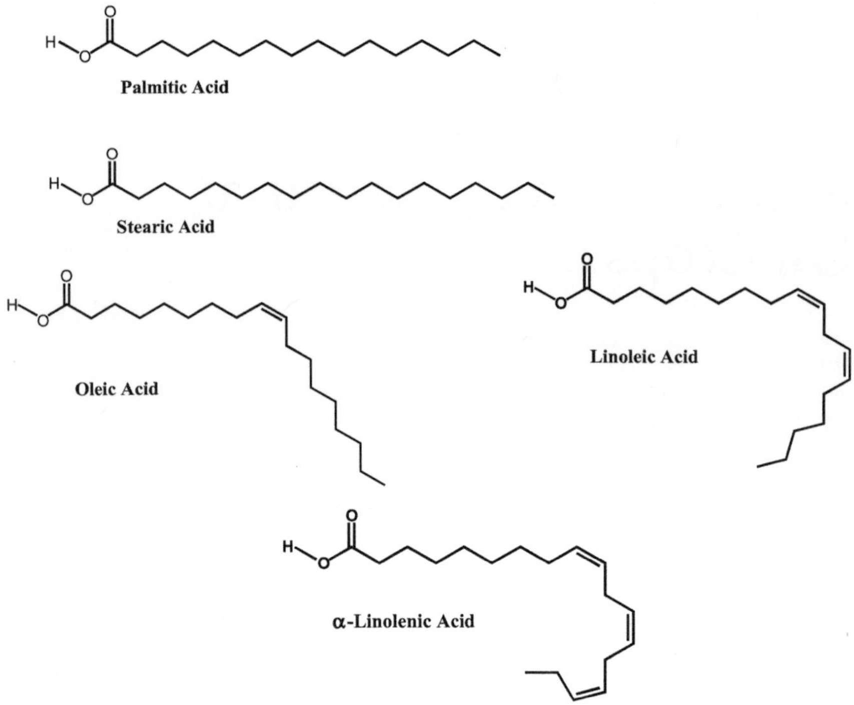

FIGURE 1.1 Fatty acids present in vegetable oils used primarily for food.

WHAT IS AN INDUSTRIAL OIL CROP?

While any seed oil can fill certain nonfood applications, there are hundreds of seed oils containing a different complement of fatty acids that impart physical and chemical properties making the oil and associated fatty acids especially useful for industrial and other nonfood uses. These differences may lie in chain length, degree or nature of unsaturation, or the presence of unusual functional groups on the fatty acid chain. While some of these oils are also valued for food, such as coconut (*Cocos nucifera*) (see chapter: Coconut (*Cocos nucifera*)) and palm kernel oils with high laurate (12:0) content (Fig. 1.2), and linseed oil with high linolenate content (see chapter: Flax (*Linum usitatissimum* L.)), there are other oils with no nutritive value such as castor oil (see chapter: Castor (*Ricinus communis* L.)), which is composed of 90% ricinoleate (12-OH 18:1Δ^9; Fig. 1.2), and tung oil (*Aleurites fordii*) (see chapter: Tung (*Vernicia fordii* and *Vernicia montana*)), which is composed of 80% eleostearate (18:3$\Delta^{9cis,11trans,13trans}$; Fig. 1.2). These oils are uniquely valuable for industrial applications due to their chemical structure. This book will provide descriptions and uses of the most important and widely used oils mentioned above and examine progress in expanding production of some industrial oil crops.

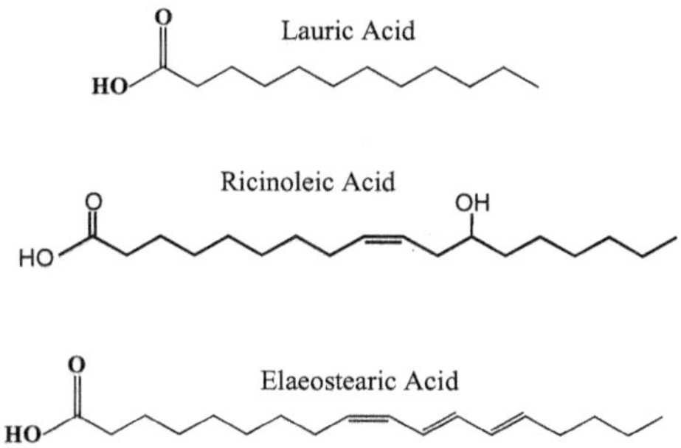

FIGURE 1.2 Some fatty acids present in seed oils used for industrial applications.

TABLE 1.1 World Oil Consumption, 2014[a]

Region	Oil Consumption, 1 Million Barrels/Day
Asia Pacific	30
Africa	3
Middle East	9
Europe and Eurasia	19
South and Central America	7
North America	23

[a]*http://www.bp.com/en/global/corporate/about-bp/energy-economics/statistical-review-of-world-energy/review-by-energy-type/oil/oil-production.html.*

WHY DO WE CARE ABOUT INDUSTRIAL SEED OILS?

More than 33 billion barrels of petroleum are used annually worldwide (Table 1.1). While the primary use of petroleum is in fuel and energy production, up to 30% is used for a range of products including polymers, lubricants, detergents, resins, paints, plasticizers and coatings, cosmetics, inks, solvents, and assorted other chemicals. Seed oils can provide products equal or superior to those derived from petroleum, as the conserved chemical structure of the different oils provides a more uniform starting material with less refining needed. A comparison of some products derived from petroleum with those derived from seed oils is provided in Table 1.2. A key difference of these products is the biodegradability of many of those derived from seed oils, even after chemical modifications for specific uses. This property is especially important in reducing environmental contamination of soil and water. Contamination of soil, aquifers, and waterways resulting from spills of petroleum-based products poses a long-term problem requiring great expense for cleanups or detrimental health effects for

TABLE 1.2 Comparison of Products Derived From Petroleum and Seed Oils

Petroleum Products	Seed Oil–Derived Products
Fuels—gasoline, diesel	Fuel—biodiesel
Paints and coatings	Paints and coatings with low VOC
Lubricants and greases	Biodegradable lubricants and greases
Nylon 6,6; Nylon 12,12	Nylon 11, Nylon 10,10
Detergents	Soaps and detergents
Polyurethanes	Partially biodegradable polyurethanes
Polystyrene	Polyesters
Printing inks	Biodegradable, nonsmudging inks
Hydraulic fluids	Biodegradable hydraulic fluids

local populations. Use of renewable oils such as seed oils to replace petroleum products is a major goal of research in sustainable development, and products currently derived from these oils include a broad representation of the types currently provided by petroleum. Specific applications including biodiesel (see chapter: Biodiesel and Its Properties) and polymers and coatings (see chapter: Polymeric Products Derived From Industrial Oils for Paints, Coatings, and Other Applications) will be covered in separate chapters in this volume. Other products, applications, and sources will be covered throughout this volume in chapters describing individual crops.

WHERE DO INDUSTRIAL OILS COME FROM?

While some plants produce seeds with a high starch content to provide energy for germination, some seeds store and utilize TAG for the germinating seed. These oilseeds utilize oxidation of fatty acids derived from the oil to produce energy for germination and early growth. Oils that carry uncommon fatty acids such as those described earlier are a result of an evolutionary process that may provide the seed with a selective advantage resulting from a protective effect against disease or predation. While the time scale of evolution may be very long, the process has provided a rich collection of resources containing potentially useful fatty acids. In many cases, though, the plants that produce these fatty acids are not suitable as crops. The adoption of these plants as crops is stymied by the need for growers to have a stable market and for the industrial user to have a stable supply of the crop. Moreover, additional research is usually needed to develop plants that are better adapted to cultivation, harvesting, and high yield. Ultimately, a commercial sponsor is required to enable the adoption of such crops. The challenges and economics of introducing these emerging oilseed crops are presented in Chapters Emerging Industrial Oil Crops and Successful Commercialization of Industrial Oil Crops, respectively.

HOW ARE INDUSTRIAL OIL CROPS IMPROVED?

Breeding

Traditional breeding techniques have been used to alter levels of fatty acids present in seed oil. An important example is the development of canola from rapeseed (*B. napus*), which resulted in a reduction of erucic acid ($22{:}1\Delta^{13}$) content from ~50% to <2% in the seed oil. The success of this breeding resulted from the availability of suitable germplasm that could provide a low-erucate background for incorporation into highly productive rapeseed varieties (Stefansson et al., 1961; Downey, 1964) (see chapter: *Brassica* spp. Oils). Breeding can also be used to enhance the content of specific fatty acids. The development of high-oleic safflower oil (*Carthamus tinctorius*) is an early example (Knowles, 1985), with success based on identification of a single locus associated with oleate content. Such enrichment is especially valued for industrially significant fatty acids as it reduces the cost of purifying the desired component. Traditional methods of self-crossing can require 5–10 years to develop homozygous strains to incorporate in a breeding program. However, there is technology for producing haploid strains of crops in tissue culture and then doubling their chromosomes to provide homozygous strains within 1 year, greatly accelerating the ability to breed and screen for crop improvement (see chapter: Applications of Doubled Haploidy for Improving Industrial Oilseeds).

While traditional breeding relies on screening of physical and chemical traits, the introduction of marker-based breeding expanded the ability of breeding programs to screen crosses. Molecular markers such as restriction and amplified fragment length polymorphisms are used to generate quantitative trait loci tied to desired traits. This molecular marker approach in breeding has proven especially useful evaluating crosses in tree crops such as palm that take several years to produce seed (Singh et al., 2009). The availability of economical genome sequencing introduced the application of single nucleotide polymorphisms and high-throughput sequencing in breeding programs (Wilson, 2012). However, even with such advances, breeding ultimately relies on the availability of suitable germplasm, and traditional breeding cannot introduce a new fatty acid or any other component not already present in the genetic background of the plants used in crosses.

Mutagenesis

Mutagenesis using chemical agents or radiation has also been effective in modifying fatty acid composition through alteration of the genome followed by screening and breeding (Knowles, 1985). Mutagenesis is especially useful in eliminating genes or reducing or blocking gene expression, and has been successfully used to produce high-linoleate linseed (*Linum usitatissimum*) oil (Green, 1986) (see chapter: Flax (*Linum usitatissimum* L.)). The biosynthetic map of plant fatty acid and lipid biosynthesis was developed through screening of mutagenized *Arabidopsis thaliana* (Browse and Somerville, 1991), and led to the identification and cloning of genes involved in fatty acid and oil biosynthesis. The identification of genes encoding enzymes involved in fatty acid biosynthesis provided the foundation for the application of Targeting Induced Local Lesions IN Genomes (TILLING) to modify fatty acid composition. TILLING relies on mutagenesis followed by high-throughput DNA sequence screening of the M_2 generation seeds to identify those seeds that carry stop codons in the sequence of the gene to be silenced (Henikoff and Comai, 2003). Plant selections carrying these mutated genes can then be screened directly for desired characteristics. The TILLING process thus moves most of the screening effort into the laboratory, considerably reducing the population that would otherwise have to be grown in the field for phenotypic screening.

Although not specifically a form of mutagenesis, gene suppression has also proven to be an effective means for altering fatty acid composition. Gene suppression was initially discovered as an occasional byproduct of gene overexpression, and was useful in altering fatty acid composition of *Brassica* seed oil (Knutzon et al., 1992). Usually, the correct or "sense" strand of DNA is transcribed to provide the mRNA that will be translated into the desired protein, but in such gene-suppressed plants, the wrong or "antisense" strand of DNA is transcribed, limiting protein synthesis from the sense mRNA. Direct use of antisense technology led to the discovery that expression of small pieces of RNA can suppress gene expression (Smith et al., 2000). This RNA silencing technology has been used to develop crops ranging from tomatoes with delayed softening during ripening to high-oleic acid soybeans (Frizzi and Huang, 2010) (see chapter: Genetic Transformation of Crops for Oil Production).

GENETIC ENGINEERING OF FATTY ACID BIOSYNTHESIS

Years of biochemical characterization of fatty acid and oil biosynthesis provided the foundation for rapid advances in identification and cloning of genes involved in these processes. In order to modify the fatty acid composition of seed oil through genetic engineering, it is necessary to understand the processes involved in oil biosynthesis. A combination of biochemical characterization of enzymes involved in fatty acid and oil biosynthesis plus generation of mutants in lipid biosynthesis provided the means to identify genes essential for directing the synthesis of the desired fatty acids and their incorporation in oil (Browse and Somerville, 1991). The widespread availability of information on genes that direct biosynthesis of certain fatty acids are key steps in oil biosynthesis providing the means to introduce novel activities and components into oilseeds (see chapter: Genetic Transformation of Crops for Oil Production).

Fatty acid biosynthesis in plants starts with acetyl-CoA carboxylation to malonyl-CoA through the catalytic action of the acetyl-CoA carboxylase (ACCase) followed by a series of condensation reactions that result in production of long chain saturated fatty acids. Malonyl-CoA is converted to the corresponding acyl carrier protein (ACP) derivative, malonyl-ACP, and the first condensation is initiated by the keto-acyl synthase (KAS) III using acetyl-CoA to initiate the condensation with malonyl-ACP. This condensation cycle is followed by six additional condensations with malonyl-ACP catalyzed by KAS I to yield palmitoyl-ACP. This may be followed by an additional condensation catalyzed by KAS II to yield stearoyl-ACP, as depicted in Fig. 1.3. A fatty acid thioesterase can then release free palmitate or stearate which are converted to acyl-CoA derivatives and incorporated into TAG and other glycerolipids. The stearoyl-ACP can also be desaturated to oleoyl-ACP, the oleate released by thioesterase action and converted to oleoyl-CoA. The oleoyl-CoA is then incorporated into glycerolipids for incorporation into TAG or further modification. These reactions leading to palmitate, stearate, and oleate occur in the plastids, separate from reactions leading to oil biosynthesis. Given the dependence of fatty acid production on malonyl-CoA production (to provide malonyl-ACP), ACCase is thought to play a regulatory role in fatty acid production and oil biosynthesis (Weselake et al., 2009).

The first commercial plant engineered to produce an industrial useful fatty acid was high-laurate *B. napus* (Voelker et al., 1992). Plants that make oils containing medium chain fatty acids such as California bay laurel (*Umbellularia californica*) and coconut have a fatty acid thioesterase that catalyzes the release of fatty acyl chains from the fatty acid synthesis complex after four or five condensation steps carried out by the catalytic action of KAS I. When the

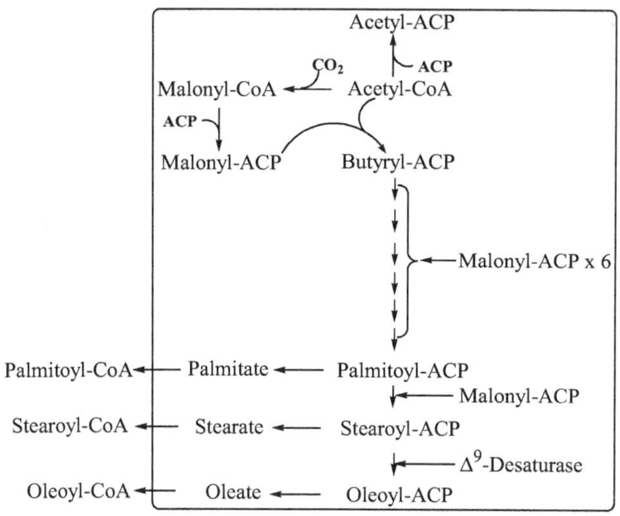

FIGURE 1.3 Pathway for fatty acid biosynthesis in plants.

gene for such a thioesterase was inserted and expressed in *B. napus*, the result was a canola seed with oil biosynthesis rerouted to produce oil containing over 50% laurate.

GENETIC ENGINEERING OF OIL BIOSYNTHESIS

The introduction of genes that modify the fatty acid composition of oil have been essential to identifying specific steps in the oil biosynthetic pathway (TAG assembly) that control the amount and the composition of the oil produced by the seed. The engineering required to develop the high-laurate *B. napus* (Voelker et al., 1996) made it clear that simple insertion of a gene to alter fatty acid biosynthesis or modify a fatty acid (Broun and Somerville, 1997) is not sufficient to significantly alter the fatty acid composition of the oil. For example, in *B. napus*, insertion of the coconut lyso-phosphatidate acylCoA acyltransferase (*LPAAT*) gene was required for acylation of the *sn*-2 position of TAG, thereby increasing laurate content (Voelker et al., 1996) because expression of the California bay laurel thioesterase alone in *B. napus* resulted in acylation of only the *sn*-1 and *sn*-3 positions of TAG.

The pathway for TAG assembly from plastid-generated acyl-CoAs (Fig. 1.3) is presented in Fig. 1.4. The Kennedy pathway for TAG assembly was driven by thermodynamically favored reactions, namely acylation using acyl-CoA or ATP to form thioesters.

Considerable research on oil biosynthesis has demonstrated additional layers of complexity in TAG assembly (Bates et al., 2013). The acyl-CoA can be elongated on the ER or incorporated into glycerolipids or phospholipids. TAG can be synthesized from DAG via the acyl-CoA-dependent DGAT. The LPCAT incorporates an acyl group from acyl-CoA into the sn-2 position to form PtdC, an important component of membranes and a substrate for desaturation or fatty acid modification of the sn-2 fatty acyl. The sn-2 acyl-group on PC can be released by PLA2 and the free fatty acid converted to acyl-CoA or the sn-2 acyl-group can be enzymatically transferred by PDAT to DAG, forming TAG. In each case, the other product is LPC which is then available for re-acylation by LPCAT. The glycerol backbone is derived from sn-glycerol-3-phosphate which is subject to sequential acylations to produce phosphatidic acid. Phosphatidic acid phosphatase catalyzes the dephosphorylation of phosphatidic acid to produce sn-1, 2-DAG (DAG). DAG can also be converted to phosphatidylcholine (PtdC) via the action of DAG:CPT or (PDCT). PtdC can also be converted to DAG by PDCT. The LPCAT can catalyze acyl-exchange between the sn-2 position of PUFA-enriched PtdC and the acyl-CoA pool, as facilitated by the ACBP. Both PLA2 and PDAT have been implicated in removal of unusual FAs from the PtdC for incorporation into TAG. The DGAT activity is unique to TAG biosynthesis, and has long been thought to be a gatekeeper for oil biosynthesis as it diverts the products leading to phospholipid production for membrane biosynthesis into TAG (Weselake et al., 2009). Increased expression of DGAT

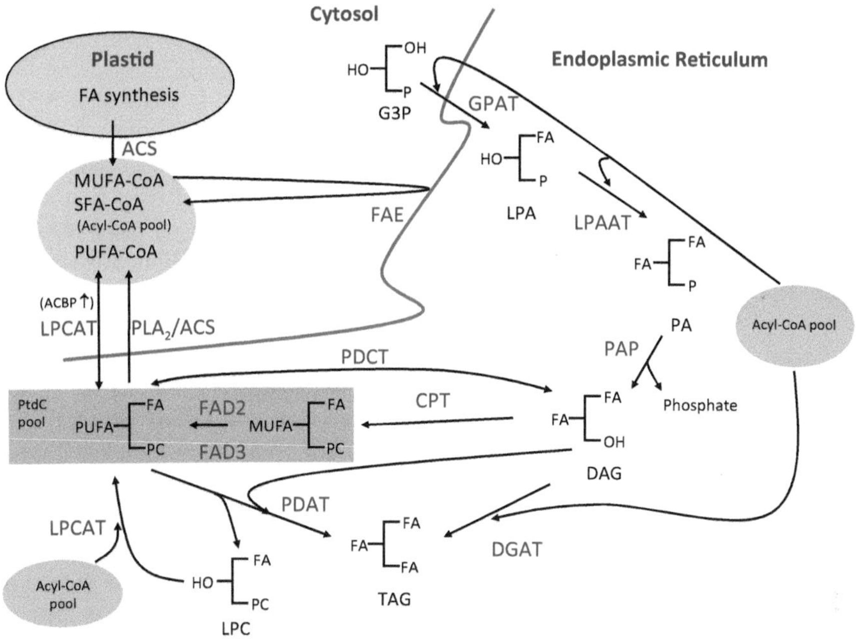

FIGURE 1.4 Generalized scheme for triacylglycerol (TAG) biosynthesis in developing seeds of oleaginous plants. *MUFAs*, monounsaturated fatty acids; *SFA*, saturated fatty acids; *CoAs*, acyl-Coenzyme As; *FAE*, fatty acid elongase; *ER*, endoplasmic reticulum; *G3P*, *sn*-glycerol-3-phosphate; *PAP*, phosphatidic acid phosphatase; *PA*, phosphatidic acid; *DAG*, *sn*-1, 2-diacylglycerol; *DGAT*, acyl-CoA:diacylglycerol acyltransferase; *PtdC*, phosphatidylcholine; *CPT*, *sn*-1,2-diacylglycerol: cholinephosphotransferase; *PDCT*, phosphatidylcholine:diacylglycerol cholinephosphotransferase; *FAD1* and *FAD2*, fatty acid desaturases 1 and 2; *PUFA*, polyunsaturated fatty acid; *PLA2*, phospholipase A2; *ACS*, acyl-CoA synthetase; *LPCAT*, acyl-CoA:*lyso*-phosphatidylcholine acyltransferase; *ACBP*, low molecular mass soluble acyl-CoA-binding protein; *PDAT*, phospholipid:diacylglycerol acyltransferase; *LPC*, lyso-phosphatidylcholine; *GPAT*, acyl-CoA:*sn*-glycerol-3-phosphate acyltransferase; *LPA*, *lyso*-phosphatidic acid; *LPAAT*, acyl-CoA:*lyso*-phosphatidic acid acyltransferase; *PC*, phosphocholine.

activities has been closely tied to increased oil production in some oilseeds (Liu et al., 2012). Additional detail is available at http://lipidlibrary.aocs.org/content.cfm?ItemNumber=40314.

Chapters Castor (*Ricinus communis* L.), *Brassica* spp. Oils, Flax (*Linum usitatissimum* L.), Crambe (*Crambe abyssinica*), Camelina *(Camelina sativa)*, Tung (*Vernicia fordii* and *Vernicia montana*), Emerging Industrial Oil Crops, and Engineering Oil Accumulation in Vegetative Tissue provide additional descriptions of oil biosynthesis including TAG incorporating uncommon fatty acids and production of oil by vegetative tissue. Recent developments in plant fatty acid and oil biosynthesis can be found in the Lipid Library (http://lipidlibrary.aocs.org/Biochemistry/content.cfm?ItemNumber=40300&navItemNumber=19219) maintained by AOCS.

TECHNICAL AND SOCIAL ISSUES RELATED TO PLANT GENETIC ENGINEERING

As mentioned, the earliest commercialized oil crop genetically engineered (GE) primarily to produce an industrial oil product was high-laurate *B. napus* (Voelker et al., 1992). The oil was developed for use by the surfactant and detergent industry, and the crop represented the first source of laurate available from a temperate crop that had suitable agronomic characteristics. Because the seed meal remaining after oil extraction was to be used as animal feed, the crop required regulatory approval for use in food. Although the crop was not a commercial success, it provided a model for the regulatory process in developing industrial crops and also an example of the difficulty in achieving commercial success.

Two incidents involving GE industrial crops pointed to some inherent weaknesses in the regulatory system and resulted in increased oversight of industrial GE crop introduction (McKeon, 2003; Lemaux, 2008). The first incident involved StarLink corn, a variety engineered to produce the insect resistance protein CrY9c, because this protein had some properties consistent with allergens, the corn was approved only for animal feed or industrial use. Since animal feed is cheaper than human food, some unscrupulous or unknowing producers and users of this corn sold it as food for humans. This problem resulted in a 7% decline in corn prices for at least a year and reduced acceptance of U.S. corn exports in a number of countries (Carter and Smith, 2007). Moreover, this incident altered U.S. regulation of GE crop introductions, requiring that all such crops, even those solely for industrial use, be evaluated for risk to humans if the product were consumed (McKeon, 2003). The second incident involved soybeans that were planted in a field used in the previous year to grow ProdiGene's GE corn that had been engineered to produce an industrial protease. Residual corn seed that germinated the following season in the soybean field was harvested with the soybeans, thereby contaminating the soy crop with the GE corn. The end result was that, upon discovery of the contamination, the harvested soybeans co-mingled with the contaminated soybeans were destroyed, resulting in a multi–million dollar loss and additional calls for increased regulation of GE crops (Fox, 2003).

These two incidents served as a warning to the agencies that regulate the plant biotechnology industry, leading to increased oversight in the granting of applications for introducing transgenic crops. It is now assumed that oilseeds engineered to produce industrial oil will produce a seed meal that may be used as food or feed, thus entering the food supply even if it is not intended to be used as food or. As a result of this possibility, even these byproducts of transgenic oilseeds must pass the substantial equivalence test for regulatory approval of the crop (Stewart and McLean, 2004).

There are currently 27 countries that produce GE food crops and 36 countries have approved GE crops for environmental release or use as food or feed (James, 2014). The major oilseed GE crops planted include soy, corn, canola (low–erucic acid rapeseed), and cotton. Although controversies regarding GE crops remain, eg, labeling of products derived from these crops, the regulations governing their production and use in food and feed have presumably been met in the countries that have approved their use. In most cases, an extensive regulatory oversight has been developed to assure safety of introduced GE crops. For example, the United States has three regulatory agencies that oversee approval of GE crops. The US Department of Agriculture Animal and Plant Health Inspection Service (USDA-APHIS) ensures that the plant is not invasive. The Environmental Protection Agency (EPA) addresses concerns that any plants carrying biopesticides or requiring application of pesticides or herbicides will not cause environmental problems. The Food and Drug Administration (FDA) ensures the crop is substantially equivalent to the unmodified crop or, if there are changes resulting in a significant difference, eg, different fatty acid composition, that the alteration does not pose a hazard for human or animal consumption (McKeon, 2003). The requirement that such crops are substantially equivalent to the non-GE crops is especially important when considering the crop for food and feed, to ensure that introduced genes do not reduce the safety or nutritive value of the crop. A number of specific concerns regarding GE crops are listed in Table 1.3.

The common practice of including antibiotic resistance as a marker for selection led to concerns that wholesale antibiotic resistance would be acquired by gut bacteria leading to disastrous untreatable infections. Although it is unlikely that these genes would survive the digestive process, the introduction of recombinase systems that can specifically excise these selection genes obviates the issue (Srivastava and Ow, 2015). Also, nonantibiotic selection systems are available, and antibiotic selection genes can be introduced at unlinked loci that segregate out from the traits of interest.

In most cases, GE plants will include genes that express proteins or enzymes; therefore, the expressed protein can potentially introduce a new allergen. As a result, the proteins and products resulting from introduced genes must be demonstrated to be readily digestible and comply with food safety regulations.

TABLE 1.3 Issues Related to Genetically Engineered Crops
Introduction of antibiotic resistance
"Random" insertion of gene
Undesirably or unpredictably altered composition
Harmful effects on beneficial insects
Outcrossing
Introduction of allergen
Crops for both food and industrial applications
End uses of residues
Identity preservation

The required testing for susceptibility to pepsin digestion serves as one indicator that the GE plant will not introduce a new allergen, but additional criteria including evaluation of the protein structure and allergenicity of the source of the protein are also used (Ladics, 2008).

The outcrossing of GE crop traits to native germplasm has occurred in some cases and introduction of herbicide resistance to closely related species that are invasive weeds is a major concern (Warwick et al., 2009). Engineered crops can be planted with a buffer zone of nonmodified crops for containment of any outcrossing to native germplasm or related weeds and to reduce the development of insect resistance to biopesticides introduced in the GE crop. Buffer zones are also used for non-GE crop plants such as canola and rapeseed, where outcrossing can occur over some distance resulting in seed with undesirable composition.

The economics of producing GE oilseeds is beyond the scope of this text. One key issue related to economics, however, is identity preservation. The investment in research and development of a transgenic oilseed is often considerable. It is generally recognized that only crops to be grown on a large scale or are of very high value will be profitable for the seed developer. Furthermore, the transgenic crop products with the industrial oil trait(s) must be kept separate from other related seed in order to maintain its premium value. There are protocols developed for the process of identity preservation that are currently applied to crops such as organic and/or non-GE soybeans (Bullock and Desquilbet, 2002).

CONCLUSIONS

There is considerable recognized value for development of oilseed crops that provide biomaterials and feedstocks to the chemical industry. Oleochemicals already make an important contribution to biomaterial production in the form of surfactants, lubricants, paints, coatings, polymers, and biofuels. Successful development of approaches to genetically engineer crops to produce oils containing industrially useful fatty acids will aid the introduction of additional oil-derived biomaterials and provide renewable resources that can reduce the need for petrochemical production.

REFERENCES

Bates, P.D., Stymne, S., Ohlrogge, J., 2013. Biochemical pathways in seed oil synthesis. Curr. Opin. Plant Biol. 16, 358–364.

Broun, P., Somerville, C., 1997. Accumulation of ricinoleic, lesquerolic, and densipolic acids in seeds of transgenic arabidopsis plants that express a fatty acyl hydroxylase cDNA from castor bean. Plant Physiol. 113, 933–942.

Browse, J., Somerville, C., 1991. Glycerolipid synthesis: biochemistry and regulation. Annu. Rev. Plant Physiol. Plant Mol. Biol. 42, 467–506.

Bullock, D.S., Desquilbet, M., 2002. The economic of non-GMO segregation and identity preservation. Food Policy 27, 81–99.

Carter, C.A., Smith, A., 2007. Estimating the market effect of a food scare: the case of genetically modified StarLink corn. Rev. Econ. Stat. 89, 522–533.

Downey, R.K., 1964. A selection of *Brassica campestris* L. containing no erucic acid in its seed oil. Can. J. Plant Sci. 44, 295.

Fox, J.L., 2003. Puzzling industry response to ProdiGene fiasco. Nat. Biotech. 21, 3–4.

Frizzi, A., Huang, S., 2010. Tapping RNA silencing pathways for plant biotechnology. Plant Biotech. J. 8, 655–677.

Green, A.G., 1986. A mutant genotype of flax (*Linum usitatissimum* L.) containing very low levels of linolenic acid in its seed oil. Can. J. Plant Sci. 66, 499–503.

Henikoff, S., Comai, L., 2003. Single-nucleotide mutations for plant functional genomics. Annu. Rev. Plant Biol. 54, 375–401.

James, C., 2014. Global Status of Commercialized Biotech/GM Crops: 2013. ISAAA Brief 46–2013: Executive Summary www.isaaa.org.

Knowles, P.F., 1985. Genetics and breeding of oil crops. In: Robbelen, G., Downey, R.K., Ashri, A. (Eds.), Oil Crops of the World. McGraw-Hill, New York, pp. 260–282.

Knutzon, D.S., Thompson, G.A., Radke, S.E., Johnson, W.B., Knauf, V.C., Kridl, J.C., 1992. Modification of Brassica seed oil by antisense expression of a stearoyl-acyl carrier protein desaturase gene. Proc. Nat. Acad. Sci. U.S.A. 89, 2624–2628.

Ladics, G.S., 2008. Current codex guidelines for assessment of potential protein allergenicity. Food Chem. Toxicol. 46, S20–S23.

Lemaux, P.G., 2008. Genetically engineered plants and foods: a scientist's analysis of the issues. Ann. Rev. Plant Biol. 59, 771–812.

Liu, Q., Siloto, R.M.P., Lehner, R., Stone, S.J., Weselake, R.J., 2012. Acyl-CoA:diacylglycerol acyltransferase: molecular biology, biochmeistry and biotechnology. Prog. Lipid Res. 51, 350–377.

McKeon, T.A., 2003. Genetically modified crops for industrial products and processes and their effects on human health. Trends Food Sci. Technol. 14, 229–241.

Singh, R., et al., 2009. Mapping quantitative trait loci for fatty acid composition in an interspecific cross of oil palm. BMC Plant Biol. 9, 114.

Smith, N.A., Singh, S.P., Wang, M.-B., Stoutjesdijk, P.A., Green, A.P., Waterhouse, P.M., 2000. Total silencing by intron-spliced hairpin RNAs. Nature 407, 319–320.

Srivastava, V., Ow, D.W., 2015. Simplifying transgene locus structure through Cre-lox recombination. In: Mysore, K.S., Senthil-Kumar, M. (Eds.), Plant Gen Silencing: Methods in Molecular Biology, vol. 1287. Springer Science+Business Media, New York, pp. 95–103. http://dx.doi.org/10.1007/978-1-4939-2453-0_6.

Stefansson, B.R., Hougen, F.W., Downey, R.K., 1961. Note on the isolation of rape plants with seed oil free from erucic acid. Can. J. Plant Sci. 41, 218–219.

Stewart, P.A., McLean, W., 2004. Fear and hope over the third generation of agricultural biotechnology: analysis of public response in the Federal Register. AgBioForum 7, 133–141.

Voelker, T.A., et al., 1992. Fatty acid biosynthesis redirected to medium chains in transgenic oilseed plants. Science 257, 72–74.

Voelker, T.A., Hayes, T.R., Cranmer, A.M., Turner, J.C., Davies, H.M., 1996. Genetic engineering of a quantitative trait: metabolic and genetic parameters influencing the accumulation of laurate in rapeseed. Plant J. 9, 229–241.

Warwick, S.I., Beckie, H.J., Hall, L.M., 2009. Gene flow, invasiveness, and ecological impact of genetically modified crops. Ann. N.Y. Acad. Sci. 1168, 72–99.

Weselake, R.J., Taylor, D.C., Rahman, M.H., Shah, S., Laroche, A., McVetty, P.B.E., Harwood, J.L., 2009. Increasing the flow of carbon into seed oil. Biotech. Adv. 27, 866–878.

Wilson, R.F., 2012. The role of genomics and biotechnology in achieving global food security for high-oleic vegetable oil. J. Oleo Sci. 61, 357–367.

Chapter 2

Biodiesel and Its Properties

Gerhard Knothe
U.S. Department of Agriculture, Agricultural Research Service, National Center for Agricultural Utilization Research, Peoria, IL, United States

INTRODUCTION

Biodiesel (Knothe et al., 2010; Mittelbach and Remschmidt, 2004) is a biogenic alternative to conventional diesel fuel (DF) obtained from vegetable oils, animal fats, or other materials consisting largely of triacylglycerols (triglycerides). Reacting an oil or fat with an alcohol, usually methanol, in the presence of a catalyst, commonly sodium methoxide, affords the corresponding mono-alkyl esters. These mono-alkyl esters are defined as biodiesel. Glycerol is obtained as a co-product. Fig. 2.1 depicts the principle of the transesterification reaction.

While the suitability of any material as fuel, including biodiesel, can be influenced by contaminants arising from production or other sources, the nature of the major fuel components ultimately determines the fuel properties. Some of the properties included in standards can be traced to the structure of the fatty esters comprising biodiesel. Since biodiesel consists of fatty acid esters, not only the structure of the fatty acids but also that of the ester moiety derived from the alcohol can influence the fuel properties of biodiesel. Furthermore, the aforementioned mono-alkyl esters that comprise biodiesel are a mixture corresponding in its fatty acid profile to that of the parent oil or fat from which it is produced with each ester component contributing to the properties of the fuel.

The properties of a biodiesel fuel that are determined by the structure of its component fatty esters include ignition quality, cold flow, oxidative stability, viscosity, and lubricity. The present work discusses the influence of the structure of fatty esters on these properties. Not all of these properties have been included in biodiesel standards, although all of them are essential to the proper functioning of the fuel. This article begins, however, with brief summaries on the historical background, production, and analysis of biodiesel.

Industrial Oil Crops. http://dx.doi.org/10.1016/B978-1-893997-98-1.00002-6

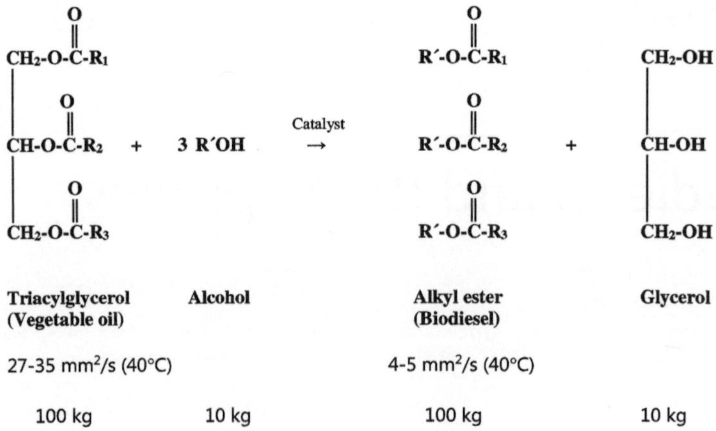

FIGURE 2.1 The transesterification reaction. R_1, R_2, and R_3 is a mixture of various fatty acid chains. The alcohol used for producing biodiesel is usually methanol ($RN{=}CH_3$).

HISTORICAL BACKGROUND

The use of vegetable oils or their derivatives in diesel engines is nearly as old as the diesel engine itself. While the engine was developed in the 1890s by Rudolf Diesel (1858–1913), the apparently first recorded run of a diesel engine using a vegetable oil as fuel, peanut oil, occurred at the World Exposition in Paris in 1900 at the request of the French government. The French government at that time was interested in developing alternative fuels for its European colonies that could be obtained or produced locally. Diesel himself made remarks supporting the concept of using vegetable oils as fuel in later writings as well as carrying out some experiments himself and describing the first test in 1900 (Diesel, 1912). Through the 1940s, numerous references can be found that describe the use of vegetable oils as diesel fuel. Probably under current aspects of using biodiesel, the most remarkable document in this connection is a Belgian patent granted in 1937 to C.G. Chavanne of the University of Brussels in which the synthesis of ethyl esters of vegetable oils for fuel use is discussed (Chavanne, 1937). A related publication details the production and use of ethyl esters in the Belgian Congo and in Belgium, including the probably first use of biodiesel in commercial buses (van den Abeele, 1942). Interest in alternative fuels waned after approximately 1945 and this lack of interest continued into the 1970s. In the United States, a few reports from The Ohio State University and Georgia Institute of Technology, after approximately the end of World War II, through the 1970s, apparently little interest existed in vegetable oil–based fuels, or alternative fuels in general.

The energy crises of the 1970s then spawned renewed interest in alternative fuels, among them vegetable oil–based fuels. In this connection, the alkyl esters of vegetable oils were "rediscovered" with the probably first reports dealing with the production of methyl esters of sunflower oil (Bruwer et al., 1980a,b). A variety of legislative and regulatory incentives in many countries around the world assisted in bringing biodiesel to market beginning in the mid to late 1990s and accelerating a few years later. Most biodiesel, however, is used in blends with petrodiesel, these blends often being termed "BXX" in which "XX" denotes the percentage level of biodiesel blended with the petrodiesel. The successful introduction and commercialization of biodiesel in many countries worldwide has been accompanied by the development of standards to ensure high product quality and instill user confidence. The first biodiesel standard was developed in Austria in 1991 as ÖNORM C1190. Major biodiesel standards that have served as guidelines for other standards around the world are ASTM D6751 (ASTM, American Society for Testing and Materials) and the European standard EN 14214 (CEN). Table 2.1 is a list of standards that have been mostly developed for application to biodiesel. Table 2.2 contains fuel property specification from the standards ASTM D6751 and EN 14214.

BIODIESEL PRODUCTION

The most common process for the production of biodiesel is the transesterification of the vegetable oil (or other oil or fat) with an alcohol, most commonly methanol, in the presence of a catalyst. Methanol is the most commonly used alcohol as it is the least expensive alcohol. As mentioned, glycerol is obtained as a co-product. This transesterification reaction with methanol is straightforward as it can be carried out at 60–65°C, slightly below the boiling point of methanol, with a molar ratio of alcohol:oil = 6:1 (Freedman et al., 1984). In case of the use of an alcohol other than methanol, the reaction temperature is raised accordingly. The reaction is usually complete after 1 h, with most of the conversion occurring during the first 10 min. Base-catalyzed transesterification is usually considerably faster than acid-catalyzed transesterification (Freedman et al., 1984). Catalyst sodium methoxide (or other alkoxide corresponding to the alcohol used) is preferable compared to hydroxide because the water-forming reaction

$$XOH + ROH \rightarrow ROX + H_2O \ (X = Na, K; R = alkyl)$$

is not possible when using methoxide. The transesterification reaction yielding biodiesel needs to be as free of water as possible to prevent the formation of salts of fatty acids (soap). Likewise, the free fatty acids should be kept to a minimum. For feedstocks high in free fatty acids, notably often used cooking oils or animal fats but also some vegetable oils, a two-step conversion is suitable in which the free fatty acids are esterified first by acid-catalyzed esterification and then the triacylycerols transesterified to the alkyl esters (Canakci and Van Gerpen, 1999).

The transesterification reaction mixture to give methyl esters consists of two phases at the outset of the reaction, methanol and vegetable oil, and

TABLE 2.1 Major Standards (ASTM and EN) Related to Biodiesel

D975	Diesel Fuel	EN 590	Diesel Fuel
D6584	Total monoglycerides, total diglycerides, total triglycerides, and free and total glycerin in B-100 by Gas Chromatography	EN 141078	FAME in diesel fuel by IR
D6751	Biodiesel blend stock (B100) for middle distillate fuels	EN 14103	Ester and linolenic acid ester content
D7321	Particulate contamination of B100 and biodiesel blends by laboratory filtration	EN 14104	Acid value
D7371	Biodiesel (FAME) content in diesel fuel by mid infrared spectroscopy	EN 14105	Free and total glycerol, acylglycerols by GC
D7398	Boiling range distribution of FAME in the range from 100 to 615°C by GC	EN 14106	Free glycerol in the range 0.005–0.07% (m/m)
D7462	Oxidation stability of B100 and blends of biodiesel with diesel fuel	EN 14107	Phosphorus by ICP
D7467	B6–B20 blends	EN 14108	Sodium content by AAS
D7501	Determination of fuel filter blocking potential of B100 by cold soak filtration test (CSFT)	EN 14109	Potassium content by AAS
D7591	Free and total glycerin in biodiesel blends by anion exchange chromatography	EN 14110	Methanol content by GC
D7797	FAME content in jet fuel using flow analysis by FTIR	EN 14111	Iodine value
D7806	FAME content of a blend of biodiesel and petroleum-based diesel Fuel oil using mid-IR	EN 14112	Oxidation stability

TABLE 2.1 Major Standards (ASTM and EN) Related to Biodiesel—cont'd

		EN 14213	FAME as heating fuels
		EN 14331	FAME analysis by LC/GC
		EN 14538	Ca and Mg by ICP-OES
		EN 15751	Modified oxidation stability
		EN 15779	Determination of polyunsaturated (≥4 double bonds) in FAME by GC

two phases at the end, methyl esters and glycerol, although in case of higher esters there exists an increased tendency toward the formation of emulsions. The methyl ester and glycerol can generally be easily separated after the reaction. The methyl ester phase can be washed with water (temperature of the wash water may be slightly elevated) to remove remaining glycerol and catalyst. Minor components remaining in the biodiesel after completion of all steps include starting material (triacylglycerols), intermediates (di- and monoacylglycerols), glycerol, catalyst, and extraneous materials. These minor components can significantly influence biodiesel fuel properties, probably the most prominent examples being cold flow and oxidative stability.

The development of other catalysts or catalytic systems as well as variations of the transesterification reaction has received considerable attention by the research community in recent years. Countless other catalysts, the enumeration of which transcends this article, are to address issues with the conventional transesterification reaction such as increasing tolerance for water, reducing problems associated with the use of other alcohols or easy removal of the catalyst. Most notably, heterogeneous catalysts or catalyst systems as well as enzymatic catalysis have been researched extensively. Some review articles that deal with conventional transesterification or alternative catalytic systems are Ma and Hanna (1999), Lotero et al. (2005), Di Serio et al. (2008), Fjerbaek et al. (2009), Andrade et al. (2011), Abbaszaadeh et al. (2012), He and Van Gerpen (2012b), Kouzu and Hidaka (2012), Motasemi and Ani (2012), Veljkovic et al. (2012), Davison et al. (2013), Hama and Kondo (2013), Kralisch et al. (2013), Narwal and Gupta (2013), and Ramachandran et al. (2013).

TABLE 2.2 Fuel Property Specifications in Biodiesel Standards

Property	Unit	ASTM D6751	EN 14214
Cetane number	–	47 min	51 min
Kinematic viscosity at 40°C	mm^2/s	1.9–6.0	3.5–5.0
Oxidation stability at 110°C	h	3 min	8 min
Cloud point	°C	Report	–
Cold soak filterability	s	200 or 360	–
Density	kg/m^3	–	860–900
Flash point	°C	93 min	101 min
Free glycerol	% Mass	0.02	0.02
Total glycerol	% Mass	0.24	0.25
Linolenic acid content		–	12% (m/m)
FAME ≥4 double bonds		–	1% (m/m)
Sodium	µg/g (ppm)	5 max, combined	5 mg/kg combined
Potassium			
Calcium	µg/g (ppm)	5 max, combined	5 mg/kg combined
Magnesium			
Sulfur	% Mass (ppm)	0.0015 or 0.05[a]	10 mg/kg
Phosphorus	% Mass	0.001 max	4 mg/kg
Acid value	mg KOH/g	0.5 max	0.5 max
Water content	% Volume	0.05 max	500 mg/kg
Alcohol content	% Mass	0.2 or flash point 130°C max	0.2% (m/m)

[a]*Depends on grade of petrodiesel fuel to be blended with.*

Other approaches to modifying the transesterification reaction include the use of a solvent to achieve a one-phase system or in situ transesterification in which the oil is not removed from the original feedstock, rather the feedstock containing the oil directly subjected to the transesterification reaction. Such procedures entail changes to the reaction such as a higher stoichiometric ratio of alcohol to oil or different reaction temperatures. Overall, the conventional transesterification reaction discussed earlier is the procedure that still is most commonly applied as the alternative procedures also have issues such as higher reaction temperatures, higher pressure, extended reaction times, reduced yield, catalyst cost and availability, etc.

Generally, the most common sources of biodiesel are classical commodity oils such as soybean, rapeseed (canola), and palm oils. The use of these oils as prevalent feedstock varies by region of the world; thus, in Europe, rapeseed (as low-erucic variety similar to canola) is the predominant feedstock, while in the United States, soybean oil is the most common and in tropical countries palm oil is used. Other oils such as sunflower oil, however, are also used. As the major cost factor in biodiesel production is the feedstock, low-cost sources such as animal fats and used cooking oils are also relatively commonly employed. Because, biodiesel derived from all of these feedstocks taken together does not suffice to replace the whole petrodiesel market and because of the perceived competition between food and fuel use, significant emphasis is being placed on identifying and developing additional sources of biodiesel. Alternative, usually nonedible oils have been promoted under these aspects, with notable examples being jatropha (Foidl et al., 1996; Wood, 2005; Rashid et al., 2010) and pongamia (Scott et al., 2008), which is also known as karanja. That unexpected problems may arise with alternative feedstocks has been discussed for jatropha (Kant and Wu, 2012). Some other alternative feedstocks that have been discussed in the literature include camelina (Moser and Vaughn, 2010a), coriander (Moser and Vaughn, 2010b), cuphea (Knothe et al., 2009), desert date (Chapagain et al., 2009), kenaf (Knothe et al., 2013), meadowfoam (Moser et al., 2010), moringa (Rashid et al., 2008), pennycress (Moser et al., 2009), rubber seed (Ikwuagwu et al., 2000; Ramadhas et al., 2005), seashore mallow (Moser et al., 2013), and thespesia (Rashid et al., 2011). A potential feedstock that is currently being intensively researched is algae. The background is, besides avoiding the food versus fuel issue, the claimed high production potential of algae, claimed to exceed that of any vegetable oil. On the other hand, critics state that the claimed high production potential is vastly exaggerated. Numerous review articles deal with the production of algal oils for fuel purposes under a variety of aspects (Aguirre et al., 2013; Chisti, 2007; Clarens and Colosi, 2010; Foley et al., 2011; Greenwell et al., 2010; Hu et al., 2008; Knothe, 2011; Kröger and Müller-Langer, 2012; Leite et al., 2013; Petkov et al., 2012; Pienkos and Darzins, 2009; Scott et al., 2010; Simionato et al., 2013; van Beilen, 2010; Williams and Laurens, 2010).

BIODIESEL ANALYSIS

As mentioned, biodiesel can contain various minor components (contaminants) as a result of the transesterification reaction but also through carry-over from the feedstock or through contact with other material after production. Most of the significant fuel quality issues are addressed in biodiesel standards. Besides analyzing the final product, monitoring the transesterification reaction has found attention in the literature.

Remaining feedstock as well as the intermediates and remaining glycerol in biodiesel are addressed by gas chromatographic (GC) methods in standards, the GC methods themselves being standards. The GC methods are ASTM D6584 and EN 14105. They are based on prior literature (Freedman et al., 1986; Plank and Lorbeer, 1995). These methods determine bound, free, and total glycerol, bound glycerol being defined as mono-, di-, and triacylglycerols. The sum of bound glycerol with free glycerol, the co-product of the transesterification reaction, gives what is usually termed total glycerol. The European biodiesel standard EN 14214 also calls for determination of the individual acylglycerol species, while ASTM D6751 only calls for free and total glycerol. The free and total glycerol limits in standards are set low as these materials can either negatively influence combustion and/or cold flow properties.

Other fuel quality parameters include limits on the heteroelements sodium, potassium, sulfur, phosphorus, calcium, and magnesium. The presence of these heteroelements can also negatively influence combustion and poison catalytic exhaust emissions controls as well as contribute to the formation of sediments when storing the fuel. These elements can arise in biodiesel as remaining transesterification catalysts, carry over from feedstocks such as animal fats having had contact with other animal parts such as meat and bones or sulfur-containing glucosinolates in rapeseed.

The monitoring of the transesterification reaction has been described by using methods such as infrared (IR), near infrared (NIR), and nuclear magnetic resonance (NMR). Finally, numerous methods have also been developed to determine the amount of biodiesel in blends with petrodiesel with IR-based methods being the most commonly applied. Some reviews on the issue of biodiesel analysis by the various methods mentioned are Mittelbach (1996), Holcapek et al. (2001), Foglia et al. (2004), Knothe (2006), Monteiro et al. (2008), Lepri et al. (2011), He and Van Gerpen (2012a), and Zhang (2012). Several standards also deal with the issues of biodiesel analysis and determination of blend levels (Table 2.1). It may be noted that one standard also deals with the determination of biodiesel in jet fuel as concerns exist that biodiesel may carry over from pipelines into jet fuel with the less favorable cold flow properties of biodiesel negatively affecting the jet fuel. The analysis of biodiesel for the properties attributed largely to the fatty esters is discussed more extensively in the section on biodiesel properties.

BIODIESEL PROPERTIES

As methanol is commonly used in its production, biodiesel is often referred to as the fatty acid methyl esters (FAME) of a vegetable oil or animal fat or other triacylglycerol-containing material. However, both moieties, the fatty acid chain and the alcohol functionality, contribute to the overall properties of a fatty ester and therefore the properties imparted by other alcohols (yielding fatty acid alkyl esters; FAAE) that could be used for producing biodiesel should be considered. Accordingly, both structural moieties will be discussed. Table 2.3 lists fuel properties of neat methyl esters of fatty acids while Table 2.4 gives properties of esters other than methyl. Properties that will be discussed here are the cetane number (CN), cold flow, oxidative stability, kinematic viscosity, and lubricity.

Cetane Number and Combustion

The dimensionless CN is a prime indicator of DF quality as it is a descriptor for the ignition behavior of the fuel. Generally, the higher the CN, the shorter the ignition delay time, ie, the time that passes between injection of the fuel into the cylinder and the onset of ignition and vice versa. Shorter ignition delay times are generally desirable.

The determination of CN is governed by standards that have been established worldwide. Such standards include ASTM D613 in the United States and, internationally, the International Organization for Standardization (ISO) standard ISO 5165. A long straight-chain hydrocarbon, hexadecane ($C_{16}H_{34}$; trivial name cetane, thus the cetane scale), is the high-quality standard on the cetane scale with an assigned CN of 100. Highly branched 2,2,4,4,6,8,8-heptamethylnonane (HMN, also $C_{16}H_{34}$), a compound with poor ignition quality, is the low-quality standard with a CN of 15 on the cetane scale. The two reference compounds on the cetane scale show that the CN decreases with decreasing chain length and increasing branching. Aromatic compounds, which occur in significant amounts in petrodiesel, have low CNs but their CNs increase with increasing size of *n*-alkyl side chains (Puckett and Caudle, 1948; Clothier et al., 1993). Compounds with $CN > 100$ or $CN < 15$ have been identified on the cetane scale. The standard ASTM D975 for petrodiesel requires a minimum CN of 40 while the standards for biodiesel prescribe a minimum of 47 in ASTM D6751 or 51 in EN 14214.

As the traditional standard for CN determination consumes a significant amount of fuel sample, an apparatus termed the Ignition Quality Tester™ (IQT™) was developed, which also offers the benefit of greater ease of use. It forms the basis of another standard, ASTM D6890, in which a derived cetane number (DCN) is obtained that correlates with the CN obtained by the traditional standard. Many CN data given here in Tables 2.3 and 2.4 were obtained as DCNs using the IQT™ (Knothe et al., 2003).

Generally, the CNs of compounds with lower CNs, such as the more unsaturated fatty compounds, show better agreement in the literature than those of compounds with higher CNs. The cause of this is the nonlinear relationship between the ignition delay time and the CN (Knothe et al., 2003). Thus, small changes at shorter ignition delay times result in greater changes in CN than at longer ignition delay times.

TABLE 2.3 Properties of Fatty Acid Methyl Esters

Trivial Name of Acid; Systematic Name; Acronym[a]	m.p. (°C)[b]	Oxidative Stability (h)[c]	Cetane No.[d]	Kinematic Viscosity (40°C; mm^2/s)[e]
Saturated				
Caprylic; octanoic; 8:0	–37.4	>24	39.75; 33.6 (98.6)	1.19
Capric; decanoic; 10:0	–13.5	>24	51.6; 47.2 (98.1)	1.72
Lauric; dodecanoic; 12:0	4.3	>24	66.7; 61.4 (99.1)	2.43
Myristic; tetradecanoic; 14:0	18.1	>24		3.30
Palmitic; hexadecanoic; 16:0	28.5	>24	85.9; 74.5 (93.6)	4.38
Stearic; octadecanoic; 18:0	37.7	>24	101; 86.9 (92.1)[f]	5.85
Arachidic; eicosanoic; 20:0	46.4			
Behenic; docosanoic; 22:0	53.2			
Lignoceric; tetracosanoic; 24:0	58.6			
Unsaturated				
Palmitoleic; 9(Z)-hexadecenoic; 16:1	–34.1	2.11	51.0, 56.6	3.67
Petroselenic; 6(Z)-octadecenoic; 18:1	–1.0			4.64

Oleic; 9(Z)-octadecenoic; 18:1	−20.2	2.79	55; 59.3	4.51
Linoleic; 9Z,12Z-octadecadienoic; 18:2	−43.1	0.94	38.2; 42.2	3.65
Linolenic; 9Z,12Z,15Z-octadecatrienoic; 18:3	−57/-52	0.00	22.7; 20.4	3.14
Gondoic; 11Z-eicosenoic; 20:1	−7.8			5.77
Arachidonic; 5(Z),8(Z),11(Z),14(Z)-eicosatetraenoic; 20:4		0.09	29.6	3.11
Erucic (13Z-docosenoic); 22:1	−3.05			7.33
4(Z),7(Z),10(Z),13(Z),16(Z),19(Z)-docosahexaenoic; 22:6		0.07	24.35	2.97

[a]*The numbers denote the number of carbons and double bonds. For example, in oleic acid, 18:1 stands for 18 carbons and 1 double bond.*
[b]*Melting point data from Knothe and Dunn (2009).*
[c]*Oxidative stability data from Knothe (2008).*
[d]*Cetane numbers from Klopfenstein (1985), Knothe et al. (2003), and Knothe (2008); CN with numbers in parentheses indicates purity (%) of the material used for CN determinations as given in Klopfenstein (1985).*
[e]*Kinematic viscosity data from Knothe and Steidley (2005a, 2011).*

TABLE 2.4 Properties of Fatty Acid Alkyl Esters Other Than Methyl

Fatty Acid Ester[a]	m.p.[b] (°C)	Oxidative Stability (h)[c]	Cetane No.[d]	Kinematic Viscosity (40°C; mm^2/s)[e]
8:0				
Ethyl ester	−44.7	>24	42.2	
Propyl ester	−45.7		39.6 (98.7)	
Butyl ester	−43.3			
10:0				
Ethyl ester	−20.4	>24	54.5; 51.2 (99.4)	1.87
Propyl ester	−21.8		52.9 (98.0)	2.30
Butyl ester	−23		54.6 (98.6)	2.60
12:0				
Ethyl ester	−1.8			2.63
Propyl ester	−4.35			3.04
Butyl ester	−6.5			3.39
Iso-butyl ester				3.48
14:0				
Ethyl ester	12.5		66.9 (99.3)	3.52
Propyl ester	9.2			4.05
Iso-propyl ester	5.6			3.91
Butyl ester				4.47
Iso-butyl ester				4.65
16:0				
Ethyl ester	23.2		93.1	4.57

TABLE 2.4 Properties of Fatty Acid Alkyl Esters Other than Methyl—cont'd

Fatty Acid Ester[a]	m.p.[b] (°C)	Oxidative Stability (h)[c]	Cetane No.[d]	Kinematic Viscosity (40°C; mm^2/s)[e]
Propyl ester	20.3		85.0	5.30
Iso-propyl ester	13–4		82.6	5.20
Butyl ester	16.1		91.9	6.49
2-Butyl ester	22.5, 28.9		84.8	6.02
Iso-butyl ester			83.6	
18:0	71		61.7	
Ethyl ester	33.0		76.8; 97.7	5.92
Propyl ester	28.1		69.9; 90.9	6.78
Iso-propyl ester	25.6		96.5	7.59
Butyl ester			80.1; 92.5	
2-Butyl ester			97.5	
Iso-butyl ester			99.3	
16:1				
Ethyl ester	−36.65			
Propyl ester	−52.6			
Butyl ester	−54.1			
18:1				
Ethyl ester	−20.3	2.68	53.9; 67.8	4.78
Propyl ester	−30.5		55.7; 58.8	5.44

Continued

TABLE 2.4 Properties of Fatty Acid Alkyl Esters Other than Methyl—cont'd

Fatty Acid Ester[a]	m.p.[b] (°C)	Oxidative Stability (h)[c]	Cetane No.[d]	Kinematic Viscosity (40°C; mm^2/s)[e]
Iso-propyl ester	−34.8		86.6	5.69
Butyl ester			59.8; 61.6	
2-Butyl ester			71.9	
Iso-butyl ester			59.6	
18:2				
Ethyl ester	−56.7		37.1; 39.6	4.25
Propyl ester	−51.5		40.6; 44.0	4.39
Butyl ester			41.6; 53.5	4.80
18:3				
Ethyl ester	−61.7		26.7	3.42
Propyl ester			26.8	
Butyl ester				
22:1				
Ethyl ester	−10.5			

[a]For definition of fatty acid acronyms, see Table 2.1.
[b]Melting point data from Knothe and Dunn (2009).
[c]Oxidative stability data from Knothe (2008).
[d]Cetane numbers from Klopfenstein (1985), Knothe et al. (2003), and Knothe (2008); CN with numbers in parentheses indicates purity (%) of the material used for CN determinations as given in Klopfenstein (1985).
[e]Kinematic viscosity data from Knothe and Steidley (2005a,b, 2011).

For petrodiesel, higher CNs were correlated with reduced nitrogen oxides (NO_x) exhaust emissions (Ladommatos et al., 1996). Despite the inherently relatively high CNs of fatty compounds, NO_x exhaust emissions usually increase slightly when operating a diesel engine on biodiesel (EPA, 2002). The connection between the structure of fatty esters and exhaust emissions was investigated (McCormick et al., 2001; Knothe et al., 2006) by studying the exhaust emissions caused by enriched FAAE as fuel. NO_x exhaust emissions reportedly increase with increasing unsaturation and decreasing chain length, which can also lead to a connection with the CNs of these compounds. Particulate emissions, on the other hand, were hardly influenced by the aforementioned structural factors. The relationship between the CN and engine emissions is complicated by many factors including the technology level of the engine with older, lower injection pressure engines generally being more sensitive to CN, with increased CN causing significant reductions in the NO_x emissions due to shorter ignition delay times, and the resulting lower average combustion temperatures. New exhaust emissions control technologies such as selective catalytic reduction largely eliminate the effect of compound structure on exhaust emissions as they leave the tailpipe.

Probably the first CN test of what is now known as biodiesel was conducted with palm oil ethyl esters (van den Abeele, 1942; Knothe, 2001). This material had a high CN, a result confirmed by later studies on many other vegetable oil–based DFs and individual fatty compounds. The influence of compound structure on CNs of fatty compounds has been discussed in more recent literature (Harrington, 1986) with the predictions made in that report being confirmed by practical cetane tests (Klopfenstein, 1985; Freedman et al., 1990; Knothe et al., 1997a, 2003). In summary and as indicated earlier, the results are that CNs decrease with increasing unsaturation and increase with increasing chain length (ie, uninterrupted CH_2 moieties). Highly polyunsaturated FAME with C_{20} and C_{22} chain lengths as found in many algal oils therefore have low CN (Knothe, 2012). However, branched esters derived from alcohols such as *iso*-propanol have CNs competitive with methyl or other straight-chain alkyl esters (Knothe et al., 2003). Thus, one long straight chain suffices to impart a high CN even if the other moiety is branched. Branched esters are of interest because they exhibit improved low-temperature properties.

The CN of mixtures such as biodiesel can be approximated by the equation (Knothe, 2012)

$$\mathrm{CN_{mix}} = \sum A_c \times \mathrm{CN_c} \qquad [2.1]$$

with CN_{mix} being the CN of the mixture (biodiesel; methyl esters), A_c is the amounts of the individual components, and CN_c is the CN of the individual components. Using the data given in Table 2.3, an example is the CN of biodiesel (methyl esters) from soybean oil (SBO), which is 50 according to Eq. [2.1] when assuming a typical fatty acid profile of soy methyl esters being 11% 16:0, 4% 18:0, 24% 18:1Δ9, 1% 18:1Δ11, 52% 18:2, and 8% 18:3 and CNs of 85.9 for 16:0, 101 for 18:0, 59.3 for both C18:1 species, 38.2 for C18:2, and 22.7 for C18:3. This result agrees well with published CN data for soy methyl esters, which are usually in the upper 40s to low 50s.

Gross heat of combustion (HG) is another fuel property indicating the suitability of fatty compounds as DF. The heats of combustion of fatty esters and triacylglycerols (Freedman and Bagby, 1989; Weast et al., 1985) are in the range of approximately 1300 to about 3500 kg cal/mol for C_8–C_{22} fatty acids and esters. HG increases with chain length. The literature value (Weast et al., 1985) for the heat of combustion of hexadecane is 2559.1 kg cal/mol (at 20°C). This value is in the same range as those of fatty compounds. HG is not included in the biodiesel standards ASTM D6751 and EN 14214 but has been included in the European standard EN 14213 when using biodiesel for heating purposes (CEN).

Cold Flow Properties

One of the major problems associated with the use of biodiesel is poor low-temperature flow properties, documented by relatively high cloud points (CP) and pour points (PP). The CP, which usually occurs at a higher temperature than the PP, is the temperature at which a fatty material becomes cloudy due to the formation of crystals and solidification of saturates. Solids and crystals rapidly grow and agglomerate, clogging fuel lines and filters and causing major operability problems. With decreasing temperature, more solids form and the material approaches the PP, the lowest temperature at which it will still flow. Saturated fatty compounds have significantly higher melting points than unsaturated fatty compounds (Tables 2.3 and 2.4) and in a mixture they crystallize at higher temperature than the unsaturates. The CP depends on the nature and amounts of the saturated fatty compounds (Imahara et al., 2006). Thus, biodiesel fuels derived from fats or oils with significant amounts of saturated fatty compounds will display higher CPs and PPs as saturated fatty compounds have significantly higher melting points than their unsaturated counterparts. A recent determination of the melting points of fatty compounds is found in Knothe and Dunn (2009).

Besides CP (ASTM D2500) and PP (ASTM D97), two test methods for the low-temperature flow properties of conventional DF exist, namely the low-temperature flow test (LTFT; used in North America; ASTM D4539), and cold filter plugging point (CFPP; used outside North America; for example the European standard EN 116) (CEN). These methods have also been used to evaluate biodiesel and its blends with No. 1 and No. 2 conventional DF. Low-temperature filterability tests were stated to be necessary because of better correlation with operability tests than CP or PP (Owen and Coley, 1995). However, for fuel formulations containing at least 10 vol% methyl esters, both LTFT and CFPP are linear functions of CP (Dunn and Bagby, 1995). Additional statistical analysis showed a strong 1:1 correlation between LTFT and CP (Dunn and Bagby, 1995).

Several approaches to the low-temperature problems of esters have been investigated including blending with conventional DF, winterization, additives, branched-chain esters, and bulky substituents in the chain. The latter approach may be considered a variation of the additive approach as the corresponding compounds have been investigated in biodiesel at additive levels. Blending of esters with conventional DF will not be discussed here.

Numerous, usually polymeric, additives were synthesized and reported mainly in the patent literature to have the effect of lowering PP or sometimes even CP, a brief overview of such additives being shown in Knothe et al. (1997b). Relatedly, the use of fatty compound-derived materials with bulky moieties in the chain (Knothe et al., 2000) at additive levels was investigated. The idea associated with these materials is that the bulky moieties in these additives would destroy the harmony of the crystallizing solids. The effect of some additives appears to be limited because they more strongly affect the PP than the CP or they had only a slight influence on CP. The CP, however, is more important than the PP for improving low-temperature flow properties (Dunn et al., 1996).

Branched esters such as *iso*-propyl, *iso*-butyl, and 2-butyl instead of methyl esters (Lee et al., 1995; Foglia et al., 1997) constitute another approach for improving the low-temperature properties of biodiesel, which is based on their lower melting points in the neat form (Table 2.4). These esters showed a lower T_{CO} (crystallization onset temperature) as determined by differential scanning calorimetry (DSC) for the isopropyl esters of SBO by 7–11°C and for the 2-butyl esters of SBO by 12–14°C (Lee et al., 1995). The CPs and PPs were also lowered by the branched-chain esters. For example, the CP of *iso*-propyl soyate was given as −9°C and that of 2-butyl soyate as −12°C (Foglia et al., 1997). In comparison, the CP of methyl soyate is 0°C (Dunn and Bagby, 1995). However, in terms of economics only, *iso*-propyl esters appear attractive as branched-chain esters although even they are more expensive than methyl esters. Branching in the ester chain does not have any negative effect on the CNs of these compounds as discussed earlier.

Winterization (Lee et al., 1996; Dunn et al., 1997) is based on the lower melting points of unsaturated fatty compounds versus saturated compounds (Table 2.3). This method removes by filtration the solids formed during cooling of the vegetable oil esters, leaving a mixture with a higher content of unsaturated fatty esters and thus with lower CP and PP. This procedure can be repeated to further reduce the CPs and PPs. Saturated fatty compounds, which have higher CNs, are among the major compounds removed by winterization. Thus, the CN of biodiesel decreases during winterization. Loss of material was reduced when winterization was carried out in the presence of cold flow improvers or solvents such as hexane and *iso*-propanol (Dunn et al., 1997).

Besides the major ester components of biodiesel contributing to its cold flow problems, minor constituents can also cause this problem (Yu et al., 1998; Dunn, 2009, 2012; He and Van Gerpen, 2012a). The most problematic species in this respect are monoacylglycerols, especially those of saturated ones, and steryl glucosides. Steryl glucosides are naturally occurring compounds in plant oils consisting of a sterol and a carbohydrate (glucose) moiety (Dunn, 2009; He and Van Gerpen, 2012a). The acylated form in which they occur can be changed to the free form during the transesterification reaction that produces biodiesel. Monoacylglycerols (monopalmitin and monostearin >70°C) and free steryl glucosides (>240°C) have high melting points so that they crystallize from biodiesel more readily under common storage conditions (Dunn, 2009; He and Van Gerpen, 2012a). Furthermore, monoglycerides in biodiesel can precipitate in different crystal forms with different solubility (Chupka et al., 2011).

Oxidative Stability

The oxidative stability of biodiesel has been the subject of considerable research as it is, besides cold flow, one of the major technical issues with biodiesel. Some recent overview articles are (Knothe, 2007; Dunn, 2008; Jain and Sharma, 2010; Xin and Saka, 2010). This issue affects biodiesel primarily during extended storage. The influence of parameters such as the presence of air, heat, traces of metal, antioxidants, and peroxides as well as the nature of the storage container was investigated in the aforementioned studies. Generally, factors such as the presence of air, elevated temperatures, or the presence of metals facilitate oxidation. Studies performed with the automated Oil Stability Index (OSI) method confirmed the catalyzing effect of metals on oxidation, however, the influence of compound structure of the fatty esters, especially unsaturation, as discussed later, was even greater (Knothe and Dunn, 2003).

In biodiesel standards, oxidative stability is usually assessed with the so-called Rancimat method in which air is bubbled through the sample at an elevated temperature (110°C) and the effluent directed into water, the conductivity of which is measured. The time at which the greatest increase in conductivity is detected is the induction time. Originally, the standard EN 14112 (CEN) was developed for studying neat FAME; however, a modified version, EN 15751 (CEN), has been developed with reduced sample volatility for the investigation of blends of biodiesel with petrodiesel. Minimum oxidative stability times prescribed in biodiesel standards are 3 h in ASTM D6751 and 8 h in EN 14214. Numerous other methods, including wet chemical ones such as acid value and peroxide value, pressurized DSC, etc., have been applied in oxidation studies of biodiesel. The peroxide value, however, is not very suitable for this purpose due to the formation of secondary oxidation products so that the peroxide value decreases, although oxidative degradation has progressed (Canakci et al., 1999).

The reason for auto-oxidation is the presence of double bonds in the chains of many fatty compounds. The auto-oxidation of unsaturated fatty compounds proceeds with different rates depending on the number and position of double bonds (Frankel, 2005). The positions allylic to double bonds are especially susceptible to oxidation. The *bis*-allylic positions in common polyunsaturated fatty acids, such as linoleic acid (double bonds at C-9 and C-12, giving one *bis*-allylic position at C-11) and linolenic acid (double bonds at C-9, C-12, and C-15, giving two *bis*-allylic positions at C-11 and C-14), are even more prone to autoxidation than allylic positions. The relative rates of oxidation given in the literature (Frankel, 2005) are 1 for oleates (methyl, ethyl esters), 41 for linoleates, and 98 for linolenates. This is essential because most biodiesel fuels contain significant amounts of esters of oleic, linoleic, or linolenic acids that influence the oxidative stability of the fuels. The species formed during the oxidation process cause the fuel to eventually deteriorate.

Induction times determined for neat fatty acid esters using the Rancimat method are also given in Tables 2.3 and 2.4. It may be noted that no unsaturated fatty ester has an induction time >3 h, which is the minimum specification in the ASTM biodiesel standard. This observation implies that antioxidant additives will almost always be necessary to attain the minimum oxidative stability specification in biodiesel standards.

Viscosity

Viscosity affects the atomization of a fuel on injection into the combustion chamber and, thereby, ultimately the formation of engine deposits. The higher the viscosity, the greater is the tendency of the fuel to cause such problems. The viscosity of a transesterified oil (ie, biodiesel) is about 1 order of magnitude lower than that of the parent oil. High viscosity is a major reason why neat vegetable oils have been largely abandoned as alternative DF. Kinematic viscosity has been included in most biodiesel standards and can be determined by standards such as ASTM D445 or ISO 3104. The effect on viscosity of blending biodiesel and petrodiesel was also investigated (Tat and Van Gerpen, 1999) and an equation was derived that allows calculation of the viscosity of such blends.

The prediction of viscosity of fatty materials has received considerable attention in the literature. Viscosity increases with chain length (number of carbon atoms) and with increasing degree of saturation. This holds also for the alcohol moiety as the viscosity of ethyl esters is slightly higher than that of methyl esters. Factors such as double bond configuration influence viscosity (*cis* double bond configuration giving a lower viscosity than *trans*), while double bond position affects viscosity less (Knothe and Steidley, 2005a). Branching in the ester moiety, however, has little or no influence on viscosity, again showing that this is a technically promising approach for improving low-temperature properties without significantly affecting other fuel properties. While often dynamic viscosity has been determined for biodiesel and its components, here only kinematic viscosity values are given as kinematic viscosity is prescribed in biodiesel standards.

Similar to the equation for calculating the CN of mixtures, the kinematic viscosity of a mixture of fatty esters can be determined by the equation (Knothe and Steidley, 2011)

$$\nu_{\text{mix}} = \sum A_C \times \nu_C \qquad [2.2]$$

with ν_{mix} being the kinematic viscosity of the biodiesel sample (mixture of FAAE), A_C the amounts of the individual esters, and ν_C is the kinematic viscosity of the individual compounds in the mixture. As an example, the kinematic viscosity (40°C) of soy methyl esters in the literature has usually been reported

~4.0–4.1 mm^2/s. This value agrees well with the application of this equation using the kinematic viscosity values given in Table 2.3 to a typical fatty acid profile of soy methyl esters being 11% 16:0, 4% 18:0, 24% 18% 18:1 Δ9, 1% 18:1 Δ11, 52% 18:2, and 8% 18:3.

Lubricity

With the advent of low-sulfur petroleum-based DFs, the issue of DF lubricity had become increasingly important. Desulfurization of conventional petrodiesel fuel reduces or eliminates the inherent lubricity of this fuel (Nikanjam and Henderson, 1993; Wang and Reynolds, 1994; Tucker et al., 1994), which is essential for proper functioning of vital engine components such as fuel pumps and injectors. It may be noted that oxygen- and nitrogen-containing species are the lubricity-imparting components to petrodiesel (Barbour et al., 2000), confirmed by lubricity studies on neat C-3 compounds with OH, NH_2, and SH groups (Knothe and Steidley, 2005a,b). Numerous studies on the lubricity of biodiesel or fatty compounds have shown the beneficial effect of these materials on the lubricity of petrodiesel, particularly low-sulfur petrodiesel fuel with poor lubricity. An advantage of biodiesel compared to lubricity-enhancing additives is that biodiesel inherently possesses fuel properties competitive and compatible with petrodiesel, which those additives usually do not possess. Thus, adding biodiesel at low levels (1–2%) restores the lubricity to low-sulfur petroleum-derived DF. On the other hand, neat free fatty acids, monoacylglycerols, and glycerol possess better lubricity than neat esters (Knothe and Steidley, 2005b; Hillion et al., 1999) because of their free OH groups as OH groups also enhance lubricity in fatty acid chains (Knothe and Steidley, 2005a,b; Goodrum and Geller, 2005), with lubricity also improving slightly with chain length and the presence of double bonds. Overall, an order of oxygenated moieties enhancing lubricity ($COOH > CHO > OH > CO OCH_3 > C{=}O > C{-}O{-}C$) was obtained from studying various oxygenated C_{10} compounds (Knothe and Steidley, 2005a,b). Adding commercial biodiesel improves the lubricity of low-sulfur petrodiesel more than neat fatty esters. This indicates that biodiesel components other than the methyl esters are primarily responsible for lubricity enhancement at low biodiesel blend levels. Adding glycerol, however, to a neat ester and then adding this mixture at low blend levels to low-lubricity petrodiesel does not improve petrodiesel lubricity, but adding polar compounds such as free fatty acids or monoacylglycerols improves the lubricity of low-level blends of esters in low-lubricity petrodiesel. Thus, some materials, namely free fatty acids and monoacylglycerols, often considered undesirable constituents because of their impairment of low-temperature or other properties of biodiesel, are largely responsible for the lubricity of low-level blends of biodiesel in ultra-low sulfur petrodiesel (Knothe and Steidley, 2005a,b). Thus, commercial biodiesel is required at a level of 1–2% in low-lubricity petrodiesel, which exceeds the typical additive level, to attain the lubricity-imparting additive level of biodiesel contaminants in petrodiesel.

The commonly applied procedure for lubricity testing of petrodiesel and biodiesel is the high-frequency reciprocating rig as described in standards such as ASTM D6079 or ISO 12156. Lubricity has not been included in biodiesel standards despite the favorable behavior of biodiesel versus petrodiesel with respect to this fuel property.

Perspectives and Conclusions

An evaluation of biodiesel fuel properties as discussed here can be used to formulate what may be termed a "designer" biodiesel fuel (Knothe, 2008, 2009). Oleic acid enrichment with linoleic acid as the secondary fatty acid has been suggested with the additional aspect that such a fatty acid composition has useful organoleptic properties (Bringe, 2005). Overall, an evaluation of the properties suggests that, when using methyl esters, methyl palmitoleate or methyl decanoate would be compounds to target for enrichment in biodiesel (Knothe, 2008). Accordingly, an oil moderately enriched in palmitoleic acid, macadamia nut oil, was investigated for its fuel properties but the moderate level of palmitoleic acid and the presence of some high-melting C20 and C22 esters led to properties typical for biodiesel from other oils (Knothe, 2010). Biodiesel derived from cuphea oil highly enriched in decanoic acid (approximately 65%) possesses improved properties including a low CP of −9 to −10°C (Knothe et al., 2009), besides showing a distillation curve that more closely resembles that of petrodiesel than that of biodiesel typically derived from vegetable oils (Lovestead et al., 2010; Fisher et al., 2010). The genetic engineering of crops to enrich medium-chain fatty acids such as decanoic acid in oils has been discussed (Dehesh, 2001). Biodiesel from field pennycress oil with a fatty acid profile consisting mainly of erucic acid and C18:1 unsaturated fatty acids also exhibited a lower CP of −10°C (Moser et al., 2009).

Overall, five approaches exist for improving biodiesel fuel properties (Knothe, 2009) as indicated in the discussion here. These approaches are (1) the use of additives, (2) the use of esters other than methyl, and changing the fatty acid profile by (3) either physical procedures, (4) genetic modification, and (5) using feedstocks with inherently differently fatty acid profiles. An approach has been discussed, taking into account the fuel properties as well as the economic and political considerations briefly addressed in the present article (Pinzi et al., 2009). Algae can be modified to increase lipid production (Roessler et al., 1994; Dunahay et al., 1996) and the fatty acid profile of algae can apparently be tailored, depending on various parameters employed for their cultivation. Recently, microbiological approaches to biodiesel fuels with modified fatty ester profiles have been reported that do not use classic lipid feedstocks, instead using carbohydrates (Kalscheuer et al., 2006; Keasling et al., 2007; Wackett, 2008; Stöveken and Steinbüchel, 2008).

DISCLAIMER

Mention of trade names or commercial products in this publication is solely for the purpose of providing specific information and does not imply recommendation or endorsement by the U.S. Department of Agriculture. USDA is an equal opportunity provider and employer.

REFERENCES

Abbaszaadeh, A., Ghobadian, B., Omidkhah, M.R., Najafi, G., 2012. Current biodiesel production technologies: a comparative review. Energy Convers. Manag. 63, 138–148.

Aguirre, A.-M., Bassi, A., Saxena, P., 2013. Engineering challenges in biodiesel production from microalgae. Crit. Rev. Biotechnol. 33 (3), 293–308.

Andrade, J.E., Pérez, A., Sebastian, P.J., Eapen, D., 2011. A review of bio-diesel production processes. Biomass Bioenergy 35, 1008–1020.

ASTM (American Society for Testing and Materials), West Conshohocken, PA. Various Standards: D97 Standard Test Method for Pour Point of Petroleum Products; D445 Standard Test Method for Kinematic Viscosity of Transparent and Opaque Liquids (and Calculation of Dynamic Viscosity); D613 Standard Test Method for Cetane Number of Diesel Fuel Oil; D975 Standard Specification for Diesel Fuel Oils; D2500 Standard Test Method for Cloud Point of Petroleum Products; D6079 Standard Test Method for Evaluating Lubricity of Diesel Fuels by the High Frequency Reciprocating Rig (HFRR); D4539 Standard Test Method for Filterability of Diesel Fuels by Low Temperature Flow Test (LTFT); D6751 Standard Specification for Biodiesel Fuel (B100) Blend Stock for Distillate Fuels; D6890 Standard Test Method for Determination of Ignition Delay and Derived Cetane Number (DCN) of Diesel Fuel Oils by Combustion in a Constant Volume Chamber. ASTM, West Conshohocken, PA.

Barbour, R.H., Rickeard, D.J., Elliott, N.G., 2000. Understanding Diesel Lubricity. SAE Technical Papers Series 2000–01–1918.

Bringe, N.A., 2005. Soybean oil composition for biodiesel. In: Knothe, G., Van Gerpen, J., Krahl, J. (Eds.), The Biodiesel Handbook, first ed. AOCS Press, Champaign, IL, pp. 161–164.

Bruwer, J.J., van d. Boshoff, B., Hugo, F.J.C., Fuls, J., Hawkins, C., van der Walt, A.N., Engelbrecht, A., du Plessis, L.M., 1980a. The Utilization of Sunflower seed oil as a renewable fuel for diesel engines. In: National Energy Symposium. ASAE, Kansas City, MO.

Bruwer, J.J., van d. Boshoff, B., Hugo, F.J.C., du Plessis, L.M., Fuls, J., Hawkins, C., van der Walt, A.N., Engelbrecht, A., 1980b. Sunflower seed oil as an extender for diesel fuel in agricultural tractors. In: Symposium of South African Institute of Agricultural Engineering.

Canakci, M., Van Gerpen, J., 1999. Biodiesel production via acid catalysis. Trans. ASAE 42, 1203–1210.

Canakci, M., Monyem, A., Van Gerpen, J., 1999. Accelerated oxidation processes in biodiesel. Trans. ASAE 42, 1565–1572.

CEN (European Committee for Standardization), Brussels, Belgium. Various Standards: EN 116 Diesel and Domestic Heating Fuels - Determination of Cold Filter Plugging Point. EN 14112 Fat and Oil Derivatives. Fatty Acid Methyl Esters (FAME). Determination of Oxidation Stability (Accelerated Oxidation Test); EN 14213 Heating Fuels. Fatty Acid Methyl Esters (FAME). Requirements and Test Methods; EN 14214 Automotive Fuels - Fatty Acid Methyl Esters (FAME) for Diesel Engines - Requirements and Test Methods, EN 15751 Automotive Fuels. Fatty Acid Methyl Ester (FAME) Fuel and Blends with Diesel Fuel. Determination of Oxidation Stability by Accelerated Oxidation Method. CEN, Brussels, Belgium.

Chapagain, B.P., Yehoshua, Y., Wiesman, Z., 2009. Desert date (*Balanites aegyptiaca*) as an arid lands sustainable bioresource for biodiesel. Bioresour. Technol. 100, 1221–1226.

Chavanne, C.G., August 31, 1937. Procédé de transformation d'huiles végétales en vue de leur utilization comme carburants (Procedure for the transformation of vegetable oils in view of their use as fuels) (Belgian Patent 422,877).

Chisti, Y., 2007. Biodiesel from microalgae. Biotechnol. Adv. 25 (3), 294–306.

Chupka, G.M., Yanowitz, J., Chiu, G., Alleman, T.L., McCormick, R.L., 2011. Effect of saturated monoglyceride polymorphism on low-temperature performance of biodiesel. Energy Fuels 25, 398–405.

Clarens, A., Colosi, L., 2010. Putting algae's promise into perspective. Biofuels 1 (6), 805–808.

Clothier, P.Q.E., Aguda, B.D., Moise, A., Pritchard, H., 1993. How do diesel-fuel ignition improvers work? Chem. Soc. Rev. 22, 101–108.

Davison, T.J., Okoli, C., Wilson, K., Lee, A.F., Harvey, A., Woodford, J., Sadhukhan, J., 2013. Multiscale modelling of heterogeneously catalysed transesterification reaction process: an overview. RSC Adv. 3 (18), 6226–6240.

Dehesh, K., 2001. How can we genetically engineer oilseed crops to produce high levels of medium-chain fatty acids? Eur. J. Lipid Sci. Technol. 103, 688–697.

Diesel, R., 1912. The diesel oil-engine and its industrial importance particularly for Great Britain. Proc. Inst. Mech. Eng. 179–280.

Di Serio, M., Tesser, R., Pengmei, L., Santacesaria, E., 2008. Heterogeneous catalysts for biodiesel production. Energy Fuels 22 (1), 207–217.

Dunahay, T.G., Jarvis, E.E., Dais, S.S., Roessler, P.G., 1996. Manipulation of microalgal lipid production using genetic engineering. Appl. Biochem. Biotechnol. 57–58, 223–231.

Dunn, R.O., 2008. Antioxidants for improving storage stability of biodiesel. Biofuels, Bioprod. Biorefin. 2, 304–318.

Dunn, R.O., 2009. Effects of minor constituents on cold flow properties and performance of biodiesel. Prog. Energy Combust. Sci. 35, 481–489.

Dunn, R.O., 2012. Effects of monoacylglycerols on the cold flow properties of biodiesel. J. Am. Oil Chem. Soc. 89, 1509–1520.

Dunn, R.O., Bagby, M.O., 1995. Low-temperature properties of triglyceride-based diesel fuels: transesterified methyl esters and petroleum middle distillate/ester blends. J. Am. Oil Chem. Soc. 72, 895–904.

Dunn, R.O., Shockley, M.W., Bagby, M.O., 1996. Improving the low-temperature properties of alternative diesel fuels; vegetable-oil derived methyl esters. J. Am. Oil Chem. Soc. 73, 1719–1728.

Dunn, R.O., Shockley, M.W., Bagby, M.O., 1997. Winterized Methyl Esters from Soybean Oil: An Alternative Diesel Fuel with Improved Low-temperature Flow Properties. SAE Technical Paper Series 971682.

Environmental Protection Agency (EPA), October 2002. A Comprehensive Analysis of Biodiesel Impacts on Exhaust Emissions. U.S. EPA Draft Technical Report 420-P-02–001 National Service Center for Environmental Publications, Cincinnati, OH. http://www.epa.gov/otaq/models/analysis/biodsl/p02001.pdf.

Fisher, B.T., Knothe, G., Mueller, C.J., 2010. Liquid-phase penetration under unsteady in-cylinder conditions: soy- and cuphea-derived biodiesel fuels versus conventional diesel. Energy Fuels 24, 5163–5180.

Fjerbaek, L., Christensen, K.V., Norddahl, B., 2009. A review of the current state of biodiesel production using enzymatic transesterification. Biotechnol. Bioeng. 102 (5), 1298–1315.

Foglia, T.A., Nelson, L.A., Dunn, R.O., Marmer, W., 1997. Low-temperature properties of alkyl esters of tallow and grease. J. Am. Oil Chem. Soc. 74, 951–955.

Foglia, T.A., Jones, K.C., Nuñez, A., Phillips, J.G., Mittelbach, M., 2004. Comparison of chromatographic methods for the determination of bound glycerol in biodiesel. Chromatographia 60 (5–6), 305–311.

Foley, P.M., Beach, E.S., Zimmerman, J.B., 2011. Algae as a source of renewable chemicals: opportunities and challenges. Green Chem. 13, 1399–1405.

Foidl, N., Foidl, G., Sanchez, M., Mittelbach, M., Hackel, S., 1996. *Jatropha curcas* L. as a source for the production of biofuel in Nicaragua. Bioresour. Technol. 58 (1), 77–82.

Freedman, B., Pryde, E.H., Mounts, T.L., 1984. Variables affecting the yields of fatty esters from transesterified vegetable oils. J. Am. Oil Chem. Soc. 61, 1638–1643.

Freedman, B., Kwolek, W.F., Pryde, E.H., 1986. Quantitation in the analysis of transesterified soybean oil by capillary gas chromatography. J. Am. Oil Chem. Soc. 63, 1370–1375.

Freedman, B., Bagby, M.O., 1989. Heats of combustion of fatty esters and triglycerides. J. Am. Oil Chem. Soc. 66, 1601–1605.

Freedman, B., Bagby, M.O., Callahan, T.J., Ryan III, T.W., 1990. Cetane Numbers of Fatty Esters, Fatty Alcohols and Triglycerides Determined in a Constant Volume Combustion Bomb. SAE Technical Paper Series 900343.

Frankel, E.N., 2005. Lipid Oxidation, second ed. The Oily Press, PJ Barnes & Associates, Bridgwater, England.

Goodrum, J.W., Geller, D.P., 2005. Influence of fatty acid methyl esters from hydroxylated vegetable oils on diesel fuel lubricity. Bioresour. Technol. 96, 851–855.

Greenwell, H.C., Laurens, L.M.L., Shields, R.J., Lovitt, R.W., Flynn, K.J., 2010. Placing microalgae on the biofuels priority list: a review of the technological challenges. J. R. Soc. Interface 7, 703–726.

Hama, S., Kondo, A., 2013. Enzymatic biodiesel production: an overview of potential feedstocks and process development. Bioresour. Technol. 135, 386–395.

Harrington, K.J., 1986. Chemical and physical properties of vegetable oil esters and their effect on diesel fuel performance. Biomass 9, 1–17.

He, B.B., Van Gerpen, J.H., 2012a. Analyzing biodiesel for contaminants and moisture retention. Biofuels 3 (3), 351–360.

He, B., Van Gerpen, J.H., 2012b. Application of ultrasonication in transesterification processes for biodiesel production. Biofuels 3 (4), 479–488.

Hillion, G., Montagne, X., Marchand, P., 1999. Methyl esters of plant oils used as additives or organic fuel. Ol., Corps Gras, Lipides 6, 435–438.

Holcapek, M., Jandera, P., Fischer, J., 2001. Analysis of acylglycerols and methyl esters of fatty acids in vegetable oils and in biodiesel. Crit. Rev. Analyt. Chem. 31 (1), 53–56.

Hu, Q., Sommerfeld, M., Jarvis, E., Ghirardi, M., Posewitz, M., Seibert, M., Darzins, A., 2008. Microalgal triacylglycerols as feedstocks for biofuel production: perspectives and advances. Plant J. 54 (4), 621–639.

Ikwuagwu, O.E., Ononogbu, I.C., Njoku, I.U., 2000. Production of biodiesel using rubber (*Hevea brasiliensis* (Kunth. Muell.)) seed oil. Ind. Crops Prod. 12, 57–62.

Imahara, H., Minami, E., Saka, S., 2006. Thermodynamic study on cloud point of biodiesel with its fatty acid composition. Fuel 85, 1666–1670.

Jain, S., Sharma, M.P., 2010. Stability of biodiesel and its blends: a review. Renew. Sustainable Energy Rev. 14, 667–678.

Kalscheuer, R., Stölting, T., Steinbüchel, A., 2006. Microdiesel: *Escherichia coli* engineered for fuel production. Microbiology 152, 2529–2536.

Kant, P., Wu, S., 2012. The extraordinary collapse of jatropha as a global biofuel. Environ. Sci. Technol. 45 (17), 7114–7115.

Keasling, J.D., Hu, Z., Somerville, C., Church, G., Berry, D., Friedman, L., Schirmer, A., Brubaker, S., del Cardayré, S.B., November 29, 2007. Production of Fatty Acids and Derivatives Thereof. WO/2007/136762 http://www.wipo.int/pctdb/en/wo.jsp?WO=2007136762.

Klopfenstein, W.E., 1985. Effect of molecular weights of fatty acid esters on cetane numbers as diesel fuels. J. Am. Oil Chem. Soc. 62, 1029–1031.

Knothe, G., Bagby, M.O., Ryan III, T.W., 1997a. Cetane Numbers of Fatty Compounds: Influence of Compound Structure and of Various Potential Cetane Improvers. SAE Technical Paper Series 971681.

Knothe, G., 2001. Historical perspectives on vegetable oil-based diesel fuels. INFORM 12, 1103–1107.

Knothe, G., 2006. Analyzing biodiesel: standards and other methods. J. Am. Oil Chem. Soc. 83 (10), 823–833.

Knothe, G., 2007. Some aspects of biodiesel oxidative stability. Fuel Proc. Technol. 88, 669–677.

Knothe, G., 2008. "Designer" biodiesel: optimizing fatty ester composition to improve fuel properties. Energy Fuels 22, 1358–1364.

Knothe, G., 2009. Improving biodiesel fuel properties by modifying fatty ester composition. Energy Environ. Sci. 2, 759–766.

Knothe, G., 2011. A technical evaluation of biodiesel from vegetable oils versus algae. Will algae-derived biodiesel perform? Green Chem. 13, 3048–3065.

Knothe, G., 2012. Fuel properties of highly polyunsaturated fatty acid methyl esters. Prediction of fuel properties of algal biodiesel. Energy Fuels 26, 5265–5273.

Knothe, G., Dunn, R.O., Bagby, M.O., 1997b. Biodiesel: the use of vegetable oils and their derivatives as alternative diesel fuels. ACS Symp. Ser. 666, 172–208 (Fuels and Chemicals from Biomass).

Knothe, G., Dunn, R.O., Shockley, M.W., Bagby, M.O., 2000. Synthesis and characterization of some long-chain diesters with branched or bulky moieties. J. Am. Oil Chem. Soc. 77, 865–871.

Knothe, G., Matheaus, A.C., Ryan III, T.W., 2003. Cetane numbers of branched and straight-chain fatty esters determined in an ignition quality tester. Fuel 82, 971–975.

Knothe, G., Dunn, R.O., 2003. Dependence of oil stability index of fatty compounds on their structure and concentration and presence of metals. J. Am. Oil Chem. Soc. 80, 1021–1026.

Knothe, G., Steidley, K.R., 2005a. Kinematic viscosity of biodiesel fuel components and related compounds. Influence of compound structure and comparison to petrodiesel fuel components. Fuel 84, 1059–1065.

Knothe, G., Steidley, K.R., 2005b. Lubricity of components of biodiesel and petrodiesel. The origin of biodiesel lubricity. Energy Fuels 19, 1192–1200.

Knothe, G., Krahl, J., Van Gerpen, J. (Eds.), 2010. The Biodiesel Handbook, second ed. AOCS Press, Urbana, IL, USA.

Knothe, G., Sharp, C.A., Ryan III, T.W., 2006. Exhaust emissions of biodiesel, petrodiesel, neat methyl esters, and alkanes in a new technology engine. Energy Fuels 20, 403–408.

Knothe, G., Dunn, R.O., 2009. A comprehensive evaluation of the melting points of fatty acids and esters determined by differential scanning calorimetry. J. Am. Oil Chem. Soc. 86, 843–856.

Knothe, G., Cermak, S.C., Evangelista, R.L., 2009. Cuphea oil as source of biodiesel with improved fuel properties caused by high content of methyl decanoate. Energy Fuels 23, 1743–1747.

Knothe, G., 2010. Biodiesel derived from a model oil enriched in palmitoleic acid, macadamia nut oil. Energy Fuels 24, 2098–2103.

Knothe, G., Steidley, K.R., 2011. Kinematic viscosity of fatty acid methyl esters: prediction, calculated viscosity contribution of esters with unavailable data, and carbon–oxygen equivalents. Fuel 90, 3217–3224.

Knothe, G., Razon, L.F., Bacani, F.T., 2013. Kenaf oil methyl esters. Ind. Crops Prod. 49, 568–572.

Kouzu, M., Hidaka, J.-s., 2012. Transesterification of vegetable oil into biodiesel catalyzed by CaO: a review. Fuel 93, 1–12.

Kralisch, D., Staffel, C., Ott, D., Bensaid, S., Saracco, G., Bellantoni, P., Loeb, P., 2013. Process design accompanying life cycle management and risk analysis as a decision support tool for sustainable biodiesel production. Green Chem. 15 (2), 463–477.

Kröger, M., Müller-Langer, F., 2012. Review on possible algal-biofuel production processes. Biofuels 3 (3), 333–349.

Ladommatos, N., Parsi, M., Knowles, A., 1996. The effect of fuel cetane improver on diesel pollutant emissions. Fuel 75, 8–14.

Lee, I., Johnson, L.A., Hammond, E.G., 1995. Use of branched-chain esters to reduce the crystallization temperature of biodiesel. J. Am. Oil Chem. Soc. 72, 1155–1160.

Lee, I., Johnson, L.A., Hammond, E.G., 1996. Reducing the crystallization temperature of biodiesel by winterizing methyl soyate. J. Am. Oil Chem. Soc. 73, 631–636.

Leite, G.B., Abdelaziz, A.E.M., Hallenbeck, P.C., 2013. Algal biofuels: challenges and opportunities. Bioresour. Technol. 145, 134–141.

Lepri, F.G., Chaves, E.S., Vieira, M.A., Ribeiro, A.S., Curtius, A.J., De Oliveira, L.C.C., De Campos, R.C., 2011. Determination of trace elements in vegetable oils and biodiesel by atomic spectrometric techniques. A review. Appl. Spectrosc. Rev. 46 (3), 173–204.

Lotero, E., Liu, Y., Lopez, D.E., Suwannakarn, K., Bruce, D.A., Goodwin Jr., J.G., 2005. Synthesis of biodiesel via acid catalysis. Ind. Eng. Chem. Res. 44 (14), 5353–5363.

Lovestead, T.M., Windom, B.C., Bruno, T.J., 2010. Investigating the unique properties of cuphea-derived biodiesel fuel with the advanced distillation curve method. Energy Fuels 24, 3665–3675.

Ma, F., Hanna, M.A., 1999. Biodiesel production: a review. Bioresour. Technol. 70, 1–15.

McCormick, R.L., Graboski, M.S., Alleman, T.L., Herring, A.M., 2001. Impact of biodiesel source material and chemical structure on emissions of criteria pollutants from a heavy-duty engine. Environ. Sci. Technol. 35, 1742–1747.

Mittelbach, M., 1996. Diesel fuel derived from vegetable oils, VI: specifications and quality control of biodiesel. Bioresour. Technol. 56, 7–11.

Mittelbach, M., Remschmidt, C., 2004. Biodiesel - the Comprehensive Handbook. M. Mittelbach, Graz, Austria.

Monteiro, M.R., Ambrozin, A.R.P., Lião, L.M., Ferreira, A.G., 2008. Critical review on analytical methods for biodiesel characterization. Talanta 77 (2), 593–605.

Moser, B.R., Knothe, G., Vaughn, S.F., Isbell, T.A., 2009. Production and evaluation of biodiesel from field pennycress (*Thlaspi arvense* L.) oil. Energy Fuels 23, 4149–4155.

Moser, B.R., Vaughn, S.F., 2010a. Evaluation of alkyl esters from *Camelina sativa* oil as biodiesel and blend components in ultra-low sulfur diesel fuel. Bioresour. Technol. 101, 646–653.

Moser, B.R., Vaughn, S.F., 2010b. Coriander seed oil methyl esters as biodiesel fuel: unique fatty acid composition and excellent oxidative stability. Biomass Bioenergy 34, 550–558.

Moser, B.R., Knothe, G., Cermak, S.C., 2010. Biodiesel from meadowfoam (*Limnanthes alba* L.) seed oil: oxidative stability and unusual fatty acid composition. Energy Environ. Sci. 3, 318–327.

Moser, B.R., Dien, B.S., Seliskar, D.M., Gallagher, J.L., 2013. Seashore mallow (*Kosteletzkya pentacarpos*) as a salt-tolerant feedstock for production of biodiesel and ethanol. Renew. Energy 50, 833–839.

Motasemi, F., Ani, F.N., 2012. A review on microwave-assisted production of biodiesel. Renew. Sustainable Energy Rev. 16 (7), 4719–4733.

Narwal, S.K., Gupta, R., 2013. Biodiesel production by transesterification using immobilized lipase. Biotechnol. Lett. 35 (4), 479–490.

Nikanjam, M., Henderson, P.T., 1993. Lubricity of Low Sulfur Diesel Fuels. SAE Technical Paper Series 932740.

Owen, K., Coley, T., 1995. Automotive Fuels Reference Book, second ed. Society of Automotive Engineers, Warrendale, PA.

Petkov, G., Ivanova, A., Iliev, I., Vaseva, I., 2012. A critical look at the microalgae biodiesel. Eur. J. Lipid Sci. Technol. 114, 103–111.

Pienkos, P.T., Darzins, A., 2009. The promise and challenges of microalgal-derived biofuels. Biofuels, Bioprod. Biorefin. 3 (4), 431–440.

Pinzi, S., Garcia, L.L., Lopez-Gimenez, F.J., Luque de Castro, M.D., Dorado, G., Dorado, M.P., 2009. The ideal vegetable oil-based biodiesel composition: a review of social, economical and technical implications. Energy Fuels 23, 2325–2341.

Plank, C., Lorbeer, E., 1995. Simultaneous determination of glycerol, and mono-, di-, and triglycerides in vegetable oil methyl esters by capillary gas chromatography. J. Chromatogr. A 697, 461–468.

Puckett, A.D., Caudle, B.H., 1948. Ignition Qualities of Hydrocarbons in the Diesel Fuel Boiling Range. U.S. Bureau of Mines Information Circular No. 7474, p. 14.

Ramachandran, K., Suganya, T., Nagendra Gandhi, N., Renganathan, S., 2013. Recent developments for biodiesel production by ultrasonic assist transesterification using different heterogeneous catalyst: a review. Renew. Sustainable Energy Rev. 22, 410–418.

Ramadhas, A.S., Muraleedharan, C., Jayaraj, S., 2005. Performance and emission evaluation of a diesel engine fueled with methyl esters of rubber seed oil. Renew. Energy 30 (12), 1789–1800.

Rashid, U., Anwar, F., Moser, B.R., Knothe, G., 2008. *Moringa oleifera* oil: a possible source of biodiesel. Bioresour. Technol. 99, 8175–8179.

Rashid, U., Anwar, F., Jamil, A., Bhatti, H.N., 2010. *Jatropha curcas* seed oil as a viable source for biodiesel. Pak. J. Bot. 42 (1), 575–582.

Rashid, U., Anwar, F., Knothe, G., 2011. Biodiesel from Milo (*Thespesia populnea* L.) seed oil. Biomass Bioenergy 35, 4034–4039.

Roessler, P.G., Brown, L.M., Dunahay, T.G., Heacox, D.A., Jarvis, E.E., Schneider, J.C., Talbot, S.G., Zeiler, K.G., 1994. Genetic engineering approaches for enhanced production of biodiesel fuel from microalgae. ACS Symp. Ser. 566, 255–270 (Enzymatic conversion of biomass for fuels production).

Scott, P.T., Preglj, L., Chen, N., Hadler, J.S., Djordjevic, M.A., Grasshoff, P.M., 2008. *Pongamia pinnata*: an untapped resource for the biofuels industry of the future. Bioenergy Res. 1 (1), 2–11.

Scott, S.A., Davey, M.P., Dennis, J.S., Horst, O., Howe, C.J., Lea-Smith, D.J., Smith, A.G., 2010. Biodiesel from algae: challenges and prospects. Curr. Opin. Biotechnol. 21 (3), 277–286.

Simionato, D., Basso, S., Giacometti, G.M., Morosinotto, T., 2013. Optimization of light use efficiency for biofuel production in algae. Biophys. Chem. 182, 71–78.

Stöveken, T., Steinbüchel, A.A., 2008. Bacterial acyltransferases as an alternative for lipase-catalyzed acylation for the production of oleochemicals and fuels. Angew. Chem. Int. Ed. Engl. 47, 3688–3694.

Tat, M.E., Van Gerpen, J.H., 1999. The kinematic viscosity of biodiesel and its blends with diesel fuel. J. Am. Oil Chem. Soc. 76, 1511–1513.

Tucker, R.F., Stradling, R.J., Wolveridge, P.E., Rivers, K.J., Ubbens, A., 1994. The Lubricity of Deeply Hydrogenated Diesel Fuels The Lubricity of Deeply Hydrogenated Diesel Fuels. The Swedish Experience. SAE Technical Paper Series 942016.

van Beilen, J.B., 2010. Why microalgal biofuels won't save the internal combustion machine. Biofuels, Bioprod. Biorefin. 4 (1), 41–52.

van den Abeele, M., 1942. L'huile de palme, matière première pour la preparation d'un carburant lourd utilizable dans les moteurs à combustion interne (Palm oil as raw material for the production of a heavy motor fuel). Bull. Agr. Congo Belge 33, 3–90 Chem. Abstr. 38:28051 (1944).

Veljkovic, V.B., Avramovic, J.M., Stamenkovic, O.S., 2012. Biodiesel production by ultrasound-assisted transesterification: state of the art and the perspectives. Renew. Sustainable Energy Rev. 16 (2), 1193–1209.

Wackett, L.P., 2008. Biomass to fuels via microbial transformations. Curr. Opin. Chem. Biol. 12, 187–193.

Wang, J.C., Reynolds, D.J., 1994. The Lubricity Requirement of Low Sulfur Diesel Fuels. SAE Technical Paper Series 942015.

Weast, R.C., Astle, M.J., Beyer, W.H., 1985–1986. Handbook of Chemistry and Physics, 66th ed. CRC Press, Boca Raton, FL, pp. D-272–D-278.

Williams, P.J.B., Laurens, L.M.L., 2010. Microalgae as biodiesel & biomass feedstocks: review & analysis of the biochemistry, energetics & economics. Energy Environ. Sci. 3 (5), 554–590.

Wood, P., 2005. Out of Africa. Could jatropha vegetable oil be Europe's biodiesel feedstock? Refocus 6 (4), 40–44.

Xin, J., Saka, S., 2010. Test methods for the determination of biodiesel stability. Biofuels 1, 275–289.

Yu, L., Lee, I., Hammond, E.G., Johnson, L.A., Van Gerpen, J.H., 1998. The influence of trace components on the melting point of methyl soyate. J. Am. Oil. Chem. Soc. 75, 1821–1824.

Zhang, W.-B., 2012. Review on analysis of biodiesel with infrared spectroscopy. Renew. Sustainable Energy Rev. 16 (8), 6048–6058.

Chapter 3

Polymeric Products Derived From Industrial Oils for Paints, Coatings, and Other Applications

Douglas G. Hayes
Department of Biosystems Engineering and Soil Science, University of Tennessee, Knoxville, TN, United States

Marie-Josée Dumont
Department of Bioresource Engineering, McGill University, Ste-Anne-de-Bellevue, QC, Canada

INTRODUCTION

Industrial oilseeds are viable feedstocks for the preparation of polymeric materials, particularly thermosetting polymers: liquid-phase polymers that are cured irreversibly via heat, irradiation, or chemical reactions to form an insoluble polymeric network. Paints, coatings, adhesives, foams, and gels are commonly thermosets. Biobased feedstocks are receiving increasing attentions as replacements for petroleum in the preparation of thermosets and other polymers due to the reduced environmental impact associated with their derivation (no mining involved, approximately neutral in the net production of the greenhouse gas, CO_2) and increased cost competitiveness.

This chapter focuses on the preparation of thermosets, polyesters, and other polymers from industrial oilseeds. Nature has provided a few examples of plant oils that possess multiple functional groups needed for polymer synthesis, such as castor (*Ricinus communis*), lesquerella (*Lesquerella fendleri*), and vernonia (*Vernonia galamensis*) oils, enriched in —OH and epoxide-functionalized fatty acids: ricinoleic, lesquerolic, and vernolic acid, respectively (Table 3.1). Many common plant seed oils (eg, soybean, cottonseed, corn, soybean, safflower, sunflower, canola, jatropha, and olive oils) are enriched in C_{16}—C_{18} saturated and mono- and diunsaturated fatty acids, such as palmitic (16:0), oleic (18:1–9c), and linoleic (18:2–9c,12c) acids and lesser amounts of α-linolenic acid (18:3–9c,12c,15c); however, linseed (flaxseed), camelina (*Camelina sativa*),

Industrial Oil Crops. http://dx.doi.org/10.1016/B978-1-893997-98-1.00003-8

TABLE 3.1 Fatty Acids Encountered in Industrial Seed Oils That Are Commonly Used in Polymerization Reactions

Fatty Acyl Species, Common Name	Molecular Structure	Seed Oil Source
Polyunsaturated Fatty Acids		
γ-Linolenic acid	18:3–6c,9c,12c	Evening primrose (*Oenothera biennis*), borage (*Borago officinalis*)
Eleostearic acid	18:3–9c,11t,13c	Tung (*Vernicia fordii*), Chinese melon (*Momordica charantia* L.)
α-Linolenic acid	18:3–9c,12c,15c	Flax (linseed; *Linum usitatissimum*), Camelina (*Camelina sativa*), perilla (*Perilla frutescens*), Conophor (*Tetracarpidium conophorum*)
Licanic acid	18:3–9t,11,13t, 4-oxy	Oiticica (*Licania rigida*)
Long-Chain Fatty Acids (≥20 Carbons)		
Eicosaenoic acid	20:1–11c	Jojoba (*Simmondsia chinensis*)
Eicosaenoic acid	20:1–5c (and 22:2–5c,13c)	Meadowfoam (*Limnanthes alba*)
Erucic acid	22:1–13c	Crambe (*Crambe abyssinica*), high erucic acid rapeseed (*Brassica napus*), pennycress (*Thlaspi arvense* L.); jojoba, meadowfoam
Hydroxy Fatty Acids		
Dimorphecolic acid	S-18:2–10t,12t —OH-9	Dimorphotheca (*Dimorphotheca pluvialis*)
Ricinoleic acid	R-18:1–9c, —OH-12	Castor (*Ricinus communis*)
Isoricinoleic acid	S-18:1–12c, —OH-9	*Wrightia tinctoria, Wrightia coccinea*
Lesquerolic acid	R-20:1–11c, —OH-14	Lesquerella (*Lesquerella fendleri*)
Epoxy Fatty Acids		
Vernolic	(+)-(12S,13R)-Epoxy-18:1–9c	Vernonia (*Vernonia galamensis*)

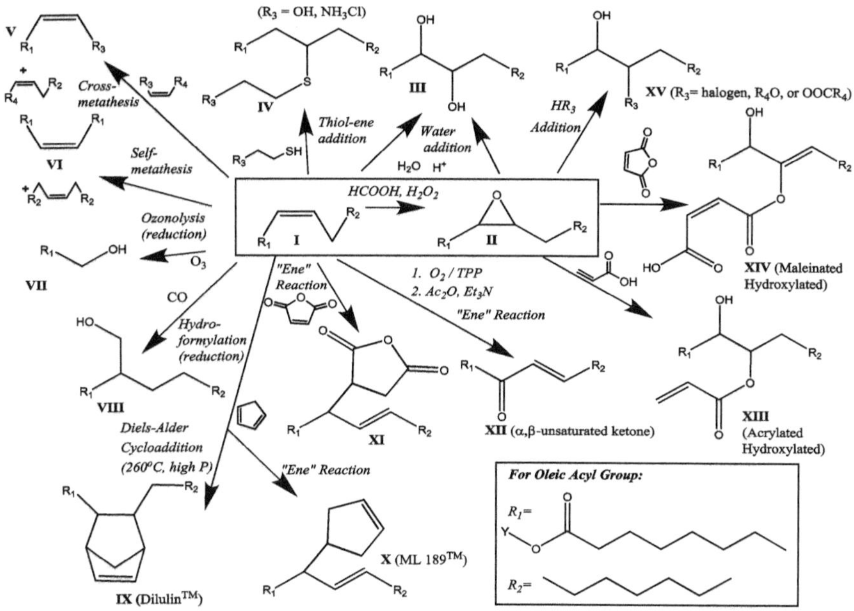

FIGURE 3.1 Modification of double bonds in fatty acyl groups to produce functional groups useful for polymerization. Starting material (**I**) is a double bond on a fatty acyl group (eg, for the oleic acyl group, R_1 and R_2 are defined in the lower right corner, where *Y* would equal CH_3 for methyl oleate, for example). Symbols: *TPP*, meso-tetraphenylporphyrin, *Ac_2O*, acetic anhydride, and *Et_3N*, triethylamine.

perilla (*Perilla frutescens*), and conophor (*Tetracarpidium conophorum*) oils possess a relatively large amount of the latter compared to other common seed oils (5–10%). To make use of free fatty acids (FFAs), fatty acid methyl esters (FAME), and triacylglycerols (TAG) from the above-mentioned common oils as monomers, additional functional groups must be incorporated, such as hydroxy, epoxy, or maleinate groups (Fig. 3.1). The hydroxy and epoxy groups can in turn serve as sites to add other functional groups (Fig. 3.1). Oils enriched in polyunsaturated fatty acids, such as α- or γ-linolenic acid or eleostearic acid, possess additional utility; for instance, they have been used as wood varnishes and in flooring for centuries. Particularly attractive are those oils enriched in conjugated double bonds (eg, tung oil, which contains eleostearic and acyl groups), which can undergo additional reactions such as Diels–Alder.

In the last 50 years, scientists have developed novel polymerization approaches, leading to new polymeric materials and applications, with the pace of new discoveries growing. This chapter provides an overview of this body of research, focusing on reactions involving covalent modification of double bonds to prepare monomers (particularly on the formation of polyols for employment as monomers for polyurethane synthesis), direct polymerization reactions between polyunsaturated acyl groups (eg, Diels–Alder, free radical, metathesis, and cationic polymerizations), and reactions unique to hydroxy and epoxy acid–containing oils. The reader is also referred to other outstanding reviews on this topic (Lu and Larock, 2009; Montero de Espinosa and Meier, 2011; Wool and Khot, 2001; Xia et al., 2013). Scientific discoveries have led to several commercially available biobased monomers and polymers, listed in Table 3.2.

Examples of bifunctional derivatives formed from fatty acids or their methyl esters that are useful monomers are given in Table 3.3.

PREPARATION OF NEW MONOMERS FROM UNSATURATED AND POLYUNSATURATED FATTY ACIDS

As outlined in Fig. 3.1, the double bond is a useful moiety for modification of fatty acyl groups (as components of TAG, FFAs, or FAME) for their conversion into monomeric units useful for polymerization. Several of the reactions that involve formation of polyols (ie, the formation of **II, III, VII, VIII**, and **XV**) will be discussed in a separate section below. Other, less common, modifications of fatty acids at double bond positions are reviewed elsewhere (Behr and Gomes, 2010).

Thiol-ene Reactions

Thiol-ene reactions (**I→IV**, Fig. 3.1) involve reactions between C=C double bonds and thiol groups, forming a thioether linkage. The reaction proceeds via a free radical mechanism, initiated by ultraviolet radiation or radical initiators (reviewed in Lligadas et al., 2013; Tueruenc and Meier, 2013) This reaction follows anti-Markovnikov behavior. Terminal double bonds (eg, 10-undecenoic acid, a derivative of castor oil, Table 3.3) are more reactive than nonterminal bonds (eg, *cis*-C_9 double bond of oleic acid). Thiol-ene reactions can also be used to form random block copolymers between α,ω-dithiols and molecules possessing double bonds at the omega position (eg, allyl 10-undecenaote) (Lluch et al., 2010). The resultant thiol-ene product, oligomeric diol, was a useful telechelic, used in the synthesis of biobased polyurethane (Lluch et al., 2010).

Alder-*ene* Reactions

Also important for modification of C=C double bonds are *Alder-ene* ("ene") reactions. Reviewed elsewhere (Lligadas et al., 2013), the *ene* reaction utilizes an alkene and an "enophile" as reactants, with the latter referring to molecule containing multiple double bonds that are frequently conjugated with electron withdrawing groups (eg, C=O and C=N). Common enophiles for modification of double bonds in fatty acyl groups include cyclopentadiene and dicarboxylic acids or their anhydrides (eg, formaldehyde, fumaric acid, and maleic anhydride). Reactions are operated at elevated temperatures (>180°C) using Lewis acid catalysts, such as $AlCl_3$, $SnCl_4$, $TiCl_4$, and alkylaluminum halides.

TABLE 3.2 Commercially Available Polymeric Products From Industrial Oilseeds

Product Name	Manufacturer	Application
Epoxidized Oils		
HiBond™ (linseed)	Polar Industries (Fisher Branch, MB, Canada)	Paints, plasticizers, adhesives, coatings, polyols (polyurethanes), thermosets
Vikoflex® (linseed, soybean)	Arkema (King-of-Prussia, PA, USA)	
PlastiSoy™ epoxidized soybean oil (ESO)	CHS (Inver Grove Heights, MN, USA)	
ESO	Makwell (Maharashtra, India), The Chemical Company (Jamestown, RI, USA), MultiPlus (Nonthaburi, Thailand), PolyMar Enterprises (Irvine, CA, USA), FMC (Philadelphia, PA, USA)	
Ebecryl 860 acrylated ESO	Advent International (Boston, MA, USA)	Surface coatings
Heloxy™ flexibilizers: castor oil and dimer acid glycidyl ether	Momentive Specialty Chemicals, Columbus, OH, USA	Epoxy resins, adhesives
ERISYS™ GE-35H castor oil and dimer acid glycidyl ethers	Emerald Performance Materials, Moorestown, NJ, USA	Epoxy resins, adhesives
Biobased Polyurethanes		
Minwax® (Tung polyurethanes)	Miniwax (Upper Saddle River, NJ USA)	Wood finish
Polymerized Tung oil	Sutherland Welles (North Hyde Park, VT, USA)	Coatings
BiOH® polyols	Cargill (Minneapolis, MN, USA)	
Biobased polyurethanes	Synthetic Natural Polymers (Durham, NC, USA)	

Continued

TABLE 3.2 Commercially Available Polymeric Products From Industrial Oilseeds—Cont'd

Product Name	Manufacturer	Application
Agrol® biobased polyols (soy, castor)	BioBased Technologies (Springdale, AR, USA)	
Lupranol® Balance 50 (ethoxylated castor-based polyol)	BASF (Ludwigshafen, Germany)	
Renuva® biobased polyols	Dow Chemical (Midland, MI, USA)	
Linseed Oil—Cyclopentadiene Polymers		
Dilulin™ (via Diels–Alder)	Cargill (Minneapolis, MN, USA)	Drying oil
ML189 (via *ene* reaction)	Archer Daniels Midland (Decatur, IL, USA)	Varnish, enamel, aluminum paint, reinforced oil
Estolides		
LubriGreen® (Coco Estolide™)	Biosynthetic Technologies (Irvine, CA, USA)	Lubricants
Vorite® polymerized castor oil	Vertellus (Indianapolis, IN, USA)	Sealant, adhesive
Zenigloss® isostearic acid-grafted copolymer of castor oil and succinic acid	Zenitech (Toronto, ON, Canada)	Emollient, lip gloss agent, personal care applications
Polyglycerol polyrinoleate	Fraken Biochemicals (Qingdao, China) Spell Organics (Gurgaon, India)	Food emulsifier/texture and viscosity control (eg, chocolate)
Polyamides		
Rilsan® Polyamide 11 (from 11-aminoundecanoic acid, derived from castor oil)	Arkema (King of Prussia, PA, USA)	Electrical cables, automotive, and pneumatic and hydraulic hose

TABLE 3.3 Monomers Derived From Industrial Oilseeds

Monomer Name	Monomer Molecular Structure	Fatty Acid Source	Method of Production	References
Diacids				
Azaelic acid	$HOOC(CH_2)_7COOH$	Oleic acid	Ozonolysis	
Sebacic acid	$HOOC(CH_2)_8COOH$	Ricinoleic acid	Alkali, 250°C	Ogunniyi (2006)
Brassylic acid	$HOOC(CH_2)_{11}COOH$	Erucic acid	Ozonolysis	Thames et al. (1998)
2-Undecendiotic acid methyl ester	$CH_3OOCCH{=}CH(CH_2)_7COOCH_3$	Methyl oleate	Cross-metathesis with methyl acrylate	
1,26-Hexacosanediodic acid	$HOOC(CH_2)_{24}COOH$	Erucic acid	Self-metathesis (reduction)	Vilela et al. (2012)
ω-Hydroxy or ω-Amino Acids				
Methyl 9-hydroxynonanoate	$HO(CH_2)_8COOCH_3$	Methyl oleate	Ozonolysis	Liu et al. (2008)
ω-Hydroxyundecanoic acid	$HO(CH_2)_{10}COOH$	Ricinoleic acid	Alkali fusion	Mutlu and Meier (2009)
ω-Aminoundecanoic acid	$HOOC(CH_2)_{10}NH_2$	Ricinoleic acid	HBr + undecenoic acid; NH_3	Ogunniyi (2006)
Fatty Acids With Terminal Unsaturation				
Methyl 9-dodecenate	$CH_3OOC(CH_2)_7CH{=}CH_2$	Methyl oleate	Cross metathesis with ethylene	Rybak et al. (2008)
10-Undecenoic acid	$HOOC(CH_2)_8CH{=}CH_2$	Ricinoleic acid	Pyrolysis	Van der Steen and Stevens Christian (2009)
Others				
1,26-Hexacosane-1,26-diol	$HO(CH_2)_{26}OH$	Erucic acid	Self-metathesis (reduction)	Vilela et al. (2012)
Mono- and di-lactones of 9-hydroxy nonanoic acid	$HO[(CH_2)_8COO]_nH$ ($n = 1–2$)	Methyl 9-hydroxynonanoate	$HfCl_4$	Liu et al. (2008)
1-Monoricinolein	$HOCH_2CH(OH)CHOOC(CH_2)_7CH{=}CHCH_2CH(OH)(CH_2)_7CH_3$	Castor oil	Hydrolysis or glycerolysis	Ogunniyi (2006)
18:2–9c,11t	$HOOC(CH_2)_8CH{=}CHCH{=}CH(CH_2)_5CH_3$	Ricinoleic acid	Dehydration	

During the reaction, the double bond of the alkene donor undergoes an allylic shift, and the allylic hydrogen atom of the alkene migrates to a double bond of the enophile (a 1,5-hydrogen shift).

Common *ene* reactions conducted for unsaturated acyl groups involve cyclopentadiene and maleic anhydride as enophiles, producing **X** and **XI**, respectively, Fig. 3.1 (Eren et al., 2003). The *ene* reaction between linseed oil and cyclopentadiene forms a commercially available polymer, ML189 (Table 3.2; (Xia and Larock, 2010)).

Cádiz et al. developed a procedure to prepare α,β-unsaturated ketones from methyl oleate or high-oleic sunflower oil via an *ene* reaction (Montero de Espinosa et al., 2008). The reaction occurs in two steps. The first step, photoperoxidation, is conducted under high pressure in the presence of ultraviolet radiation and meso-tetraphenylporphyrin (TPP), using methylene chloride as solvent. Photoperoxidation produces a peroxy group at position C_9 of oleic acid, and promotes a shift of the *cis*-C_9 double bond to *trans*-C_{10}. In the second step, acetic anhydride (Ac_2O) is added in the presence of triethylamine, Et_3N, to convert the peroxy group into an α,β-unsaturated ketone (Fig. 3.1, **I→XII**). The triketone derivative that is formed from *ene* conversion of high oleic sunflower oil is subsequently reacted with diaminodiphenylmethane (DMM), a common cross-linker, perhaps serving as a substitute of amine-cured epoxidized soybean oil, via an aza-Michael reaction (Montero de Espinosa et al., 2008). In a subsequent report, the same group replaced step two of the above-mentioned protocol with reduction via $NaBH_4$ catalyst to yield α,β-unsaturated alcohols (secondary allylic alcohols), which were subsequently esterified with acrylic acyl groups and shown to be highly reactive in free radical polymerization (Montero de Espinosa et al., 2009).

Diels–Alder Reactions

The Diels–Alder reaction, a [4+2] cycloaddition occurring between a conjugated diene and an alkene ("dienophiles") at an elevated temperature, is a well-known method for monomer preparation in oleochemistry. Fig. 3.2 depicts the Diels–Alder reaction between eleostearic acid methyl ester and acrylic acid, the dienophile (Huang et al., 2010). The *trans*-C_{11} and *cis*-C_{13} bonds of the eleostearic acyl group participate in the reaction with acrylic acid, forming a six-member ring between C_{11}—C_{14} of the former and the two carbons of the latter. The two carboxylic acid groups of the product readily serve as reaction sites for polymerization with polyamines (Huang et al., 2010). The Diels–Alder reaction can also be applied to petrochemically derived conjugated dienes and the double bonds of fatty acyl groups, with the latter serving as dienophile. An example is the synthesis of Dilulin™ (Table 3.2), involving the reaction between linseed oil and cyclopentadiene (**I→IX** of Fig. 3.1), forming a bicyclic-derivatized TAG structure. The extent of derivatization for Dilulin™ is one bicyclic group per TAG (Henna et al., 2008). The Diels–Alder, thiol-ene, and *ene* reactions are

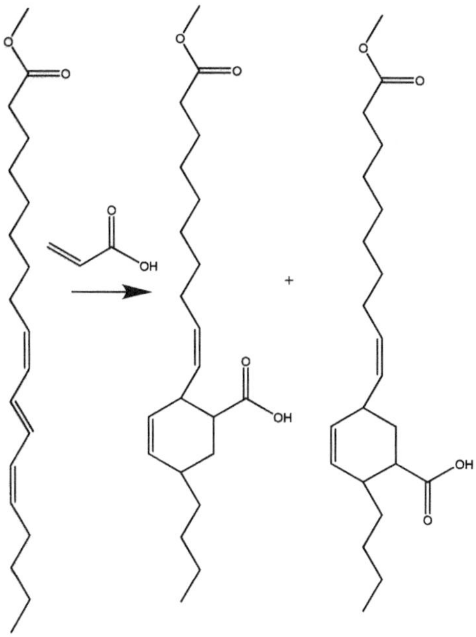

FIGURE 3.2 Diels–Alder reaction between eleostearic (18:3–9c,11t,13c) and acrylic acids. *From Huang, K., Xia, J., Yang, X., Li, M., Ding, H., 2010. Properties and curing kinetics of C_{21}-based reactive polyamides as epoxy-curing agents derived from tung oil. Polym. J. (Tokyo, Jpn.) 42, 51–57.*

all examples of "click chemistry," defined as quick and reliable reactions that involve the joining of smaller molecules to prepare a larger molecule.

The Diels–Alder reaction is also involved in the synthesis of dimer acid from oleic acid or plant oils. Dimer acids typically consist of a broad distribution of highly branched 36-carbon dicarboxylic acids that contain cyclohexene groups. They are formed using clay catalysts at a temperature above 200°C, and are manufactured worldwide for employment in polyamide resins (Breuer, 2007; Heidbreder et al., 1999). The hydrogenation of dimer acids yields dimer diols, which are useful as polyols in the preparation of polyurethanes (Heidbreder et al., 1999).

Of the fatty acyl groups listed in Table 3.1, only dimorphecolic eleostearic and licanic are conjugated. Conjugated FFA, FAME, and TAG can readily be formed via isomerization of nonconjugated sources enriched in polyunsaturated fatty acyl groups using metallic compounds (eg, ruthenium- or platinum-based), bases, or acid as catalysts (Knoerr et al., 1995; Larock, 2003; Larock et al., 2001). Oils that are commonly subject to conjugation reactions include linseed, soybean, and corn. Also, dehydration of ricinoleic acid at ~250°C in the presence of acidic catalysts produces a mixture enriched in conjugated linoleic acid, 18:2-9c,11t, and18:2-9c,12t (Mutlu and Meier, 2009; Ogunniyi, 2006). Related to this topic is conjugated linoleic acid, typically a mixture of linoleic acid isomers, with18:2-9c,11t and 18:2-10t,12c being the most prominent species, produced via alkaline isomerization of linoleic acid (Berdeaux et al., 1997; Fernie, 2003; Moreno et al., 2012). Conjugated linoleic acid is typically utilized as a dietary supplement, due to its anticarcinogenicity and ability to lower body fat, atherosclerosis, and diabetes (Berdeaux et al., 1997; Fernie, 2003; Moreno et al., 2012). It is a potential source of conjugated fatty acids for polymer synthesis.

Metathesis

Metathesis, reviewed elsewhere (Meier, 2009; Montero de Espinosa and Meier, 2012), has been a particularly useful method for modifying unsaturated oleochemicals in recent years. Although the technique was developed in the 1970s, its widespread usage started approximately 10 years as a result of novel ruthenium-based "second generation" Grubbs and Hoveyda-Grubbs catalysts, leading to the award of the 2005 Nobel Prize in Chemistry to Chauvin, Grubbs, and Schrock. In simple terms, this reaction involves the exchange of substituents attached to C=C bonds between two molecules. In cross-metathesis, two different molecules are involved with the exchange. For example, cross-metathesis between oleic acid and an unsaturated alcohol or amine would yield terminally functionalized ω-hydroxy and ω-amino fatty acids, respectively, both of which are useful monomers (cf. Fig. 3.1, **I→V**, R_3=OH or NH_2, respectively). Cross-metathesis of ethylene (R_3=R_4=H, Fig. 3.1) and methyl oleate would yield fatty acids with terminally unsaturated methyl 9-dodecenoate (Table 3.3). Self-metathesis entails an exchange between two molecules of the same species (**I→VI**, Fig. 3.1). Self-metathesis of unsaturated FFA (or FAME) would yield diacid (or diacid dimethyl ester), used as monomer for nylon preparation, with alkene formed as co-product. Self-metathesis of unsaturated TAG would produce cross-link formation between TAG molecules. Hydroxy fatty acids (Table 3.1) undergoing self-metathesis would produce α,ω-diols.

Acrylated and Maleated Epoxidized Oils

The incorporation of acrylic acid into TAG results in multifunctional monomers useful for preparation of thermosets by free radical polymerization. The esterification of acrylic acid onto hydroxy acids, such as ricinoleic or lesquerolic acid, or their TAG sources, castor and lesquerella oil, respectively, is a straight forward chemical conversion utilizing acid chlorides, or biocatalytic using lipases to carry out the acylation (Brister et al., 2000; Thames et al., 1998). Acrylated epoxidized soybean oil (AESO) (Fig. 3.1, **I→II→XIII**) is readily prepared by mixing the two reactants, often in the presence of styrene as dilutent (reviewed in Lu and Larock, 2009; Wool and Khot, 2001). AESO is produced industrially as Ebercryl 860 (Table 3.2) and used in surface coatings. AESO-styrene thermosets have been combined with natural fibers such as cellulose or flax to prepare biocomposites (reviewed in Lu and Larock, 2009) Acrylated epoxidized linseed oil has potential utility as a coating for wood (Wuzella et al., 2012). Maleic anhydride can be added to AESO to form cross-links through ester bond formation with the free hydroxyls and/or unreacted epoxide groups. Alternatively, half-esters can be formed from maleic anhydride via attachment directly onto hydroxy or epoxy acyl groups of hydroxylated or epoxidized TAG, with the typical degree of modification being 2–2.3 maleates per TAG (Fig. 3.1, **I→II→XIV**; reviewed in Lu and Larock, 2009; Wool and Khot, 2001) Maleated TAG can form cross-links with hydroxyls or epoxides on other TAG molecules. Another route for incorporating maleic anhydride for cross-linking is to produce half-esters by reacting maleic anhydride with the glycerol hydroxyls of monoacylglycerol (MAG), the latter obtained via glycerolysis of TAG (Can et al., 2001). These monomers were used in the synthesis of rigid thermosets for liquid molding resins (Can et al., 2001). The tuning of the molecular structure of AESO and maleinated TAG or MAG to achieve desired mechanical strength properties, and their applications (eg, composites with glass or natural fibers), is reviewed elsewhere (Wool and Khot, 2001).

PREPARATION OF POLYOLS FROM UNSATURATED FATTY ACIDS

Polyurethanes are a versatile class of polymers that can be tailored for numerous applications (typically as thermosets): foam (both rigid and flexible; eg, in mattresses, furniture, and insulation), automotive parts, sealants, elastomers, shoe soles, flooring, high-performance adhesives, hoses, and coatings, to name a few. Polyurethanes compose approximately 5% of the global polymer market. Their applications are anticipated to reach 18 million metric tons by 2018 (Nohra et al., 2013; Pfister et al., 2011). Polyurethanes are formed through reactions between molecules possessing two or more isocyanate (—N=C=O) groups and polyols in the presence of catalysts, forming a urethane (NH—COO—) linkage between two monomeric units. The number of isocyanate monomers typically used for polyurethane synthesis is much smaller than the number of polyol monomers. Oleochemicals have been used primarily as biobased substitutes for petroleum-derived polyols. The preparation of such polyols is the topic of this section. The preparation of biobased isocyanates and biobased polyurethanes via other synthetic routes is reviewed elsewhere (Nohra et al., 2013; Pfister et al., 2011).

Nature has provided a few hydroxy fatty acyl oils that are useful for polyol synthesis. Castor oil has important industrial applications due to its ricinoleic acid content (12-hydroxy-9-octadecenoic acid), which comprises over 89% of its overall fatty acid composition (Ogunniyi, 2006). Ricinoleic acid contains a secondary hydroxyl group within its structure which renders castor oil suitable for many applications such as adhesives and coatings, paints and varnishes, polyurethane films, elastomers and foams, among others (Ogunniyi, 2006).

Lesquerella oil contains between 54% and 60% of the hydroxy acid lesquerolic acid (and minor amounts of auricolic, R-20:2–11c,17c, OH-14, and densipolic, R-18:2–9c,15c, OH-12, acids), which can be utilized as a polyol (Hayes et al., 1995a). Other sources of hydroxy acids include dimorphotheca (*Dimorphotheca pluvialis*) oil, containing 60% of the conjugated hydroxy fatty acid, dimorphecolic acid (Cuperus et al., 1996; Derksen et al., 1992), and *Wrightia* sp. seed oil, containing 76.1% isoricinoleic acid (Ahmad and Lie Ken Jie, 2008; Siddiqi et al., 1980). In addition, vernonia oil is a natural source of epoxy fatty acids, containing 72–81% vernolic acid. The epoxy group is easily converted to a secondary hydroxyl group (Grinberg et al., 1994; Singh et al., 1993). The structures of the hydroxy and epoxy acids described above are given in Table 3.1.

Other vegetable oils must be changed into polyols by modifying C=C double bond(s) contained in their unsaturated fatty acyl group(s) to introduce hydroxyl functionality. This section covers chemical routes to hydroxylate unsaturated TAGs. The hydroxylation reactions involving air oxidation and enzymatic reactions are not covered here.

Transesterification With Polyols

Polyols can be synthesized through transesterification of vegetable oils with other polyols or alcohols, producing fatty acid polyol mono- and di-esters. This technique was found to effectively increase the hydroxyl number (OH number) of castor oil. For example, it was found that the OH number of castor oil (160 mg KOH/g) could be increased significantly when transesterified with pentaerythritol (250 mg/g KOH) (Valero and Gonzalez, 2012) or glycerol (441 mg KOH/g) (Veronese et al., 2011). Other vegetable oils from palm, rapeseed, olive and soybean, among others, were successfully converted into polyols through transesterification (Lligadas et al., 2010). Another route to increasing the OH number is to transform vegetable oils to MAG via glycerolysis or hydrolysis.

Epoxidation of TAG and Ring-Opening of Epoxide Groups

The hydroxylation of TAG through epoxidation followed by a ring opening reaction may lead to different polyol structures, depending on the reagent used. The first step of this synthesis, the oxidation of alkenes contained in fatty acyl groups of TAG by peracetic acid (ie, **I→II** in Fig. 3.1) is one of the most common routes yielding epoxidized TAG that mimic vernonia oil in structure (Sharmin et al., 2007; Tan and Chow, 2010). Generally, the reaction yield is greater than 90% (Guo et al., 2000; Javni et al., 2000; Zlatanic et al., 2004, 2002). Other reactions that involve the use of dioxiranes (Sonnet et al., 1995), aldehydes and molecular oxygen (Kuo and Chou, 1987), or hydrogen peroxide with a catalyst (Crivello and Narayan, 1992) are found in the literature. Immobilized lipases can also catalyze epoxidation of unsaturated fatty acid esters in the presence of H_2O_2 (Hagstroem et al., 2011; Lu et al., 2010).

Epoxidized oils, particularly linseed and soybean (ELSO and ESO, respectively), produced at 200,000 t annually (Gunstone, 2004), are commercially available (Table 3.2). In addition to serving as polyurethane monomers, ESO and ELSO have numerous applications in coatings: plasticizers and stabilizers for vinyl plastics (particularly polyvinylchloride), paints, coatings, nonvolatile reactive solvents for surface coatings, and ultraviolet-curable resins, to name a few (Hutchinson, 2002). ELSO, when combined with a diacid cross-linker, is a potential replacement for polyvinylchloride in flooring applications (Carter et al., 2008) and as an epoxy resin in the preparation of "green" composites (Samper et al., 2012).

The second, ring-opening, step can be achieved by several pathways. The epoxy group can be converted by: (1) alcohol-mediated ring-opening reaction (Fig. 3.1, **II→XV**; R_3=OR_4, where R_4 is an n-alkyl group; (Dai et al., 2009; Guo et al., 2000; Hu et al., 2002; Javni et al., 2000; Lozada et al., 2009; Wang et al., 2009; Zlatanic et al., 2004, 2002)), (2) catalytic hydrogenation (Campanella et al., 2009) (Fig. 3.1, **II→III**), or (3) hydroxylation with an acid (Caillol et al., 2012; Sharmin et al., 2007) (Fig. 3.1, **II→XV**; R_3=Cl or Br) or with the use of sodium cyanoborohydride and boron trifluoride (Zhao et al., 2008). Interestingly, the use of hydrochloric and hydrobromic acids (Guo et al., 2000) yield halogenated polyols. It was found that the highest yield (100%) and functionality are obtained with hydrogen bromide. However, the brominated polyols are highly viscous at room temperature and therefore challenging to process.

In general, the conversion yield of epoxides into polyols is above 75% (Zlatanic et al., 2004). All of these reactions lead to polyols with secondary hydroxyl groups. This leaves the remainder of the carbon–carbon chain dangling, allowing it to act as a plasticizer. Furthermore, reactions such as epoxidation and hydroxylation using an alcohol leave alkyl ether groups within the polyol structure, which increases stearic hindrance and may limit the yield of further reactions.

Recently, new poly(ether ester) polyols were synthesized by epoxidation, followed by hydroxylation and transesterification with 1,3-propanediol and 1,2-propanediol (Kong et al., 2011, 2012) (Fig. 3.3). This reaction has the peculiarity of yielding polyols which do not contain the glycerol backbone. The hydroxyl number of these polyols ranges from 270 to 320 mg KOH/g.

Hydroformylation–Hydrogenation Reaction

Hydroformylation (or *oxo* synthesis) is an important industrial process. This reaction allows for the attachment of an aldehyde (—CH=O) group across a C=C bond through the latter's reaction with synthetic gas (carbon monoxide and hydrogen). The formation of the alcohol group is achieved through subsequent reduction of the aldehyde (Figs. 3.1 (**I→VIII**) and 3.4). The hydroformylation

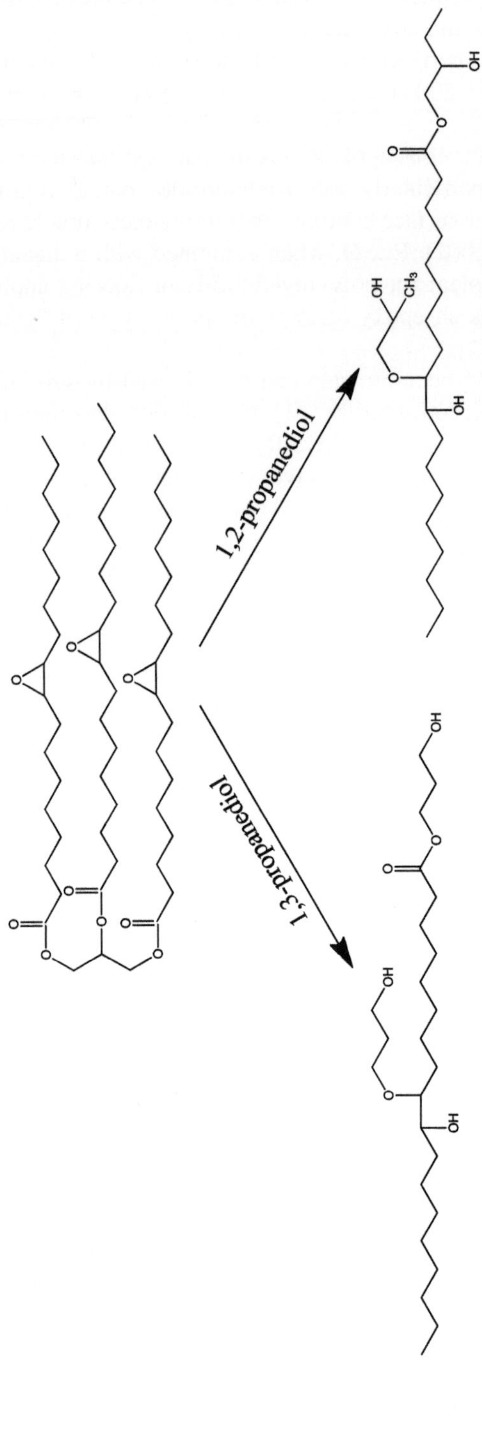

FIGURE 3.3 Poly(ester ether)polyols derived from epoxidized soybean oil and propanediols (Kong et al., 2012).

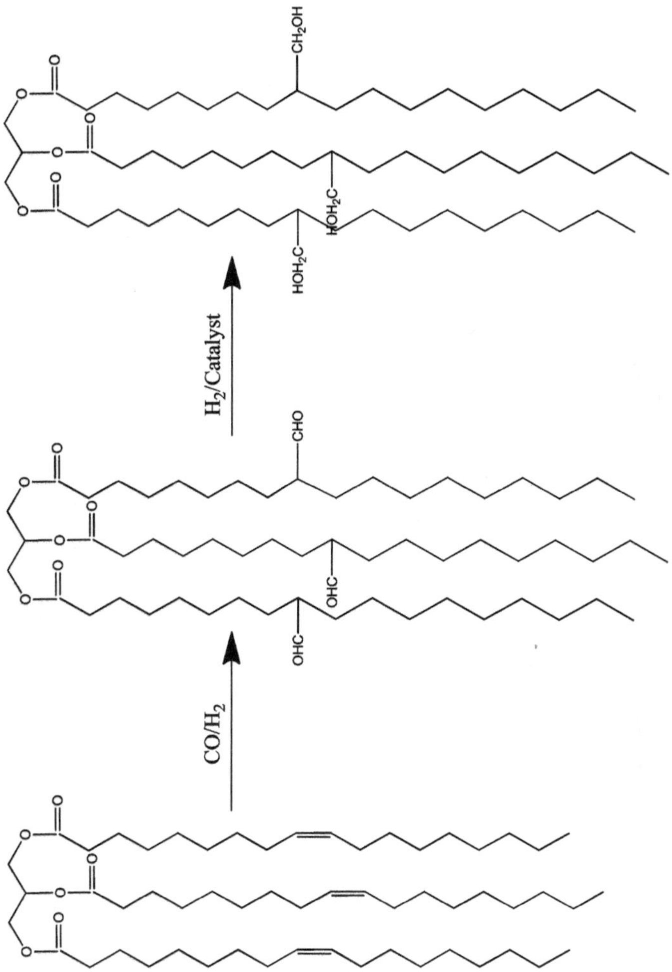

FIGURE 3.4 Hydroformylation–hydrogenation reaction (Petrovic et al., 2008).

reaction requires the use of an organometallic catalyst, where cobalt and rhodium catalysts are the most commonly used. The former has the advantage of being low cost compared to the latter but has limited efficiency (Petrovic, 2008). Rhodium is, therefore, the preferred catalyst because it generally shows a higher activity and selectivity (Behr et al., 2008). For example, a study demonstrated that the functionality and reaction yield of soy polyols was decreased by 52% and 30%, respectively, when cobalt catalyst was used instead of rhodium catalyst (Guo et al., 2002).

The hydroformylation–hydrogenation reaction generates primary hydroxyl groups, which are located in the middle of the main carbon chain and are more reactive than the secondary midchain hydroxyl groups of ricinoleic and lesquerolic acid. Similar to the epoxidation and ring-opening reactions discussed previously in this section, polyols are produced that possess long alkyl chains (Guo et al., 2002, 2006).

Ozonolysis–Hydrogenation

The ozonolysis–hydrogenation reaction yields polyols with terminal, hydroxyls. The primary hydroxyl groups have a reaction rate 3–3.3 times faster than secondary hydroxyl groups when reacting with isocyanates for the formation of polyurethanes (Lligadas et al., 2010). This is of importance because physical properties such as the degree of swelling and the cross-linking density are influenced by the hydroxyl group reactivity (Zlatanic et al., 2002). The first step of the reaction, ozonolysis, yields aldehydes and carbonyl oxides via the cycloaddition reaction of ozone to the C=C double bond of unsaturated fatty acids (Omonov et al., 2011). The product of ozonolysis (ozonide) can then be hydrolyzed as a side-reaction, producing carboxylic acid (Bailey, 1958; Narine et al., 2007). This is an obstacle to the subsequent hydrogenation step since the carboxylic acid formed cannot be converted into alcohols. However, the amount of carboxylic acid in the resulting polyol pool is not so high as to prevent the formation of polyurethanes (Kong et al., 2007; Narine et al., 2007; Omonov et al., 2011; Petrovic et al., 2005).

Polyols can be synthesized by ozonolysis in the presence of a diol (Tran et al., 2005) by the use of sodium borohydride as a reducing agent (Petrovic et al., 2005) under an alkaline medium (Guo et al., 2000), or by reduction of the ozonide through hydrogenation using Raney nickel as catalyst (Narine et al., 2007; Yue and Narine, 2007) (Figs. 3.1 (**I→VII**) and 3.5). Of note, as shown in the Fig. 3.5, the double bonds at the C_9 position are the ones that preferably undergo conversion when 18:2 and α-18.3 are ozonized. One advantage of the ozonolysis–hydrogenation reaction is the formation of short chain alcohol co-products, which have industrial applications. For example, hexanol is used in the perfume industry (Kandra and Wagner, 1998) whereas 1,3-propanediol is used as a chain extender (Miao et al., 2013; Rashmi et al., 2013).

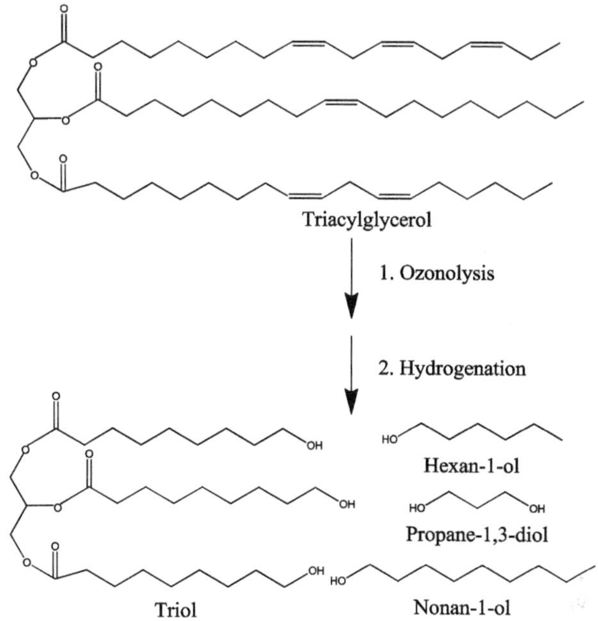

FIGURE 3.5 Ozonolysis–hydrogenation reaction (Narine et al., 2007).

Cyclotrimerization

The literature reports a few studies of the transformation of fatty acids into aromatic triols (Dumont et al., 2010; Lligadas et al., 2007; Yue and Narine, 2008) and hexa-ols (Song and Narine, 2008). These polyols, synthesized from free fatty acids, are novel because their structure has the peculiarity of containing a benzene ring (Fig. 3.6). So far, these conversions have been performed with oleic and erucic acid as starting materials. The aromatic triol synthesis pathway involves a bromination, dehydrobromination, esterification, transition metal–catalyzed cyclotrimerization, and reduction steps. Depending on the pathway, these reactions yield a mixture of aromatic symmetric and asymmetric triols (Fig. 3.6A and B) in a proportion of 17% and 61%, respectively (Yue and Narine, 2008). The yield of 1,3,5-(9-hydroxynonyl)benzene (Fig. 3.6C) is 70%, and the yield of the aromatic hexa-ols (Fig. 3.6D) is 94%. These polyols have the advantage of being highly functional and may confer high mechanical properties due to the benzene ring. However, these reactions are done through several steps requiring palladium on carbon, an expensive catalyst, as well as numerous expensive solvents. Therefore, these polyols are not likely to be utilized for commodity applications.

Employment of Thiol-ene Reaction

Polyols can also be prepared via the thiol-ene reaction applied to polyunsaturated oils (Fig. 3.1, **I→IV**, with R_3=OH). For example, as shown in Fig. 3.7,

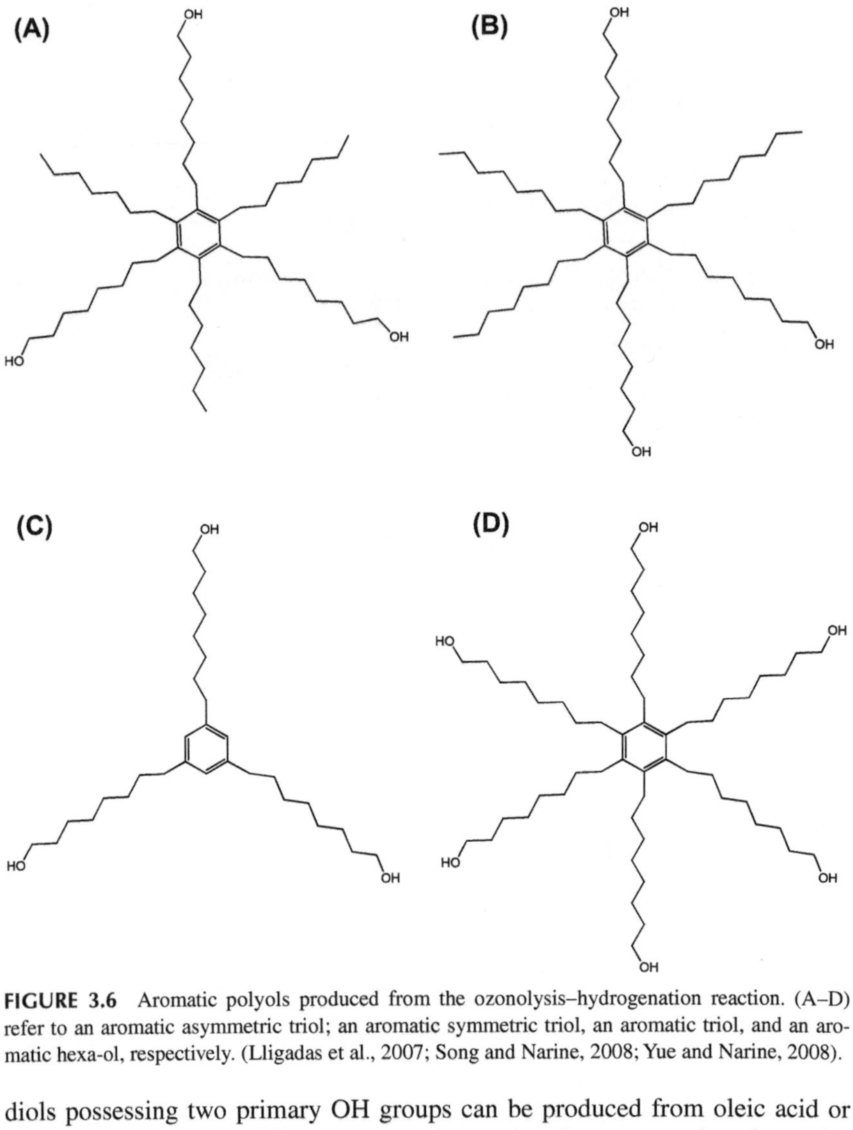

FIGURE 3.6 Aromatic polyols produced from the ozonolysis–hydrogenation reaction. (A–D) refer to an aromatic asymmetric triol; an aromatic symmetric triol, an aromatic triol, and an aromatic hexa-ol, respectively. (Lligadas et al., 2007; Song and Narine, 2008; Yue and Narine, 2008).

diols possessing two primary OH groups can be produced from oleic acid or methyl oleate by two different approaches, using 2-mercaptoethanol as thiol (Gonzalez-Paz et al., 2011). For one route, oleic acid is esterified with allyl alcohol (2-propenol) to yield an ester with two double bonds. Subsequently, the thiol-ene reaction is applied to both double bonds of the ester, yielding a diol. Alternatively, oleic acid methyl ester and 2-mercaptoethanol undergo the thiol-ene reaction, producing thiolation at the C_9 position of oleic acid; subsequently, the ester bond is reduced into a primary OH group using $LiAlH_4$.10-Undecenoic acid and its methyl ester have also successfully been converted into diols via the approaches illustrated in Fig. 3.7 (Gonzalez-Paz et al., 2011).

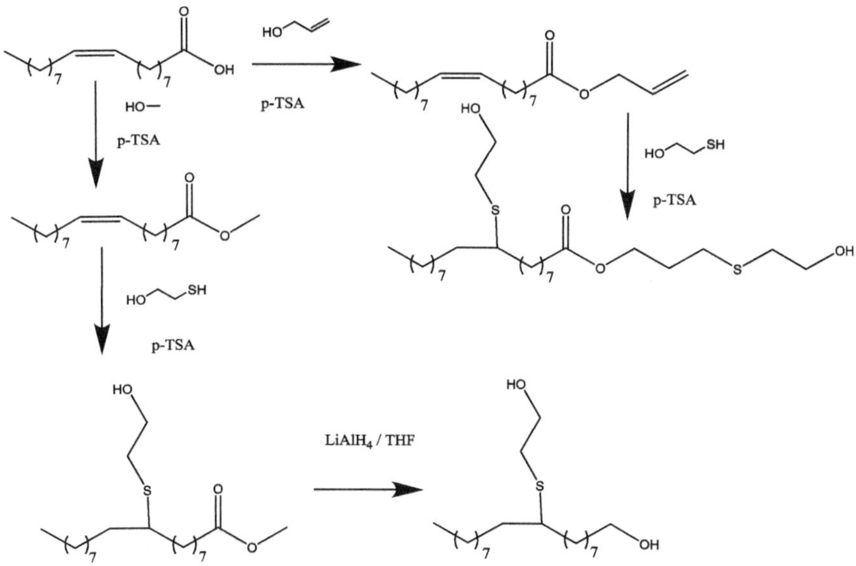

FIGURE 3.7 Polyols prepared from oleic acid or its methyl ester and 2-mercaptoethanolvia thio-ene addition. *p*-TSA refers to *p*-toluenesulfonic acid. *From Gonzalez-Paz, R.J., Lluch, C., Lligadas, G., Ronda, J.C., Galia, M., Cadiz, V., 2011. A Green approach toward oleic- and undecylenic acid-derived polyurethanes. J. Polym. Sci. Part A Polym. Chem. 49, 2407–2416.*

CONJUGATED OILS AS COATINGS: DIRECT FREE RADICAL POLYMERIZATION

Oils enriched in polyunsaturated fatty acids, particularly linseed, tung, and Chinese melon oils, have been used as coatings, particularly as varnishes for wood-based furniture and products, linoleum floor covers, and printing inks and are commonly referred to as "drying oils." The oils form a thin film coating through auto-oxidation, followed by free radical polymerization. Auto-oxidation is initiated by the dehydrogenation (oxidation) of carbon adjacent to double bonds (particularly, hydrogen abstraction from the allylic group between two double bonds) via O_2, forming conjugated and oxidized free radicals, hydroperoxides, and perepoxides (Fig. 3.8), as described in detail elsewhere (Narine and Kong, 2005; Xia et al., 2013). The latter two produce cross-linking between acyl chains via —O—O—, or peroxide, linkages. Tung-based coatings cure more rapidly than linseed oil since it major constituent, eleostearic acid (Table 3.1), has conjugated double bonds, which are more reactive than the nonconjugated double bonds in linseed oil's major fatty acyl component, α-linolenic acid. Also, the peroxidation mechanism differs between tung and linseed oils. As a result, enhanced cross-linking and branching of the resultant polymer occur for the former (Hutchinson, 2002). Therefore, linseed oil is typically combined with a chemical drying accelerator, such as an oil-soluble metal salt. To further accelerate the drying step, tung and linseed-based oils are heated to >200°C,

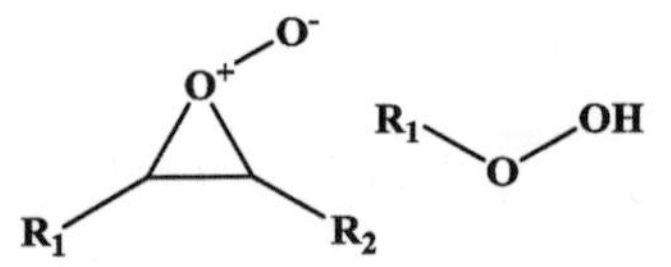

FIGURE 3.8 Molecular structure of perepoxide and hydroperoxide groups. Groups R_1, R_2 are arbitrary.

leading to the partial formation of cross-links via free radical polymerization and Diels–Alder reactions ("bodied oils"). Conjugated oils readily undergo free radical polymerization with acrylonitrile, divinylbenzene, or dicyclopentadiene, using common initiators such as azobisisobutyronitrile (AIBN), to produce transparent thermosets possessing a broad range of mechanical strength properties (reviewed in Lu and Larock, 2009; Xia et al., 2013) A modification of this approach is reacting (per)oxidized linseed or soybean oil with acrylates (eg, methyl methacrylate), taking advantage of the free radical produced from the seed oil, resulting in a polymethylmethacrylate backbone with grafted groups from the oxidized oils (Cakmakli et al., 2005, 2004). Alternatively, cross-linkers such as maleic anhydride can be added to the thermally polymerized oil (Zovi et al., 2011).

Although drying oils have been traditionally used for alkyd resins used in paints, their utility has decreased due to the decreasing popularity of oil-based paints (Hutchinson, 2002). However, due to this trend, more water-soluble "semi-drying oils," rich in α-linoleic acid (eg, soybean and sunflower) are encountered in alkyd resins today (Hutchinson, 2002).

For the more abundant and less expensive mono- and di-unsaturated fatty acyl groups and their oils, free radical polymerization is difficult to achieve; hence, several methods have been developed to incorporate acrylic groups to enhance their reactivity, and maleic acyl groups to enable cross-linking (described earlier).

CATIONIC POLYMERIZATION OF UNSATURATED OILS

An alternative reaction to free radical polymerization is cationic polymerization (reviewed in Lu and Larock, 2009; Ronda et al., 2011; Xia et al., 2013) The latter reduces the formation and entrapment of air bubbles and cracks. The reaction is initiated by strong Brønsted–Lowry and Lewis acids, such as $AlCl_3$, $TiCl_4$, $ZnCl_2$, and $SnCl_4$, with $BF_3{-}O(CH_3)_2$ being the most common. As shown in Fig. 3.9 for the latter initiator, initiation begins with its reaction with water, yielding BF_3OH_2, which readily reacts with an alkene group, yielding a cation. During propagation, the newly produced cation can react with a double bond, resulting in formation of a new covalent bond and transfer of the cationic charge (Fig. 3.9). Reactions occur at high temperatures (typically >100°C),

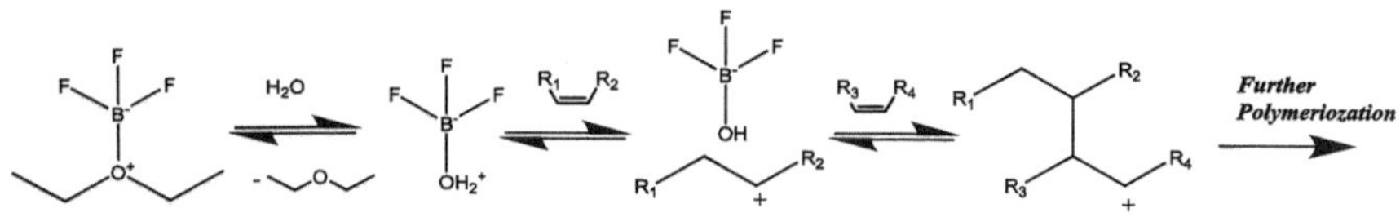

FIGURE 3.9 Mechanism of cationic polymerization. Groups R_1–R_4 are arbitrary.

frequently heated using microwaves (Ronda et al., 2011). The polymers that form are highly cross-linked and therefore possess high mechanical strength. The best-performing cationic thermosets have been produced via copolymerization of polyunsaturated oils (eg, tung, soy, or conjugated linseed) with divinylbenzene, styrene, and dicyclopentadiene. The cationic polymer products have been useful in preparing biocomposites with glass fibers and nanocomposites with layered silicates (Lu and Larock, 2009). In recent work, linseed oil–cyclopentadiene conjugates Dilulin™ and ML189 (described earlier and in Fig. 3.1 and Table 3.2) have been successfully polymerized via a cationic route with dicyclopentadiene (Xia et al., 2009).

METATHESIS POLYMERIZATION OF UNSATURATED OILS

In addition to producing novel monomers via self- and cross-metathesis (cf. Table 3.3), metathesis can be used for direct polymerization of vegetable oils. Meier and co-workers have developed acyclic diene metathesis (ADMET) for the preparation of linear polymers via a step-growth mechanism (reviewed in Meier, 2009; Montero de Espinosa and Meier, 2012). A molecule containing two double bonds, preferably at the terminal positions (α,ω-dienes; eg, 10-undecenoyl 10-undecenoate) undergoes self-metathesis, with ethylene produced as co-product. The type of polymer obtained depends in part on the nature of the monomer; for instance, ADMET polymerization of 10-undecenoyl 10-undecenoate would yield a lipophilic polyester. ADMET is terminated through addition of a chain-stopper, such as methyl acrylate. Chain-terminators serve as end groups for the polymer. The timing of their addition can be used to control the degree of polymerization. Polymeric chain terminators would yield ABA block copolymers.

To obtain a branched polymer, acyclic triene metathesis (ATMET) can be used. As shown in Fig. 3.10, self-metathesis of a polyunsaturated TAG (the "triene;" ie, triolein in the figure) would yield a hyperbranched polymer devoid of cross-links (Biermann et al., 2010). Of note, through employment of methyl acrylate as chain stopper, the resultant polymer will possess polymerizable carboxylic acid methyl ester end groups that can undergo further polymerization.

The research groups of Larock and Meier have developed and applied ring opening metathesis polymerization (ROMP) for the production of highly cross-linked polymers (reviewed in Lu and Larock, 2009; Montero de Espinosa and Meier, 2011). ROMP possesses an inherent mechanism similar to that of living

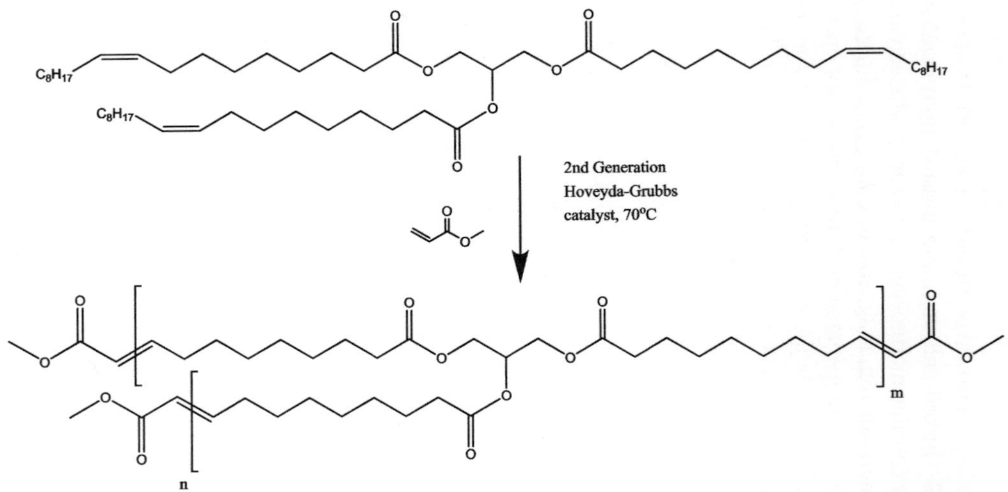

FIGURE 3.10 Acyclic triene metathesis (ATMET) of triolein, using methyl acrylate as chain stopper. *Adapted from Biermann, U., Metzger, J.O., Meier, M.A.R., 2010. Acyclic triene metathesis oligo- and polymerization of high oleic sun flower oil. Macromol. Chem. Phys. 211, 854–862.*

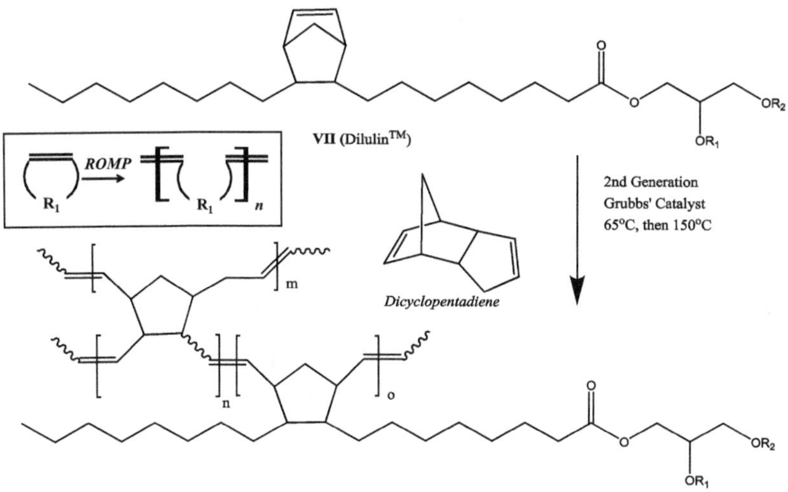

FIGURE 3.11 Ring-opening metathesis polymerization (ROMP) of Dilulin™ and dicylopentadiene. Inset: mechanism of ROMP. *Adapted from Henna, P.H., Kessler, M.R., Larock, R.C., 2008. Fabrication and properties of vegetable-oil-based glass fiber composites by ring-opening metathesis polymerization. Macromol. Mater. Eng. 293, 979–990.*

polymerization. The typical reaction, described in the inset of Fig. 3.11, is a homopolymerization (ie, self-metathesis) that often uses a norbornene derivative as monomer.

The ROMP reaction involving the norbornene-based co-monomers Dilulin™ and dicyclopentadiene is shown in Fig. 3.11 (Henna et al., 2008). For this reaction, therefore, both self- and cross-metathesis occur. As observed in the figure, the double bond of Dilulin™ is opened, yielding a cyclopentane moiety with two polymerizable double bonds attached to it as a monomeric unit. ROMP opens both double bonds of cyclopentadiene, yielding a cyclopentane group attached to four double bonds as monomeric unit, thereby allowing for enhanced cross-linking. The highly cross-linked polymers that formed possessed high glass transition temperatures that made them suitable for the preparation of biobased composites with glass fibers (Henna et al., 2008).

VULCANIZATION OF POLYUNSATURATED OILS

Plant oils (eg, soybean, rapeseed, crambe, jojoba, and meadowfoam oils, or their mixtures) react with sulfur (or S_2Cl_2), producing materials known as factices. Sulfur atoms serve as cross-linking agents between double bonds located on different TAG molecules. This process has been used for over a century (Erhan and Kleiman, 1990a,b, 1993). Factices are used as processing aids and property modifiers in rubber, and impart enhanced ozone resistance.

ESTOLIDES AND POLYCONDENSATION PRODUCTS FROM HYDROXY FATTY ACIDS

Castor and ricinoleic acid are multifunctional molecules that can serve as monomers in the synthesis of polyesters. Ricinoleic acid is well known to form homopolymers, serving as members of the "estolide" family. Estolides, oligoesters of hydroxy fatty acids and their derivatives, have potential utility as lubricant materials, sealant, adhesive, emollient, lip gloss agents, and in personal care applications (reviewed in Isbell, 2011; Zerkowski, 2008). Estolides occur in nature as well, including several species of the genus *Lesquerella* (reviewed in Hayes et al., 1995a). They readily form via polycondensation of hydroxy acids such as ricinoleic acid and lesquerolic acid, or their TAG, in the absence of catalyst at ≥175°C in vacuo (12–24 h) (Isbell and Cermak, 2002) or via lipases (Bodalo et al., 2008, 2009; Hayes, 1996; Hayes and Kleiman, 1995; Kelly and Hayes, 2006). They have been reported to occur during the isolation of dimorphotheca oil via mechanical expelling (Hayes et al., 1995b). Estolides are also produced from unsaturated fatty acids and oils (eg, oleic acid, meadowfoam oil fatty acids, soybean oil, and crambe oil) via sulfuric or perchloric (mineral) acid-catalysis at ~50–60°C for 12–24 h (reviewed in Isbell, 2011; Zerkowski, 2008). The typical average degree of polymerization achieved via the catalytic or enzymatic route varies between 2 and 5, with the average position of the internal ester bonds described as a distribution centered at the double bond position of the starting material that spans two carbons from the double bond position. The mechanism of the acid-catalyzed estolide formation entails formation of a carbocation at a double bond position, which is attacked by the COOH unit of another molecule.

In order for the estolide's physical properties to be optimal (eg, low pour points for lubrication), both the COOH and —OH end groups of estolides should be "capped" by alcohols and fatty acids, respectively, through formation of ester bonds (Isbell, 2011; Zerkowski, 2008). When using a fatty acyl source that contains saturated fatty acyl groups for acid-catalyzed reactions involving polyunsaturated oils, or non–hydroxy fatty acyl groups for the case of polycondensation in the absence of acid catalyst, these groups readily esterify to the free —OH end group of the estolide oligomer (Isbell, 2011; Kelly and Hayes, 2006; Zerkowski, 2008). Hayes and co-workers attached 10-undecenoic acid onto the —OH terminus of ricinoleic acid estolide to yield oligomer possessing more reactive end groups, allowing for its further chemical modification (Hayes et al., 2012). The attachment of alcohols or α,ω-diols to estolides can be readily achieved catalytically (eg, via use of *p*-toluenesulfonic acid) (Isbell, 2011; Zerkowski, 2008) or biocatalytically (Hayes and Kleiman, 1995). Hayes synthesized a star polymer potentially useful as a lubricant additive consisting of estolides covalently attached via their carboxylic acid end groups to polyols that contained primary OH groups (pentaerythritol, dimer diol, and trimethylolpropane) through solvent-free biocatalysis (Kelly and Hayes, 2006). Máximo and coworkers prepared polyglycerol polyricinoleate, an important food emulsifier in food products and processing (eg, in chocolates and as a greasing agent for tin; see Table 3.2 for industrial producers) using a similar enzymatic approach, with the resultant product meeting European Union standards for food additives (Bodalo et al., 2009; Gomez et al., 2011). The enzymatic approach is a desirable "green" alternative to the high-temperature approach used in current industrial practice (Wilson et al., 1998).

Other examples of the direct use of ricinoleic or lesquerolic acid, or their TAG, have been reported. Polyesters of castor oil and succinic acid, a biorefinery platform chemical, have potential utility in personal care products (O'Lenick and LaVay, 2002). Copolymers of lactic acid and ricinoleic acid, and castor oil esterified to polylactic acid have utility in drug delivery systems (Sokolsky-Papkov et al., 2009). Thames and co-workers prepared alkyds containing lesquereolic acid–trimethylolethane monoester and phthalic acid as comonomers through performing glycerolysis of lesquerella oil, followed by the addition of phthalic anhydride, with the resultant material used in the preparation of polyurethanes (Thames et al., 1994).

Ricinoleic and lesquerolic acid, their methyl esters, and their sources (castor and lesquerella oils, respectively) can be readily converted into new difunctional derivatives, or conjugated polyunsaturated fatty acids, useful for polymerization, many of which are given in Table 3.3 and have been discussed earlier: 10-undecenoic acid, conjugated linoleic acid (18:2–9c,11t), ω-hydroxyundecanoic acid and ω-amino-undecanoic acid (the latter used to prepare Polyamide 11: Table 3.2), and sebacic acid (reviewed in Mutlu and Meier, 2009). Monoricinolein, obtained by hydrolysis or glycerolysis of castor oil, is a useful polyol for polyurethane synthesis (Mutlu and Meier, 2009). The C_{12}—OH groups of castor oil and ricinoleic acid have been used to prepare new monomers through covalent attachment of new groups to them via ester bond formation. Henna and Larock attached carboxylic-functionalized norbornene derivatives to castor oil, and used the resultant product as a monomer for ROMP, producing rubbery thermosets that were stable above 400°C (Henna and Larock, 2007). Domb and coworkers prepared diacids by reacting ricinoleic acid with maleic or succinic anhydride, producing half-esters of the latter. The diacid derivatives were subsequently used to prepare biodegradable polyanhydrides (Teomim et al., 1999). Castor and lesquerella oil's —OH groups can also be readily esterified to acrylic acid and related compounds (eg, hydroxyethyl methacrylate), yielding monomers for free radical polymerization that are particularly useful for vinyl and vinyl acrylic latexes used in architectural coatings (Brister et al., 2000; Mutlu and Meier, 2009). Glycidyl ether–functionalized castor oil, which is commercially available (Table 3.2), is commonly used in epoxy resins for coatings and adhesives, to impart enhanced flexibility and impact and thermal shock resistance (Thames et al., 2000).

CONCLUSIONS

Vegetable oils are important feedstocks for the preparation of biobased polymers, particularly thermosets. Their double bonds are of particular importance for incorporating functional groups useful for free radical polymerization and incorporation of cross-linking groups. Polyunsaturated oils such as linseed and tung, and particularly the latter due to its conjugated double bonds, have been used as drying oils for paints and coatings for centuries, with their chemical modification for enhancing the properties of the coatings. Metathesis is becoming an increasingly important reaction for preparing monomers and polymers from unsaturated oils, due in part to its versatility. Of particular importance is the preparation of polyurethanes from vegetable oils and their ethoxylated forms, particularly to prepare polyols. Several different approaches for synthesizing biobased polyols have been reviewed herein. Depending on the synthetic chemical route chosen, the polyols will exhibit different molecular weight and hydroxyl numbers that will influence the viscosity and therefore the processability of the final product. It is expected that, in the future, the synthesis of polyols will become environmentally friendlier as new chemistries (eg, click chemistry) are developed in order to reduce the use of harmful chemical agents. Oils rich in hydroxy and epoxy fatty acids, such as castor and vernonia oils, respectively, have been used in polymer synthesis for several decades, and will continue to be attractive sources of monomers, particularly in the preparation of new co-polyesters.

REFERENCES

Ahmad, I., Lie Ken Jie, M.S.F., 2008. Oleochemicals from isoricinoleic acid (*Wrightia inctoria* seed oil). Ind. Eng. Chem. Res. 47, 2091–2095.

Bailey, P.S., 1958. The reactions of ozone with organic compounds. Chem. Rev. 58, 925–1010.

Behr, A., Gomes, J.P., 2010. The refinement of renewable resources: new important derivatives of fatty acids and glycerol. Eur. J. Lipid Sci. Technol. 112, 31–50.

Behr, A., Westfechtel, A., Gomes, J.P., 2008. Catalytic processes for the technical use of natural fats and oils. Chem. Eng. Technol. 31, 700–714.

Berdeaux, O., Christie, W.W., Gunstone, F.D., Sebedio, J.L., 1997. Large-scale synthesis of methyl *cis*-9, *trans*-11-octadecadienoate from methyl ricinoleate. J. Am. Oil Chem. Soc. 74, 1011–1015.

Biermann, U., Metzger, J.O., Meier, M.A.R., 2010. Acyclic triene metathesis oligo- and polymerization of high oleic sun flower oil. Macromol. Chem. Phys. 211, 854–862.

Bodalo, A., Bastida, J., Maximo, M.F., Montiel, M.C., Gomez, M., Murcia, M.D., 2008. Production of ricinoleic acid estolide with free and Immobilized lipase from *Candida rugosa*. Biochem. Eng. J. 39, 450–456.

Bodalo, A., Bastida, J., Maximo, M.F., Montiel, M.C., Gomez, M., Ortega, S., 2009. Screening and selection of lipases for the enzymatic production of polyglycerol polyricinoleate. Biochem. Eng. J. 46, 217–222.

Breuer, T.E., 2007. Dimer acids. In: Seidel, A. (Ed.), Kirk-Othmer Encyclopedia of Chemical Technology, fifth ed. John Wiley and Sons, Hoboken, NJ, USA.

Brister, E.H., Johnston, T., King, C.L., Thames, S.F., 2000. New monomers from vegetable oils. In: Havelka, K.O., McCormick, C.L. (Eds.), Specialty Monomers and Polymers. ACS Symposium Series, vol. 755. American Chemical Society, Washington, DC, pp. 159–169.

Caillol, S., Desroches, M., Boutevin, G., Loubat, C., Auvergne, R., Boutevin, B., 2012. Synthesis of new polyester polyols from epoxidized vegetable oils and biobased acids. Eur. J. Lipid Sci. Technol. 114, 1447–1459.

Cakmakli, B., Hazer, B., Tekin, I.O., Coemert, F.B., 2005. Synthesis and characterization of polymeric soybean oil-g-methyl methacrylate (and n-butyl methacrylate) graft copolymers: biocompatibility and bacterial adhesion. Biomacromolecules 6, 1750–1758.

Cakmakli, B., Hazer, B., Tekin, I.O., Kizgut, S., Koksal, M., Menceloglu, Y., 2004. Synthesis and characterization of polymeric linseed oil grafted methyl methacrylate or styrene. Macromol. Biosci. 4, 649–655.

Campanella, A., Bonnaillie, L.M., Wool, R.P., 2009. Polyurethane foams from soyoil-based polyols. J. Appl. Polym. Sci. 112, 2567–2578.

Can, E., Kusefoglu, S., Wool, R.P., 2001. Rigid, thermosetting liquid molding resins from renewable resources. I. Synthesis and polymerization of soybean oil monoglyceride maleates. J. Appl. Polym. Sci. 81, 69–77.

Carter, D.T., Stansfield, N., Mantle, R.J., France, C.M., Smith, P.A., 2008. An investigation of epoxidised linseed oil as an alternative to PVC in flooring applications. Ind. Crops Prod. 28, 309–319.

Crivello, J.V., Narayan, R., 1992. Epoxidized triglycerides as renewable monomers in photoinitiated cationic polymerization. Chem. Mater. 4, 692–699.

Cuperus, F.P., Boswinkel, G., Derksen, J.T.P., 1996. The processing of new oilseed crops – an economic evaluation. J. Am. Oil Chem. Soc. 73, 1635–1640.

Dai, H.H., Yang, L.T., Lin, B., Wang, C.S., Shi, G., 2009. Synthesis and characterization of the different soy-based polyols by ring opening of epoxidized soybean oil with methanol, 1,2-ethanediol and 1,2-propanediol. J. Am. Oil Chem. Soc. 86, 261–267.

Derksen, J.T.P., Muuse, B.G., Cuperus, F.P., Van Gelder, W.M.J., 1992. New seed oils for oleochemical industry: evaluation and enzyme-bioreactor mediated processing. Ind. Crops Prod. 1, 133–139.

Dumont, M.J., Kong, X.H., Narine, S.S., 2010. Polyurethanes from benzene polyols synthesized from vegetable oils: dependence of physical properties on structure. J. Appl. Polym. Sci. 117, 3196–3203.

Eren, T., Kuesefoglu, S.H., Wool, R., 2003. Polymerization of maleic anhydride-modified plant oils with polyols. J. Appl. Polym. Sci. 90, 197–202.

Erhan, S.M., Kleiman, R., 1990a. Meadowfoam oil factice and its performance in natural rubber mixes. Rubber World 203, 33–36.

Erhan, S.M., Kleiman, R., 1990b. Vulcanized meadowfoam oil. J. Am. Oil Chem. Soc. 67, 670–674.

Erhan, S.M., Kleiman, R., 1993. Factice from oil mixtures. J. Am. Oil Chem. Soc. 70, 309–311.

Fernie, C.E., 2003. Conjugated linoleic acid. In: Caballero, B., Trugo, L., Finglas, P.M. (Eds.), Encyclopedia of Food Sciences and Nutrition, second ed. Academic Press, Waltham, MA, USA, pp. 1581–1587.

Gomez, J.L., Bastida, J., Maximo, M.F., Montiel, M.C., Murcia, M.D., Ortega, S., 2011. Solvent-free polyglycerol polyricinoleate synthesis mediated by lipase from *Rhizopus arrhizus*. Biochem. Eng. J. 54, 111–116.

Gonzalez-Paz, R.J., Lluch, C., Lligadas, G., Ronda, J.C., Galia, M., Cadiz, V., 2011. A green approach toward oleic- and undecylenic acid-derived polyurethanes. J. Polym. Sci. Part A Polym. Chem. 49, 2407–2416.

Grinberg, S., Kolot, V., Mills, D., 1994. New chemical derivatives based on *Vernonia galamensis* oil. Ind. Crops Prod. 3, 113–119.

Gunstone, F.D., 2004. The Chemistry of Oils and Fats. CRC Press, Boca Raton, FL, USA.

Guo, A., Cho, Y.J., Petrovic, Z.S., 2000. Structure and properties of halogenated and nonhalogenated soy-based polyols. J. Polym. Sci. Part A Polym. Chem. 38, 3900–3910.

Guo, A., Demydov, D., Zhang, W., Petrovic, Z.S., 2002. Polyols and polyurethanes from hydroformylation of soybean oil. J. Polym. Environ. 10, 49–52.

Guo, A., Zhang, W., Petrovic, Z.S., 2006. Structure-property relationships in polyurethanes derived from soybean oil. J. Mater. Sci. 41, 4914–4920.

Hagstroem, A.E.V., Toernvall, U., Nordblad, M., Hatti-Kaul, R., Woodley, J.M., 2011. Chemo-enzymatic epoxidation-process options for improving biocatalytic productivity. Biotechnol. Prog. 27, 67–76.

Hayes, D.G., 1996. The catalytic activity of lipases toward hydroxy fatty acids – a review. J. Am. Oil Chem. Soc. 73, 543–549.

Hayes, D.G., Kleiman, R., 1995. Lipase-catalyzed synthesis and properties of estolides and their esters. J. Am. Oil Chem. Soc. 72, 1309–1316.

Hayes, D.G., Kleiman, R., Phillips, B.S., 1995a. The triglyceride composition, structure, and presence of estolides in the oils of *Lesquerella* and related species. J. Am. Oil Chem. Soc. 72, 559–569.

Hayes, D.G., Kleiman, R., Weisleder, D., Adlof, R.O., Cuperus, F.P., Derksen, J.T.P., 1995b. Occurrence of estolides in processed *Dimorphotheca pluvialis*. Ind. Crops Prod. 4, 295–301.

Hayes, D.G., Mannam, V.K., Ye, R., Zhao, H., Ortega, S., Montiel, M.C., 2012. Modification of oligo-ricinoleic acid and its derivatives with 10-undecenoic acid via lipase-catalyzed esterification. Polymers (Basel, Switz.) 4, 1037–1055.

Heidbreder, A., Hofer, R., Grutzmacher, R., Westfechtel, A., Blewett, C.W., 1999. Oleochemical products as building blocks for polymers. Fett/Lipid 101, 418–424.

Henna, P.H., Kessler, M.R., Larock, R.C., 2008. Fabrication and properties of vegetable-oil-based glass fiber composites by ring-opening metathesis polymerization. Macromol. Mater. Eng. 293, 979–990.

Henna, P.H., Larock, R.C., 2007. Rubbery thermosets by ring-opening metathesis polymerization of a functionalized castor oil and cyclooctene. Macromol. Mater. Eng. 292, 1201–1209.

Hu, Y.H., Gao, Y., Wang, D.N., Hu, C.P., Zu, S., Vanoverloop, L., Randall, D., 2002. Rigid polyurethane foam prepared from a rape seed oil based polyol. J. Appl. Polym. Sci. 84, 591–597.

Huang, K., Xia, J., Yang, X., Li, M., Ding, H., 2010. Properties and curing kinetics of C_{21}-based reactive polyamides as epoxy-curing agents derived from tung oil. Polym. J. (Tokyo, Jpn.) 42, 51–57.

Hutchinson, G.H., 2002. Traditional and new uses for vegetable oils in the surface coatings and allied industries. Surf. Coat. Int. Part B 85, 1–8.

Isbell, T.A., 2011. Chemistry and physical properties of estolides. Grasas Aceites (Sevilla, Spain) 62, 8–20.

Isbell, T.A., Cermak, S.C., 2002. Synthesis of triglyceride estolides from lesquerella and castor oils. J. Am. Oil Chem. Soc. 79, 1227–1233.

Javni, I., Petrovic, Z.S., Guo, A., Fuller, R., 2000. Thermal stability of polyurethanes based on vegetable oils. J. Appl. Polym. Sci. 77, 1723–1734.

Kandra, L., Wagner, G.J., 1998. Pathway for the biosynthesis of 4-methyl-1-hexanol volatilized from petal tissue of *Nicotiana sylvestris*. Phytochemistry 49, 1599–1604.

Kelly, A.R., Hayes, D.G., 2006. Lipase-catalyzed synthesis of polyhydric alcohol-poly(ricinoleic acid) ester star polymers. J. Appl. Polym. Sci. 101, 1646–1656.

Knoerr, W., Daute, P., Gruetzmacher, R., Hoefer, R., 1995. Development of new fields of application for linseed oil. Fett Wiss. Technol. 97, 165–169.

Kong, X.H., Liu, G.G., Curtis, J.M., 2011. Characterization of canola oil based polyurethane wood adhesives. Int. J. Adhes. Adhes. 31, 559–564.

Kong, X.H., Liu, G.G., Curtis, J.M., 2012. Novel polyurethane produced from canola oil based poly(ether ester) polyols: synthesis, characterization and properties. Eur. Polym. J. 48, 2097–2106.

Kong, X.H., Yue, J., Narine, S.S., 2007. Physical properties of canola oil based polyurethane networks. Biomacromolecules 8, 3584–3589.

Kuo, M.C., Chou, T.C., 1987. Kinetics and mechanism of the catalyzed epoxidation of oleic-acid with oxygen in the presence of benzaldehyde. Ind. Eng. Chem. Res. 26, 277–284.

Larock, R.C., 2003. The conjugation of natural oils. Lipid Technol. 15, 58–61.

Larock, R.C., Dong, X., Chung, S., Reddy, C.K., Ehlers, L.E., 2001. Preparation of conjugated soybean oil and other natural oils and fatty acids by homogeneous transition metal catalysis. J. Am. Oil Chem. Soc. 78, 447–453.

Liu, G., Kong, X., Wan, H., Narine, S., 2008. Production of 9-hydroxynonanoic acid from methyl oleate and conversion into lactone monomers for the synthesis of biodegradable polylactones. Biomacromolecules 9, 949–953.

Lligadas, G., Ronda, J.C., Galia, M., Cadiz, V., 2007. Polyurethane networks from fatty-acid-based aromatic triols: synthesis and characterization. Biomacromolecules 8, 1858–1864.

Lligadas, G., Ronda, J.C., Galia, M., Cadiz, V., 2010. Plant oils as platform chemicals for polyurethane synthesis: current state-of-the-art. Biomacromolecules 11, 2825–2835.

Lligadas, G., Ronda, J.C., Galia, M., Cadiz, V., 2013. Monomers and polymers from plant oils via click chemistry reactions. J. Polym. Sci. Part A Polym. Chem. 51, 2111–2124.

Lluch, C., Ronda, J.C., Galia, M., Lligadas, G., Cadiz, V., 2010. Rapid approach to biobased telechelics through two one-pot thiol-ene click reactions. Biomacromolecules 11, 1646–1653.

Lozada, Z., Suppes, G.J., Tu, Y.C., Hsieh, F.H., 2009. Soy-based polyols from oxirane ring opening by alcoholysis reaction. J. Appl. Polym. Sci. 113, 2552–2560.

Lu, H., Sun, S., Bi, Y., Yang, G., Ma, R., Yang, H., 2010. Enzymatic epoxidation of soybean oil methyl esters in the presence of free fatty acids. Eur. J. Lipid Sci. Technol. 112, 1101–1105.

Lu, Y., Larock, R.C., 2009. Novel polymeric materials from vegetable oils and vinyl monomers: preparation, properties, and applications. ChemSusChem 2, 136–147.

Meier, M.A.R., 2009. Metathesis with oleochemicals: new approaches for the utilization of plant oils as renewable resources in polymer science. Macromol. Chem. Phys. 210, 1073–1079.

Miao, S.D., Zhang, S.P., Su, Z.G., Wang, P., 2013. Synthesis of bio-based polyurethanes from epoxidized soybean oil and isopropanolamine. J. Appl. Polym. Sci. 127, 1929–1936.

Montero de Espinosa, L., Meier, M.A.R., 2011. Plant oils: the perfect renewable resource for polymer science? Eur. Polym. J. 47, 837–852.

Montero de Espinosa, L., Meier, M.A.R., 2012. Olefin metathesis of renewable platform chemicals. Top. Organomet. Chem. 39, 1–44.

Montero de Espinosa, L., Ronda, J.C., Galia, M., Cadiz, V., 2008. A new enone-containing triglyceride derivative as precursor of thermosets from renewable resources. J. Polym. Sci. Part A Polym. Chem. 46, 6843–6850.

Montero de Espinosa, L., Ronda, J.C., Galia, M., Cadiz, V., 2009. A new route to acrylate oils: crosslinking and properties of acrylate triglycerides from high oleic sunflower oil. J. Polym. Sci. Part A Polym. Chem. 47, 1159–1167.

Moreno, M., Gomez, M.V., Cebrian, C., Prieto, P., de la Hoz, A., Moreno, A., 2012. Sustainable and efficient methodology for CLA synthesis and identification. Green Chem. 14, 2584–2594.

Mutlu, H., Meier, M.A.R., 2009. Castor oil as a renewable resource for the chemical industry. Eur. J. Lipid Sci. Technol. 112, 10–30.

Narine, S.S., Kong, X., 2005. Vegetable oils in production of polymers and plastics. Bailey's Ind. Oil Fat Prod. 6, 279–306 (Sixth ed.).

Narine, S.S., Yue, J., Kong, X.H., 2007. Production of polyols from canola oil and their chemical identification and physical properties. J. Am. Oil Chem. Soc. 84, 173–179.

Nohra, B., Candy, L., Blanco, J.-F., Guerin, C., Raoul, Y., Mouloungui, Z., 2013. From petrochemical polyurethanes to biobased polyhydroxyurethanes. Macromolecules (Washington, DC, U.S.A.) 46, 3771–3792.

O'Lenick, A.J., LaVay, C., June 2002. Castor polyesters for personal care. Cosmet. Toiletries 59–62.

Ogunniyi, D.S., 2006. Castor oil: a vital industrial raw material. Bioresour. Technol. 97, 1086–1091.

Omonov, T.S., Kharraz, E., Curtis, J.M., 2011. Ozonolysis of canola oil: a study of product yields and ozonolysis kinetics in different solvent systems. J. Am. Oil Chem. Soc. 88, 689–705.

Petrovic, Z.S., 2008. Polyurethanes from vegetable oils. Polym. Rev. 48, 109–155.

Petrovic, Z.S., Guo, A., Javni, I., Cvetkovic, I., Hong, D.P., 2008. Polyurethane networks from polyols obtained by hydroformylation of soybean oil. Polym. Int. 57, 275–281.

Petrovic, Z.S., Zhang, W., Javni, I., 2005. Structure and properties of polyurethanes prepared from triglyceride polyols by ozonolysis. Biomacromolecules 6, 713–719.

Pfister, D.P., Xia, Y., Larock, R.C., 2011. Recent advances in vegetable oil-based polyurethanes. ChemSusChem 4, 703–717.

Rashmi, B.J., Rusu, D., Prashantha, K., Lacrampe, M.F., Krawczak, P., 2013. Development of water-blown bio-based thermoplastic polyurethane foams using bio-derived chain extender. J. Appl. Polym. Sci. 128, 292–303.

Ronda, J.C., Lligadas, G., Galia, M., Cadiz, V., 2011. Vegetable oils as platform chemicals for polymer synthesis. Eur. J. Lipid Sci. Technol. 113, 46–58.

Rybak, A., Fokou, P.A., Meier, M.A.R., 2008. Metathesis as a versatile tool in oleochemistry. Eur. J. Lipid Sci. Technol. 110, 797–804.

Samper, M.D., Fombuena, V., Boronat, T., Garcia-Sanoguera, D., Balart, R., 2012. Thermal and mechanical characterization of epoxy resins (ELO and ESO) cured with anhydrides. J. Am. Oil Chem. Soc. 89, 1521–1528.

Sharmin, E., Ashraf, S.M., Ahmad, S., 2007. Epoxidation hydroxylation acrylation and urethanation of *Linum usitatissimum* seed oil and its derivatives. Eur. J. Lipid Sci. Technol. 109, 134–146.

Siddiqi, S.F., Ahmad, F., Saleem Siddiqi, M., Osman, S.M., 1980. *Wrightia coccinea* seed oil: a rich source of isoricinoleic acid. Chem. Ind. (London) 115–116.

Singh, C., Kapur, N., Kaul, B.L., 1993. *Vernonia galamensis* (Cass.) oil – a potential new source of epoxidized triglyceride oil. Res. Ind. 38, 152–153.

Sokolsky-Papkov, M., Shikanov, A., Ezra, A., Vaisman, B., Domb, A.J., 2009. Fatty acid-based biodegradable polymers: synthesis and applications. In: Celina, M.C., Wiggins, J.S., Billingham, N.C. (Eds.), Polymer Degradation and Performance. ACS Symposium Series, vol. 1004. American Chemical Society, Washingtin, DC, pp. 60–69.

Song, D., Narine, S.S., 2008. Synthesis and characterization of highly functionalized symmetric aromatic hexa-ol intermediates from oleic acid. Chem. Phys. Lipids 155, 43–47.

Sonnet, P.E., Lankin, M.E., McNeill, G.P., 1995. Reactions of dioxiranes with selected oleochemicals. J. Am. Oil Chem. Soc. 72, 199–204.

Tan, S.G., Chow, W.S., 2010. Biobased epoxidized vegetable oils and its greener epoxy blends: a review. Polym.-Plast. Technol. Eng. 49, 1581–1590.

Teomim, D., Nyska, A., Domb, A.J., 1999. Ricinoleic acid-based biopolymers. J. Biomed. Mater. Res. 45, 258–267.

Thames, S.F., Bautista, M.O., Watson, M.D., Wang, M.D., 1994. Application of lesquerella oil in industrial coatings. In: Fishman, M.L., Friedman, R.B., Huang, S.J. (Eds.), Polymers from Agricultural Coproducts. ACS Symposium Series, vol. 575. American Chemical Society, Washington, DC, pp. 212–222.

Thames, S.F., Blanton, M.D., Mendon, S., Subramanian, R., Yu, H., 1998. Surfactants and fatty acids: plant oils. Biopolym. Renewable Resour. 249–280.

Thames, S.F., Yu, H., Subramanian, R., 2000. Cationic ultraviolet curable coatings from castor oil. J. Appl. Polym. Sci. 77, 8–13.

Tran, P., Graiver, D., Narayan, R., 2005. Ozone-mediated polyol synthesis from soybean oil. J. Am. Oil Chem. Soc. 82, 653–659.

Tueruenc, O., Meier, M.A.R., 2013. The thiol-ene (click) reaction for the synthesis of plant oil derived polymers. Eur. J. Lipid Sci. Technol. 115, 41–54.

Valero, M.F., Gonzalez, A., 2012. Polyurethane adhesive system from castor oil modified by a transesterification reaction. J. Elastomers Plast. 44, 433–442.

Van der Steen, M., Stevens, C.V., 2009. Undecylenic acid: a valuable and physiologically active renewable building block from castor oil. ChemSusChem 2, 692–713.

Veronese, V.B., Menger, R.K., Forte, M.M.D., Petzhold, C.L., 2011. Rigid polyurethane foam based on modified vegetable oil. J. Appl. Polym. Sci. 120, 530–537.

Vilela, C., Silvestre, A.J.D., Meier, M.A.R., 2012. Plant oil-based long-chain C_{26} monomers and their polymers. Macromol. Chem. Phys. 213, 2220–2227.

Wang, C.S., Yang, L.T., Ni, B.L., Shi, G., 2009. Polyurethane networks from different soy-based polyols by the ring opening of epoxidized soybean oil with methanol, glycol, and 1,2-propanediol. J. Appl. Polym. Sci. 114, 125–131.

Wilson, R., Van Schie, B.J., Howes, D., 1998. Overview of the preparation, use and biological studies on polyglycerol polyricinoleate (PGPR). Food Chem. Toxicol. 36, 711–718.

Wool, R.P., Khot, S.N., 2001. Bio-based resins and natural fibers. In: Donaldson, S.L., Miracle, D.B. (Eds.), ASM Handbook, Volume 21 (Composites). ASM International, Materials Park, OH, USA, pp. 184–193.

Wuzella, G., Mahendran, A.R., Mueller, U., Kandelbauer, A., Teischinger, A., 2012. Photocrosslinking of an acrylated epoxidized linseed oil: kinetics and its application for optimized wood coatings. J. Polym. Environ. 20, 1063–1074.

Xia, Y., Henna, P.H., Larock, R.C., 2009. Novel thermosets from the cationic copolymerization of modified linseed oils and dicyclopentadiene. Macromol. Mater. Eng. 294, 590–598.

Xia, Y., Larock, R.C., 2010. Vegetable oil-based polymeric materials: synthesis, properties, and applications. Green Chem. 12, 1893–1909.

Xia, Y., Quirino, R.L., Larock, R.C., 2013. Bio-based thermosetting polymers from vegetable oils. J. Renewable Mater. 1, 3–27.

Yue, J., Narine, S.S., 2007. Separation and quantification of vegetable oil based polyols by high performance liquid chromatography with evaporative light scattering detection. J. Am. Oil Chem. Soc. 84, 803–807.

Yue, J., Narine, S.S., 2008. Synthesis of aromatic triols and triacids from oleic and erucic acid: separation and characterization of the asymmetric and symmetric isomers. Chem. Phys. Lipids 152, 1–8.

Zerkowski, J.A., 2008. Estolides: from structure and function to structured and functionalized. Lipid Technol. 20, 253–256.

Zhao, H.P., Zhang, J.F., Sun, X.S., Hua, D.H., 2008. Syntheses and properties of cross-linked polymers from functionalized triglycerides. J. Appl. Polym. Sci. 110, 647–656.

Zlatanic, A., Lava, C., Zhang, W., Petrovic, Z.S., 2004. Effect of structure on properties of polyols and polyurethanes based on different vegetable oils. J. Polym. Sci. Part B Polym. Phys. 42, 809–819.

Zlatanic, A., Petrovic, Z.S., Dusek, K., 2002. Structure and properties of triolein-based polyurethane networks. Biomacromolecules 3, 1048–1056.

Zovi, O., Lecamp, L., Loutelier-Bourhis, C., Lange, C.M., Bunel, C., 2011. A solventless synthesis process of new UV-curable materials based on linseed oil. Green Chem. 13, 1014–1022.

Chapter 4

Castor (*Ricinus communis* L.)

Thomas A. McKeon
United States Department of Agriculture, Agricultural Research Service, Western Regional Research Center, Albany, CA, United States

INTRODUCTION

The castor plant has been cultivated for thousands of years, providing a useful source of medicine, lamp fuel, and lubricant long before petroleum came into wide use (Copley et al., 2005). The oil content of castor seed ranges from 45% to 60%, with most cultivars closer to 50% oil content. The presence of the hydroxy fatty acid ricinoleate (Fig. 4.1) [(9Z, 12R)-12-hydroxyoctadec-9-enoate] in the oil at levels up to 90% imparts unique physical and chemical properties that give the oil unique value as a chemical feedstock. This midchain hydroxy group results in hydrogen bonding among the triacylglycerols (TGs) that carry the 2 or 3 ricinoleoyl chains (Lin et al., 2003). The intermolecular interaction results in viscosity that is significantly greater than that of other commodity seed oils, providing superior properties for use in lubricants and greases. The susceptibility of the hydroxyl group to substitution or elimination provides a wide range of derivatives. The chemical interaction of the hydroxy group separated by one methylene from a carbon–carbon double bond supports several cleavage reactions that provide a collection of useful chemicals (Fig. 4.2). Products derived from castor oil have been in use for centuries, long before "biobased" was a buzzword or requirement for meeting sustainable product regulations (Copley et al., 2005; Mutlu and Meier, 2010).

Castor was previously grown and processed in a number of nations, but castor cultivation is currently limited to a few countries as major producers, including India, the major exporter of castor oil, China, Brazil, and Thailand. Castor is a minor crop in about 20 countries including Mexico, Colombia, Central America, Pakistan, Iran, and Turkey (FAOSTAT, 2015). Approximately 400 ha of castor are grown in the United States (Campbell et al., 2014). In the United States, the loss of castor cultivation has been explained as resulting from the loss of agricultural price supports, the increased cost of energy needed to detoxify the seed meal, and the availability of more profitable crops (Severino et al., 2012; McKeon et al., 2014).

Industrial Oil Crops. http://dx.doi.org/10.1016/B978-1-893997-98-1.00004-X

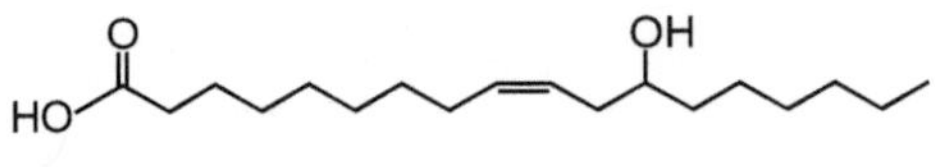

FIGURE 4.1 Ricinoleic acid.

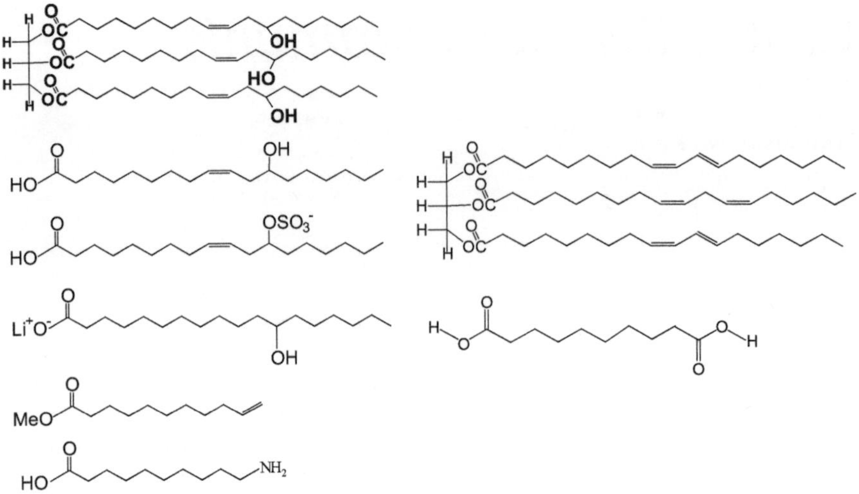

FIGURE 4.2 Biomaterials from castor oil. From top left: castor oil; ricinoleic acid; ricinoleyl-sulfate; lithium grease (lithium hydroxystearate); 10-undecylenic acid; 11-amino-undecanoic acid. From top right: dehydrated castor oil; sebacic acid.

Because of its commercial importance and unique biochemistry, castor remains a crop of great interest. With a relatively small genome (350 Mb), it was a leading candidate for sequencing (Chan et al., 2010). Numerous genes were identified, including 28 for the protein toxin ricin and related proteins, 110 genes related to oil production, and numerous genes related to pest resistance (Chan et al., 2010). Knowledge of the sequence coupled with breeding programs will enable significant improvement in agronomic performance, oil production, yield, and stress and pest resistance, as well as reduction or elimination of ricin content. There continues to be an effort to produce a substitute for castor oil in transgenic plants, and this effort is aided by the availability of extensive genome information from castor.

AGRONOMY

The castor plant is adapted to a wide range of climates. It is thought to originate in Ethiopia and grows as a perennial tree or shrub in tropical and semitropical climates. However, it will grow even in northern parts of the temperate zone as an annual crop, producing castor seed as long as the growing season lasts 4–5 months needed to set seed and mature. Despite the considerable variation in characteristics such as plant height, leaf and stem color, dehiscence, etc., it is a monotypic species classified in the family *Euphorbiaceae*, and only one species of *Ricinus* has been confirmed (Weiss, 2000; Anjani, 2012). The so-called *Ricinus sanguineus* appears to be a colorful red variant with no documentation to justify a difference in species name. The castor plant is monoecious and self-fertile, bearing numerous seed pods that develop on a raceme. The seedpods contain three seeds, and a single raceme may bear over 100 pods. Most castor plants produce multiple racemes, although some produce most of the seed on the primary raceme, while still other castor cultivars may produce racemes that each bear a significant number of seed pods over an extended growing season. The cover illustration displays a castor plant bearing seedpods on racemes.

While castor grows on agriculturally marginal lands and will produce some seed and oil without agricultural inputs, it benefits considerably from the inclusion of fertilizer. Growth of castor depletes soil nitrogen, with estimates of nitrogen required for optimal production ranging from 40 to 65 kg N/1000 kg seed/ha (Weiss, 2000; Reddy and Matcha, 2010). Soil nitrogen is best supplied as inorganic nitrogen derived from ammonium or nitrate (Severino et al., 2012), but nitrogen is often supplied by incorporating seedpods and seed meal in the soil. During plant growth and seed production, phosphorus is removed from soil at 9 kg/1000 kg seed/ha, potassium at 16 kg/1000 kg seed/ha, calcium at 6 kg/1000 kg seed/ha, and magnesium at 5 kg/1000 kg seed/ha (Weiss, 2000; Severino et al., 2012). The plant also requires trace minerals including sulfur, boron, and zinc (Severino et al., 2012).

Castor plants produce a deep tap root and extensive root system, enabling the plants to absorb water from deep within the soil and endure drought conditions under which other crops would be severely damaged (Weiss, 2000). With water available in early development, the seeds germinate, and some seed production can be obtained from dry cropping with little or no irrigation provided. For example, castor is often planted after the rainy or monsoon season in India, and may receive no additional irrigation. Although dry cropping does produce castor seed, yields are greatly enhanced with adequate irrigation (Ramanjaneyulu et al., 2013). In one published study, castor seed yields were increased greater than10-fold, from 232 to 2758 kg/ha, by increasing the water provided through irrigation and rainfall from 188 to 897 mm (Severino and Auld, 2013).

Because castor is somewhat resistant to drought, it has also been hypothesized to be resistant to saline conditions (Lima-Neto et al., 2014). While five cultivars tested were able to germinate and grow when irrigated with 100 mM NaCl (10 dS/m) (Rodrigues et al., 2014), the cultivars Memphis, Brigham, and Hale showed increased seedling emergence. However, all cultivars grown under high salinity showed height reductions of 40–50% and fruit dry weight reductions of 55–78% (Sun et al., 2013). It has previously been pointed out that salinity tolerance of castor is dependent on the cultivar. Of the commercial cultivars tested by Sun et al. (2013), Memphis and Energia appear to be the most tolerant to salt, with no apparent leaf damage and low sodium content of the plant.

A key factor that can affect yield is the planting density. There is considerable variation in plant height and bushiness, with some castor plants producing many racemes and other cultivars produce most of their seed on a single raceme with a limited number of additional racemes. As a result, the planting density can range from 4200 plants/ha for tall cultivars to 70,000 plants/ha for a dwarf variety (Severino et al., 2012). In addition to plant characteristics, environmental conditions and agricultural practices are considered in determining the appropriate planting density (Soratto et al., 2012). The cited study compared densities of 25,000/ha with 55,000/ha obtained by varying the distance between rows and between plants for the higher-density plantings. The seed yield at the greater density was up to 40% higher, and with nearly even planting distances, 0.4 m between plants and 0.45 m between rows, yields were 10–20% higher than with other high-density arrangements (Soratto et al., 2012).

CASTOR BREEDING

As the only commercial source of hydroxy fatty acid, breeding efforts for castor are concerned with yield of seed, increased oil content, synchronization of maturity, early maturity, pest resistance, dwarfing, indehiscence, adaptability to mechanical harvesting, and elimination of the toxin ricin and allergens from the seed. The importance of breeding castor for industrial purposes has long been recognized (White, 1918). A key component of breeding programs is the availability of germplasm representing wide genetic diversity. Germplasm repositories contain a diverse population of castor selections that have proven useful in breeding programs (Anjani, 2012; Severino et al., 2012). While these repositories contain castor collected from every growing region, the genetic diversity of the *Ricinus* genus is thought to be low (Allan et al., 2008). Genetic crosses with selection, hybridization, chemical and radiation mutagenesis, and genetic modification have all been applied to castor improvement, thus expanding the genetic diversity to some extent.

In castor growing regions, the castor plants grown vary from semiwild accessions to hybrids bred for specific traits. Each castor raceme bears male (staminate) and female (pistillate) flowers, with the male flowers on the lower portion of the raceme and the female flowers above. Proportions of male to female flowers vary and can be affected by a number of conditions (Severino et al., 2012; Tan et al., 2013) including temperature, hormone treatment, or wounding, as well as cultivar variation. Cultivars that produce only staminate or pistillate flowers are naturally desired for breeding hybrid castor, and there are several such cultivars (Tan et al., 2013).

Since the main use of the castor plant is castor oil production, traits that enhance the oil content are of greatest interest. Dwarfing can provide an enhanced yield of oil per unit area because more photosynthate is put into seed production than stem production, similar to wheat developed during the Green Revolution. Moreover, high-density plantings are possible, thus increasing production (Severino et al., 2012). There is variation in the arrangement of seed capsules on the raceme as well, with compact racemes carrying many more capsules than loosely packed racemes, resulting in a concomitant increase in seed production for the compact racemes (Anjani, 2010). Although seed size can vary considerably, there is as yet no strict correlation between seed size and oil content (Severino et al., 2012). Another desirable trait is nonbranching castor. Such plants are of value for mechanical harvesting. A breeding program designed to develop unbranched castor was able to enrich a population to 11% unbranched and provide a source of material for further development (Baldanzi and Pugliesi, 1998).

Castor is a very hardy plant and in some areas an invasive weed. It is susceptible to numerous insect and microbial pests, but castor germplasm is available for resistance to many of these pests. One key to successful breeding is the ability to screen large numbers of plants quickly. In the case of leafminer resistance, Anjani et al. (2010) found that resistance was correlated with high phenolic content as reflected by purple leaf coloration. Over 100 castor plant accessions, of 1979 tested, proved to be resistant to Fusarium (*Fusarium oxysporum* f. Sp. *ricini*) wilt based on their limited susceptibility to wilting in a contaminated field (Anjani et al., 2014).

The availability of the castor genome sequence (Chan et al., 2010) has provided the identity of over 120 genes involved in pest resistance (Chan et al., 2010; Anjani, 2012). This knowledge coupled with the development of amplified fragment polymorphisms and single sequence repeats as molecular markers underlies the effort to incorporate molecular breeding into castor breeding programs.

Two key features of castor seed are the ricin content of the endosperm and the ricinoleate content of the seed oil. There are no known null plants for either component, although there is one castor mutant that produces oil with low ricinoleate content (14% vs 90%) (Barros et al., 2004). Elimination of the toxin is an obvious desired outcome (Auld et al., 2001), and breeding efforts have been effective in producing castor seed cultivars with reduced ricin content (Auld et al., 2003; McKeon et al., 2014). The genome sequencing of castor revealed the presence of 28 genes related to ricin and RCA, a less toxic tetrameric homologue of ricin (Chan et al., 2010). The effort to remove all these genes through breeding or mutagenesis is daunting, and it would take years of backcrosses to incorporate each lost gene. However, gene suppression through genetic manipulation (McKeon et al., 2002; Sujatha et al., 2008) appears to be a viable approach for eliminating the expression of ricin and RCA genes.

Biotechnological approaches to breeding castor include in vitro propagation in tissue culture and plant regeneration. Somaclonal propagation maintains specific traits (Tan et al., 2013), and there are numerous approaches to regenerating castor plants from tissue culture (Ahn et al., 2007; Sujatha et al., 2008). There has been very limited success to date in genetic transformation of the castor plant (McKeon and Chen, 2003; Sujatha and Sailaja, 2005). Sujatha et al. (2008) have reported development of castor plants resistant to foliage feeders by expression of inserted genes coding for *Cry1A* and *Cry1E* proteins, with successful plant regeneration in culture. Further application of this technique to incorporate useful traits not available from castor germplasm as well as gene suppression of ricin and allergen will be of great benefit in expanding the production of castor.

RICIN AND RCA

A major problem inhibiting widespread cultivation of castor is the presence of noxious proteins in the seed meal residue remaining after oil extraction. These proteins include the toxin ricin, the homologue *Ricinus communis* hemagglutinin (RCA), and a collection of allergens including 2S albumin as the major component of the allergen fraction. Efforts to completely eliminate these proteins from the crop have proven unsuccessful. However, some progress has been made in addressing the problems these proteins pose.

Castor seed and the residue remaining after oil extraction have long been recognized as toxic. A recent report suggested that as far back as 22,000 BC, castor seed was used as an arrow poison (d'Errico et al., 2012a). The presence of ricinoleate and its trans isomer on wooden poison applicators is the basis for this conclusion, although this interpretation remains controversial (Evans, 2012; d'Errico et al., 2012b). Ricin is known as a weapon of assassination, most notoriously used to kill Georgi Markov, a Bulgarian dissident living in London. The ricin was carried by a small ball delivered by a needle shot from an umbrella (Knight, 1979). Ricin has long been a concern as a potential bioweapon, and the appearance of a simple method for producing ricin in planning material captured from Al Qaeda intensified this concern (Pita, 2009). Use of ricin as a poison has captured the fancy of novelists and dramatists, and it was a featured poison in the television series "Breaking Bad." Despite the well-earned reputation of ricin as a slow-acting and easily concealed poison, its attempted uses have led some to consider it to be much less a threat than feared (Blum, 2014).

Although the allergic and toxic responses to castor meal were originally thought to be from the same component, ricin was characterized separately from the allergenic material as a toxic albumin in 1881 (Ritthausen, 1881), and the protein was crystallized to purity in 1948 (Kunitz and McDonald, 1948). Since that time, the modes of action of ricin and the homologous agglutinin RCA have been identified. Ricin is a dimeric glycoprotein composed of a toxic A-chain and a lectin B-chain linked by disulfide bonds, and classified as a type 2 ribosomal-inhibiting protein (type 2 RIP). The A-chain is an N-glycosidase that inactivates ribosomes by depurinating a specific adenylate residue from 28S ribosomal RNA, leading to inactivation of eukaryotic protein synthesis (Endo et al., 1987). The B-chain, which contains two galactose-binding sites, binds specifically to cell surface glycoproteins or glycolipids and facilitates the movement of A-chain into cells (Sphyris et al., 1995). RCA is a tetramer consisting of two ricin-like dimers held together by noncovalent bonds (Roberts et al., 1985). RCA is less toxic than ricin but causes agglutination of red blood cells in mammals (Roberts et al., 1992;

Lord et al., 1994). This activity may explain some of the final symptoms of crude ricin poisoning which include hemorrhaging.

Ricin is toxic to humans at an estimated 1–5 μg/kg of body weight on intravenous, peritoneal, or inhalation delivery. However, the level for oral toxicity is much higher, with the toxic dose in the range of 5 mg/kg (He et al., 2010a; Worbs et al., 2011). The amount of ricin in seeds varies within a range of 1.2–6% of seed weight (McKeon et al., 2014). While there is concern about the potential use of ricin in bioterror attacks, growing and harvesting castor is generally safe, with few incidents reported over many years of castor cultivation (Apen et al., 1967). There are some reports of ricin toxicity due to exposure of workers during processing of castor seed (Apen et al., 1967), and these were distinguished from allergic responses based on symptoms, with allergic responses arising quickly after exposure and the onset of symptoms from exposure to ricin delayed for several hours. The exposure occurred when the castor meal was produced as a fine powder, for example to produce animal feed or fertilizer. Exposure can be prevented by using extraction methods that eliminate powder formation. There are numerous reports of death from ricin by accidental ingestion, assassination, and murder (Worbs et al., 2011), and the major militaries, intelligence services, and some terrorist organizations have considered ricin a possible weapon of mass destruction or assassination (Knight, 1979; Pita, 2009). During World War II, several US companies were involved in preparing ricin, known in secret documents as Agent W (Stewart, 2006). Although ricin is considerably more toxic than RCA, a crude toxin containing both ricin and RCA appears to be more toxic than ricin alone, suggesting a synergistic effect of the two in causing toxicity (He et al., 2010a).

The complete primary sequences of ricin and RCA proteins have been determined (Yoshitake et al., 1978; Funatsu et al., 1979) and later deduced from the nucleotide sequence of cloned cDNAs and genomic clones (Roberts et al., 1985; Lamb et al., 1985; Halling et al., 1985). Both ricin and RCA genes encode a precursor protein containing an N-terminal signal sequence, an A chain, a 12 amino acid linking region and a B chain (Roberts et al., 1985; Lamb et al., 1985; Tregear and Roberts, 1992). The A chains of ricin (31kD) and RCA differ in 18 of their 267 residues, with 93% identity at the amino acid level, whereas the B chains (for ricin, 31.5 kD) differ in 41 of 262 residues with 84% identical (Roberts et al., 1985). Formerly, ricin and RCA were thought to be encoded by a multigene family composed of approximately eight members (Halling et al., 1985; Tregear and Roberts, 1992), at least three of which were nonfunctional (Tregear and Roberts, 1992). However, the sequencing of the castor genome demonstrated the presence of 28 copies of ricin-related genes and pseudogenes (Chan et al., 2010). Expression of ricin and RCA genes is tissue specific and developmentally regulated so that mature ricin and RCA accumulate only in the endosperm of castor seeds (Roberts and Lord, 1981; Tregear and Roberts, 1992; Chen et al., 2005). Although the biological functions of ricin and RCA remain a mystery, it is generally believed that they have a defense role against predators.

RICIN DETECTION

Due especially to concerns about the use of ricin as a bioweapon, numerous approaches for ricin detection and quantitation have been developed. These methods fall into four basic categories: activity/toxicity, chemical instrumentation, immunodetection, and indirect detection.

The gold standard for ricin detection has been mouse toxicity (Zhan and Zhou, 2003), but even with this reduced number of mice needed to determine the LD_{50}, the lethal dose for 50% of the population treated, animal experiments are an expensive and time-consuming process. The in vivo inhibition of protein biosynthesis in cell cultures (Rasooly and He, 2012) and in vitro inhibition of protein synthesis (He et al., 2008) have proven to be reliable methods for detecting ricin without the need for animal testing and biopsy. When Vero cells transduced with adenovirus carrying a gene for green fluorescent protein are exposed to ricin, the loss of protein synthesis is measured by reduction in fluorescence. This method can detect 200 pg ricin/mL (Rasooly and He, 2012). An alternative method of similar sensitivity measures loss of luminescence in ricin-treated samples in an in vitro protein biosynthesis system with luciferase mRNA added (He et al., 2008). This method relies directly on the enzymatic action of the ricin A subunit, namely the N-glycosidase releasing an adenine from 28S RNA resulting in loss of ribosome function. As such, the assay is effective in detecting both type 1 and type 2 RIPs (type 1 RIP has no lectin subunit—eg, curcin from Jatropha: see Chapter 11). Another assay uses Vero cells and monitors the number of viable cells remaining after treatment with ricin or *R. communis* agglutinin (Pauly et al., 2012). The specificity of these in vivo and in vitro systems were assured by demonstrating reversal of the ricin effect after adding antibody to ricin, resulting in inactivation of ricin activity. Another assay that relies on detection of released adenine from RNA has also been developed and is effective in detecting both types of RIP (Sturm and Schramm, 2009). The method uses an enzyme-coupled reaction that converts released adenine to ATP, which in turn causes emission of luminescence from firefly luciferase.

The detection of ricin directly by instrumental methods has been limited to matrix-assisted laser desorption ionization-time of flight mass spectrometry (MALDI-TOF MS) detection (Darby et al., 2001; Brinkworth et al., 2009). Darby et al. (2001) separated ricin from other material present in samples using size exclusion chromatography and then hydrolyzed the isolated ricin with trypsin. The hydrolysate was analyzed by MALDI-TOF MS and ricin peptides identified by comparison of mass ions with an online database and comparison of identified peptides with those from a reference sample obtained from ricin. Brinkworth et al. (2009) generated and measured the intact ricin ion as opposed to peptide fragments. Since the presence of readily ionized substances such as salts in the sample can overwhelm the detection of ionized ricin, the sample was concentrated and ultrafiltered to eliminate material of MW below 30 kDa. The method detected ricin at levels of 4 μg/mL.

The most widely used and sensitive approaches for specific detection of ricin have been based on immunodetection. Early work used radial immunodiffusion (RID) using the distance required for formation of an immunoprecipitin band between cylindrical wells in gel plates containing ricin+RCA and ricin antibody, with the distance from the sample well to the precipitin line inversely related to ricin/RCA concentration in the sample. Although this was not a very sensitive technique, it was reproducible and proved effective in identifying and breeding low ricin castor varieties (Pinkerton et al., 1999; Auld et al., 2003). Development of monoclonal antibodies to ricin provided the means for specific detection of ricin and the agglutinin (Brandon and Hernlem, 2009) without interference from the carbohydrate portion of the protein that can result in the detection of nonricin proteins. Recently, monoclonal antibodies that distinguish ricin and RCA have been reported (Brandon et al., 2016). Monoclonal antibodies to ricin have been effective for detecting ricin in milk (Brandon et al., 2014) as well as other foods, and have aided the breeding effort of castor by identifying cultivars with low levels of ricin (McKeon et al., 2014).

The most sensitive assay for ricin was based on immunoPCR (He et al., 2010b). The method used the ricin sample bound to a 96-well plate, which then bound streptavidin-labeled antibody to ricin. A biotinylated strand of DNA 340-bp long was then captured by the streptavidin, the DNA enzymatically released, and then amplified and quantified by real-time PCR (RT-PCR) in a separate plate. This assay is 10-fold to 1000-fold more sensitive than ELISA assays in detecting ricin in food samples, depending on food matrix effects that influence the sensitivity of the assay. As a result of its sensitivity, it was possible for the first time to monitor levels of ricin in the bloodstream of mice treated with ricin (He et al., 2010b).

Methods for indirect detection of ricin analyzed samples for components that are present in castor seed extracts, in particular castor DNA and the alkaloid ricinine (Fig. 4.3). These approaches were concerned with crude ricin preparations as described in a "terrorist handbook" (Darby et al., 2001). Although ricin is not a major component of these crude preparations, they do pose some threat and represent a concern due to their description in such a handbook.

Castor DNA was specifically detected using a DNA primer that corresponds to the gene for ricin. While simple PCR is effective in detecting castor DNA, by using RT-PCR, the level of DNA could be measured to obtain an indication of the level of contamination (He et al., 2007). This method has been applied to

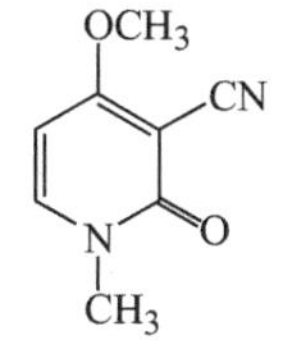

FIGURE 4.3 Ricinine.

food samples spiked with crude castor seed extracts and procedures for detection in the presence of high protein, fat, and other lipid components. Although the method detected low levels of DNA, it cannot distinguish between contamination with castor seed or from any other part of the castor plant.

The alkaloid ricinine is present throughout the castor plant from the time of early germination to seed maturation (Waller and Nakazawa, 1963). It has been used as an indirect means for detecting ricin in crude ricin preparations (Darby et al., 2001), but is more commonly used to detect ricin poisoning (eg, Hamelin et al., 2012), as it is eliminated from the body in urine and is readily identified by GC/MS or LC/MS.

CASTOR SEED ALLERGEN

Castor seed meal has long been recognized as a potent allergen. Most reactions to castor meal are allergic responses. Many allergic reactions to various stages of castor oil processing have been reported, including townsfolk living near a castor processing plant (Ordman, 1955), dockworkers carrying green coffee in burlap bags that had previously been used for castor meal fertilizer (Figley and Rawling, 1950), and processing plant workers (reviewed in Apen et al., 1967). While safety standards of oilseed processing plants are much higher now, even in more recent times, individuals may still be exposed and react to castor dust— eg, as a component of some organic fertilizers (Baur et al., 1998). Initial research on identification of the allergenic fraction of castor seed meal led to isolation of a protein fraction designated CB-1A (Spies and Coulson, 1943). It was later determined that several proteins in this fraction reacted to IgE obtained from a group of patients with severe allergic responses to castor meal. The main allergen was identified as 2S albumin because 96% of the IgEs reacted with that protein, although a lower proportion of the IgE from the patients bound to several other proteins (Thorpe et al., 1988).

Based on fractionation studies, 2S albumin proteins were identified as the primary allergenic components of the allergenic CB-1A fraction from castor meal (Youle and Huang, 1978a; Machado and Silva, 1992). The amino acid sequencing of one of these castor 2S albumins led to the cloning of a full-length cDNA and isolation of a genomic clone (Sharief and Li, 1982; Irwin et al., 1990; Irwin and Lord, 1990). Sequence analysis of these clones indicated a castor 2S albumin gene with a 776-bp open reading frame, encoding a precursor protein of 258 amino acids. The precursor produces two heterodimeric 2S albumin proteins, RicC1 (Sharief and Li, 1982) and RicC3 (Machado and Silva, 1992; Silva et al., 1996; Bashir et al., 1998). The main function of castor 2S albumins is storage protein to provide the germinating seed with an adequate supply of amino acids, for use by the seedling before the onset of photosynthesis (Youle and Huang, 1978b). Castor 2S albumin sequences are homologous to amylase inhibitor and trypsin inhibitor genes in plant (Irwin et al., 1990; Bashir et al., 1998). Whether the castor 2S albumins are solely a source of amino acids for the germinating seed or if their putative inhibitory function as insect feeding deterrents is confirmed, the function of this protein remains to be seen. The 2S albumin present in a number of other plants—eg, ragweed and soy—is also a major allergen due to the size, disulfide cross-linking of the dimer structure, and possible protease inhibitor activity of the protein (Chen et al., 2004).

PROCESSING

Commercially grown castor averages 50% oil content, with some cultivars producing up to 60% oil. Thus, simple pressing is effective for extracting oil from the castor seed, followed by solvent extraction of the press cake to maximize recovery of oil. Due to the high content of hydroxy fatty acid, castor oil is soluble in methanol, ethanol, and acetone, although hexane is commonly used to extract it. After pressing, the oil requires drying to eliminate water and prevent deterioration of the oil. The dried press oil or solvent-extracted oil are further refined by filtration to remove particulate material, additional settling and degumming to remove phospholipid, neutralization to remove free fatty acid, and bleaching with activated clay or activated carbon to remove pigments and deodorization (http://www.castoroil.in/extraction/extraction.html).

Supercritical carbon dioxide (SC-CO_2) is a solvent that has been used for oil extraction, especially of oils containing reactive fatty acids that may be altered by solvent extraction. In laboratory-scale experiments, castor seeds ground to a fine powder were extracted, and fatty acids (FAs) from the oil were directly converted to methyl esters using SC-CO_2 containing 7% methanol in a bed of immobilized lipase (Turner et al., 2004). However, extraction of castor seed with SC-CO_2 was not effective in extracting castor oil, as yields were at best 9.29% (Danlami et al., 2015).

CASTOR OIL BIOCHEMISTRY AND MOLECULAR BIOLOGY

Castor oil is a product of great interest to plant lipid scientists, as it is an established commodity oil with a significant market and a cost ranging from two to three times that of soybean oil. Since it is available from very few countries, it represents a useful commercial target for production in a transgenic plant and also a long-term intellectual challenge for understanding how the castor seed can produce an oil that is 90% ricinoleate in seeds containing up to 60% oil. Understanding the basis for the regulation of oil production in plants is also a major research interest, and castor has served as a benchmark for high oil content in an annual oilseed crop.

As a result of its high ricinoleate content, castor oil is composed mainly of triglycerols (TGs) carrying three ricinoleates per glycerol molecule, or approximately 70%. Lin et al. (2003) and Lin and Arcinas (2007) have characterized the other TG components of castor oil, initially identifying 16 TG species. Almost all the residual TG incorporates two ricinoleoyl residues, with oleate, stearate, palmitate, linoleate, linolenate, and the hydroxy FA lesquerolate as the third acyl chain on the TG. The authors determined that most of the nonhydroxylated FA is present in the *sn*-2 position with >90% regiospecificity (Lin and Arcinas, 2007). Later work indicated the presence of numerous, albeit minor, TG-carrying dihydroxy and trihydroxy FAs in addition to ricinoleate (Lin et al., 2009; Lin and Chen, 2010).

Interest in castor oil biochemistry exceeds 50 years. In the 1960s, both the Stumpf research group at the University of California, Davis, and the Morris group at Unilever, Great Britain, carried out basic research investigating the hydroxylation reaction that converts oleate to ricinoleate (Yamada and Stumpf, 1964; Galliard and Stumpf, 1966; Morris, 1967). These early biochemical developments led to additional biochemical characterization of the pathway, followed by cloning, expression, and characterization of genes identified as being involved in the castor oil pathway. These genes, in turn, have been expressed in transgenic oilseeds in order to test their importance to ricinoleate or other hydroxy FA production. The transgenic work was intended as proof-of-concept, and the results obtained from the transformed plants provided additional insight into the process of castor oil biosynthesis.

The oleoyl-12-hydroxylase enzyme proved challenging to purify (Moreau and Stumpf, 1981; Bafor et al., 1991; Richards et al., 1993; McKeon et al., 1997). However, the availability of a stable microsomal preparation from developing castor endosperm provided a means to identify some of the key biochemical steps in castor oil biosynthesis (McKeon et al., 1997; Lin et al., 1998). The initial focus of the biochemical approaches was based on the Kennedy pathway for TG biosynthesis, in which the final committed step is carried out by the acylCoA: diacylglycerol acyltransferase, the key step that differentiates TG from phospholipid biosynthesis for membranes. When ^{14}C-labeled oleoylCoA was added to the castor microsomes, it was rapidly taken up into the phosphatidylcholine (PC) fraction and then hydroxylated (McKeon et al., 1997). This data indicated the importance of lysophosphatidylcholine acyltransferase (LPCAT) in supplying the PC substrate for hydroxylation. While added ^{14}C-labeled *sn*-2-oleoyl-PC was only hydroxylated to a limited extent (Lin et al., 1996), use of the substrate allowed additional characterization of the pathway for TG biosynthesis. Among the labeled products detected were ^{14}C-ricinoleoyl-PC, ^{14}C-ricinoleate as a free FA, and ^{14}C-ricinoleoyl-TG (Lin et al., 1998). The presence of labeled ricinoleate confirmed previous results that a phospholipase A2 releases the ricinoleate from the PC product (Bafor et al., 1991). Comparison of the incorporation of the main FAs present in castor oil indicated a sixfold preference for incorporating ricinoleate into TGs versus the hydroxylase substrate oleate (Lin et al., 2002). These results suggested that the acyl-CoA–dependent diacylglycerol acyltransferase (DGAT), the final step in the Kennedy pathway for TG biosynthesis, is responsible for the high specificity for inserting ricinoleate into castor TG, while the hydroxylase is responsible for the high level of ricinoleate produced. These results are depicted in the Kennedy pathway for castor oil biosynthesis (Fig. 4.4).

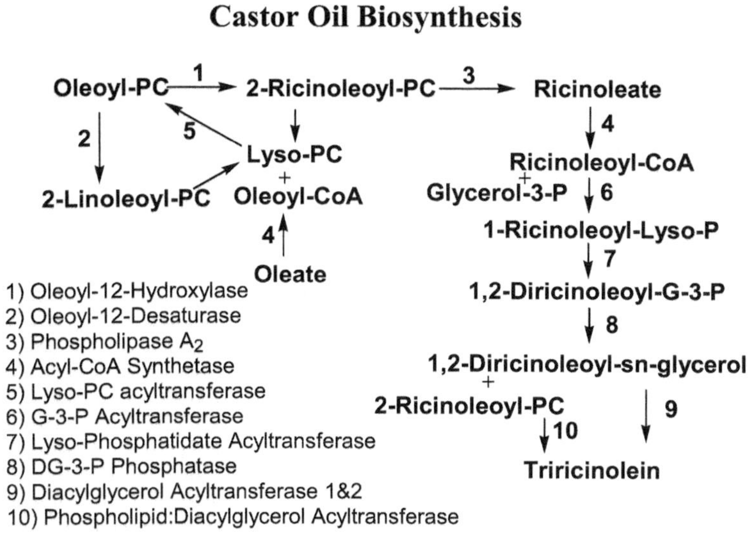

FIGURE 4.4 Kennedy pathway for castor oil biosynthesis.

While the Kennedy pathway helps to explain the unique features of castor oil biosynthesis, work by Mancha and Stymne (1997) indicated that the picture is more complex. When they repelleted, mixed, and incubated castor microsomes that had been labeled with either ^{14}C-glycerol-3-phosphate or ^{14}C-ricinoleoyl-CoA, they found that label was transferred from triricinolein (TG with three ricinoleoyl groups) to produce additional TG bearing one or two ricinoleoyl groups, indicating the presence of an acyl–exchange reaction (Mancha and Stymne, 1997). Moreover, additional experiments with castor endosperm tissue slices indicated the presence of two pools of labeled TG–one that remained stable in the oil bodies, and the other in the microsomal fraction composed of monoricinoleins and diricinoleins. An additional enzyme involved in TG biosynthesis in castor was identified as phospholipid:diacylglycerol acyltransferase (PDAT). This enzyme is present in castor microsomes and carries out an acyl-CoA-independent acylation of diacylglycerol (DG) from the acyl group on the *sn*-2 position of PC (Dahlqvist et al., 2000). The authors demonstrated that castor PDAT has an approximately eight fold preference for transferring ricinoleate vs oleate from PC to DG and thus provides an additional means to produce ricinoleoyl-TG. The simplified pathway (Fig. 4.4) illustrates the enzymatic steps that have been demonstrated in vitro to be involved in castor oil biosynthesis.

With increasing availability of methods for cloning and identifying genes for the enzymes involved in FA and oil biosynthesis, the biochemical results logically led to cloning genes for these enzymes, expressing them in microbial systems, and characterizing their activity, either in isolated form or in vivo in the microbial expression system.

The cDNA for the oleoyl-12-hydroxylase gene was the first and logically most important gene cloned (van de Loo et al., 1995). Based on the hypothesis that the hydroxylation reaction is analogous to, or the first step in, the desaturation reaction, the authors proposed that the hydroxylase would share sequence elements in common with FA desaturases. Using this approach, hundreds of cDNAs from developing castor seed were sequenced, prospective hydroxylase cDNAs expressed in tobacco seed, and the seed oil assayed for hydroxy FAs. Although ricinoleate production was low, 0.1%, it was sufficient to show that the hydroxylase had been cloned and successfully expressed in a transgenic plant. However, oilseeds transformed to express the gene for the castor oleoyl-12-hydroxylase (FAH gene) produce much less than the 90% present in castor oil, with most transgenic plants producing less than 20% hydroxy FA content in oil (Broun and Somerville, 1997; van Erp et al., 2015). As a result, it was generally accepted that other enzymes involved in castor oil biosynthesis would be required to increase production of ricinoleate. One logical choice was the castor DGAT (later designated RcDGAT1), which was cloned and expressed in yeast (He et al., 2004a). Although the transcription of the gene is maximal before the onset of oil deposition in the castor endosperm, the expression of RcDGAT1 protein coincides with oil biosynthesis (He et al., 2004b). It is a highly active DGAT, but it has only a twofold preference for acylating 1,2-diricinoleoyl glycerol (He et al., 2004a; McKeon and He, 2015). The RcDGAT2 was later cloned and expressed at the correct size in yeast, but no activity was detected (McKeon et al., 2007). However, another research group was able to express an active RcDGAT2 in yeast and demonstrated that it had high specificity for incorporating ricinoleate into TG (Kroon et al., 2006). Although the specific activity of the RcDGAT2 is very low compared with the RCDGAT1, it appears to be the key enzyme in determining high ricinoleate content. Based on the activity comparison and the presence of RcDGAT1 protein throughout oil biosynthesis, it appears likely that DGAT1 is involved in mass production of TG that is tailored to triricinolein TG by RcDGAT2 (McKeon and He, 2015). Furthermore, expression of RCDGAT1, RcDGAT2, and PDAT genes support the idea that all three of the corresponding enzymes are involved in accumulation of high levels of ricinoleate in TG (Dauk et al., 2009; Li et al., 2010). In some plants such as coconut, the lysophosphatidate acyltransferase (LPAT) is a key enzyme in the incorporation of specific FAs in the *sn*-2 position of the acylglycerol (Arroyo-Caro et al., 2013a). Two LPAT genes cloned from castor encode the corresponding enzymes RcLPAT2 and RcLPATB, both expressed during castor endosperm development. The LPAT2 has some preference for ricinoleoyl-CoA as the substrate for acylating *sn*-1 ricinoleoyl glycerol 3-phosphate. Although it is not a strong preference, it supports the idea that castor enzymes involved in oil biosynthesis have evolved to interact with ricinoleoyl-lipids. These researchers have also characterized a castor LPCAT, the enzyme previously shown to rapidly incorporate oleate from oleoyl-CoA into the *sn*-2 position of lyso-PC for hydroxylation by the oleoyl-12-hydroxylase (McKeon et al., 1997).

The expressed gene produces an active LPCAT which preferentially uses oleoyl-CoA or palmitoleoyl-CoA as the acyl donor (Arroyo-Caro et al., 2013b).

The results derived from cloning and biochemical characterization have resulted in a variety of hypotheses as to which combination of genes would lead to high ricinoleate incorporation in transgenic plants. While these experiments can be seen as proof of concept, they have also led to new hypotheses and greater understanding of the problems associated with high ricinoleate production. As noted, expression of the oleoy-12-hydroxylase (FAH12), the key enzyme in ricinoleate production, in transgenic plants leads only to low levels of hydroxy FAs, less than 20% (Broun and Somerville, 1997; van Erp et al., 2015). To date, the most successful approaches have produced seeds containing up to 30% hydroxy FA. Burgal et al. (2008) demonstrated that a non-C18 FA elongating *Arabidopsis* strain, when transformed to incorporate both the RcDGAT2 and FAH12, produced nearly 30% ricinoleate, with the transgenic plant that expressed both genes producing levels of triricinolein up to 16 times that of the transgenic expressing only the FAH12 gene. Expression of PDAT can also affect ricinoleate incorporation into TG. Three PDAT genes cloned from castor, RcPDAT1A, RcPDAT1B, and RcPDAT2, were coexpressed with FAH12 in the same *Arabidopsis* strain. Only the transformant expressing PDAT1A had elevated levels of hydroxy FA, from 17% in the plant expressing FAH12 only to 27% in the plant incorporating FAH12 plus PDAT1A (van Erp et al., 2011). However, the FAH12 and the double transgenic plants produced less oil than the wild-type seed. Incorporation of the RcDGAT2 gene led to a slight increase in % hydroxy FA, but a more significant increase in the overall amount of hydroxy FA (van Erp et al., 2011), indicating a significant role for DGAT2 in enhancing oil production in castor. Similar results were obtained by Kim et al. (2011). A gene for the "exchange" enzyme phosphatidylcholine diacylglycerol cholinephosphotransferase (PDCT) enriched the content of hydroxy FA when coexpressed with FAH12, compared with *Arabidopsis* expressing only FAH12 (Hu et al., 2012).

At this time, the challenge of producing hydroxy FA in transgenic plants seems limited to <35%. There are several explanations for the current limitation. There are two obvious obstacles to high ricinoleate in transgenic plants. One is that castor has evolved to produce high levels of ricinoleate, and it would be expected that the enzymes involved in oil biosynthesis coevolved to better utilize hydroxy FA lipid substrates, while the TG pathway enzymes in transgenic plants have not. As described above, a number of enzymes in the castor oil pathway show some preference, and in some cases high specificity, for ricinoleate-containing substrates. The second major obstacle is competition from the corresponding enzymes in the transgenic plant. This is supported by the *Arabidopsis* line expressing FAH12, RCDGAT2, RcPDAT1a, with inactivated endogenous AtDGAT1 and blocked expression of AtPDAT1. This line produced 33% hydroxy FA (van Erp et al., 2015), indicating that endogenous enzymes do interfere with hydroxy FA incorporation into TG in transgenic plants.

Since the hydroxy FA may be "detected" as an oxidized form of the endogenous FAs and excised by the plant's "editing function" for degradation, Beta-oxidation is another factor in preventing accumulation of hydroxy FA. This is particularly the case when ricinoleate accumulates in membrane PC (Bates and Browse, 2011; Bates et al., 2014). As a result of this editing process, the accumulation of hydroxy FA is limited. Moreover, the total content of oil may also be reduced. Effects on limiting FA biosynthesis in transgenic plants expressing FAH12 have also been observed as a result of acetyl-CoA carboxylase (ACCase) inhibition, perhaps through feedback inhibition (Bates et al., 2014). Coexpression of RcPDAT1 or RcDGAT2 in these plants results in reversion of the transgenic to near normal levels of FA synthesis, as the corresponding enzymes effectively move ricinoleate into TG, reducing the availability of ricinoleate for inhibiting ACCase.

While it may seem doubtful that any transgenic plant will replace the castor plant in terms of high yield of triricinolein or its equivalent, it remains an important challenge in understanding how any plant can make such a high level of oil at what is a technical grade of purity—90% of ricinoleate. While results of experiments to date suggest it may be necessary to eliminate all the TG biosynthetic genes in a transgene and replace them with the corresponding genes from castor. It will be interesting to see if there are genetic shortcuts that elicit high levels of ricinoleate in a transgenic plant oil.

CASTOR OIL PRODUCTION AND APPLICATIONS

While castor oil is well known as a powerful laxative, this medicinal use of the oil is relatively minor, <1%. Beyond this infamous application of castor oil, it is an important feedstock used by the chemical industry to produce a wide array of materials, many of which are superior to equivalent products derived from petroleum. Since castor oil is composed 85–90% of the hydroxy FA ricinoleic acid (Fig. 4.1), castor oil is considered a source of ricinoleate at a technical grade of purity. As mentioned previously, the presence of the hydroxy group in proximity to the double bond imparts physical, chemical, and even physiological activities that make castor oil such a uniquely useful commodity and feedstock.

Castor Oil Production

Castor is grown in over 20 nations (http://faostat3.fao.org/download/Q/QC/E), with India, Brazil, and China the leading producers of castor seed (Oil World, 2015). Table 4.1 presents the castor seed production for these countries for the last 3 years. The major castor oil producers also include India, Brazil, and China (Table 4.1), accounting for 613,000 of the 645,000 tons of castor oil produced in 2014 (Oil World, 2015). India is a net exporter of oil (447,000 tons in 2014), accounting for over 90% of castor oil exports, and the European Union, United

TABLE 4.1 Production of Castor Seed and Castor Oil, 2012–14

Country	Seed 2014[a]	Seed 2013[a]	Seed 2012[a]	Castor Oil 2014[b]	Castor Oil 2013[b]	Castor Oil 2012[b]
India	1250	1300	1280	563	585	576
China	71	105	145	31	46	63
Brazil	44	25	60	19	11	27
Thailand	13	14	14			
Other countries	60	60	59	32	32	32

[a]*Total seed crushed, exported, and imported, 1000 tonnes.*
[b]*Total oil production, 1000 tonnes.*

States, and China are major importers, accounting for 84% of imported castor oil (Oil World, 2015).

Castor oil is principally composed of TGs, of which 70% are triricinoleins, with most of the remainder being diricinoleate esters containing an additional acyl group derived from oleate, linoleate, linolenate, palmitate, stearate, and lesquerolate (14-hydroxy cosa-*cis*-11-enoate) (Lin et al., 2003). Other minor constituents include estolides, diacylglycerols, and free FAs as well as sterols.

There are literally hundreds of uses for castor oil and ricinoleate (Naughton, 1974; Naughton and Vignolo, 1992; Caupin, 1997; Ogunniyi, 2006; McKeon and Lin, 2007; Mutlu and Meier, 2010), but many are of limited application due to low production of the oil at only 645,000 tons in 2014 (Oil World, 2015), compared with soy oil with production greater than 42 million tons in 2013 (FAOSTAT, 2015). The remainder of this chapter describes major uses of the oil and its derivatives (Figs. 4.1, 4.2, 4.5, 4.6, and 4.7). An extensive description of castor oil chemistry is provided by Naughton and Vignolo (1992).

Polymers

Also see chapter 3 "Polymeric Products Derived from Industrial Oilseed for Paints, Coatings, and Other Applications" by Hayes and Dumont for additional detail on polymers derived from castor oil.

Castor oil is an important renewable source of monomers for use in the production of polymers. Castor oil as a monomer bears 2 or 3 ricinoleate moieties, making it a natural polyol. With the presence of a double bond and a free hydroxyl group, there are two reactive sites for polymerization. Ricinoleic acid derived from castor oil also has a free carboxyl group providing an additional reactive site for building polymers.

As a natural polyol, castor oil readily forms polymers when reacted with cross-linking reagents. These reagents can include dibasic carboxylic acids such as sebacic acid, or diisocyanates—eg, toluene diisocyanate. The diisocyanates are probably the most prominent cross-linking agents, as their reaction with castor oil produces polyurethanes (Fig. 4.5) (Naughton, 1974). The polyurethanes produced are elastomers that find uses in applications including foams, electrical insulators, automotive belts, O-rings, hoses, and coatings (Yeganeh and Mehdizadeh, 2003). One manufacturer of mattresses, Magniflex (Allessanderx S.p.a., Prato, Italy), produces a polyurethane foam from castor oil, with the claim of no solvent outgassing and foam with superior breathability for heat dissipation during sleep. The value of these applications can be seen in the fact that limited availability of castor oil led to development of chemically derived polyols from other fats and oils after epoxidation and ring opening (Scholnick et al., 1968). Due to increasing demand for TG polyols derived from natural sources and limited castor oil supplies, soybean oil–based polyols have achieved increasing popularity as replacements for petroleum-derived polyols, a market exceeding 1 million tons, with an estimated market for soy polyols of 400,000 t (de Guzman, 2003). Reliable and increased production of castor oil could certainly be accommodated in the supply chain for polyurethanes.

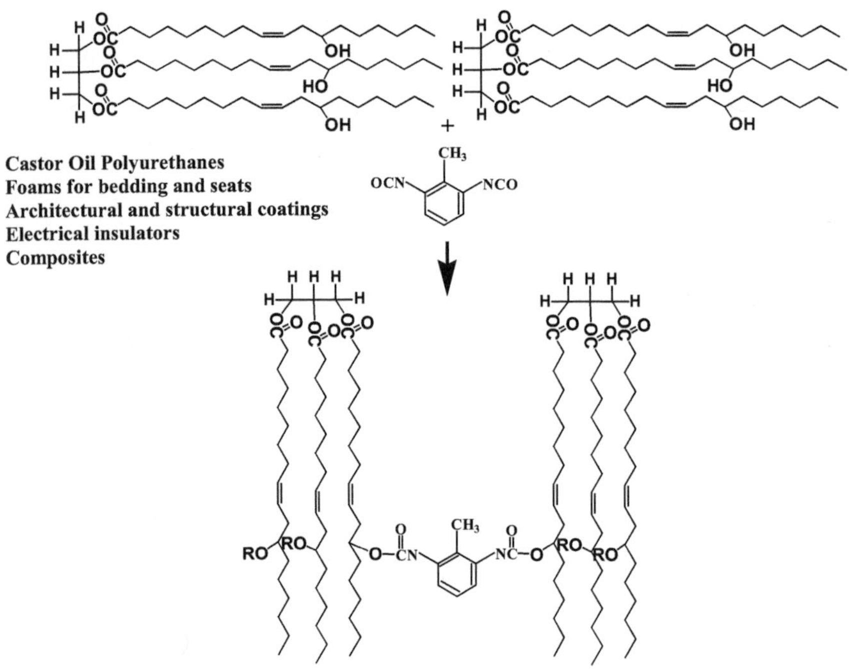

FIGURE 4.5 Polyurethanes from castor oil.

Given the three functional groups on ricinoleate, it is also a useful monomer for producing polymers. When heated in vacuo, a linear oligomer averaging 5 ricinoleate residues forms (Wilson et al., 1998). This oligomer is combined with polyglycerol to form polyglycerol polyricinoleate (PGPR), an emulsifier with food uses (Wilson et al., 1998). Both oligoricinoleate synthesis and esterification of the oligoricinoleates to polyalcohols can be carried out by the catalytic action of lipases (Kelly and Hayes, 2006).

Methyl ricinoleate, formed by transesterification of castor oil followed by vacuum distillation to attain higher purity, can undergo several reactions to yield useful monomers (Figs. 4.6 and 4.7). The first description of heat-induced cleavage of ricinoleate appeared in an 1827 report that identified one of the products as "oenanthol," later identified as heptaldehyde (reviewed by Hargreaves and Owen, 1947). Later research on cracking of castor oil indicated that heating above 400°C led to production of heptaldehyde and undecylenic (10-undecenoic) acid, with higher temperatures up to 550°C maximizing production of the aldehyde and undecylenic acid. The pyrolysis of castor oil also produces acrolein from glycerol as well as a number of unidentified side products (Vernon and Ross, 1936). Use of methyl ricinoleate obviated the formation of the toxic acrolein gas when using a castor oil feedstock (Han et al., 1996). While undecenoate can be used in polymer production (Kulkarni et al., 1968), the major application is conversion to 11-amino undecanoate by a series of chemical conversions (reviewed by Naughton, 1974) (Fig. 4.6). This monomer is used to produce an 11-carbon polyamide (PA 11) via polycondensation (Aelion, 1961) known as

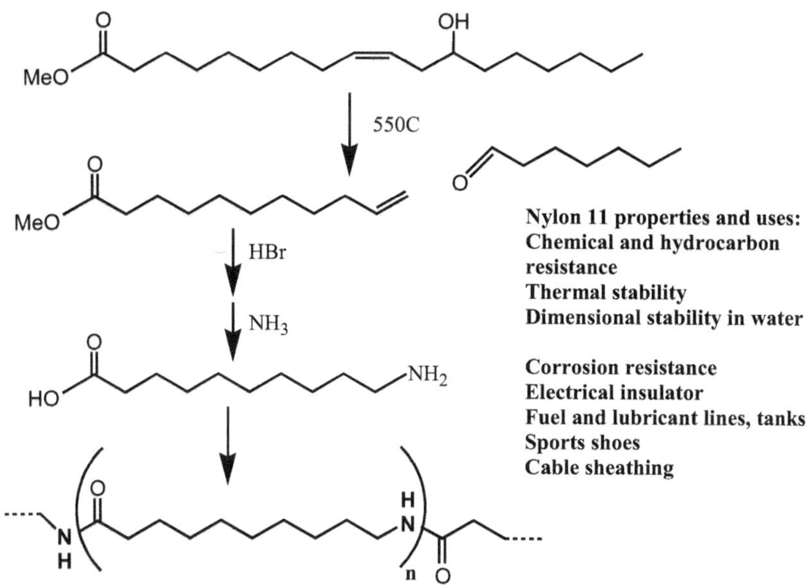

FIGURE 4.6 11-Amino undecanoic acid from methyl ricinoleate.

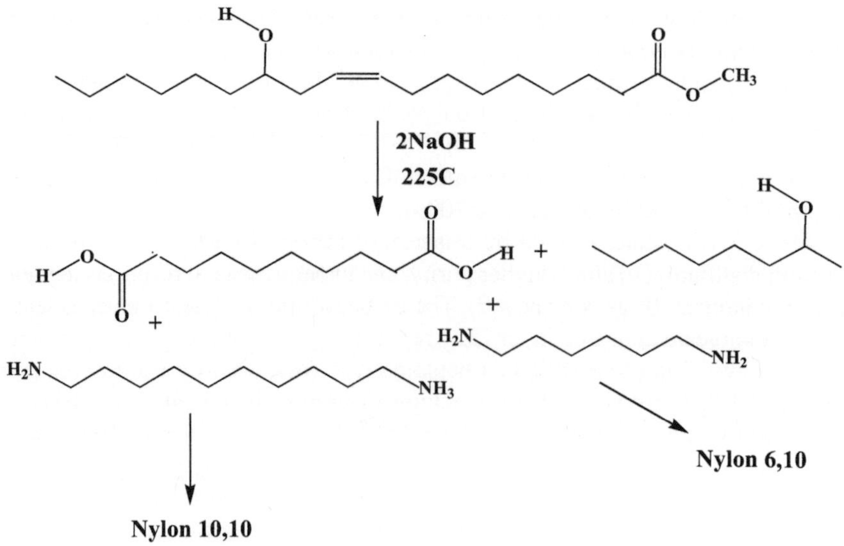

FIGURE 4.7 Sebacic acid from methyl ricinoleate.

Rilsan (Arkema), PA11, or Nylon 11 (Plastemart, 2011). This polyamide has many properties that make it useful for applications very different from the original polyamide Nylon 66 developed by DuPont. Since PA 11 has a lower melting point than Nylon 66, it is a thermoplastic that can be molded, extruded, or rotomolded (Borg et al., 2009). It does not swell or shrink when exposed to water and is highly resistant to chemicals and solvents, and thus finds uses in fuel and lubricant lines, oil and gas pipelines, cable sheathing, and numerous other engineering applications (Caupin, 1997; Borg et al., 2009). As a coating, it protects metals from chemical corrosion or impact. The heptanal produced during the pyrolytic reaction is used in flavoring and perfumes and can be oxidized to produce heptanoic acid for use in lubricants or structured fats in pharmaceutical applications (Borg et al., 2009).

Heating methyl ricinoleate with two saponifying equivalents of sodium hydroxide in the presence of 1-octanol or 2-octanol at 185–195°C leads to production of 10-hydroxy decanoate and methyl hexyl ketone (Hargreaves and Owen, 1947; Diamond and Applewhite, 1967). The 10-hydroxy-decanoate can be polymerized by heating with catalyst for ester formation, or by lipase action to condense the monomers into polyesters (Mahapatro et al., 2004). The enzymatic approach eliminates possible side reactions caused by heating. The ketone co-product is used in flavorings and perfumes.

Heating methyl ricinoleate in excess alkali as above but at 250°C results in production of sebacic acid and 2-octanol (Hargreaves and Owen, 1947; Diamond et al., 1965). Sebacic acid, a 10-carbon dibasic acid, is used in production of polyamide 10,10 (PA10,10), a polymer with properties similar to those of PA11 (Naughton, 1974; Caupin, 1997). Both PA 11 and PA10,10 are important plastics for use in engineering applications due to their resistance to corrosives and organic solvents, as well as their strength, thermoplastic properties, and ability to withstand exposure to water without changing shape or size (Plastemart, 2011). Sebacic acid is also used in production of PA 6,10, which is similar to PA 10,10, with better heat resistance but increased water and moisture uptake (Borg et al., 2009; Spini, 2013). Other uses for sebacic acid include ester plasticizers, jet lubricants, and the production of hydraulic fluids, paints, and adhesives (Johnson, 1984; Borg et al., 2009). The coproduct octanol has many uses including diester plasticizers, ester lubricants, cosmetics, and as a precursor for caprylic acid in production of dietary capryloyl triacylglycerols TGs (Borg et al., 2009).

A recent publication demonstrated the bioconversion of ricinoleate to undecylenic acid and to sebacic acid, eliminating the pyrolytic step or the alkali fusion, respectively, as synthetic steps to produce these useful monomers (Song et al., 2014). Industrial implementation of this approach could result in lower energy input and additional environmental benefit from production and utilization of these castor oil–based polymers.

Interpenetrating polymer networks (IPNs) are combinations of polymers in which one polymer is formed in the presence of another previously formed polymer, or in which formation of the polymers is initiated at the same time (a simultaneous interpenetrating network) (Sperling, 1994). There are varying reasons for producing IPNs, but generally speaking, the properties of IPNs differ significantly and are an improvement over the properties that a simple blending of two polymers would yield (Sperling, 1994). There are numerous examples of castor oil–based IPNs in the literature, including those involving castor oil-derived polyurethanes and other polymers such as polystyrene (Valero et al., 2009). Composites are a type of IPN classified as semi-IPNs (Sperling, 1994). Composites include a linear or branched polymer such as cellulose or polyacrylamide, minerals, etc., around which the polymer forms, usually to increase the mechanical strength of the composite. Polyurethanes and polyamides derived from castor oil are often used in such composites (Borg et al., 2009; Mosiewicki and Aranguren, 2013).

Plasticizers

Plasticizers are materials added to polymers to alter properties for improved uses and processing. A major effect of plasticizers is reduction of the glass transition temperature, the temperature at which a material loses its brittle structure and becomes more flexible or moldable. Plasticizers may also lower the melting point to make the polymer more easily processed or molded, alter the general appearance, or increase the tensile strength of the polymer. There are numerous other possible effects of plasticizers, including increased impact strength, flexural strength, stretchability, ductility, etc. (Wypych, 2012a). While a number of natural products were originally used as plasticizers, many were displaced by less expensive petroleum-derived plasticizers—eg, phthalate esters. Plasticizers derived from natural products were not entirely displaced by phthalate esters, and they have enjoyed a resurgence as concern about the biological effects of phthalate plasticizers has led to increased interest in biocompatible and physiologically safe plasticizers (Matsumoto et al., 2007).

Castor oil was one of the earliest recorded plasticizers (Wypych, 2012b), added to cellulose nitrate to allow shaping of the latter into films and lacquers. Castor oil and its derivatives are still used as plasticizers for a number of polymers. Castor oil itself is used in polyvinylacetate, cellulose nitrate, and rubber, whereas epoxidized castor oil, ricinoleate, and epoxidized ricinoleate are used as plasticizers in other plastics—eg, polyvinylchloride (Wypych, 2012a). In addition, alkyl diesters of sebacic acid, a product derived from castor oil, are also used in a broad array of polymers (Wypych, 2012a).

Soaps, Detergents, and Surfactants

By definition, any oil or fat can be turned into a soap via saponification, although the FA composition of the oil will determine the effectiveness and applications for the soap. Saponified castor oil is occasionally used as a component of commercial soaps because it imparts antimicrobial activity and increases moisturization (Johnson, 2007). Castor oil soaps can also make soap bars translucent. Sulfation of castor oil with fuming sulfuric acid produced a sulfated castor oil known as Turkey red oil, one of the first synthetic surfactants produced (Naughton, 1974). Chemical modification of the hydroxy group with a variety of acylating and alkylating reagents has led to production of ionic and nonionic surfactants with a range of properties, depending on the chemical modification (Caupin, 1997).

In 1834, FF Runge treated olive oil with concentrated sulfuric acid to obtain a sulfonated product for use in textile dyeing. This product and those derived from other oils were referred to as "Turkey red oils," as they were used to hasten the process of dyeing cotton to impart a bright red color known as Turkey red. Later work in the 1870s reported the treatment of castor oil with concentrated sulfuric acid to produce an oil that was transparent and more suitable for use as a dyeing mordant than previous sulfated oils (Parks, 1931). The principal product was shown to be an oil bearing the ester sulfate of ricinoleate, though a number of side products were also present (Grun and Woldenberg, 1909). In addition to their past use in dyeing, castor oil sulfates continue to be used in tanning, proving to interact more effectively with collagen fibrils than sodium dodecyl sulfate (Kronick and Cooke, 1998). Alkyl ricinoleate sulfates are useful emulsifiers with properties similar to those of sodium dodecyl sulfate (SDS), including their surface tension effects, emulsification, foaming action, and critical micelle concentration, but exhibiting higher tolerance to calcium ion (Karuna et al., 2005). In the same study, only the dodecyl ester of 12-hydroxystearate sulfate matched the emulsifying properties of SDS.

Ethoxylated derivatives of castor oil are also useful as industrial, cosmetic, and food emulsifiers for use in detergents, lubricating and cutting oils, hydraulic fluids, and hydraulic fracturing (fracking). Ethoxylation can proceed by heating castor oil and ethylene oxide in the presence of acid or base or via SN_2 displacement of bromine from hydrobrominated castor oil or ricinoleate by the sodium base of polyethylene oxide monomethyl ethers, allowing controlled production of ethoxylated castor oil (El-Azab et al., 2010).

Lubricants

The presence of the hydroxy groups on castor oil results in extensive hydrogen bonding, leading to a very high viscosity oil (230 mPa s at 40°C) compared with nonhydroxy oils such as soy (31 mPa s at 40°C) (Quinchia et al., 2014). Thus, castor oil can be used to modify the viscosity of lubricants, and it has been used since ancient times as a lubricant. Both the high and low temperature performances of a lubricant are important, as these factors impose physical limits on the useful range and the applications in which the lubricant can be used. The cloud point of a liquid lubricant is the temperature at which the liquid becomes hazy and may interfere with flow or clog filters (Dwivedi and Sharma, 2014). The cloud point of castor oil is −20.8°C, one of the lowest cloud points among unmodified seed oils (Dwivedi and Sharma, 2014), due to its low saturated FA content and high ricinoleate content that block crystal formation (Quinchia et al., 2014). Castor oil can be modified to produce lubricants with even lower cloud point—eg, castor oil estolide has a cloud point of −54°C (Cermak et al., 2006).

The major use of castor oil in lubricants is in the form of lithium grease (Fig. 4.2), in which the castor oil is hydrogenated and hydrolyzed with lithium hydroxide to form the lithium soap of hydroxystearate. Lithium 12-hydroxystearate is an effective thickening agent that is usually blended with mineral oil (Naughton and Vignolo, 1992; Meng and Jie, 1998). Greases and other lubricants must interact effectively with the substrates they are coating, retaining their structure to remain bound for a reasonable span of time to prevent friction of lubricated parts. The lithium soap and mineral oil blend together to make a colloidal system that is thixotropic (Meng and Jie, 1998). In addition to serving as a thickener, the soap adds oxidative stability, resistance to shearing, and high temperature performance. The carboxylate and hydroxy groups aid in interaction with metal surfaces, while the hydrocarbon chain of the hydroxystearate helps to maintain the friction-reduction of the grease. Lithium greases are blended with different components to obtain desired properties, such as a higher dropping point, the temperature at which the grease components either separate or become liquid (Silver and Stanley, 1974). With increased emphasis on using biodegradeable greases and lubricants, grease blends of lithium hydroxystearate soaps with castor and other seed oils have demonstrated properties similar to those of mineral oil–based greases (Garcia-Zapateiro et al., 2014).

Castor oil, ricinoleate, and sebacic esters derived from castor may be used to alter properties of lubricants. Esters of castor oil and ricinoleate, including estolides formed with other FAs, are used as viscosity enhancers for lubricant blends in making greases (Cermak et al., 2006; Garcia-Zapeteiro et al., 2014).

Paints and Coatings

Also see chapter 3 "Polymeric Products Derived from Industrial Oilseed for Paints, Coatings, and Other Applications" by Hayes and Dumont.

Dehydration of castor oil produces a drying oil containing linoleate and conjugated linoleate, mostly as the 9-*cis*, 11-*trans* isomer (Priest and von Mikusch, 1940; Naughton, 1974) (Fig. 4.2). The proportion of conjugated linoleate varies with the method of dehydration (Grummitt and Marsh, 1953). Methods for dehydration include heating in the presence of mineral acids, acid anhydrides, alumina, or cation exchange resins, with each method providing different levels of conjugation in the product (Grummitt and Marsh, 1953; Naughton and Vignolo, 1992). Microwave heating in the presence of sodium bisulfate and sodium bisulfite also yields dehydrated castor oil (DCO) containing up to 43% conjugated linoleate (Nezihe et al., 2011). Complete conversion of the ricinoleate in castor oil to conjugated linoleate can be accomplished through heating in the presence of phosphoric acid (54% conjugated linoleate) followed by isomerization with a rhodium catalyst to DCO with 87% conjugated linoleate (Villeneuve et al., 2005).

At one time, DCO accounted for 5–6% of all drying oils used in coatings (Grummitt and Marsh, 1953), representing over one-third of the use of castor oil in the United States (Terrill, 1950). It dries rapidly, providing a nonyellowing, waterproof coating with drying properties that are easier to control than those of tung oil and with faster bodying (a process that increases viscosity by heat or chemical treatment to improve the coating texture) than that of linseed oil (Terrill, 1950). Currently, drying oils derived from petroleum have replaced DCO in many applications, although the low volatile organic compound output from DCO and other seed oils is valued in meeting regulations related to air pollution and worker exposure. Paint and coating applications for DCO include resins modified for specific applications such as weatherproofing or enamels (Majumdar et al., 1998; Jovicic et al., 2008).

As described earlier, castor oil is used to produce polyurethanes, and these may be used both in water- and oil-based coatings (Hofer et al., 1997). In addition, although castor oil is not a drying oil, it has been found as a component of Ripolin house paints (Dredge et al., 2013). These paints were developed in the late 19th century and were considered high-quality architectural coatings. Based on correspondence and studio photos, it is known that artists including Picasso and Picabia used these paints in some of their works. The Dadaist Picabia even produced a tribute to castor oil in "Portrait a l'huile de ricine," even though no evidence of castor oil was detected in the portrait itself (Umland and Sudhalter, 2008).

Fuel

Use of castor oil as a fuel for lamps in Egypt was documented over 2400 years ago by Herodotus, and analysis of lamps from a later age has identified castor oil as one of the fuels used in the lamps (Copley et al., 2005). Although castor oil may rarely be used as a lamp fuel now, it is still burned in the form of lubricant in two-stroke engines and as fuel for diesel engines.

Two-stroke gasoline engines are power sources for transportation including scooters and motorcycles, as well as recreational and landscaping equipment. Especially for these uses, the characteristics of the engine, including simple design and repair, high power output, and low cost (Singh, 2011) are key factors in the popularity of the engine. However, they are a major source of pollution, as the engines produce smoky emissions laden with unburned hydrocarbons and carbon monoxide. Two-stroke engines are usually lubricated by the addition of the lubricating oil to the fuel, so that both the fuel and oil are burned during operation of the engine. Esterified and acetylated castor oil is soluble in hydrocarbon fuel, and when added to gasoline it reduced smoke pollution from the engine by 50–70% (Singh, 2011).

With growing concern over petroleum use, alternative fuels derived from renewable materials have experienced increased interest. One such fuel is biodiesel which uses vegetable oils or FA alkyl esters derived from oils and fats (see chapter: Biodiesel and Its Properties by Knothe). FA methyl esters (FAMEs) represent the major source of biodiesel. In addition to reducing CO_2 output from petroleum sources, biodiesel can improve lubricity and cetane value of the fuel, important measures of performance for diesel fuels. As a result of the oxygen content of FAMEs, biodiesel also burns cleaner and reduces soot output from the engine. Since FAMEs improve lubricity, they can eliminate the need for the sulfur compounds added to diesel to improve lubricity, thereby reducing SO_2 pollution (Goodrum and Geller, 2005). Methyl esters derived from castor oil and from pure methyl ricinoleate are significantly better lubricity agents than FAME derived from nonhydroxylated FAs (Geller and Goodrum, 2004; Goodrum and Geller, 2005). While added lubricity from as little as 0.2% castor FAME points to the potential of castor in improving diesel fuel performance, additions at the level of the usual blends of biodiesel with diesel up to 20% are detrimental (Canoira et al., 2010; Knothe et al., 2012). Castor oil FAMEs exhibit unacceptably high kinematic viscosity, poor cetane number, and higher water content than biodiesel derived from nonhydroxy FAMEs and do not meet ASTM standards (Knothe et al., 2012). Blends of castor oil meet accepted engineering standards for biodiesel—eg, ethyl esters (20%) with 20% ethanol and 60% petroleum diesel (da Silva et al., 2013), or 10% castor oil FAME in petroleum diesel (Berman et al., 2011). However, the ASTM standard for diesel fuel requires that the pure biodiesel meets the same ASTM standards as diesel fuel (Berman et al., 2011); therefore, castor oil blends can only be used where standards differ from those of ASTM. Ultimately, the value of castor oil to the chemical industry, the limited supply of the oil, and its high price are major limitations for its use in biodiesel (Scholz and da Silva, 2008).

Medicinal, Cosmetic, and Food Applications

Castor oil is best known for its laxative properties. It is still used today, despite the availability of milder laxatives, and also serves in medical studies evaluating diarrhea and bowel function. The laxative effect is directly related to ricinoleate released after lipolysis of castor oil. Other hydroxy FAs including 12-hydroxy stearate, lesquerolate, 10-hydroxy stearate, and ricinelaidate do not induce smooth muscle contractions in the intestine (Stewart et al., 1975). The basis for the laxative effect of ricinoleate is physiological. Both castor oil and ricinoleate affect several signaling pathways including NO, eicosanoids, and platelet activating factor (reviewed in McKeon et al., 2006). Castor oil has also been used in birthing, as it is said to induce or promote contractions, and recent evidence using mice has shown that ricinoleate acts as an agonist of prostanoid receptors. Deletion of these receptors removed the effect of ricinoleate-induced smooth muscle contractions in the uterus and intestines (Tunaru et al., 2012).

The emulsifier Cremophor EL is an ethoxylated castor oil derivative that has been widely used in drug delivery (Gelderblom et al., 2001). Although the antitumor activity of paclitaxel from yew was reported in 1971, it took over a decade before a formulation could deliver it intravenously due to solubility problems. Cremophor EL in combination with ethanol was found to be suitable, and Cremophor EL was used in formulations for delivery of a number of pharmaceuticals (Gelderblom et al., 2001). Serious medical complications were observed in many patients when Cremophor EL–paclitaxel emulsions were delivered and were thought to be due to Cremophor EL. However, approaches were developed to mitigate the medical complications (Gelderblom et al., 2001). Later work indicated that the medical problems observed might have been due to paclitaxel, and the Cremophor EL formulation continues to be the most widely used delivery system for paclitaxel (Li et al., 2015).

The value of castor oil in cosmetics lies in its high viscosity and miscibility with polar and nonpolar compounds, allowing it to bind organic and mineral pigments as well as to interact with a range of other cosmetic components. As a result, it is a common component of lipstick, ranging in concentration from 15% to 81%, and was found in over half the lipsticks analyzed, 492 of 962 (Johnson, 2007). Castor oil is also considered noncomedogenic compared with most seed oils, as it does not clog pores and is not a skin irritant (Johnson, 2007).

Derivatives of castor oil also have value in cosmetics. Zinc ricinoleate is useful as a deodorant both in personal deodorant applications and in treating animal waste (Kuhn et al., 2000). Although ricinoleate salts are derived from base hydrolysis of castor oil, thereby eliminating any trace of protein allergens, in some individuals they do cause an allergic reaction (Magerl et al., 2001; Johnson, 2007). Other derivatives of castor oil are used for both textural and antimicrobial properties. Hydroxystearate derived from hydrogenation and hydrolysis of castor oil is similar to wax and provides a skin feel similar to the more expensive cocoa butter. It also serves as an antifungal agent. Similarly, undecylenic acid and its salts have antifungal activity and are used in cosmetic creams as well as topical antifungals (McKeon et al., 2006; Johnson, 2007).

Castor oil is generally considered to have no food value, as it causes diarrhea in doses of 200 mg/kg body weight. Both castor oil and related derivatives are allowed at levels up to 500 ppm in the final food product when used as a mold release for hard candy, in food packaging, and in other applications where ingestion from food products may occur (Johnson, 2007).

The emulsifier PGPR described in the Polymer section is used in the chocolate industry to replace cocoa butter and improve handling. Chocolate containing 32% cocoa butter and 0.1% PGPR has similar flow properties to chocolate containing 35% cocoa butter (Afoakwa et al., 2007), allowing replacement of some cocoa butter with a minor amount of PGPR. PGPR also allows processing and tempering of chocolate at reduced temperature (Schantz and Rohm, 2005) and may thereby help to reduce fat bloom of chocolate (Bastida-Rodriguez, 2013). Since small amounts of PGPR can replace significant amounts of cocoa butter without affecting physical properties of the chocolate product, some cocoa butter can be retained for use in cosmetics, enhancing profitability for the manufacturer. Because it is a replacement ingredient for cocoa butter, and despite the fact that it is metabolized and considered safe at regulated levels (Wilson et al., 1998), use of PGPR in the chocolate industry has engendered consumer complaints about it serving as an artificial additive to a beloved product.

MISCELLANEOUS USES OF THE CASTOR PLANT

As mentioned in the sections on ricin and allergen in seed meal, there are serious issues with using castor meal in animal feed. It is generally considered suitable for feed use if the allergen and toxin have been inactivated, usually by heat and hydrolytic treatment. Over the years, a number of treatments have been used to inactivate the ricin and allergen (Gardner et al., 1960; Mottola et al., 1968; Fuller et al., 1971; Horton and Williams, 1989; Fernandes et al., 2012). Ricin has also been inactivated by proteolytic action (McKeon et al., 2013) and reactive processing, in which the seed is extracted with strong base and methanol to generate FAMEs (Dubois et al., 2013). However, if not carefully or conscientiously carried out, the reported ricin and allergen inactivation is not always effective, resulting in animal death or sickness (Worbs et al., 2011; Akande et al., 2015). In short, commercialization of castor meal for animal feed would likely require that it be tested and certified as ricin and allergen free. Under conditions where these noxious proteins are inactivated, it appears that castor seed meal could be a useful feed supplement (Akande et al., 2015) In Nigeria, castor seed is fermented by a collection of *Bacillus* species to produce the fermented bean product ogiri, and it is apparently safe to consume, probably due to extensive heating prior to fermentation and bacterial proteolysis during fermentation (Odunfa, 1985).

In addition to seed meal, the husks of the fruit that bears castor seed is a significant coproduct of castor oil production. Castor meal is sold and used as fertilizer, as it is rich in nitrogen, 7.54% N, which is 10 times that of steer manure (Lima et al., 2011). The authors found that castor meal in excess of 4% prevented germination and caused plant growth inhibition or death, but a mixture of 4.5% meal and 5.5% husk promoted plant growth. While the meal is rich in N and P, the husk is rich in K, providing a balanced soil amendment that also adds organic material to the soil.

As described earlier in the chapter, the castor plant grows well in soil with small amounts of salt present. It has also been shown to grow in soils contaminated with nickel, chromium, copper, manganese (Gupta and Sinha, 2007), and arsenate (Melo et al., 2009). Because castor has a deep and extensive root system, it is possible to use the plant for phytoremediation of soils contaminated with heavy metal ion from mining and manufacturing. Preliminary experiments indicate that castor can also be used for phytoremediation of cesium ion, a process of considerable potential value as the Cs^{137} ion is a major contaminant in nuclear reactor accidents.

There is a considerable background in applications of the castor plant for folk medicine. While the use of castor oil as a purgative is widely known, application of other parts of the plant are documented (Scarpa and Guerci, 1982; Rana et al., 2012). As described in these reviews, the leaves of the plant are said to be useful in healing and reducing inflammation. Other remedies include ingestion of the seed for smallpox and leprosy (Scarpa and Guerci, 1982), although it is not known if such treatments are effective. Some effects of castor plant extracts are imputed to the ricinine present throughout the plant. Both reviews indicate that leaf extracts have antifungal and antimicrobial activity, and ricinine itself is effective in killing nematodes, ants, termites, and army worms (Scarpa and Guerci, 1982; Bullangpoti et al., 2011).

SUMMARY

Castor oil is a unique natural product used in many industries and many types of products. With its 85–90% content of the hydroxy FA ricinoleic acid, castor oil provides this FA at the equivalent of a technical grade of purity, a rare natural phenomenon. The uses of castor oil in industry are limited by the availability of the oil. The presence of a midchain hydroxyl group and the polyol nature of castor oil provide nearly endless possibilities for chemists to develop new applications for the oil. More widespread cultivation of the castor plant will expand utilization of castor oil by industry.

ACKNOWLEDGMENTS

The author would like thank USDA-ARS, USDA-IFAFS, Dow Chemical, and the Department of Energy for supporting aspects of his research on castor. He would also like to express thanks to his coworkers, collaborators, and coauthors on castor research.

REFERENCES

Aelion, R., 1961. Nylon 6 and related polymers. Ind. Eng. Chem. 53, 826–828.

Afoakwa, E.O., Paterson, A., Fowler, M., 2007. Factors influencing rheological and textural qualities of chocolate – a review. Trends Food Sci. Technol. 18, 290–298.

Ahn, Y.-J., Vang, L., McKeon, T.A., Chen, G.Q., 2007. High-frequency plant regeneration through adventitious shoot formation in casstor (*Ricinus communis* L.). In Vitro Cell. Dev. Biol. Plant 43, 9–15.

Akande, T.O., Odunsi, A.A., Akinfala, E.O., 2015. A review of nutritional and toxicological implications of castor bean (*Ricinus communis* L.) meal in animal feeding systems. J. Anim. Physiol. Anim. Nutr. http://dx.doi.org/10.1111/jpn. 12360.

Allan, G., Williams, A., Rabinowitz, P.D., Chan, A.P., Ravel, J., Keim, P., 2008. Worldwide genotyping of castor bean (*Ricinus communis* L.) germplasm using AFLPs and SSRs. Genet. Resour. Crop Evol. 55, 365–378.

Anjani, K., 2010. Extra-early maturing germplasm for utilization in castor improvement. Ind. Crops Prod. 31, 139–144.

Anjani, K., 2012. Castor genetic resources: a primary gene pool for exploitation. Ind. Crops Prod. 35, 1–14.

Anjani, K., Pallavi, M., Sudhakara Babu, S.N., 2010. Biochemical basis of resistance to leafminer in castor (*Ricinus communis* L.). Ind. Crops Prod. 31, 192–196.

Anjani, K., Raoof, M.A., Desai, A.G., 2014. Evaluation of world castor (*Ricinus communis* L.) germplasm for resistance to Fusarium wilt (*Fusarium oxysporum* f.sp. *ricini*). Eur. J. Plant Pathol. 139, 567–578.

Apen, E.M., Cooper, W.C., Horton, R.J.M., Scheel, L.D., 1967. Health Aspects of Castor Bean Dust. Public Health Service Publication No. 999-AP-36, pp. 1–132.

Arroyo-Caro, J.M., Chileh, T., Kazachkov, M., Zou, J., Lopez-Alonso, D., Garcia-Maroto, F., 2013a. The multigene family of lysophosphatidate acyltransferase (LPAT) – related enzymes in *Ricinus communis*. Cloning and molecular characterization of two *LPAT* genes that are expressed in castor seeds. Plant Sci. 199–200, 29–40.

Arroyo-Caro, J.M., Chileh, T., Kazachkov, M., Zou, J., Lopez-Alonso, D., Garcia-Maroto, F., 2013b. Molecular characterization of a lysophosphatidylcholine acyltransferase gene belonging to the MBOAT family in *Ricinus communis* L. Lipids 48, 663–674.

Auld, D.L., Rolfe, R.D., McKeon, T.A., 2001. Development of castor with reduced toxicity. J. New Seeds 3, 61–69.

Auld, D.L., Pinkerton, S.D., et al., 2003. Registration of TTU-LRC castor germplasm with reduced levels of ricin and RCA 120. Crop Sci. 43, 746–747.

Bafor, M., Smith, M.A., Jonsson, L., Stobart, K., Stymne, S., 1991. Ricinoleic acid biosynthesis and triacylglycerol assembly in microsomal preparations from developing castor-bean (*Ricinus communis*) endosperm. Biochem. J. 280, 507–514.

Baldanzi, M., Pugliesi, C., 1998. Selection for non-branching in castor, *Ricinus communis* L. Plant Breed. 117, 392–394.

Barros, P.R., Haro, A., Munoz, J., Martinez, J.M.F., 2004. Isolation of a natural mutant in castor with high oleic/low ricinoleic acid content in the oil. Crop Sci. 44, 76–80.

Bashir, M.E.H., Hubatsch, I., Leinenbach, H.P., Zeppezauer, M., Panzani, R.C., Hussein, I.H., 1998. RicC1 and RicC3, the allergenic 2S albumin storage proteins of *Ricinus communis*: complete primary structures and phylogenetic relationships. Int. Arch. Allergy Immunol. 115, 73–82.

Bastida-Rodriguez, J., 2013. The food additive polyglycerol polyricinoleate (E-476): structure, applications, and production methods. ISRN Chem. Eng. 2013, 21.:124767. http://dx.doi.org/10.1155/2013/124767.

Bates, P.D., Browse, J., 2011. The pathway of triacylglycerol synthesis through phosphatidylcholine in Arabidopsis produces a bottleneck for the accumulation of unusual fatty acids in transgenic seeds. Plant J. 68, 387–399.

Bates, P.D., Johnson, S.R., et al., 2014. Fatty acid synthesis is inhibited by inefficient utilization of unusual fatty acids for glycerolipid assembly. Proc. Natl. Acad. Sci. U.S.A. 111, 1204–1209.

Baur, X., Chen, Z., Hurter, T., 1998. Asthma and rhinoconjunctivitis caused by castor bean dust. Pneumologie 52, 539–540.

Berman, P., Nizri, S., Wiesman, Z., 2011. Castor oil biodiesel and its blends as alternative fuel. Biomass Bioenergy 35, 2861–2866.

Blum, D., June 7, 2014. How did ricin become the poison of choice for incompetent schemers? Los Angeles Times. http://www.latimes.com/opinion/op-ed/la-oe-blum-ricin-20140608-story.html (July 2, 2015).

Borg, P., Le, G., Lebrun, S., Pees, B., 2009. Example of industrial valorisation of derivative products of castor oil. Ol. Corps Gras Lipides 16, 211–214. http://dx.doi.org/10.1051/ocl.2009.0276 (July 29, 2015).

Brandon, D.L., Hernlem, B.J., 2009. Development of monoclonal antibodies specific for ricinus agglutinins. Food Agric. Immunol. 20, 11–22.

Brandon, D.L., Korn, A.M., Yang, L.L., 2014. Immunosorbent analysis of ricin contamination in milk using colorimetric, chemiluminescent and electrochemiluminescent detection. Food Agric. Immunol. 25, 160–172.

Brandon, D.L., McKeon, T., Patfield, S., Kong, Q., He, X., 2016. Analysis of castor by ELISAs that distinguish ricin and *Ricinus communis* agglutinin (RCA). J. Am. Oil Chem. Soc. http://dx.doi.org/10.1007/s11746-015-2773-6(in press).

Brinkworth, C.S., Pigott, E.J., Bourne, D.J., 2009. Detection of intact ricin in crude and purified extracts from castor beans using matrix-assisted laser desorption ionization mass spectrometry. Anal. Chem. 81, 1529–1535.

Broun, P., Somerville, C., 1997. Accumulation of ricinoleic, lesquerolic, and densipolic acids in seeds of transgenic *Arabidopsis* plants that express a fatty acyl hydroxylase cDNA from castor bean. Plant Physiol. 113, 933–942.

Bullangpoti, V., Khumrungsee, N., Pluempanupat, W., Kainoh, Y., Saguanpong, U., 2011. Toxicity of ethyl acetate extract and ricinine from *Jatropha gossypifolia* senescent leaves against *Spodoptera exigua* Hubner (Lepidoptera: Noctuidae). J. Pestic. Sci. 36, 260–263.

Burgal, J., Shockey, J., Lu, C., Dyer, J., Larson, T., Graham, I., Browse, J., 2008. Metabolic engineering of hydroxy fatty acid production in plants: RcDGAT2 drives dramatic increases in ricinoleate levels in seed oil. Plant Biotechnol. J. 6, 819–831.

Campbell, D.N., Rowland, D.L., Schnell, R.W., Ferrell, J., Wilkie, A.C., 2014. Developing a castor (*Ricinus communis* L.) production system in Florida, U.S.: evaluating crop phenology and response to management. Ind. Crops Prod. 53, 217–227.

Canoira, L., Galean, J.G., Alcantara, R., Lapuerta, M., Garcia-Contreras, R., 2010. Fatty acid methyl esters (FAMEs) from castor oil: production process assessment and synergistic effects in its properties. Renewable Energy 35, 208–217.

Caupin, H.J., 1997. Products from castor oil: past, present, and future. In: Gunstone, F.D., Padley, F.B. (Eds.), Lipid Technologies and Applications. Marcel Dekker Inc., New York, pp. 787–795.

Cermak, S.C., Brandon, K.B., Isbell, T.A., 2006. Synthesis and physical properties of estolides from lesquerella and castor fatty acid esters. Ind. Crops Prod. 23, 54–64.

Chan, A.P., Crabtree, J., et al., 2010. Draft genome sequence of the oilseed species *Ricinus communis*. Nat. Biotechnol. 28, 951–956.

Chen, G.Q., He, X., Liao, L.P., McKeon, T.A., 2004. 2S albumin gene expression in castor plant (*Ricinus communis* L.). J. Am. Oil Chem. Soc. 81, 867–872.

Chen, G.Q., He, X., McKeon, T.A., 2005. A simple and sensitive assay for distinguishing the expression of ricin and *Ricinus communis* agglutinin genes in developing castor seeds (*R. communis* L.). J. Agric. Food Chem. 53, 2358–2361.

Copley, M.S., Bland, H.A., Rose, P., Horton, M., Evershed, R.P., 2005. Gas chromatographic, mass spectrometric and stable carbon isotopic investigations of organic residues of plant oils and animal fats employed as illuminants in archaeological lamps from Egypt. Analyst 130, 860–871.

Dahlqvist, A., et al., 2000. Phospholipid:diacylglycerol acyltransferase: an enzyme that catalyzes the acyl-CoA-independent formation of triacylglycerol in yeast and plants. Proc. Natl. Acad. Sci. U.S.A. 97, 6487–6492.

Danlami, J.M., Zaini, M.A.A., Arsad, A., Yunus, M.A.C., 2015. A parametric investigation of castor oil (*Ricinus communis* L.) extraction using supercritical carbon dioxide via response surface optimization. J. Taiwan Inst. Chem. Eng. 53, 32–39.

d'Errico, F., Backwell, L., et al., 2012a. Early evidence of San material culture represented by organic artifacts from Border Cave, South Africa. Proc. Natl. Acad. Sci. U.S.A. 109, 13214–13219.

d'Errico, F., Backwell, L., et al., 2012b. Reply to Evans: use of poison remains the most parsimonious explanation for Border Cave castor bean extract. Proc. Natl. Acad. Sci. U.S.A. 109, E3291–E3292.

da Silva, N.L., Santander, C.M.G., Rueda, S.M.G., Maciel, M.R.W., Filho, R.M., 2013. Characterization of blend properties of castor biodiesel and bioethanol. Ind. Eng. Chem. Res. 52, 15504–15508.

Darby, S.M., Miller, M.L., Allen, R.O., 2001. Forensic determination of ricin and the alkaloid ricinine from castor bean extracts. J. Forensic Sci. 46, 1033–1046.

Dauk, M., Lam, P., Smith, M.A., 2009. The role of diacylglycerol acyltransferase-1 and phospholipid:diacylglycerol acyltransferase-1 and -2 in the incorporation of hydroxy fatty acids into triacylglycerol in *Arabidopsis thaliana* expressing a castor bean oleate 12-hydroxylase gene. Botany 87, 552–560.

de Guzman, D., September 22, 2003. Soy polyols in polyurethanes emerge as niche market. Chem. Market Rep. 264 (9), 14.

Diamond, M.J., Applewhite, T.H., 1967. Alkaline cleavage of hydroxy unsaturated fatty acids and derivatives. II. 10-hydroxydecanoic acid from ricinoleates and 1,10-decandiol from ricinoleyl alcohol. J. Am. Oil Chem. Soc. 44, 656–658.

Diamond, M.J., Binder, R.G., Applewhite, T.H., 1965. Alkaline cleavage of hydroxy unsaturated fatty acids. I. Ricinoleic and lesquerolic acid. J. Am. Oil Chem. Soc. 42, 882–884.

Dredge, P., Schilling, M.R., Gautier, G., Mazurek, J., Learner, T., Wuhrer, R., 2013. Lifting the lids off Ripolin: a collection of paint from Sidney Nolan's studio. J. Am. Inst. Conserv. 52, 213–226.

Dubois, J.L., Piccirilli, A., Magne, J., He, X., 2013. Detoxification of castor meal through reactive seed crushing. Ind. Crops Prod. 43, 194–199.

Dwivedi, G., Sharma, M.P., 2014. Potential and limitation of straight vegetable oils as engine fuel – an Indian perspective. Renewable Sustainable Energy Rev. 33, 316–322.

El-Azab, W.I.M., Aiad, I., Azzam, E.M.S., Gad, E.A.M., 2010. Molecular, surface, and thermodynamic properties of nonionic surfactants based on castor oil. J. Dispersion Sci. Technol. 31, 1150–1156.

Endo, Y., Mitsui, K., Motizuki, M., Tsurugi, K., 1987. The mechanism of action of ricin and related toxic lectins on eukaryotic ribosomes. The site and the characteristics of the modification in 28 S ribosomal RNA caused by the toxins. J. Biol. Chem. 262, 5908–5912.

Evans, A.A., 2012. Arrow poisons in the paleolithic? Proc. Natl. Acad. Sci. U.S.A. 109, E3290.

FAOSTAT, 2015. http://faostat3.fao.org/download/Q/QC/E (Production, Crops, Castor oil seed) November 28, 2015.

Fernandes, K.V., Deus de Oliveira, N., et al., 2012. Simultaneous allergen inactivation and detoxification of castor bean cake by treatment with calcium compounds. Braz. J. Med. Biol. Res. 45, 1002–1010.

Figley, K.D., Rawling, F.F.A., 1950. Castor bean: an industrial hazard as a contaminant of green coffee dust and used burlap bags. J. Allergy 21, 545–553.

Fuller, G., Walker, H.G., Mottola, A.C., Kuzmicky, D.D., Kohler, G.O., Vohra, P., 1971. Potential for detoxified castor meal. J. Am. Oil Chem. Soc. 48, 616–618.

Funatsu, G., Kimura, M., Funatsu, M., 1979. Primary structure Ala chain of ricin D. Agric. Biol. Chem. 43, 2221–2224.

Galliard, T., Stumpf, P.K., 1966. Fat metabolism in higher plants: XXX. Enzymatic synthesis of ricinoleic acid by a microsomal preparation from developing *Ricinus communis* seeds. J. Biol. Chem. 241, 5806–5812.

Garcia-Zapeteiro, L.A., Valencia, C., Franco, J.M., 2014. Formulation of lubricating greases from renewable basestocks and thickener agents: a rheological approach. Ind. Crops Prod. 54, 115–121.

Gardner, H.K., D'Aquin, E.L., Koltun, S.P., McCourtney, E.J., Vix, H.L.E., Gastrock, E.A., 1960. Detoxification and deallergenization of castor beans. JAOCS 37, 142–148.

Gelderblom, H., Verweij, J., Nooter, K., Sparreboom, A., 2001. Cremophor EL: the drawbacks and advantages of vehicle selection for drug formulation. Eur. J. Cancer 37, 1590–1598.

Geller, D.P., Goodrum, J.W., 2004. Effects of specific fatty acid methyl esters on diesel fuel lubricity. Fuel 83, 2351–2356.

Goodrum, J.W., Geller, D.P., 2005. Influence of fatty acid methyl esters from hydroxylated vegetable oils on diesel fuel lubricity. Bioresour. Technol. 96, 851–855.

Grummitt, O., Marsh, D., 1953. Alternative methods for dehydrating castor oil. J. Am. Oil Chem. Soc. 30, 21–25.

Grun, A.D., Woldenberg, M., 1909. Investigation of the essential constituent of Turkey red oil and its derivatives. J. Am. Chem. Soc. 31, 490–506.

Gupta, A.K., Sinha, S., 2007. Phytoextraction capacity of the plants growing on tannery sludge dumping sites. Bioresour. Technol. 98, 1788–1794.

Halling, K.C., Halling, A.C., Murray, E.E., Ladin, B.F., Houston, L.L., Weaver, R.F., 1985. Genomic cloning and characterization of a ricin gene from *Ricinus communis*. Nucleic Acids Res. 13, 8019–8033.

Hamelin, E.I., Johnson, R.C., Osterloh, J.D., Howard, D.J., Thomas, J.D., 2012. Evaluation of ricinine, a ricin biomarker, from a non-lethal castor bean ingestion. J. Anal. Toxicol. 36, 660–662.

Han, G., Liu, Z., Yao, S., Yan, R., 1996. Study of reaction kinetics in pyrolysis of methyl ricinoleate. JAOCS 73, 1109–1112.

Hargreaves, G.H., Owen, L.N., 1947. Reactions of carbinols in the presence of alkali. Part II> the scission of ricinoleic acid. J. Chem. Soc. 753–756. http://dx.doi.org/10.1039/JR9470000753.

He, X., Turner, C., Chen, G.Q., Lin, J.-T., McKeon, T.A., 2004a. Cloning and characteization of a cDNA encoding diacylglycerol acyltransferase from castor bean. Lipids 39, 311–318.

He, X., Chen, G.Q., Lin, J.-T., McKeon, T.A., 2004b. Regulation of diacylglycerol acyltransferase in developing seeds of castor. Lipids 39, 865–871.

He, X., Carter, J.M., Brandon, D.L., Cheng, L.W., McKeon, T.A., 2007. Application of a real time polymerase chain reaction method to detect castor toxin contamination in fluid milk and eggs. J. Agric. Food Chem. 55, 6897–6902.

He, X., Lu, S., Cheng, L.W., Rasooly, R., Carter, J.M., 2008. Effect of food matrices on the biological activity of ricin. J. Food Prot. 73, 2053–2058.

He, X., McMahon, S., Henderson, T.D., Griffey, S.M., Cheng, L.W., 2010a. Ricin toxicokinetics and its sensitive detection in mouse sera or feces using immuno-PCR. PLos One 5 (9), e12858. http://dx.doi.org/10.1371/journal.pone.0012858.

He, X., McMahon, S., McKeon, T.A., Brandon, D.L., 2010b. Development of a novel immuno-PCR assay for detection of ricin in ground beef, liquid chicken egg, and milk. J. Food Prot. 695–700.

Hofer, R., Daute, P., Grutzmacher, R., Westfechel, A., 1997. Oleochemical polyols—a new raw material source for polyurethane coatings and floorings. J. Coat. Technol. 69, 65–72.

Horton, J., Williams, M.A., 1989. A cooker-extruder for deallergenation of castor bean meal. J. Am. Oil Chem. Soc. 66, 227–231.

Hu, Z., Ren, Z., Lu, C., 2012. The phosphatidylcholine diacylglycerol cholinephosphotransferase is required for efficient hydroxy fatty acid accumulation in transgenic Arabidopsis. Plant Physiol. 158, 1944–1954.

Irwin, S.D., Keen, J.N., Findlay, J.B.C., Lord, J.M., 1990. The *Ricinus communis* 2S albumin precursor: a single preproprotein may be processed into two different heterodimeric storage proteins. Mol. Gen. Genet. 222, 400–408.

Irwin, S.D., Lord, J.M., 1990. Nucleotide sequence of a *Ricinus communis* 2S albumin precursor gene. Nucl. Acids Res. 18, 5890.

Johnson, R.W., 1984. The chemistry of dibasic and polybasic fatty acids. J. Am. Oil Chem. Soc. 61, 241–246.

Johnson, W., 2007. Final report on the safety assessment of *Ricinus communis* (castor) seed oil, hydrogenated castor oil, glyceryl ricinoleate, glyceryl ricinoleate SE, ricinoleic acid, potassium ricinoleate, sodium ricinoleate, zinc ricinoleate, cetyl ricinoleate, ethyl ricinoleate, glycol ricinoleate, isopropyl ricinoleate, methyl ricinoleate, and octyldodecyl ricinoleate. Int. J. Toxicol. 26 (Suppl. 3), 31–77.

Jovicic, M.C., Radicevic, R.Z., Budinski-Simendic, J.K., 2008. Curing of alkyds based on semi-drying oils with melamine resin. J. Therm. Anal. Calorim. 94, 143–150.

Karuna, M.S.L., Reddy, J.R.C., Rao, B.V.S.K., Prasad, R.B.N., 2005. Lipase-mediated synthesis of alkyl ricinoleates and 12-hydroxy stearates and evaluation of the surfactant properties of their sulfated sodium salts. J. Surfactants Deterg. 8, 271–276.

Kelly, A.R., Hayes, D.H., 2006. Lipase-catalyzed synthesis of polyhydric alcohol-poly(ricinoleic acid) ester star polymers. J. Appl. Polym. Sci. 101, 1646–1656.

Kim, H.U., Lee, K.R., Go, Y.S., Jung, J.H., Suh, M.C., Kim, J.B., 2011. Endoplasmic reticulum-located PDAT1-2 from castor bean enhances hydroxy fatty acid accumulation in transgenic plants. Plant Cell Physiol. 52, 983–993.

Knight, B., 1979. Ricin – a potent homicidal poison. Br. Med. J. 1, 350–351.

Knothe, G., Cermak, S.C., Evangelista, R.L., 2012. Methyl esters from vegetable oils with hydroxy fatty acids: comparison of lesquerella and castor methyl esters. Fuel 96, 535–540.

Kronick, P.L., Cooke, P., 1998. Destabilization of collagen in hide and leather by anionic surfactants. II. Calorimetry of the reaction of collagen with sulfates. J. Polym. Sci. 36, 805–813.

Kroon, J.T.M., Wei, W., Simon, W.J., Slabas, A.R., 2006. Identification and functional expression of a type 2 acyl-CoA:diacylglycerol acyltransferase (DGAT2) in developing castor bean seeds which has high homology to the major triglyceride biosynthetic enzyme of fungi and animals. Phytochemistry 67, 2541–2549.

Kuhn, H., Muller, F., Peggau, J., Zekorn, R., 2000. Mechanism of the odor-absorption effect of zinc ricinoleate. A molecular dynamics computer simulation. J. Surfactants Deterg. 3, 335–343.

Kulkarni, N.G., Krishnamurti, N., Chatterjee, P.C., Aggarwal, J.S., 1968. Polymerizable monomers from castor oil. I. Vinyl monomers based on undecenoic acid. J. Am. Oil Chem. Soc. 45, 465–467.

Kunitz, M., McDonald, M.R., 1948. Isolation of crystalline ricin. J. Gen. Physiol. 32, 25–31.

Lamb, F.I., Roberts, L.M., Lord, J.M., 1985. Nucleotide sequence of cloned cDNA coding for preproricin. Eur. J. Biochem. 148, 265–270.

Li, R., Yu, K., Hildebrand, D.F., 2010. DGAT1, DGAT2, and PDAT expression in seeds and other tissues of epoxy and hydroxy fatty acid accumulating plants. Lipids 45, 145–157.

Li, Y., Chen, N., Palmisano, M., Zhou, S., 2015. Pharmacologic sensitivity of paclitaxel to its delivery vehicles drives distinct clinical outcomes of paclitaxel formulations. Mol. Pharm. 12, 1308–1317.

Lima-Neto, M.C., Lobo, A.K.M., Martins, M.O., Fontinele, A.V., Silveira, J.A.G., 2014. Dissipation of excess photosynthetic energy contributes to salinity tolerance: a comparative study of salt-tolerant *Ricinus communis* and salt-sensitive *Jatropha curcas*. J. Plant Physiol. 171, 23–30.

Lima, R.L.S., Severino, L.S., Sampaio, L.R., Sofiatti, V., Gomes, J.A., Beltrao, N.E.M., 2011. Blends of castor meal and castor husks for optimized use as organic fertilizer. Ind. Crops Prod. 33, 364–368.

Lin, J.T., McKeon, T.A., Goodrich-Tanrikulu, M., Stafford, A.E., 1996. Characterization of oleoyl-12-hydroxylase in castor microsomes using the putative substrate, 1-acyl-2-oleoyl-sn-glycero-3-phosphocholine. Lipids 31, 571–577.

Lin, J.-T., Woodruff, C.L., Lagouche, O.J., McKeon, T.A., Stafford, A.E., Goodrich-Tanrikulu, M., Singleton, J.A., Haney, C.A., 1998. Biosynthesis of triacylglycerols containing ricinoleate in castor microsomes using 1-acyl-2-oleoyl-sn-glycero3-phosphocholine as the substrate of oleoyl-12-hydroxylase. Lipids 33, 59–69.

Lin, J.-T., Chen, J.M., Liao, L.P., McKeon, T.A., 2002. Molecular species of acylglycerols incorporating radiolabeled fatty acids from castor (*Ricinus communis* L.) microsomal incubations. J. Agric. Food Chem. 50, 5077–5081.

Lin, J.-T., Turner, C., Liao, L.P., McKeon, T.A., 2003. Identification and quantification of the molecular species of acylglycerols in castor oil by HPLC using ELSD. J. Liq. Chromatogr. Relat. Technol. 26, 773–780.

Lin, J.-T., Arcinas, A., 2007. Regiospecific analysis of diricinoleoylacylglycerols in castor (*Ricinus communis* L.) oil by electrospray ionization-mass spectrometry. J. Agric. Food Chem. 55, 2209–2216.

Lin, J.-T., Arcinas, A., Harden, L.A., 2009. Identification of acylglycerols containing dihydroxy fatty acids in castor oil by mass spectrometry. Lipids 44, 359–365.

Lin, J.-T., Chen, G.Q., 2010. Acylglycerols containing trihydroxy fatty acids in castor oil and the regiospecific quantification of triacylglycerols. J. Am. Oil Chem. Soc. 87, 1371–1379.

Lord, J.M., Roberts, L.M., Roberts, J.D., 1994. Ricin: structure, mode of action, and some current applications. FASEB J. 8, 201–208.

Machado, O.L.T., Silva Jr., J.G., 1992. An allergenic 2S storage protein from *Ricinus communis* seeds which is a part of the 2S albumin precursor predicted by cDNA data. Braz. J. Med. Biol. Res. 25, 567–582.

Magerl, A., Heiss, R., Frosch, P.J., 2001. Allergic contact dermatitis from zinc ricinoleate in a deodorant and glyceryl ricinoleate in a lipstick. Contact Dermatitis 44, 119–121.

Mahapatro, A., Kumar, A., Gross, R.A., 2004. Mild, solvent-free ω-hydroxy acid polycondensations catalyzed by *Candida antarctica* Lipase B. Biomacromolecules 5, 63–68.

Majumdar, S., Kumar, D., Nirvan, Y.P.S., 1998. Acrylate grafted dehydrated castor oil alkyd – a binder for exterior paints. J. Coat. Technol. 70, 27–33.

Mancha, M., Stymne, S., 1997. Remodeling of triacylglycerols in microsomal preparations from developing castor bean (*Ricinus communis* L.) endosperm. Planta 203, 51–57.

Matsumoto, M., Hirata-Koizumi, M., Ema, M., 2007. Potential adverse effects of phthalic acid esters on human health: a review of recent studies on reproduction. Regul. Toxicol. Pharmacol. 50, 37–49.

McKeon, T.A., Lin, J.-T., Goodrich-Tanrikulu, M., Stafford, A.E., 1997. Ricinoleate biosynthesis in castor microsomes. Ind. Crops Prod. 6, 383–389.

McKeon, T.A., Lin, J.-T., Chen, G., 2002. Developing a safe source of castor oil. Inform 13, 381–385.

McKeon, T.A., Chen, G.Q., 2003. Transformation of *Ricinus communis*, the Castor Plant. U.S. Patent 6,620,986 B1.

McKeon, T.A., Turner, C., He, X., Chen, G., Lin, J.T., 2006. Hydroxy fatty acids. In: Shahidi, F. (Ed.), Nutraceutical and Specialty Lipids and Their Co-products. CRC Press, Boca Raton.

McKeon, T.A., Chen, G.Q., He, X., Ahn, Y.-J., Lin, J.-T., 2007. The enzymology of castor oil biosynthesis. In: Janick, J., Whipkey, A. (Eds.), Issues in New Crops and New Uses. ASHS Press, Alexandria, VA, pp. 101–104.

McKeon, T.A., Lin, J.-T., 2007. Castor oil products. In: Gunstone, F.D., Harwood, J.L., Djkstra, A.J. (Eds.), The Lipid Handbook. CRC Press, Boca Raton, pp. 632–636.

McKeon, T.A., Shim, K.B., He, X., 2013. Reducing the toxicity of castor seed meal through processing treatments. Biocatal. Agric. Biotechnol. 2, 159–161.

McKeon, T.A., Auld, D., Brandon, D.L., Leviatov, S., He, X., 2014. Toxin content of commercial castor cultivars. J. Am. Oil Chem. Soc. 91, 1515–1519.

McKeon, T.A., He, X., 2015. Castor diacylglycerol acyltransferase type 1 (DGAT1) displays greater activity with diricinolein than Arabidopsis DGAT1. Biocatal. Agric. Biotechnol. 4, 276–278.

Melo, E.E.C., Costa, E.T.S., Guilherme, L.R.G., Faquin, V., Nascimento, C.W.A., 2009. Accumulation of arsenic and nutrients by castor bean plants grown on an As-enriched nutrient solution. J. Hazard. Mater. 168, 479–483.

Meng, Y., Jie, Z., 1998. A rheological model for lithium lubricating grease. Tribol Int. 31, 619–625.

Moreau, R.A., Stumpf, P.K., 1981. Recent studies of the enzymic synthesis of ricinoleic acid by developing castor beans. Plant Physiol. 67, 672–676.

Morris, L.J., 1967. The mechanism of ricinoleic acid biosynthesis in *Ricinus communis* seeds. Biochem. Biophys. Res. Commun. 29, 311–315.

Mosiewicki, M.A., Aranguren, M.I., 2013. A short review of novel biocomposites based on plant oil precursors. Eur. Polym. J. 49, 1243–1256.

Mottola, A.C., Hendrickson, A.P., O'Connell, D.E., Palter, R., Kohler, G.O., 1968. Pilot plant deactivation of castor meal allergen. Lime process. J. Agric. Food Chem. 16, 725–729.

Mutlu, H., Meier, M.A.R., 2010. Castor oil as a renewable resource for the chemical industry. Eur. J. Lipid Sci. Technol. 112, 10–30.

Naughton, F.C., 1974. Production, chemistry, and commercial applications of various chemicals from castor oil. J. Am. Oil Chem. Soc. 51, 65–71.

Naughton, F.C., Vignolo, R.L., 1992. The Chemistry of castor Oil and Its Derivatives and Their Applications. Int. Castor Oil Assoc., Westfield, NJ, 391 references, pp. 162.

Nezihe, A., Elif, D., Ozlem, Y., Tuncer, E.A., 2011. Microwave heating application to produce dhydrated castor oil. Ind. Eng. Chem. Res. 50, 398–403.

Odunfa, S.A., 1985. Microbiological and toxicological aspects of fermentation of castor oil seeds for *ogiri* production. J. Food Sci. 50, 1758–1759.

Ogunniyi, D.S., 2006. Castor oil: a vital industrial raw material. Bioresour. Technol. 97, 1086–1091.

Oil World, 2015. www.oilworld.de/statistics. (6/22/2015).

Ordman, D., 1955. An outbreak of bronchial asthma in South Africa, affecting more than 200 persons, caused by castor bean dust from an oil-processing factory. Int. Arch. Allergy Appl. Immunol. 7, 10–24.

Parks, L.R., 1931. The chemistry of turkey red dyeing. J. Phys. Chem. 35, 488–510.

Pauly, D., Worbs, S., Kirchner, S., Shatohotina, O., Dorner, M.B., Dorner, B.G., 2012. Real-time cytoxicity assay for rapid and sensitive detection of ricin from complex matrices. PLos One 7 (4), e35360. http://dx.doi.org/10.1371/journal.pone.0035360.

Pinkerton, S.D., Rolfe, R., Auld, D.L., Ghetie, V., Lauterbach, B.F., 1999. Selection of castor for divergent concentrations of ricin and *Ricinus communis* agglutinin. Crop Sci. 39, 353–357.

Pita, R., 2009. Toxin weapons: from World War I to jihadi terrorism. Toxin Rev. 28, 219–237.

Plastemart, 2011. New Grades of Polyamide Based on Naturally Occurring Feedstocks. http://www.plastemart.com/technical-market-reports-plastics-industry.asp. June 14, 2011 (July 29, 2015).

Priest, G.W., von Mikusch, J.D., 1940. Composition and analysis of dehydrated castor oil. Ind. Eng. Chem. 32, 1314–1319.

Quinchia, L.A., Delgado, M.A., Reddyhoff, T., Gallegos, C., Spikes, H.A., 2014. Tribological studies of potential oil-based lubricants containing environmentally friendly viscosity modifiers. Tribol. Int. 69, 110–117.

Ramanjaneyulu, A.V., Reddy, A.V., Madhavi, A., 2013. The impact of sowing date and irrigation regime on castor (*Ricinus communis* L.) seed yield, oil quality characteristics and fatty acid composition during post-rainy season in South India. Ind. Crops Prod. 44, 25–31.

Rana, M., Dhamija, H., Prashar, B., Sharma, S., 2012. *Ricinus communis* L. – a review. Int. J. PharmTech Res. 4, 1706–1711.

Rasooly, R., He, X., 2012. Sensitive bioassay for detection of biologically active ricin in food. J. Food Prot. 75, 951–954.

Reddy, K.R., Matcha, S.K., 2010. Quantifying nitrogen effects on castor bean (*Ricinus communis* L.) development, growth, and photosynthesis. Ind. Crops Prod. 31, 185–191.

Richards, D.E., Taylor, R.D., Murphy, D.J., 1993. Localization and possible substrate requirement of the oleate 12-hydroxylase of developing *Ricinus communis* seeds. Plant Physiol. Biochem. 31, 89–94.

Ritthausen, H., 1881. Krystallinische eiweisskorper aus verschiedenen oelsamen. J. Prakt. Chem. 23, 481–486.

Roberts, L.M., Lord, J.M., 1981. Protein biosynthetic capacity in the endosperm tissue of ripening castor bean seeds. Planta 152, 420–427.

Roberts, L.M., Lamb, F.I., Pappin, D.J., Lord, J.M., 1985. The primary sequence of *Ricinus communis* agglutinin. Comparison with ricin. J. Biol. Chem. 260, 15682–15686.

Roberts, L.M., Tregear, J.W., Lord, J.M., 1992. Molecular cloning of ricin. Targeted Diagn. Ther. 7, 81–97.

Rodrigues, C.R.F., Silva, E.N., Moura, R.d.M., dos Anjos, D.C., Hernandez, F.F.F., Viegas, R.A., 2014. Physiological adjustment to salt stress in *R. communis* seedlings is associated with a probable mechanism of osmotic adjustment and a reduction in water lost by transpiration. Ind. Crops Prod. 54, 233–239.

Scarpa, A., Guerci, A., 1982. Various uses of the castor oil plant (*Ricinus communis* L.) a review. J. Ethnopharmacol. 5, 117–137.

Schantz, B., Rohm, H., 2005. Influence of lecithin-PGPR blends on the rheological properties of chocolate. Lebensm. Wiss. Technol. 38, 41–45.

Scholnick, F., Saggese, E.J., Wrigley, A.N., Ault, W.C., Monroe, H.A., Zubillaga, M., 1968. Urethane foams from animal fats IV. Rigid foams from epoxidized glycerides. J. Am. Oil Chem. Soc. 45, 76–77.

Scholz, V., da Silva, J.N., 2008. Prospects and risks of the use of castor oil as a fuel. Biomass Bioenergy 32, 95–100.

Severino, L.S., Auld, D.L., et al., 2012. A review on the challenges for increased production of castor. Agron. J. 104, 853–880.

Severino, L.S., Auld, D.L., 2013. Seed yield and yield components of castor influenced by irrigation. Ind. Crops Prod. 49, 52–60.

Sharief, F.S., Li, S.S., 1982. Amino acid sequence of the small and large subunits of seed storage proteins from *Ricinus communis*. J. Biol. Chem. 257, 14753–14759.

Silva Jr., J.G., Machado, O.L., Izumi, C., Padovan, J.C., Chait, B.T., Mirza, U.A., Greene, L.J., 1996. Amino acid sequence of a new 2S albumin from *Ricinus communis* which is part of a 29-kDa precursor protein. Biochem. Biophys. 336, 10–18.

Silver, H.B., Stanley, I.R., 1974. The effect of the thickener on the efficiency of load-carrying additives in greases. Tribol. Int. 7, 113–118.

Singh, A.K., 2011. Castor oil-based lubricant reduces smoke emission in two-stroke engines. Ind. Crops Prod. 33, 287–295.

Song, J.-W., Lee, J.-H., Bornscheuer, U.T., Park, J.-B., 2014. Microbial synthesis of medium-chain α,ω-dicarboxylic acids and ω-amincarboxylic acids from renewable long-chain fatty acids. Adv. Synth. Catal. 356, 1782–1788.

Soratto, R.P., Souza-Schlick, G.D., Fernandes, A.M., Zanotto, M.D., Crusciol, C.A.C., 2012. Narrow row spacing and high plant population to short height castor genotypes in two cropping seasons. Ind. Crops Prod. 35, 244–249.

Sperling, L.H., 1994. Interpenetrating polymer networks: an overview. In: Klempner, D., Sperling, L.H., Utracki, L.A. (Eds.), Interpenetrating Polymer Networks, Advances in Chemistry. American Chemical Society, Washington, DC, pp. 3–38.

Sphyris, N., Lord, J.M., Wales, R., Roberts, L.M., 1995. Mutational analysis of the Ricinus lectin B-chains. Galactose-binding ability of the 2 gamma subdomain of *Ricinus communis* agglutinin B-chain. J. Biol. Chem. 270, 20292–20297.

Spies, J.R., Coulson, C.J., 1943. Isolation and properties of an active protein-polysaccharide fraction, CB-1A, from castor beans. J. Am. Chem. Soc. 65, 1720–1725.

Spini, E., 2013. Polyamide 6, 10: a polymer with a smaller environmental footprint for high-tech applications. Proc. Eurotech 2013, 95–100.

Stewart, C.E., 2006. Possible biotoxins. In: Weapons of Mass Casualties and Terrorism Response Handbook. Chapter 13. Jones and Bartlett, Sudbury, MA, USA, pp. 169–171.

Stewart, J.J., Gaginella, T.S., Bass, P., 1975. Actions of ricinoleic acid and structurally related fatty acids on the gastrointestinal tract: I. Effects on smooth muscle contractility in vitro. J. Pharmacol. Exp. Ther. 195, 347–354.

Sturm, M.B., Schramm, V.L., 2009. Detecting ricin: sensitive luminescent assay for ricin A-chain ribosome depurination kinetics. Anal. Chem. 81, 2847–2853.

Sujatha, M., Reddy, T.P., Mahasi, M.J., 2008. Role of biotechnological interventions in the improvement of castor (*Ricinus communis* L.) and *Jatropha curcas* L. Biotechnol. Adv. 26, 424–435.

Sujatha, M., Sailaja, M., 2005. Stable genetic genetic transformation of castor (*Ricinus communis* L.) via *Agrobacterium tumefaciens*-mediated gene transfer using embryo axes from mature seeds. Plant Cell Rep. 23, 803–810.

Sun, Y., Niu, G., et al., 2013. Seedling emergence, growth, and leaf mineral nutrition of *Ricinus communis* L. cultivars irrigated with saline solution. Ind. Crops Prod. 49, 75–80.

Tan, M., Yan, M., Wang, L., Yan, X., Fu, C., Wang, L., 2013. Replication of pistillate plants of *Ricinus communis* L. and investigation of the sex stability and genetic variation of the somaclones. Ind. Crops Prod. 50, 50–57.

Terrill, R.L., 1950. Dehydration of castor oil. J. Am. Oil Chem. Soc. 27, 477–481.

Thorpe, S.C., Kemeny, D.M., Panzani, R.C., McGurl, B., Lord, M., 1988. Allergy to castor bean. II. Identification of the major allergens in castor bean seeds. J. Allergy Clin. Immunol. 82, 67–72.

Tregear, J.W., Roberts, L.M., 1992. The lectin gene family of *Ricinus communis*: cloning of a functional ricin gene and three lectin pseudogenes. Plant Mol. Biol. 18, 515–525.

Tunaru, S., Althoff, T.F., Nusing, R.M., Diener, M., Offermanns, S., 2012. Castor oil induces laxation and uterus contractions via ricinoleic acid activating prostaglandin EP3 receptors. Proc. Natl. Acad. Sci. U.S.A. 109, 9179–9184.

Turner, C., Whitehand, L.C., Nguyen, T., McKeon, T.A., 2004. Optimization of a supercritical fluid extraction/reaction methodology for the analysis of castor oil using experimental design. J. Agric. Food Chem. 52, 26–32.

Umland, A., Sudhalter, A., 2008. In: Umland, Sudhalter (Eds.), Picabia in Dada in the Collection of the Museum of Modern Art. The Museum of Modern Art, New York, NY, pp. 241–242.

Valero, M.F., Pulido, J.E., Ramirez, A., Cheng, Z., 2009. Simultaneous interpenetrating networks of polyurethane from pentaerythritol-modified castor oil and polystyrene: structure-property relationships. J. Am. Oil Chem. Soc. 86, 383–392.

van de Loo, F.J., Broun, P., Turner, S., Somerville, C., 1995. An oleate-12-hydroxylase from *Ricinus communis* L. is a fatty acid desaturase homolog. Proc. Natl. Acad. Sci. U.S.A. 92, 6743–6747.

van Erp, H., Bates, P.D., Burgal, J., Shockey, J., Browse, J., 2011. Castor phospholipid:diacylglycerol acyltransferase facilitates efficient metabolism of hydroxy fatty acids in transgenic Arabidopsis. Plant Physiol. 155, 683–693.

van Erp, H., Shockey, J., Zhang, M., Adhikari, N.D., Browse, J., 2015. Reducing isozyme competition increases target fatty acid accumulation in seed triacylglycerols of transgenic Arabidopsis. Plant Physiol. 168, 36–46.

Vernon, A.A., Ross, H.K., 1936. Some characteristics of the residue from the cracking of castor oil. J. Am. Chem. Soc. 58, 2430–2433.

Villeneuve, P., et al., 2005. Production of conjugated linoleic acid isomers by dehydration and isomerization of castor bean oil. J. Am. Oil Chem. Soc. 82, 261–269.

Waller, G.R., Nakazawa, K., 1963. Nicotinic acid-ricinine relationship in sterile cultures of *Ricinus communis* L. Plant Phys. 38, 318–322.

Weiss, E.A., 2000. Castor. In: Oilseed Crops (World Agriculture Series), second ed. Blackwell Science Ltd, Oxford, GB, pp. 13–51.

White, O.E., 1918. Breeding new castor beans. J. Hered. 9, 195–200.

Wilson, R., van Schie, B.J., Howes, D., 1998. Overview of the preparation, use and biological studies on polyglycerol polyricinoleate (PGPR). Food Chem. Toxicol. 36, 711–718.

Worbs, S., Kohler, K., Pauly, D., Avondet, M., Schaer, M., Dorner, M.B., Dorner, B.G., 2011. *Ricinus communis* intoxications in human and veterinary medicine – a summary of real cases. Toxins 3, 1332–1372.

Wypych, G., 2012a. Introduction. In: Wypych, G. (Ed.), Handbook of Plasticizers. ChemTec Publishing, Toronto, pp. 1–6.

Wypych, G., 2012b. Plasticizers use and selection for specific polymers. In: Wypych, G. (Ed.), Handbook of Plasticizers. ChemTec Publishing, Toronto, pp. 273–379.

Yamada, M., Stumpf, P.K., 1964. Enzymic synthesis of ricinoleic acid by extracts of developing *Ricinus communis* L. seeds. Biochem. Biophys. Res. Comm. 14, 165–171.

Yeganeh, H., Mehdizadeh, M.R., 2003. Synthesis and properties of isocyanate curable millable polyurethane elastomers based on castor oil as a renewable resource polyol. Eur. Polym. J. 40, 1233–1238.

Yoshitake, S., Funatsu, G., Funatsu, M., 1978. Isolation and sequences of peptic peptides and the complete sequence of Ile chain of ricin D. Agric. Biol. Chem. 42, 1267–1274.

Youle, R.J., Huang, A.H.C., 1978a. Albumin storage proteins in the protein bodies of castor bean. Plant Physiol. 61, 13–16.

Youle, R.J., Huang, A.H.C., 1978b. Evidence that castor bean allergens are storage proteins in the proteins bodies of castor bean. Plant Physiol. 61, 1040–1042.

Zhan, J., Zhou, P., 2003. A simplified method to evaluate the acute toxicity of ricin and ricinus agglutinin. Toxicology 186, 119–123.

Chapter 5

Brassica spp. Oils

Peter B.E. McVetty
Department of Plant Science, University of Manitoba, Winnipeg, MB, Canada

Elzbieta Mietkiewska, Tolibjon Omonov, Jonathan Curtis
Agricultural Lipid Biotechnology Program, Department of Agricultural, Food and Nutritional Sciences, University of Alberta, Edmonton, AB, Canada

David C. Taylor
National Research Council of Canada, Saskatoon, SK, Canada

Randall J. Weselake
Department of Agricultural, Food and Nutritional Sciences, University of Alberta, Edmonton, AB, Canada

INTRODUCTION

The *Brassica* genus of plants belongs to the Brassicaceae family (formerly Crucifer family). This family is characterized by plants which produce very-long-chain fatty acids (VLCFAs) in their seed oil, and secondary defense-related metabolites called glucosinolates in the whole plant (Booth and Gunstone, 2004). *Brassica* species can germinate and grow at low temperatures and are one of the few oilseeds adapted to cooler temperate agricultural zones and winter production (Booth and Gunstone, 2004). The emerging emphasis on renewable energy, chemical feed stocks, industrial oils and novel uses of vegetable oils, and the steadily growing bioeconomy will provide significant growth opportunities for industrial (nonedible) *Brassica* oils. Canola (including canola with a modified fatty acid profiles), high erucic acid ($22{:}1^{cis\Delta13}$) rapeseed (HEAR) and eventually super-high erucic acid rapeseed (SHEAR) cultivars will be developed and grown to meet emerging and expanding markets. Double-digit annual growth in demand for industrial *Brassica* oil is anticipated (Friedt and Snowdon, 2009).

This chapter begins by describing the characteristics of *Brassica* oilseed species (BOS) and some important industrial applications of the oils derived from them. Thereafter, the discussion moves to *Brassica napus* canola/HEAR cultivar development over the last several decades, outlining the contributions made by breeders to the steady incremental improvement in agronomic

Industrial Oil Crops. http://dx.doi.org/10.1016/B978-1-893997-98-1.00005-1

performance and seed quality in BOS for industrial oil applications. Currently in some parts of the world, *B. napus* seed serves as a valuable source of oil for the production of biodiesel, and this bioproduct is also discussed. Plant oils can also be chemically modified to produce useful bioproducts such as plastics, insulation and other building materials. Therefore, some attention is devoted to this burgeoning area of "green chemistry." Although breeding has been successfully used to produce *Brassica*-derived oils for industrial applications, genetic engineering of this genus to produce modified oils has been restricted to investigations using growth chambers, green houses or small field trials. The potential for increasing seed oil content and developing specialty oils in *B. napus* and other *Brassica* species, however, is vast. Prior to discussing strategies for modifying seed oil metabolism for different industrial applications, we provide a brief overview of the process of seed oil formation so as to provide the reader with adequate background information. Thereafter, we discuss the engineering of BOS to produce high oleic acid ($18{:}1^{cis\Delta9}$), high medium-chain fatty acids and SHEAR oils. The reader will come to appreciate that one of the major challenges in developing any specialty oil as industrial feedstock is to enrich that oil, as much as possible, in a particular fatty acid (Dyer et al., 2008). Uniformity is a key factor. Sometimes this happens naturally, but not necessarily in a plant with favorable agronomic characteristics. For example, castor (*Ricinus communis*) seed contains about 90% ricinoleic acid (12-OH $18{:}1^{cis\Delta9}$), which is a valuable feedstock for the production of useful polymers, but the seed also contains toxic proteins (Jaworski and Cahoon, 2003; Hartley and Lord, 2004). After discussing strategies for producing specialty oils for industrial applications, we move on to briefly discuss some recent biotechnological approaches which have been used to increase seed oil content in BOS. Indeed, increased seed oil content is important to both edible and industrial applications of *Brassica*-derived oils. Finally, we feature a discussion of *Brassica carinata* as an emerging industrial platform crop in the Brassicaceae, especially for the production of biojet fuel.

CHARACTERISTICS OF *BRASSICA* OILSEED SPECIES AND INDUSTRIAL USES OF THEIR OILS

Members of the Brassicaceae were among the first plants domesticated by man several thousand years ago. Rapeseed was cultivated in India 3000–4000 years ago, in China 2000–2500 years ago, in Japan, Greece and Italy 2000–2500 years ago, in Europe 800 years ago and only 60 years ago in North America (Li, 1980; Prakash, 1980; Gupta and Pratap, 2007; Shahidi, 1990). There are several oilseed species within the *Brassica* genus including *B. carinata* (Abyssinian "Ethiopian" mustard), *Brassica juncea* (Indian mustard), *B. napus* (oilseed rape or rapeseed), *Brassica nigra (black mustard)*, *Brassica rapa* (turnip rape or rapeseed) and *Brassica oleracea* (cabbage, kale and cauliflower grown as a vegetable, not as an oilseed). *Brassica carinata*, *B. juncea* and *B. napus* are amphidiploids combining chromosome sets of the diploid parents *B. nigra* (black mustard), *B. oleracea* (kale) and *B. rapa* (Nagaharu, 1935). *B. rapa* rapeseed is thought to have a primary center of origin in the Indian subcontinent with secondary centers of origin in Europe, the Mediterranean area and Asia (Hedge, 1976). *Boleracea oleracea* and *B. napus* rapeseed both originated in the Mediterranean area (Hedge, 1976). *Brassica nigra* and *B. juncea* both originated in the Middle East (Prakash and Hinata, 1980), while *B. carinata* originated in Northeast Africa (Downey and Röbbelen, 1989). *Brassica napus* rapeseed and *B. rapa* rapeseed have both spring and winter annual forms, while *B. juncea* and *B. carinata* are solely spring types. The winter forms of *B. napus* rapeseed and *B. rapa* rapeseed are more productive than the spring forms, but less winter hardy than winter cereals (Downey and Röbbelen, 1989). Rapeseed plant heights range from 0.5 to 2.5m. The flowers are yellow with four petals, typical of crucifer flowers; flowering is indeterminate, lasting 2–4 weeks, beginning at the lowest bud on the main raceme. *Brassica napus* is a self-compatible species displaying a high degree of self-pollination, while most *B. rapa* is self-incompatible (except for the Indian subspecies yellow sarson, which is self-compatible) (Downey and Röbbelen, 1989). Each fertilized flower produces 25 or more seeds per pod, and each plant produces 40 or more pods, resulting in multiplication rates of 1000-to-1 or more in the *Brassica* rapeseed species (Downey and Röbbellen, 1989). Spring habit *B. napus* rapeseed is grown in Canada and Northern China and matures in 90–140 days, while winter habit *B. napus* rapeseed is grown in most of Europe as well as Central and Southern China, and matures in 320–360 days. The seed of rapeseed contains over 40% oil, and the meal contains 36–44% protein. Rapeseed meal is used a fertilizer in Asia and as an animal feed in the rest of the world (Booth and Gunstone, 2004).

Brassica rapa rapeseed and *B. juncea* mustard are the common *Brassica* oilseeds in India, *B. napus* rapeseed and *B. juncea* mustard the common *Brassica* oilseeds in China, and *B. rapa* rapeseed and *B. napus* rapeseed the common *Brassica* oilseeds in Europe and North America (Gupta and Pratap, 2007; Shahidi, 1990).

Rapeseed and mustard oils were originally used for lighting in oil-burning lamps, in religious ceremonies, for medicinal purposes, cooking and frying foods, and for food seasoning (Booth and Gunstone, 2004). Rapeseed oils had limited industrial use until the invention of steam power in the early 1700s (ie, the Newcomen steam engine of 1712), when rapeseed oil was found to be an excellent lubricant for the steam and water-washed surfaces commonly found in steam engines (Gupta and Pratap, 2007).

Interest in the Brassicaceae as an oilseed for industrial oils' uses has largely been as a consequence of erucic acid content in the oil and the high protein meal left over after oil extraction (Röbbelen, 1991) Several *Brassica* oilseed species including *B. carinata*, *B. juncea*, *B. nigra*, *B. napus* (oilseed rape/rapeseed) and *B. rapa* naturally produce seed oil with medium-to-high erucic acid content and medium-to-high protein meal after oil extraction (Downey and Röbbelen, 1989). Erucic acid content in the seed oil of these species, as reported by Velasco et al. (1998), was: *B. carinata*: 29.6–51.0%; *B. juncea*: 15.5–52.3%; *B. napus*: 5.6–58.1%; and *B. rapa*: 6.5–61.5%. *B. nigra* is reported to have seed oil with an erucic acid content of 19.3–44.2% (Goering et al., 1965). Even though all of these oilseed species have some potential for development as sources of high erucic acid oil, only *B. napus* rapeseed has been successfully developed as an industrial oil species (Stefansson and Downey, 1995). More recently, low erucic acid rapeseed oil has been used for biodiesel production and dielectric fluids in transformers, while high oleic acid rapeseed oil is being developed for use in hydraulic equipment. SHEAR oil is being developed for select chemical feedstock applications.

The demand for rapeseed oil grew rapidly in the 20th century, and parallel with this, significant improvements in cultivar development, production and processing were achieved (Booth and Gunstone, 2004). Numerous types of rapeseed have been developed in the last 40 years including low erucic acid, medium erucic acid, high erucic acid and high oleic oil types. Quality improvement also focused on meal quality through the reduction of glucosinolates. Double-low rapeseed was renamed canola in Canada in the 1970s to distinguish it from rapeseed. High erucic acid, low glucosinolate rapeseed (HEAR) has also been developed and commercialized in the last 30 years.

Canola is low erucic acid (<2% erucic acid in the oil), low glucosinolate (<20 μmole glucosinolates per gram of meal at 8.5% H_2O) rapeseed (synonym double-low rapeseed) first developed in Canada by Dr. Stefansson at the University of Manitoba (Stefansson and Kondra, 1975). Both traits result in a high-quality edible oil and animal feed. Conversion of rapeseed to canola was rapid in Canada, the European Union and Australia. *Brassica napus* rapeseed was the first species to be converted to canola, followed closely by *B. rapa* and much later, *B. juncea* (Potts et al., 2003). Canola oil is primarily used for edible purposes; however, new nonedible uses for canola have arisen in recent years, the largest of which is as a feedstock for biodiesel (Friedt and Snowdon, 2009). Numerous other uses for canola oil include chemical feedstocks, lubricants, coolants, cosmetics and printing inks (Canola Council of Canada, 2010). In addition, very high oleic acid content canola oils are used as hydraulic fluids.

HEAR is high erucic acid (minimum 50% erucic acid in the oil), low glucosinolate (<20 μmole glucosinolates per gram of meal at 8.5% H_2O) rapeseed also first developed in Canada by Dr. Stefansson at the University of Manitoba (Scarth et al., 1992). HEAR oil is used directly as a green lubricant in agricultural and forestry applications. It is also converted to erucamide or substituted amides and used as a slip-promoting agent in plastic manufacture. Alternatively, erucic acid is converted to substituted amines and used as a surfactant. It can also be split at the double-bond position to produce brassylic acid (C_{13}) and pelargonic acid (C_9), which are then used as lubricants, or as ingredients in perfumes and coatings. Additionally, brassylic and pelargonic acids are feedstocks for nylon 13-13 and nylon 9-9, respectively (Van Dyne and Blasé, 1991).

PRODUCTION OF *BRASSICA* OILSEED SPECIES

Canola Production

Canola cultivars are grown in both Europe and Canada. Spring habit (*B. napus*) canola cultivars are grown in the Canadian Prairie provinces and Northern Europe, while winter habit (*B. napus*) canola cultivars are grown primarily in Europe (Downey and Röbbelen, 1989). Australia also grows spring canola cultivars over their fall and winter periods (Potter et al., 1995). Seed yields for the spring types are 1.0–1.5 t/ha, while seed yields for winter types are 2.5–3.0 t/ha (Röbbelen, 1991). Canola is the world's third-largest source of vegetable oil (12.1 t), behind only soybean (27.9 t) and palm (27.8 t) (Friedt and Snowdon, 2009).

High Erucic Acid Rapeseed Production

High erucic acid, low glucosinolate rapeseed (HEAR) cultivars are grown in both the European Union and Canada. Winter habit HEAR cultivars are primarily grown in Europe, and spring habit HEAR cultivars are grown on the Canadian prairies. Seed yields for winter HEAR cultivars are 2.5–3.0 t/ha, while seed yields for spring HEAR cultivars are 1.0–1.5 t/ha (Röbbelen, 1991). HEAR is a major global source of vegetable-based industrial oil along with flax (*Linum usitatissimum*) and castor (*R. communis*) oils. HEAR production is at least an order of magnitude smaller than canola production globally.

BREEDING OF *BRASSICA* OILSEED SPECIES

Canola Cultivar Development

Canola was first developed in Canada in the 1970s. Canola is the dominant *Brassica* oilseed in Australia, Europe and North America, and is currently increasing in importance in Asia and South America. Innovations in canola cultivar development have appeared in spring habit germplasm mainly because of the much shorter life cycle length compared with winter habit germplasm. Initial canola cultivars in both spring and winter habit forms were open-pollinated population (OPP) types. With the successful adoption of microspore-based doubled haploidy techniques (Keller and Armstrong, 1983; Palmer and Keller, 1997), a number of canola cultivars were based on doubled haploid (DH) lines. The discovery of significant heterosis for seed yield in *B. napus* (Brandle and McVetty, 1989, 1990; Grant and Beversdorf, 1985; Sernyk and Stefansson, 1982) led to the development of synthetics, varietal associations and hybrids to exploit this heterosis (Friedt and Snowdon, 2009). Biotechnology provided the tools needed to develop novel broad-spectrum herbicide-tolerant canola cultivars. Finally, naturally occurring variation, mutagen-induced variation and genetic modification in *B. napus* led to the development of low α-linolenic ($18{:}3^{cis\Delta9,12,15}$), high oleic, high palmitic (16:0) and high lauric acid (12:0) specialty oil canola profiles.

Canola Breeding Objectives

Globally, breeding objectives include: seed yield, seed quality, local adaptation, disease and pest resistance, herbicide tolerance and development of specialty types for specific end uses.

Breeding for improved seed yield is the primary breeding objective for canola cultivar development programs. For *B. napus* canola cultivars, improved seed yield has involved, successively, selection within OPP landraces, crosses of OPPs and pedigree selection of derived families, the development and selection of pure breeding DH lines, the development of synthetics, varietal associations and finally, hybrids (Friedt and Snowdon, 2009). The development of pollination control systems which can be used to produce commercial quantities of hybrid canola seed has been a major breeding objective for the last 30 years (Friedt and Snowdon, 2009). Several pollination control systems for hybrid cultivars have been developed and evaluated in *B. napus*, predominantly with the Male Sterile Lembke (MSL) genetic male sterility (GMS) system, the *ogu* INRA cytoplasmic male sterility (CMS) system (Friedt and Snowdon, 2009) and the nuclear male sterility (NMS) system (Williams et al., 1997). Breeding for local adaptation in spring habit canola has focused on early maturity, while in winter habit canola, the focus has been on winter hardiness (Friedt and Snowdon, 2009). Breeding for local resistance to diseases and pests in canola is also important (Friedt and Snowdon, 2009). Breeding for seed quality in canola is increasingly important now that the transition to low erucic acid, low glucosinolate types in *B. napus*, *B. rapa* and *B. juncea* is complete in much of the world (Friedt and Snowdon, 2009). Improved oil content and the development of yellow-seeded *B. napus* cultivars (which would increase both oil and protein content simultaneously because of decreased mass of the hull) are major global seed-quality improvement objectives (Friedt and Snowdon, 2009). The development of different fatty acid–enriched specialty oil canola cultivars with lower levels of α-linolenic acid (Scarth et al., 1992), and more recently with higher levels of oleic acid combined with lower levels of α-linolenic acid (Friedt and Snowdon, 2009), have become major breeding objectives in Canada and the European Union.

Breeding for broad-spectrum herbicide tolerance in *B. napus* canola cultivars, initially beginning in Canada but expanding globally, is an important breeding objective in this comparatively noncompetitive crop, at least in the seedling stage. Canola cultivars tolerant to triazine, glufosinate, glyphosate, imidazolinone or bromxoynil have been successively developed and commercialized in Canada in the last 26 years (McVetty and Zelmer, 2007). The conversion to herbicide-tolerant canola cultivars in Canada is complete and is progressing in most other canola-growing areas of the world (McVetty and Zelmer, 2007).

High Erucic Acid Rapeseed Cultivar Development

Several spring habit HEAR cultivars were developed in Denmark in the 1980s and 1990s by Danisco. The OPP HEAR cultivars Industry, Sheila and Shirely were grown primarily in Scandinavia and the United Kingdom. The breeding objectives used to develop these HEAR cultivars included increased seed yield, improved seed quality (higher oil and protein content and higher erucic acid concentration), adaptation to local growing conditions, and enhanced disease and pest resistance. These cultivars were replaced in the early 2000s by much-higher-yielding winter hybrid HEAR cultivars bred by NPZ Lembke. The breeding objectives used to develop the winter hybrid HEAR cultivars were similar. The exploitation of heterosis for seed yield was accomplished through the use of the MSL GMS pollination control system in which female lines (C-lines) and male restorer lines (R-lines) in different heterotic pools were developed and crossed. The spring and winter HEAR cultivars used in the European Union are non–genetically engineered (GE) types, in keeping with Europe's aversion to GE cultivars of any type.

Numerous spring habit HEAR cultivars for Canadian growing conditions were developed by the University of Manitoba in the 1990s and 2000s. Hero (Scarth et al., 1991), Mercury (Scarth et al., 1994), Neptune (McVetty et al., 1996a), Venus (McVetty et al., 1996b), Castor (McVetty et al., 1998), MillenniUM 01 (McVetty et al., 1999), MillenniUM 02 (McVetty et al., 2000a) and MillenniUM 03 (McVetty et al., 2000b) were all non-herbicide-tolerant (non-GMO) OPP HEAR cultivars. Red River 1826 (McVetty et al., 2006a), Red River 1852 (McVetty et al., 2006b) and Red River 1997 (McVetty et al., 2010) are all Roundup Ready™ (GMO) OPP HEAR cultivars. The Red River series of HEAR cultivars is the world's first series of commercially successful glyphosate-tolerant (Roundup Ready™) HEAR cultivars. The University of Manitoba is also developing glufosinate ammonium–tolerant HEAR cultivars; the world's first two glufosinate ammonium–tolerant HEAR lines were supported for registration in Canada in March 2008. These are the first Liberty Link™–tolerant HEAR lines in the world to reach this development stage.

The development of hybrid HEAR cultivars is also underway at the University of Manitoba. Cuthbert et al. (2009) reported high parent heterosis for seed yield in HEAR hybrids exceeding 100%. The University of Manitoba is using the *ogu* INRA CMS system for development of hybrid HEAR cultivars. In addition, DL Seeds and the University of Manitoba are collaboratively developing hybrid HEAR cultivars using Lembke's proprietary MSL genetic pollination control system (Frauen and Paulmann, 1999).

High Erucic Acid Rapeseed Breeding Objectives

The primary breeding objectives for winter HEAR cultivars are to: increase seed yield, increase oil production, improve meal quality, increase erucic acid concentration, improve disease resistance and, for the winter habit types, improve winter hardiness and frost tolerance (Downey and Röbbelen, 1989). The objectives for breeding spring HEAR cultivars are similar. Improved weed control via development of herbicide-tolerant HEAR cultivars is also a major breeding objective. Development of MSL hybrid Roundup Ready™ HEAR cultivars and *ogu* INRA CMS hybrid Roundup Ready™ HEAR cultivars are HEAR objectives at the University of Manitoba.

Breeding for improved seed yield is the main breeding objective for HEAR cultivar development programs. For *B. napus* HEAR cultivars, this has involved, successively, selection within OPP landraces, crosses of OPPs and pedigree selection of derived families and hybrids (McVetty et al., 2009).

MSL GMS and *ogu* INRA CMS pollination control systems which can be used to produce commercial quantities of hybrid HEAR seed are being used to produce hybrid HEAR cultivars (Friedt and Snowdon, 2009; McVetty et al., 2009). The winter HEAR cultivars currently available are MSL hybrid HEAR types, which exploit the natural heterosis known to occur in *B. napus* (Brandle and McVetty, 1990). Spring hybrid HEAR cultivars will also exploit the heterosis for seed yield available in HEAR cultivars/lines (Cuthbert et al., 2009). Seed yield increases have been accompanied by simultaneous improvements in adaptation, disease resistance and seed quality for both spring and winter habit HEAR cultivars, and herbicide tolerance for spring habit HEAR cultivars (McVetty et al., 2009). Breeding for local adaptation in spring habit HEAR has focused on early maturity in Canada, while that in winter habit HEAR has focused on winter hardiness (McVetty et al., 2009; Friedt and Snowdon, 2009).

Breeding for resistance to local diseases and pests is equally as important in HEAR as it is with canola cultivars (McVetty et al., 2009; Friedt and Snowdon, 2009).

Breeding for seed quality in HEAR focuses on increases in erucic acid concentration and enhanced oil and protein content (McVetty et al., 2009; Friedt and Snowdon, 2009). The development of SHEAR with over 66% erucic acid is an additional breeding objective (Scarth et al., 1992).

The development of broad-spectrum herbicide tolerance in spring habit *B. napus* HEAR cultivars has yielded glyphosate-tolerant HEAR and glufosinate-tolerant HEAR cultivars in Canada, with Roundup Ready™ HEAR cultivars grown commercially since 2006 (McVetty et al., 2009).

Canola/High Erucic Acid Rapeseed Breeding Methods

The canola/HEAR cultivar development programs reviewed in this chapter are all based on *B. napus* L. The breeding methods used to develop a wide range of fatty acid profile types are species-specific. *Brassica napus* is a self-compatible, predominantly self-pollinated species with out-crossing rates of 2–30% (Downey and Röbbelen, 1989; Cuthbert and McVetty, 2001; Rakow and Woods, 1987). Pedigree breeding is the most common breeding method, sometimes combined with backcrossing (Downey and Röbbelen, 1989). The development of a microspore-based DH line protocol (Keller and Armstrong, 1983; Palmer and Keller, 1997) led to DH line development for new canola/HEAR cultivars. Reports of heterosis for seed yield in *B. napus* canola crosses

(Sernyk and Stefansson, 1982; Grant and Beversdorf, 1985; Brandle and McVetty, 1990) created significant interest in the development of hybrid canola cultivars. The report of significant heterosis for seed yield in HEAR hybrids (Cuthbert et al., 2009) created similar interest.

Initial hybrid canola/HEAR research focused on development of a pollination control system because *B. napus* is a perfect flowered species. A number of pollination control systems including CMS systems (McVetty, 1997), GMS systems (Sawhney, 1997) and self-incompatibility (SI) systems (McCubbin and Dickinson, 1997) have been studied for several decades. The *nap* CMS (Thompson, 1972), *pol* CMS (Fu, 1981), *mur* CMS (Hinata and Konno, 1979), *ogu* CMS (Ogura, 1968) and *ogu* INRA CMS (Pelletier et al., 1983) systems have been developed for use in hybrid canola/HEAR seed production. However, the successful development of hybrid canola/HEAR cultivars has been impeded by deficiencies in available CMS pollen-control systems (McVetty, 1997). Relative to the naturally occurring *nap* cytoplasm in *B. napus*, the *mur*, *ogu* and *pol* cytoplasms have significant biological costs (McVetty et al., 1990; Riungu and McVetty, 2004). The *nap* and *pol* CMS systems also displayed male sterility–related temperature instability (Fan and Stefansson, 1986; Burns et al., 1991). The *ogu* INRA CMS system combines the mitochondria from *Raphanus sativus* and the chloroplasts from *B. napus* in a hybrid cytoplasm (Pelletier et al., 1987). The initial absence of a restorer gene for the *ogu* INRA CMS system resulted in the development of nonrestored hybrid canola cultivars (known as varietal associations) in Europe (Booth and Gunstone, 2004). Varietal associations had widely varying seed yields correlated to the extent of insect cross-pollination occurring in the field (Booth and Gunstone, 2004). The development of restorer lines for the *ogu* INRA CMS system produced a fully functional CMS system (Heyn, 1978; Pellan-Delourme et al., 1987). GMS systems have also been developed and evaluated for use in hybrid canola/HEAR cultivar development (Sawhney, 1997). The MSL GMS pollination control system is widely used to produce hybrid canola/HEAR cultivars globally. The development of a GMS based on recombinant DNA technology created a successful pollination control system in *B. napus* based on the *barstar* and *barnase* genes, both linked to broad-spectrum herbicide tolerance (Williams et al., 1997). This NMS pollination control system is being used to produce very successful hybrid canola cultivars worldwide.

Canadian canola/HEAR cultivar breeding programs have added the development of herbicide-tolerant canola cultivars to their breeding methods. Canadian canola breeders have developed non-GE triazine–tolerant and imidolizine-tolerant canola cultivars (McVetty and Zelmer, 2007). They have also developed GE bromoxynil-, glufosinate- and glyphosate-tolerant canola cultivars either within, or in collaboration with, corporate multinational chemical and life-science companies (McVetty and Zelmer, 2007). Similarly, GE bromoxynil- and glufosinate-tolerant HEAR lines and glyphosate-tolerant HEAR cultivars have been developed at the University of Manitoba (McVetty and Zelmer, 2007).

Marker-Assisted Selection (MAS) in Canola/High Erucic Acid Rapeseed

Molecular markers are being developed and widely used in canola/HEAR breeding programs. A number of genetic maps have been developed in *B. napus*, starting with a restriction fragment length polymorphism (RFLP) map (Landry et al., 1991). Recently, an ultradense sequence–related amplified polymorphism (SRAP) (Li and Quiros, 2001) genetic recombination map for *B. napus* has been developed (Sun et al., 2007). This map has great potential for map-based cloning and within-gene molecular marker development. A review of *B. napus* mapping populations, marker systems, maps and traits studied has been published (Snowdon et al., 2006). The development of genome-specific erucic acid gene molecular markers for the two erucic acid–controlling genes in *B. napus* (Rahman et al., 2008) has permitted the selection of homozygous canola and high erucic acid genotypes in segregating generations and backcross generation progeny of canola × HEAR crosses, greatly improving breeding efficiency for the University of Manitoba HEAR breeding program.

Specialty Fatty Acid Profiles in Canola/High Erucic Acid Rapeseed

Several different fatty acid profiles have been developed in canola all of which could have both edible and industrial oil uses. These include low α-linolenic acid canola developed via mutagenesis, crossing, backcrossing and selection (Scarth et al., 1992), high oleic, low α-linolenic canola developed via mutagenesis, crossing of low α-linolenic lines, backcrossing and selection (Friedt and Snowdon, 2009) and high lauric acid canola developed using transformation technology (Voelker et al., 1996). Very high oleic acid canola (80–90% oleic acid) is required for industrial applications, because oleic acid is a chemical feedstock for further chemical synthesis where low levels of polyunsaturated fatty acids (PUFAs) are required (Lühs and Friedt, 1994).

In HEAR, the development of SHEAR has been in progress for a few decades. SHEAR will have erucic acid content of over 66% in the seed oil combined with low glucosinolates in the meal. Two approaches to develop SHEAR are being used. In the first approach, crosses of the *B. napus* parental species *B. oleracea* and *B. rapa* to produce resynthesized *B. napus* lines, followed by microspore mutagenesis to create DH lines with homozygous mutations for fatty acid profile variations, including possible SHEAR genotypes and selection are being done at the University of Manitoba (McVetty et al., 2009). In the second approach, transformation technology is being used to introduce transgenes which control erucic acid pool size and triacylglycerol (TAG) formation into HEAR and resynthesized HEAR lines (McVetty et al., 2009; Taylor et al., 1995). The role of genetic engineering in further increasing the erucic acid content of HEAR will be discussed in more depth later in this chapter. HEAR lines with 62–64% erucic acid content have been developed using these two approaches. Recently, development of the world's first transgenic SHEAR line with 72% erucic acid in the seed oil has been reported by Nath et al. (2009). This indicates that SHEAR development from HEAR is possible using genetic engineering.

BRASSICA SEED OIL AS A FEEDSTOCK FOR BIODIESEL PRODUCTION

Due to the increasing energy demands and pollution problems caused by the use of fossil fuels, it has become necessary to develop alternative fuels as renewable sources of energy. Biodiesel has been reported as a possible substitute for conventional petroleum-based diesel (Adamczak et al., 2009). According to the US Standard Specification for Biodiesel (ASTM 6751-02), biodiesel is defined as a fuel comprised of mono-alkyl esters of long-chain fatty acids derived from vegetable oils or animal fats. Currently, biodiesel is most commonly made by the alkali-catalyzed transesterification of plant oil or animal fat with methanol to produce fatty acid methyl esters (FAME) and glycerol (Lois, 2007; Piazza and Foglia, 2001). Most plant TAGs have much higher viscosity than conventional diesel, at a range of 17.3–32.9 mm^2/s compared with 1.9–4.1 mm^2/s (Knothe and Steidley, 2005). To overcome this problem, TAGs have to be converted to less viscous fatty acid esters by esterification with alcohol, most commonly methanol (Durret et al., 2008). In general, vegetable oil as a neat fuel or in blends with diesel fuel have the following advantages over diesel fuel: they produce less smoke and particulates, have higher octane values, produce lower carbon monoxide and hydrocarbon emissions, and are biodegradable and nontoxic (Durrett et al., 2008; Graboski and McCormick, 1998; Piazza and Foglia, 2001). Biodiesel is less volatile than petroleum diesel, which is reflected by its higher flash point, resulting in its safer handling and storage (Körbitz, 1995). The biggest advantages to using biodiesel today are that biodiesel is a sustainable source of liquid transportation fuels and is neutral with respect to the production of carbon dioxide.

In recent years, production of biodiesel has expanded in both the European Union and the United States. In 2006, European countries produced approximately 5.6 billion liters of biodiesel compared with 0.86 billion liters produced in the United States (Durret et al., 2008). The future development of biodiesel in Europe is mainly driven by the Renewable Energy Directive, which has a binding target of 10% biofuels in the transport sector until 2020. The main feedstock for biodiesel in Europe is still rapeseed oil, with a share of approximately 70%; 13% is covered by soybean oil, and almost 7% comes from waste frying oil and animal fat (Mittelbach, 2010). Soybean oil is the primary source of biodiesel production in the United States (Karmakar et al., 2010).

Despite an increase in overall production, biodiesel still represents a small percentage of total world diesel consumption. Limited adoption of biodiesel is attributed to several factors related to the fuel characteristics of biodiesel such as poor cold-temperature properties, higher rates of oxidation and increased emissions of nitrogen oxides compared with conventional diesel. Properties of biodiesel are dependent on the fatty acid composition of the plant oil used for its production (Ramos et al., 2009). Improving cold-temperature flow characteristics requires a fuel with low saturated fatty acid levels, whereas increasing oxidative stability and reducing nitrous oxide emissions require decreasing the amount of PUFAs (Durrett et al., 2008). Considering the fatty acid requirements, plant oils high in monounsaturated fatty acids (MUFAs) such as oleic acid and palmitoleic acid ($16{:}1^{cis\Delta9}$), and low in both saturated and PUFAs, is a good source for the production of high-quality biodiesel. As a result of experimental mutagenesis combined with conventional selection, new rapeseed varieties have been developed with oleic acid contents over 60%. For instance, oil of varieties such as "Splendor"or "Nexera" contain >75% oleic acid and <3% linoleic acid ($18{:}2^{cis\Delta9}$), and therefore the oil from these varieties is well suited for biodiesel (Friedt and Snowdon, 2009). Recently, a new mutant allele of *FAD3* was generated in *B. oleracea* via chemical mutagenesis which may prove to be useful in further lowering α-linolenic acid content in BOS carrying the C genome (Rahman et al., 2013).

The biggest factor affecting the cost of biodiesel production is the price of the input oil. The recent increase in the proportion of plant oils used for production of biodiesel has already contributed to higher vegetable oil prices (Durrett et al., 2008), not only making biodiesel production more expensive, but also affecting food prices. The conversion of plant oils to biodiesel competes with their use in foodstuffs, resulting in the often-cited "food versus fuel" debate. The current production of rapeseed oil and other vegetable oils is not sufficient to satisfy industrial needs for biodiesel. Therefore, other waste oils could be an alternative solution for biodiesel production after proper procedures for oil purification (Buczek and Czepirski, 2004; Murugesan et al., 2009). Since there is concern that a biodiesel feedstock may compete with the food supply, the production of biodiesel from different nonedible oils has been extensively investigated over the last few years. Some of these nonedible oilseed crops include jatropha (*Jatropha curcas*), karanja (*Pongamia pinnata*), tobacco (*Nicotiana tabacum* L.), rice (*Oryza sativa*), mahua (*Madhuca indica*), neem (*Azadirachta indica*), rubber tree (*Hevea brasiliensis*) and microalgae (Carlsson, 2009; Balat and Balat, 2010).

CHEMICAL MODIFICATION OF *BRASSICA* OILS FOR OTHER INDUSTRIAL APPLICATIONS

The double bonds of unsaturated fatty acids in the TAGs of vegetable oils can be modified in different ways to obtain reactive, technically useful derivatives of these oils. For example, the most widely reported and manufactured vegetable oil–based products are derivatives of epoxidized, hydroxylated or acrylated vegetable oils that find use in a wide range of industrial applications. Examples of these chemical reactions are shown in Fig. 5.1. Although the most notable

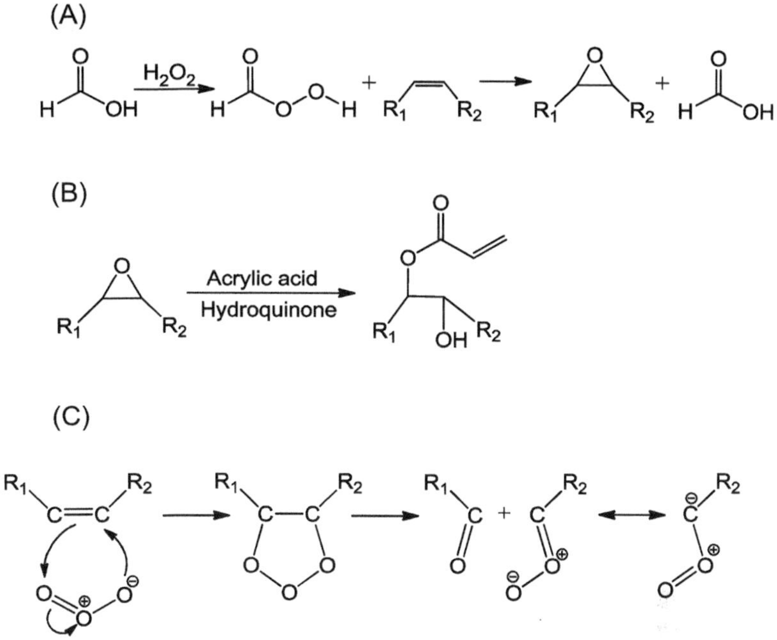

FIGURE 5.1 Chemical reactions in the modification of fatty acids. (A) in situ formation of performic acid and its reaction with double bond to facilitate an epoxide, (B) simplified scheme of acrylation of epoxy group (Wool and Sun, 2005), and (C) simplified mechanisms of ozonolysis of the double bonds.

examples of these are based on soybean (*Glycine max*) and linseed (*Linum usistassimum*) oils, these processes can easily be adapted to other oil sources including *Brassica* oils.

Epoxidation

Epoxidation of the double bonds of alkenes was firstly described by Prilezhaev (1909). This process became a commonly applied procedure for the epoxidation of unsaturated fatty acids and oils to create epoxidized vegetable oils. Generally, the epoxidation process is carried out using in situ percarboxylic acid generation in the reaction medium (Jones, 1999), formed by reacting hydrogen peroxide with formic or acetic acid. The in situ–generated performic acid in turn reacts with the double bonds of fatty acids to yield an epoxide group while regenerating formic acid, furthering the reaction (Fig. 5.1A). Consequently, a low amount of formic acid is required; the low acid content also reduces epoxide loss due to acid-catalyzed ring-opening reactions. Moreover, an essentially quantitative yield of epoxides can readily be attained. The synthesis and use of epoxidized vegetable oils in a range of applications have been reported by Earls et al. (2007), with the most widely available epoxides produced from soybean and flax oil.

Currently, epoxidized vegetable oils are produced on a commercial scale by several companies. Examples include the Vikoflex® line of products from Arkema (www.arkema-inc.com) which are produced from renewable sources including epoxidized linseed, soybean and tall oil fatty acid esters. Other examples include epoxidized linseed oil HiBond™ produced by Polar Industries, Inc. (www.polarindustries.net); epoxidized linseed oil Lankroflex L and epoxidized plasticizer/stabilizer from soybean oil Plastoflex, manufactured by Akcros Chemicals (www.akcros.com); epoxidized linseed oil Merginate ELO™ from HOBUM Oleochemicals Gmbh (www.hobum.de); epoxidized soybean oil Plasthall® ESO and Paraplex® and epoxidized linseed oil Plasthall®ELO produced by Hallstar (www.hallstar.com); plasticizer series made from epoxidized linseed oil Flexol™ LOE and from soybean oil Flexol™ EPO by Dow Chemical Company (www.dow.com); and the Epoxol® series from epoxidized soybean and linseed oils manufactured by American Chemical Service (www.acs-chem.com). These vegetable oil epoxides are used as plasticizers for PVC, additives for specialty coatings, adhesives in biocomposite production, additives in flexible polyurethane formulations, reactive diluents for UV and cationic cure applications and in intermediates in the production of surfactants, lubricants and fuel additives.

Vegetable oil epoxides can be cured in similar ways to petroleum-based epoxy derivatives using amines, isocyanates, mercaptans or anhydrides to manufacture the epoxy resins. Today, epoxy resins are a very important class of materials, widely used in many applications as insulators, as well as in the automobile, electrical and electronics industries due to their excellent chemical resistance, exceptional adhesion to a variety of materials, high tensile and compressive strength, and outstanding electrical insulation properties.

The use of a series of vegetable oils including canola oil as base fluids for industrial and automotive applications has been discussed in the literature (Erhan and Perez, 2002). These vegetable oil–based products were prepared by chemical modifications of the epoxidized vegetable oils to produce environmentally friendly hydraulic fluids, lubricants, metalworking fluids and diesel engine oils. Environmentally friendly, nontoxic and biodegradable motor oil based on canola (Johnson et al., 2002) has shown superior properties compared with synthetic and conventional motor oils.

Epoxidized oils and their methyl esters are potentially useful in making bio-based surfactants (Doll and Erhan, 2006, 2009) with excellent emulsifying/solubilizing properties for water–oil mixtures, similar to commercially available surfactants. Bio-based, biodegradable, nonionic surfactants from different vegetable oils are available from major manufacturers. For example, Clorox Company has successfully started a "green" production line of all-purpose, biodegradable, plant and mineral based cleaners and detergents under the Green Works® trademark (www.greenworkscleaners.com); Proctor & Gamble (www.pg.com) has recently introduced new natural cleaning product lines under the trademark Pure Essentials™, adding to its well-known brands; Dow

Chemical Company (www.dow.com) is producing its bio-based biodegradable, nonionic surfactants for household, industrial and institutional cleaners and textiles under the trademark Ecosurf™. There are many other examples.

Acrylation

Following epoxidation, vegetable oils can be further modified using acrylic acid to form acrylated/epoxidized oils (Fig. 5.1B). This process has been applied to different vegetable oils including canola oil (La Scala and Wool, 2005a; Wool and Sun, 2005). These oils contain incomplete acrylic double bonds which will readily undergo an addition reaction and are capable of free-radical polymerization with themselves or with a comonomer such as styrene. Acrylated/epoxidized vegetable oils are suitable for producing structural thermoset composites. It has been reported (La Scala and Wool, 2005b; Lu et al., 2005) that the acrylation of various natural oils can be carried out to different levels of acrylation, leading to the control of the curing process of thermoset polymers on their base. As examples, acrylic acid–modified epoxidized oil acrylates, acrylated polyester oligomers from Sartomer (www.sartomer.com) and acrylated epoxidized vegetable oils from Cytec Industries (www.cytec.com) are commercially available, and these are mainly produced from soybean oil. Acrylated epoxidized fatty acids from vegetable oils are used in ultraviolet light– or electron beam–curable inks to reduce the shrinkage and brittleness of the cured system. These products offer good pigment wetting, water balance and cure speeds.

Hydroxylation (Production of Polyols)

Polyols are mainly used in the production of polyurethanes. Vegetable oil–based polyols (hydroxylated oils) can be prepared by two main methods described in literature. The first is the ozonolysis of the vegetable oil to produce the aldehyde oil (Pryde et al., 1961; Fitchet et al., 2005), followed by the catalytic hydrogenation of this aldehyde oil to make a polyol. The second and more widely used method involves the epoxidation of oils (as described above) followed by a hydroxylation process. For instance, it was shown by Swern et al. (1945) that further action of the strongly acidic formic acid from the in situ epoxidation of unsaturated fatty acids results in the opening of the oxirane ring with addition of formic acid to yield hydroxy–formoxy derivatives. These can be converted to the corresponding dihydroxy derivatives via saponification reactions. Similar reactions have been described involving acid or Lewis acid catalyzed ring opening (Erhan, 2005; Soi et al., 2009), resulting in products that can contain either primary or secondary hydroxyl groups. Often, the TAG structure of the initial oil remains unaffected, and the polyol is primarily a monomer (Jones, 1999).

An alternative route is the strong acid-catalyzed reaction of a dialcohol with the epoxy ring by nucleophilic substitution (Curtis et al., 2012), leading to formation of monomers containing both primary and secondary hydroxyl groups, as well as to transesterification reactions involving loss of the glycerol moiety. Such processes for making polyols also result in other side reactions including oligomerization, so that the final polyols will have highly complex structures and a wide distribution of molecular weights.

Thus, polyols designed to have widely different structures and hydroxyl numbers can be prepared by varying the types of oils and modification routes. At present, the most common way that vegetable oil-based polyols are used is as a partial replacements for their petroleum-based counterparts in polyurethane formulations used in furniture manufacturing, bedding, automotive molded foams, carpets, coatings, sealants, adhesives, and elastomers. For example, natural oil polyol lines under trademark RENUVA™ (from soybean and high oleic canola) from Dow Chemicals (www.dow.com) are used to produce polyurethanes for adhesives, sealants, slabstock foam, molded foams, molded foam binders, coatings, flexible packaging, and footwear. The soybean oil–based polyol series BiOH® (www.bioh.com) from Cargill and Honey Bee™ polyol series produced by MCPU Polymer Engineering LLC (www.honeybee.cc) are used in a variety of polyurethane applications. A unique production line of Eco-pur Industries Inc. (www.ecopur.com.mx) is ready to offer ecological polyester biopolyols based on soybean oil and polyethylene terephthalate. Cognis (www.cognis.com) has introduced new bio-based, branched polyether/-ester polyol line under the tradename Sovermol® line. This polyol is best suited for use in coatings for high humidity, high temperature or damp conditions. Typical applications of Sovermol® include flooring, roofing, encapsulation, and composite coatings and adhesives.

Ozonolysis

Another route for modification of unsaturated TAGs involves the oxidative cleavage of double bonds using ozonolysis, a convenient and highly effective method owing to the complete reaction of ozone with the starting material (Pryde et al., 1960), to give an array of compounds including carboxylic acids, aldehydes, ketones and alcohols (Fitchet et al., 2005; McGuire et al., 2006; Omonov et al., 2011). The modification of vegetable oils, including *Brassica* oils, via oxidative cleavage of their double bonds by ozonolysis (Fig. 5.1C) has been described in the literature (Throckmorton and Pryde, 1972; Khan et al., 2004; Fitchet et al., 2005). The kinetics of ozonolysis and other important reactions occurring during the ozonolysis process have been summarized by Omonov et al. (2011). Several attempts have been made to create aldehydes as the primary products, starting from rapeseed (Fitchet et al., 2005) or canola oil (Petrovic et al., 2005) and using different reduction processes (Petrovic et al., 2005; Yue and Narine, 2007).

In addition, the direct synthesis of polyols through ozonolysis reactions involving vegetable oils has also been described in the literature (Tran et al.,

2005; Benecke et al., 2008). However, the main focus of the application of the ozonolysis process is in the manufacture of mono- and di-carboxylic acids from unsaturated fatty acids, including those derived from vegetable oils. The world's largest producer of azelaic and pelargonic acids, Emery Oleochemicals (www.emeryoleo.com), offers a diverse array of vegetable oil–derived products manufactured via the ozonolysis process.

GENETIC ENGINEERING OF *BRASSICA* OILSEED SPECIES TO PRODUCE INDUSTRIAL OILS

Overview of Seed Oil Formation

Before delving into genetic engineering strategies to produce *Brassica*-derived industrial oils, it is worthwhile to provide the reader with a brief overview of fatty acid synthesis and the assembly of fatty acyl chains into TAG, the main component of seed oil. Further details on these processes can be found at the AOCS Lipid Library (http://lipidlibrary.aocs.org/plantbio/fa_biosynth/index.htm, http://lipidlibrary.aocs.org/plantbio/tag_biosynth/index.htm) and in recent reviews (Snyder et al., 2009; Weselake et al., 2009; Taylor et al., 2011; Weselake, 2011; Snyder and Weselake, 2012; Singer et al., 2013). A schematic representation of fatty acid synthesis and TAG assembly is depicted in Fig. 5.2. Fatty acids are initially produced in the plastids of developing zygotic embryos of BOS while attached to the acyl carrier protein of the fatty acid synthase complex. In the plastid, chain elongation typically proceeds, two carbons at a time, up to a length of C18 (stearic acid; 18:0), with coproduction of some palmitoyl (16:0) chains. Malonyl-CoA, produced through the catalytic action of plastidial acetyl-CoA carboxylase, serves as a 2-carbon donor for fatty acyl chain elongation. A soluble acyl-ACP desaturase catalyzes the production of monounsaturated 18:1-ACP. This enzyme is responsible for the production of the 60% oleic acid content in regular canola oil. Acyl-ACP hydrolase (or thioesterase) action results in the liberation of fatty acid from the fatty acid synthase complex. Thereafter, free fatty acids move across the plastidial envelope and are re-esterified to coenzyme A (CoA) outside of the plastid. In HEAR, further elongation of fatty acyl chains, while esterified to CoA, can occur through four successive reactions of the enzymes in the "elongase" complex in the endoplasmic reticulum (ER). Both plastidial-derived and elongated fatty acids contribute to the cytoplasmic acyl-CoA pool, which can undergo further modification through acyl exchange with the acyl chains of phospholipid (discussed below).

TAG assembly begins with *sn*-glycerol-3-phosphate (G3P) which is produced through the catalytic action of glyceraldehyde-3-phosphate dehydrogenase acting upon glycolytically derived dihydroxyacetone phosphate. In the de novo G3P pathway (Kennedy pathway) leading to TAG, the glycerol backbone can undergo three acyl-CoA-dependent acylations catalyzed by membrane-bound acyltransferases. Prior to the final acylation, catalyzed by diacylglycerol

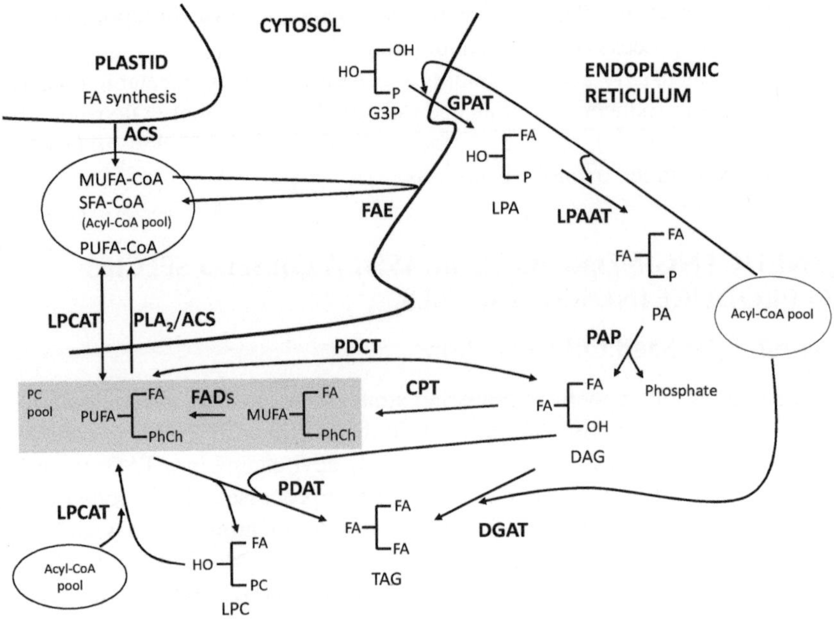

FIGURE 5.2 Simplified overview of fatty acid and triacylglycerol (TAG) biosynthesis in developing embryos of oleaginous oilseed crops. The formation of polyunsaturated fatty acids (PUFAs), catalyzed by fatty acid desaturases (FADs), occurs on phosphatidylcholine (PC) (shown in the gray box). In major oil crops, such as *Brassica napus*, linoleic acid ($18{:}2^{cis\Delta 9,12}$) and α-linolenic acid ($18{:}3^{cis\Delta 9,12,15}$) are formed from oleic acid ($18{:}1^{cis\Delta 9}$) via the sequential catalytic action of membrane-bound FAD2 and FAD3. Abbreviations for other enzymes: *ACS*, acyl-CoA synthetase; *CPT*, cholinephosphotransferase; *DGAT*, diacylglycerol acyltransferase; *FAE*, fatty acid elongase; *GPAT*, *sn*-glycerol-3-phosphate acyltransferase; *LPAAT*, lysophosphatidic acid acyltransferase; *LPCAT*, lyosphosphatidylcholine acyltransferase; *PAP*, phosphatidic acid phosphatase; *PDAT*, phospholipid:diacylglycerol acyltransferase; *PDCT*, phospholipid: diacylglycerol cholinephosphotransferase; PLA_2, phospholipase A_2. Other abbreviations: *CoA*, Coenzyme A; *DAG*, *sn*-1,2-diacylglycerol G3P, *sn*-glycerol-3-phosphate; *LPA*, lysophosphatidic acid; *LPC*, lysophosphatidylcholine; *MUFA*, monounsaturated fatty acid; *PhCh*, phosphocholine (polar headgroup of PC); *SFA*, saturated fatty acid. *Image adapted from:* http://lipidlibrary.aocs.org/plantbio/tag_biosynth/index.htm *with permission from the AOCS.*

acyltransferase (DGAT), the phosphate group is removed from the *sn*-3 position of phosphatidic acid (PA) by the catalytic action of phosphatidic acid phosphatase (PAP).

In seed oils containing PUFAs and/or unusual fatty acids (such as ricinoleic acid), however, a large proportion of the acyl chains used for TAG synthesis are actually derived from acyl chains which have been modified in the nitrogenous phospholipids (eg, phosphatidylcholine (PC)) in the ER. *sn*-1,2-Diacylglycerol (DAG) formed via PAP action can be converted to PC via the catalytic action of phospholipid:diacylglycerol cholinephosphotransferase (PDCT) and/or cholinephosphotransferase (CPT). Once the DAG skeleton is embedded in PC, it serves as a substrate for membrane-bound fatty acid desaturase (FAD) 2 and FAD3 to sequentially form linoleic ($18{:}2^{cis\Delta 9,12}$) and α-linolenic acid ($18{:}3^{cis\Delta 9,12,15}$). In some plant species, other modifications can occur on PC, leading to the formation of unusual fatty acids such as ricinoleic acid and conjugated fatty acids (eg, α-eleostearic acid). PDCT is a recently discovered enzyme which catalyzes the transfer of the phosphocholine headgroup between PUFA-enriched PC and DAG produced via de novo glycerolipid assembly (Lu et al., 2009). The products of this symmetrical reaction are DAG enriched in PUFA and PC enriched in 18:1 at the *sn*-2 position. Thus, PDCT provides a mechanism for channeling PUFAs from PC into TAG. A second possible mechanism for channeling PUFA into TAG is the reaction catalyzed by lysophosphatidylcholine acyltransferase (LPCAT). Under the right conditions, LPCAT can catalyze the exchange of fatty acyl groups at the *sn*-2 position of PC with the acyl-CoA pool, thereby enriching the acyl-CoA pool in PUFA-CoAs, which in turn would serve as substrate for the acyl-CoA-dependent reactions of the G3P pathway. In a third possible mechanism for PUFA enrichment of TAG, fatty acids can be removed from PUFA-enriched PC via the catalytic action of phospholipase A_2. The liberated PUFA could then be converted to PUFA-CoA via the catalytic action of acyl-CoA synthetase. In a fourth mechanism, phospholipid:diacylglycerol acyltransferase (PDAT) could catalyze the transfer of PUFA from PC to DAG to form TAG via an acyl-CoA-independent process. In the case of PLA_2 or PDAT action, the coproduct is lysophosphatidylcholine (LPC which), can be converted back to PC by the forward reaction catalyzed by LPCAT. A fifth means for enriching TAG in PUFAs lies in the selectivity properties of DGAT. Certain DGATs may exhibit increased preference for PUFA- or "unusual" fatty acid-enriched acyl-CoA substrates. Other possibilities for moving PUFA into TAG might include phospholipase C action to generate PUFA-enriched DAG or phospholipase D action to generate PUFA-enriched PA. Overall, the acyl-editing processes involving the interaction between the G3P pathway and membrane metabolism/turnover are complex, and there may be differences in the acyl editing processes used among different plant species.

Once TAG accumulates to a sufficient level in between the outer leaflets of the ER, oil bodies pinch off of the ER as 0.2–2 μM droplets of TAG which are surrounded by a half-unit membrane of phospholipid. In addition to a coating of phospholipid, the oil bodies are also coated with oleosins, which are proteins having a central hydrophobic region that allows the polypeptide to penetrate the TAG core of the oil body. The end regions of oleosin polypeptides, which lie along the hydrophilic head groups of the phospholipids, are believed to have a role in preventing oil bodies from coalescing. Having small oil droplets in the cytoplasm of germinating seeds of BOS is also believed to provide a greater surface area for TAG lipase action in order to facilitate effective storage lipid mobilization in support of the growth of the young seedling.

Engineering High Oleic Acid Content

As indicated earlier in this chapter, *Brassica* oil with high oleic acid content would serve as valuable feedstock for biodiesel production (Drexler et al., 2003). In addition to conventional breeding, genetic engineering is an effective approach for increasing oleic acid content in BOS. To increase oleic acid content, the gene encoding 18:1 Δ^{12} deasturase (*FAD2*) was targeted for silencing using the seed-specific cosuppression, antisense, or RNA interference (RNAi: intron-spliced hairpin) constructs (Kinney,1994; Töpfer et al., 1995; Mietkiewska et al., 2008). In *B. napus*, transgenic lines were developed with very high oleic acid up to 85% and low PUFA fraction (4–5%) using cosuppression and antisense technologies (Kinney, 1994; Stoutjesdijk et al., 2000). High oleic acid and low linoleic acid transgenic lines were also developed in *B. juncea* using antisense suppression of *FAD2* (Sivaraman et al., 2004). In the recent study described by Peng et al. (2010), an RNAi construct was developed to silence, simultaneously, the *FAD2* and *FAE1* genes in low erucic *B. napus.* This approach led to the development of transgenic lines with oleic acid content of 85%, but still below 90%. Another very successful approach to increase oleic acid content was designed for soybean. Transgenic soybean lines with oleic acid contents of >90% were developed by downregulation of two genes: *FAD2-1* encoding a Δ^{12} desaturase and *FatB* encoding palmitoyl-ACP thioesterase (Buhr et al., 2002; Damude and Kinney, 2008). Using an RNAi construct, a similar approach was applied in *B. napus* and *B. juncea* to further increase oleic acid content (Pandian et al., 2004).

Engineering Production of Medium-Chain Fatty Acids

Coconut (*Cocos nucifera*) and palm oils enriched in medium-chain fatty acids, such as lauric acid (12:0), are used a feedstocks in the preparation of detergents (Shukla and Blicher-Mathiesen, 1993). In an early genetic engineering success, *B. napus* was engineered to produce oil with about 50% lauric acid through transformation with a cDNA encoding a lauryl-ACP thioesterase from the California bay laurel (*Umbellularia californica*) (Voelker et al., 1996). Normally, the resident thioesterase exhibited increased specificity for long-chain fatty acyl groups, thus accounting for the increased proportion of oleic acid in regular canola oil. Positional analysis of fatty acyl groups in the TAGs of these transgenic lines, however, revealed very little accumulation of lauric acid at the *sn*-2 position, likely due to a limitation in the specificity of the indigenous LPAAT which catalyzes the acylation of the *sn*-2 position of lysophosphatidic acid (LPA) to generate PA in the G3P pathway leading to TAG (Snyder et al., 2009). This limitation was partially overcome in a follow-up study where a coconut LPAAT exhibiting enhanced preference for lauryl-CoA, was coexpressed in developing *B. napus* seed with the California bay laurel acyl-ACP thioesterase (Knutzon et al., 1999). In this study, *B. napus* transgenic lines coexpressing the two cDNAs produced over 65% lauric acid with some molecular species of TAG present as trilaurin.

Engineering Rapeseed Oil with Super-High Erucic Acid Content

Currently available HEAR varieties contain about 50% erucic acid in the seed oil (Nath et al., 2009). Erucic acid content in rapeseed is limited due to: (1) lack of esterification of erucic acid at the *sn*-2 position of the TAG backbone, (2) limited activity of the microsomal fatty acid elongase (FAE) complex, and (3) limited availability of the precursor, oleic acid, caused by competitive desaturation of oleic to linoleic and α-linolenic acids. A substantial increase in erucic acid content is desired to reduce processing costs for industry (Leonard, 1994; Mietkiewska et al., 2004; Scarth and Tang, 2006). Theoretically, the highest level of erucic acid that can be achieved through classical breeding is 66 mol% (Katavic et al., 2001; McVetty and Scarth, 2002). Biotechnology approaches taken to increase erucic acid content in rapeseed and other Brassicaceae are discussed below.

The substrate specificities of the acyltransferases in conjunction with the acyl-CoA pool sizes are believed to determine the fatty acid composition of seed oils and therefore must be taken into account when engineering plants for high erucic acid production (Brough et al., 1996). Similar to the enzymatic obstacle of incorporating medium-chain fatty acids at the *sn*-2 position of TAG, high erucic acid *B. napus* cannot incorporate erucic acid at the *sn*-2 position of TAG due to the inability of the resident LPAAT to acylate the *sn*-2 position of LPA with this MUFA. Exclusion of erucic acid from *sn*-2 position limits its content in rapeseed oil and prevents synthesis of trierucin (Lassner et al., 1995; Weier et al., 1997). In contrast to rapeseed, LPAATs from certain *Limnanthes* spp. were shown to incorporate erucoy1 groups into the *sn-2* position of the glycerol backbone (Brown et al., 1995). These seed-specific microsomal LPAATs appeared to be suited for increasing erucic acid content on *sn*-2 position of TAG in rapeseed oil (Frenzten, 1998). Various transgenic rapeseed lines expressing *Limnanthes* LPAATs from *Limnanthes douglasii* and *Limnanthes alba* altered the stereochemical composition of the transgenic oil; the content of erucic acid at the *sn*-2 position increased up to 41% (Frentzen, 1998). However, the alternation of the *sn*-2 fatty acid composition did not affect the total erucic acid content, likely due to redistribution of the available erucic acid previously destined for the *sn*-1 and *sn*-3 positions (Brough et al., 1996; Lassner et al., 1995), suggesting that in HEAR expressing *Limnanthes* LPAATs, the total erucic acid content was predominantly limited by the size of the erucoyl-CoA pool in the cytosol (Lühs et al., 1999; Scarth and Tang, 2006).

A more successful approach was described by Zou et al. (1997), with yeast mutant *SLC1-1* gene encoding enzyme capable of utilizing a broad range of substrates. In contrast to *Limnanthes* LPAATs, expression of yeast *SLC1-1* affected not only the stereochemical composition of seed oil, but also total erucic acid and oil content. The total content of erucic acid increased to as much as 56% compared with 45% in the control. Studies carried out with LPAAT genes showed that additional genes have to be introduced to optimize the levels of erucic acid to about the desired level of 80% for industrial applications. Erucic acid is a VLCFA synthesized by a microsomal FAE complex using oleoyl-CoA substrate from a cytoplasmic pool maintained by de novo lipid biosynthesis in plastids. Each cycle of FAE adds two carbon units to the acyl chain and involves four reactions: first, malonyl-CoA and long chain oleoyl-CoA are condensed by the action of a 3-ketoacyl-CoA synthase (KCS, often designated FAE); the resulting 3-ketoacyl-CoA is then reduced by the action of a 3-ketoacyl-CoA reductase, resulting in the synthesis of a 3-hydroxyacyl-CoA. This is dehydrated to 2-enoyl-CoA, which is then reduced by the action of a second reductase to form the elongated acyl-CoA (Barret et al., 1998; Blacklock and Jaworski, 2002; Puyaubert et al., 2005; Mietkiewska et al., 2007). It is critical to note that in the literature, the condensing enzyme and its encoding gene have been designated variably as either FAE or the more correct acronym KCS. Over the past decade, progress in understanding VLCFA biosynthesis has been achieved by cloning *KCS (FAE)* genes from different plants and performing functional expression studies. These studies provided evidence that the KCS catalyzing the condensation reaction plays a key role in determining the chain length of fatty acid products found in seed oils and the enzyme exerts strong pathway control for seed VLCFA production (Millar and Kunst, 1997; Katavic et al., 2001; Mietkiewska et al., 2004; Taylor et al., 2010a; Taylor, 2010). Millar and Kunst (1997) have shown that the introduction of *FAE1* alone in *Arabidopsis thaliana* or tobacco (*N. tabacum*) is sufficient for the production of VLCFAs, and that introduction of extra copies of *FAE1* resulted in higher levels of these fatty acids. Since that time, several *FAE* genes encoding enzymes with different acyl-CoA specificity were isolated and subsequently overexpressed in plants to increase erucic acid content in seed oil with varying degrees of success.

In the work carried out by Katavic et al. (2000, 2001), overexpression of *Arabidopsis FAE1* in HEAR *B. napus* cv. Hero resulted in an increased erucic acid proportion of 48–53% in T_3 seed in comparison with 43% in the wild-type control. This moderate increase was explained by properties of the *Arabidopsis* FAE enzyme, which prefers 18:1- over 20:1-CoA as a substrate for elongation. A more successful approach was described by Taylor et al. (2010a), wherein the investigators utilized two plant FAEs with different substrate preferences: an *Arabidopsis* FAE that prefers to elongate 18:1 to 20:1, and a *Tropaeolum majus* FAE with a preference for elongating 20:1 to 22:1 (Mietkiewska et al., 2004). Coexpression of the *Arabidopsis FAE1*+ *T. majus FAE* in *B. carinata* increased erucic acid content from 36% in the wild-type control to as high at 47% in the best transgenic lines carrying both *FAE*s. A further increase in the content of erucic acid in transgenic *B. carinata* seed was obtained by overexpression of *Crambe abyssinica FAE* (Mietkiewska et al., 2007). A higher increase in the content of erucic acid in plants carrying *C. abyssinica FAE* was attributed to the fact that it encodes a condensing enzyme more efficient in catalyzing the two-step elongation from 18:1 to 22:1 compared with other described FAEs with preference for either 18:1 or 20:1 moieties. In view of these results, overexpression of genes encoding FAEs with specificity for 18:1 and 20:1 acyl groups appears to be a promising strategy to stimulate 22:1-CoA synthesis in transgenic rapeseed lines. Functional expression studies provided strong evidence that the amounts of VLCFAs produced by the different elongase systems are primarily controlled by the expression levels of the genes encoding the KCSs. In addition, the composition of the fatty acyl-CoA pool available for elongation, and the presence and size of the neutral lipid sink, are additional important factors influencing the types and levels of VLCFAs made in particular cells (Zou et al., 1997; Lardizabal et al., 2000; Han et al., 2001).

There is some evidence that the cytosolic pool of oleoyl-CoA may limit FAE and synthesis of erucic acid (Bao et al., 1998; Domergue et al., 1999). The microsomal *FAD2* gene encoding the FAD2 Δ^{12} desaturase enzyme is responsible for more than 90% of the PUFAs in plants (Miquel and Browse, 1992). By altering the level of the expression of the *FAD2* gene, it is possible to increase the pool of 18:1 available for elongation by the miscrosomal FAE complex (Jadhav et al., 2005). There are several reports on downregulation of *FAD2* genes that has led to the reduction of PUFA levels in oilseed crops. In the study described by Kinney (1994), a cosuppression and antisense approach resulted in a PUFA reduction to 4–5% and a concomitant increase in oleic acid up to 85% in *B. napus* transgenic seed. An antisense approach taken to silence the *FAD2* gene in HEAR *B. carinata* was quite efficient in increasing the levels of 18:1 (54–130%), and as a result the level of erucic acid was increased by 5–19% compared with the wild-type control (Jadhav et al., 2005). In the following years, more efficient methods of posttranscriptional gene silencing such as ihpRNA constructs were developed and subsequently used to alter expression of *FAD2* genes, at first in model plant *Arabidopsis*, and subsequently in crop trials (Stoutjesdijk et al., 2002; Liu et al., 2002). Silencing of a *FAD2* gene with an ihpRNA construct led to further increases in the content of erucic acid in *B. carinata* to as high as 50.6%, a 26.5% proportional increase over wild-type levels of 40% (Mietkiewska et al., 2008; Fobert et al., 2008). In the same study, it was shown that a further increase in the erucic acid content to a level of up to 58% in *B. carinata* seed oil could be achieved by ihpRNA-*FAD2* silencing combined with overexpression of the *C. abyssinica FAE* (*KCS*).

In view of these results, it is clear that by altering the expression of one gene, it is impossible to increase the erucic acid level to over 66% in *Brassica* seed oil. All the bottlenecks described above should be addressed to obtain the erucic acid proportions desired by the industry. As has been shown recently, by combining alleles of *B. napus* related to low polyunsaturated oils with the transgenic coexpression of the *L. douglasii LPAT2* and the *BnFAE1* (encoding *BnKCS*), a *B. napus* line with oil containing 72% erucic acid was obtained (Nath et al., 2009). The 72% erucic acid content of *B. napus* oil achieved in this approach represents a major breakthrough.

Development of Molecular Strategies to Increase Seed Oil Content

Increasing seed oil content in BOS is obviously critical for increasing the overall supply of oil in the future for both edible and industrial applications. In Canada alone, a one percent increase in the seed oil content of canola would result in over $70M per year for the oilseed extraction and processing industry (Canola Council of Canada). Metabolic engineering strategies for increasing the seed oil content in developing seeds of oleaginous plants have recently been reviewed elsewhere (Weselake et al., 2009; Weselake, 2011; Singer et al., 2013). Although we know through plant breeding that seed oil content is a quantitative trait in *B. napus* governed by several factors (Delourme et al., 2006; Zou et al., 2010), increases in seed oil content in BOS have been achieved through the modification of specific steps in fatty acid and TAG biosynthesis, and through the intervention of other steps in carbon metabolism. In addition, certain transcription factors can potentially affect the expression of a number of genes encoding enzymes involved in lipid synthesis and other aspects of carbon metabolism (Weselake et al., 2009). Thus, altering the expression of genes encoding transcription factors represents another useful strategy for increasing seed oil content. Oleaginous crops which have been genetically engineered in this way are only at the experimental stage, and such crops have not yet been commercialized.

Studies on developing seeds from transgenic plants with specific modifications in lipid synthesis and carbon metabolism have provided valuable insight into the process of plant lipid biosynthesis. DGAT is one target in TAG assembly that has received considerable attention (Liu et al., 2012a). Overexpression of either *Arabidopsis* or *B. napus DGAT1* in *B. napus* has been shown to lead to increased seed oil content (Weselake et al., 2008; Taylor et al., 2009a; Singer et al., 2013). Top-down control analysis of fatty acid synthesis and glycerolipid assembly in developing zygotic embryos of *B. napus* indicated that 70% of the control of TAG production was associated with glycerolipid assembly (Weselake et al., 2008). In contrast, transgenic lines overexpressing *DGAT1* exhibited a decrease in control in glycerolipid assembly to 50% which correlated with the increase in the seed oil content in these lines. In further exploring the utility of DGAT in boosting seed oil content, directed evolution has been used to generate *B. napus* DGAT1 variants with enhanced activity (Siloto et al., 2009a,b). Xu et al. (2008) showed that site-directed mutagenesis of the serine^{197} moiety to an alanine in *T. majus* DGAT1 resulted in an enzyme with highly upregulated activity when expressed in yeast. Expression studies in *B. napus* utilizing the modified *DGAT1* gene are in progress. A recent study using microspore-derived cell suspension cultures of *B. napus*, which are known to accumulate TAG (Weselake and Taylor, 1999), has indicated that plastidial ACCase undergoes feedback regulation by oleoyl-ACP (Andre et al., 2012). In the light of these results and the apparent importance of DGAT in controlling storage lipid accumulation in *B. napus*, the investigators suggested that DGAT could possibly influence TAG accumulation by consuming acyl-CoA in the cytoplasm thereby stimulating export of fatty acids from the plastid and relieving inhibition of ACCase by oleoyl-ACP.

Recent in silico models of metabolism in developing zygotic embryos of *B. napus* have been useful in gaining insight into the regulation of storage reserve accumulation, carbon allocation and network plasticity (Pilalis et al., 2011; Schwender and Hay, 2012). Computational analysis has revealed that reactions involved in sugar catabolism have a greater role in influencing TAG accumulation than reactions driving TAG assembly (Schwender and Hay, 2012). Thus, the investigators suggested that it may be useful to manipulate sucrose catabolism as a possible means of increasing seed oil content. Borisjuk et al. (2013) have recently studied seed architecture in relation to metabolism in developing seeds of *B. napus* using noninvasive nuclear magnetic resonance-based imaging. Metabolic modeling predicted that the outer region of zygotic embryos was very efficient in using light energy for storage lipid biosynthesis due mainly to the contribution of the ribulose-1,5-bisphosphate carboxylase/oxygenase-bypass (Borisjuk et al., 2013; Schwender et al., 2004). Based on this metabolic insight, Borisjuk et al. (2013) suggested that manipulation of seed shape to increase surface area may represent a useful strategy to increased seed oil content in *B. napus*.

Seed oil content in *B. napus* has been observed to decline by about 10% in the later stages of seed maturation as desiccation sets in (Kelly et al., 2013). Recently, the gene family encoding SUGAR-DEPENDENT1 triacylglycerol lipase was suppressed, resulting in an increase in oil yield by 8% in *B. napus* on either a seed, plant or unit area basis (Kelly et al., 2013). Thus, this novel strategy for boosting seed oil content in *B. napus* was based on suppressing the initial step in the degradation of TAG.

In the last few years, considerable exciting research on factors influencing seed oil content in *B. napus* has been conducted by researchers in China. Recently, Hu et al. (2009, 2013) have reported that an increase in the oil body-to-seed ratio exhibits a significant highly positive correlation with seed oil content. Some of this insight came from the analysis of seed ultrastructure in *B. napus* line YN171, which exhibited a seed oil content of 64.8%. Based on structural analysis of various *B. napus* lines differing in seed oil content, Hu et al. (2012) have estimated that the maximum seed oil content of *B. napus* could reach 75%! Thus, engineering of the oil body-to-seed ratio represents another potential strategy to boost seed oil content in *B. napus*.

Liu et al. (2010) reported that two genes encoding the transcription factor WRINKLED1 are differentially expressed in *B. napus* lines varying in seed oil content. Constitutive expression of the cDNAs encoding these genes in *Arabidopsis* resulted in increases in seed oil content ranging from 10% to 20%, along with increased seed size and mass.

In another recent study, Sun et al. (2012) investigated QTL associated with high seed oil content in *B. napus* through analysis of DH lines generated from crossing a high oil content cultivar (zy036) with a low oil content cultivar (51,070). The investigation led to the identification of four candidate genes which were differentially expressed in the high versus low oil content cultivars. The investigators suggested that these candidate genes may be linked to the regulation of seed oil content and thus represent promising targets for boosting seed oil content in *B. napus*. Liu et al. (2012b) conducted differential gene expression analysis on the same two *B. napus* lines and identified *B. napus growth-regulating factor 2-like* gene (*BnaGRF2*) as a differentially expressed gene. Two forms of the gene (*BnaGRF2a* and *BnaGRF2b*) were identified in the A and C genomes, respectively. Functional studies with transgenic *Arabidopsis* constitutively overexpressing *BnaGRF2*a revealed increases in seed oil production >50% which appeared to be attributable to >20% enlargement of extended leaves and >40% improvement in photosynthetic capacity due to a boost in chlorophyll content. The same two *B. napus* cultivars differing greatly in seed oil content studied by Sun et al. (2012) were also used to demonstrate that the maternal genotype has a strong influence over seed oil content (Hua et al., 2012). Expression of the gene encoding WRINKLED1 was linked to photosynthetic activity in the silique wall. In addition, increased activity of ribulose-1,5-bisphosphate carboxylase/oxygenase in the silique was associated with increased seed oil content, suggesting that genetic manipulation of photosynthesis in the silique represents a potential biotechnological strategy for increasing seed oil content.

High temperature has long been known to result in decreased seed oil content in *B. napus* (Canvin, 1965). In order to probe the molecular basis of this effect, Zhu et al. (2012) have analyzed gene expression in two near-isogenic lines differing at a QTL region affecting seed oil content at high temperature during seed development in *B. napus*. Increased temperature resulted in the downregulation of many genes associated with red light response, photosynthesis, gibberellic acid stimulus response, translational elongations and some genes associated with lipid biosynthesis. In contrast, a higher growth temperature was associated with upregulation of *BnaABA3*, which is involved in abscisic acid (ABA) signal transduction. Given the importance of ABA in seed maturation and dormancy, *BnaABA3* upregulation may have contributed to an increased rate of seed maturation, thereby resulting in less efficient accumulation of storage reserves. These research findings could lead to the development of molecular strategies to maintain higher seed oil content in *B. napus* under conditions of elevated temperature.

Brassica carinata as a Platform Crop for Producing Industrial Oils

B. carinata (commonly called Ethiopian mustard), an amphidiploid (BBCC, $2n = 34$) formed through interspecific hybridization from diploid progenitors

B. oleracea L. (CC, 2*n* = 18) and *B. nigra* L. (BB, 2*n* = 16), is highly heat- and drought-tolerant, with good resistance to blackleg, a fungal disease, to aphids and flea beetles, and has a relatively large seed size. A genetically diverse species, it has considerable potential as an oilseed crop (Alemayehu and Becker, 2002; Ripley et al., 2006). With the current interest in biofuels and bioindustrial feedstocks, *B. carinata* is considered a suitable crop for the production of oils enriched in the specialty fatty acids required for these applications. The meal, after hydrolysis and proteolytic digestion, has potential as a source of bioactive (antioxidative, hypocholesterolemic, angiotensin metabolism-inhibiting) peptides (Pedroche et al., 2007). Because of its robust root system, *B. carinata* is a good candidate for phytoremediation in sites contaminated by heavy metals (Irtelli and Navari-Izzo, 2008; Irtelli et al., 2009).

In the area known as Palliser's Triangle in SW Saskatchewan and SE Alberta, the brown soil zone covers more than 6M acres. This is semiarid land unsuitable for canola and ideal for *B. carinata* growth (Warwick et al., 2006). It was suggested that Ethiopian mustard has good potential to become a new oilseed or protein crop for Western Canada if adapted, early maturing, high-yielding strains could be developed (Getinet et al., 1996). Subsequently, the Saskatoon Research Center of Agriculture & Agri-Food Canada (AAFC) initiated a breeding program in the mid-1990s to develop early maturing strains (Falk, 1999). While pedigree selection was the dominant breeding method used, a modification of this method best describes their approach: Open-pollinated plants within selected rows or small plots are typically selected and advanced; individuals from within selected progeny rows are threshed individually and advanced to the next generation following quality analysis. The use of mass selection for disease resistance, single seed descent and microspore culture have also been used routinely.

After breeding for earliness, several promising strains have been developed. Unselected Ethiopian mustard base populations mature 10–14 days later than canola, but on average, the new selected strains mature only 5–7 days later than most *B. napus*, a major step forward since it was achieved without sacrificing seed yield. In addition to early maturity, breeding lines have been developed with very high protein content (>35% on a whole seed basis), relatively large seed size (1000-seed weight > 3 g) and lower fiber content (compared with *B. napus* canola), and oils with both low and high erucic acid content have been attained. Low erucic acid strains (<3% erucic acid) developed by the program have C18 fatty acid profiles of 31% oleic:33% linoleic:27% linolenic acids. There is also ongoing research to reduce the level of seed glucosinolates (Márquez-Lema et al., 2008). Considerable variation has also been observed for plant phenotype, leaf form and color, pod architecture, petal color, and shattering resistance (Taylor et al., 2010a).

While immediately intended for growth in the semiarid regions of Alberta and Saskatchewan, test field trials of the most advanced *B. carinata* commercial variety have been highly successful in the Maritimes, Quebec, Ontario,

TABLE 5.1 Examples of Industrial Applications of High Erucic or High Nervonic Seed Oils

Biojet and biodiesel
Viscoelastic surfactants and high high molecular weight anionic surfactants for enhanced oil recovery
Erucamide—slip-promoting antiblocking agent in manufacture of plastic films
Nylon 13, 13 or nylon 15, 15
Polyurethanes, plastics and foams
Coatings and adhesives
Modified epoxide gels and resins
Composite materials
Cosmetic formulations
Silver behenate for film processing
Paving bed polymers
High temperature lubricants for intact triacylglycerols with very-long-chain fatty acids

Manitoba, North Dakota, Montana and even as far south as Mississippi and Florida (Dr. S. Fabijanski, Agrisoma, *pers commun*).

Because the major applications for *B. carinata* require oils high in VLCFAs and concomitantly low in PUFAs (similar to desired improvements in HEAR *B. napus* oil composition), these profiles will be the focus of our discussion of the creation of tailored industrial feedstocks (see Table 5.1). The pathway for synthesis of VLCFAs has been detailed above and will not be repeated here; similarly, the basal effects of the single gene transfers in *B. carinata* for improvement in erucic acid content have been cited. However, more recently, there have been additional transgenic *B. carinata* lines produced via gene stacking which show improvement in key VLCFAs for industrial use (Taylor, 2010).

By combining the heterologous expression of the *Crambe KCS* and coincident silencing of the indigenous *B. carinata FAD2* (Mietkiewska et al., 2008) with the coexpression of a *T. majus LPAAT2* (Taylor et al., 2010b), *B. carinata* transgenic lines with up to 64% 22:1, 85% total MUFAs and less than 10% total PUFAs have been created (Fig. 5.3) (Francis et al., unpublished data). These percentages are considerably improved compared with advanced AAFC *B. carinata* breeding line AAC A100 (data courtesy of Dr. K. Falk, AAFC Saskatoon). However, since the starting germplasm

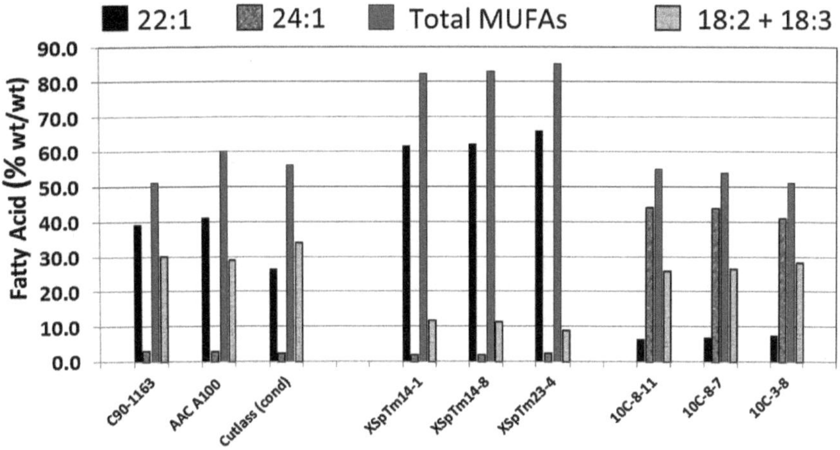

FIGURE 5.3 Composition of wild-type *Brassica carinata* breeding line C90-1163 (*courtesy of G. Rakow, Agriculture & Agri-Food Canada (AAFC) Saskatoon*) and advanced breeding line AAC A100 (*courtesy of K. Falk, AAFC Saskatoon*) and condiment mustard *cv* Cutlass, in comparison to *Brassica carinata* C90 1163 transformed with *Crambe KCS*+*Tropaeolum majus LPAT2*+silenced *B. car FAD2* (XSpTm series, T_2 generation; Francis et al., unpublished) and lines transformed with *Cardamine gracea KCS* (10C series, T_3 generation; Taylor et al., 2009c, 2010a). Fatty Acid composition is reported as % wt/wt. *Total MUFAs*, total monounsaturated fatty acids.

for insertion of the triple transgene was the less-advanced *B. carinata* C90-1163 (kindly supplied by Dr. G. Rakow, AAFC Saskatoon), the improved agronomics of the more advanced AAFC breeding lines make them the preferred base germplasm for introgression of these new traits and indeed for all future genetic engineering work.

By transgenic expression of a *KCS* gene from the crucifer *Cardamine graeca, B. carinata* lines highly enriched in nervonic acid have also been obtained (Taylor et al., 2009b,c) (see Fig. 5.3). Like erucic acid, the high nervonic oils are also of use in the production of biofuels, enhanced oil recovery (EOR) surfactants and the synthesis of bioplastics as well as cosmetics. It is particularly noteworthy that the same *B. carinata* high nervonic oil can equally find direct applications in new products for enhancing infant nutrition and treatment of neurodegenerative diseases (Taylor et al., 2009c).

In the examples of genetic engineering cited above, there were no significant changes in the agronomic performance of the transgenics, compared with nontransformed *B. carinata* host germplasm. For example, in confined field trials of the high nervonic *B. carinata* lines, the percentage of germination, date of emergence, time-to-flower; days-to-maturity and degree of lodging were all unaffected (Taylor et al., 2010a).

Two of the most promising uses of high erucic *B. carinata* oil are its conversion to biojet/biodiesel fuels and to surfactants for EOR.

Biojet and Biodiesel

In 2012, Agrisoma BioSciences launched its Resonance™ bioenergy feedstock. As shown in Table 5.2, fuels manufactured by Advanced Research Associates and UOP Honeywell from Resonance™ *B. carinata* have yielded ReadiJet™ and HRJ Renewable Green Jet™, respectively, both of which exceed ASTM specifications and can provide a 100% drop-in biojet fuel (http://www.ara.com/fuels/Why-ReadiJet.html) which can replace the crude oil–based kerosene that drives flight today. Test flight measurements conducted recently showed that emissions had essentially no sulfur, were high in hydrogen and were clean-burning, with a 25% reduction in particle number and up to a 50% reduction in black carbon using ReadiJet™ versus Jet A1 fuel. Links to test flight information are given here: http://www.altenergymag.com/news/2013/01/10/analysis-of-readijetreg-alternative-fuel-released/27751; http://www.producer.com/2012/01/jets-test-fuel-from-new-oilseed%E2%80%A9/; http://www.ara.com/fuels/PDFs/NRC-ReadiJet-flight-results.pdf; http://biomassmagazine.com/articles/8070/agrisoma-biosciences-u-s-military-contribute-to-biojet-project. While the general desired parameters and economics of seed oils of the Brassicaceae to produce biodiesel have already been described in detail above and elsewhere (Durrett et al., 2008), it is of note that Resonance™ *B. carinata* feedstock can also be converted to a 100% drop-in diesel fuel with specifications which meet, and in most cases, exceed, ASTM D 975 requirements (Table 5.3). The general desired parameters and economics of seed oils of the Brassicaceae to produce biodiesel have already been described in detail above and elsewhere (Durrett et al., 2008).

EOR Surfactants

Given the diminishing supply versus increasing global demand for petroleum-based fuels and petrochemicals, there is escalated interest in enhancing the extraction and recovery of oil from existing reservoirs (EOR). Primary oil recovery methods utilize the internal pressure from within the reservoir to extract oil, typically recovering 15–20% of the original oil in place. Once this process has been exhausted, a secondary approach employing water or a gas is used to recover additional oil, often 20–30%. However, at the point when secondary recovery becomes inefficient or ineffective, tertiary recovery methods such as EOR must be used. In EOR, various chemicals are injected into the reservoir to overcome wettability and capillary pressures, thereby recovering residual oil that would otherwise remain within the microscopic pores. Chemicals that are used include polymers, surfactants and alkali. The alkaline surfactant polymer (ASP) process uses an injection solution containing (1) an alkali to reduce adsorption and to form in situ surfactant, along with (2) one or more surfactants to lower the interfacial tension between the injection fluid and the trapped oil, and (3) a polymer to control the mobility ratio and sweep efficiency of the injected fluid. In some cases, a remarkable additional recovery of up to

TABLE 5.2 Comparison of ASTM D7566, JP-8 and Military Grade 83,133G Jet Fuel Specifications[a] With ARA ReadiJet™ and UOP Honeywell GreenJet™ Produced From Agrisoma Resonance™ *Brassica carinata* Feedstock

Spec Test/Parameter	MIL-DTL 3133G Spec	Ref Petroleum JP-8	ARA ReadiJet™ From *Brassica carinata*[b]	UOP Honeywell Green Jet™ From *Brassica carinata*[c]
Density (g/L)	775–840	804	791	760.8
Freeze point (°C) max	≤–47	–50	–52.3	–57
Flash point (°C) min	≥38	51	42.6	45
Distillation (°C)				
10% Recovery	205	182	164	175
50% Recovery	n/a	208	192	194
90% Recovery	n/a	244	226	240
End point	≤300	265	239	273
Kinematic viscosity at –20°C (cSt)	8.0 max	4.65	4.01	n/a
Total sulfur % by mass	0.3	0.019	0.002	<0.001
Aromatics (vol%)	≤25	18.8	14.3	15
Cyclo-paraffins (Mass %)	n/a	30.0	46.0	n/a

[a]*Analyses performed by Canadian DND-QETE and by the U.S. Air Force Research Laboratory, ARA and UOP.*
[b]*Manufactured by ARA's "CH" process. http://www.ara.com/fuels/Readi-Jet-Diesel-Specs.html; http://www.readifuels.com/ARA-CH-technology.htm.*
[c]*Manufactured by UOP's "HRJ" process http://www.uop.com/green-jet-fuel/.*

TABLE 5.3 Comparison of ASTM D975 Diesel Fuel Specifications with ARA ReadiDiesel™ Produced from Agrisoma Resonance™ *Brassica carinata* Feedstock

Spec Test/Parameter	ASTM D 975 Requirement	ReadiDiesel™ From *Brassica carinata*[a]
Total acid number (mg KOH/g)	0.05	0.003
Aromatics (vol%)	≤35	21.3
Olefins (vol%)	<10	5.6
Sulfur (ppm)	≤15	<1
Distillation (D2887) 90% max (°C)	356	326
Cetane Index	≥40	53
Cloud point (°C)	−5	−17.4
Flash point (°C)	≥52	69
Viscosity at 40°C, min/max, cSt	2.0–4.5	2.07
Lubricity (HFRR) at 60°C, microns	≤520	340

[a]*Manufactured by ARA's "CH" process; http://www.ara.com/fuels/Readi-Jet-Diesel-Specs.html; http://www.readifuels.com/ARA-CH-technology.htm.*

35% has been achieved (Berger, 2010; Hirasaki et al., 2011; Cohen, 2012). Currently most EOR surfactants are based on petrochemical feedstocks. However, work is ongoing to develop EOR surfactants that are partially or wholly derived from renewable agricultural crops such as *B. carinata*, affording green, biodegradable, nontoxic, environmentally friendly alternatives that are not dependent on the availability of petrochemicals (Berger, 2010; Taylor, 2010; Cohen, 2012).

As for any new crop, it is critical that it contributes positively to sustainable agriculture which has been defined as "an integrated system of plant and animal production practices having site-specific application that will, over the long term, (1) satisfy human food and fiber needs, (2) enhance environmental quality and the natural resource base upon which the agricultural economy depends, (3) make the most efficient use of nonrenewable resources and on-farm resources and integrate, where appropriate, natural biological cycles and controls, (4) sustain the economic viability of farm operations, (5) enhance the quality of life for farmers and society as a whole." (Gold, 2007).

A *B. carinata* crop platform can meet or exceed many of the targets for sustainable agriculture for the Canadian Prairies (Taylor et al., 2010a). It is well suited to the drier southern regions of Western Canada, and the addition of this new crop platform adds genetic diversity. *Brassica carinata* requires fewer inputs due to natural resistance to drought and blackleg; it is more robust in its architecture, meaning that stands are less prone to weediness; it can provide the grower with enhanced yield (kg/ha) compared with other *Brassicas* (canola) and is therefore an attractive incentive for farmers, as it could result in increased returns at the farm gate. It creates a new delivery system for bioindustrial and pharmaceutical oils that do not impact/compete with the food sector, specifically canola, minimizing any food versus fuel trade-off. Equally important, out-crossing to canola is extremely low, with typical frequencies of hybridization between *B. carinata* × *B. napus* of <0.002% (Dr. G. Séguin-Swartz, AAFC, *pers. commun.*). *Brassica carinata* oils with tailored VLCFA profiles, such as in the examples above, provide renewable, biodegradable, non–fossil fuel feedstocks for the production of biofuels, polymers, plastics, waxes, pharmaceutical and nutraceutical oils. The unique characteristics of *B. carinata* meal provide new opportunities as feedstocks for plastics and antigen delivery systems. Indeed, the utilization of the entire crop, including the biomass, storage protein and minor seed components such as tocochromanols (eg, vitamin E, a natural source of which may act as a better antioxidant compared to synthetic sources) provide greater sustainability (Taylor et al., 2010a).

As indicated by the case studies above, *B. carinata* is well suited for genetic engineering, and the generation of transgenics will play a major role in designing this crop for the delivery of bioproducts. The application of genomic tools and biotechnological methods will allow introduction of a number traits for enhanced crop performance of *B. carinata*. As an industrial crop, the potential for out-crossing of *B. carinata* with other *Brassica* species will require even further reduction, applying the principles of pollination biology to develop varieties with the appropriate pollen isolation/trapping systems. Besides the VLCFA examples discussed herein, we envision the potential for delivery of other specialty molecules, including wax esters similar to those found in *Jojoba*, and very-long-chain hydroxy fatty acids like those found in *Lesquerella* (Taylor et al., 2010a). It will be exciting to watch the evolution of *B. carinata* as a new industrial bioproducts platform crop in the years to come.

CLOSING COMMENTS

Given growing concerns about maintaining a healthy environment and the finite nature of petrochemical reserves, BOS are contributing more extensively to nonfood industrial applications. Plant breeding has resulted in great advances in HEAR for industrial applications through the recent development of hybrids. Thus far, however, genetic engineering offers the only real promise for raising erucic acid content beyond 66%, and this of course will be important in providing a more uniform feedstock for the production of polymers from rapeseed oil. Currently, chemical conversions of oil from BOS are conducted with oils composed of a mixture of fatty acids. Although high oleic acid canola has proven to be an attractive feedstock for the production of biodiesel, in the longer term, production of biodiesel from this source may not be feasible given the importance of maintaining an adequate supply of food oil for a growing global population. Subsequently, oil crops not used as a source of food oil, or other species of *Brassica* such as *B. carinata*, are being developed as new delivery systems for industrial oils with some applications in the biofuel area. Both food and nonfood applications of oil will benefit from BOS cultivars with increased seed oil content. Our increased understanding of the molecular factors affecting oil formation has the set stage for rationally designing new molecular strategies for further boosting seed oil content in BOS.

ACKNOWLEDGMENTS

RJW is grateful to Alberta Enterprise and Advanced Education, Alberta Innovates Bio Solutions, AVAC Ltd., the Canada Foundation for Innovation, the Canada Research Chairs Program and the Natural Sciences and Engineering Research Council of Canada for supporting his research on oil formation and modification in *B. napus*. DCT acknowledges the support of the Saskatchewan Agriculture Development Fund and Prairie Gold.

REFERENCES

Adamczak, M., Bornscheuer, U.T., Bednarski, W., 2009. The application of biotechnological methods for the synthesis of biodiesel. Eur. J. Lipid Sci. Technol. 111, 808–813.

Alemayehu, N., Becker, H., 2002. Genotypic diversity and patterns of variation in a germplasm material of Ethiopian mustard (*Brassica carinata* A. Braun). Genet. Resour. Crop Evol. 49, 573–582.

Andre, C., Haslam, R.P., Shanklin, J., 2012. Feedback regulation of plastidic acetyl-CoA carboxylase by 18:1-acyl carrier protein in *Brassica napus*. PNAS 109, 10107–10112.

Balat, M., Balat, H., 2010. Progress in biodiesel processing. Appl. Energy 87, 1815–1835.

Benecke, H.P., Vijayendran, B.R., Garbark, D.B., Mitchell, K.P., 2008. Low cost and highly reactive biobased polyols: a co-product of the emerging biorefinery economy. Clean – Soil Air Water 36, 694–699.

Berger, P.D., 2010. Chemically enhanced oil recovery stages a comeback. Inform 20, 682–685.

Booth, E.J., Gunstone, F.D., 2004. Rapeseed and rapeseed oil: agronomy, production and trade. In: Gunstone, F.D. (Ed.), Rapeseed and Canola Oil. CRC Press, Boca Raton, Florida, pp. 1–171.

Bao, X., Pollard, M., Ohlrogge, J., 1998. The biosynthesis of erucic acid in developing embryos of *Brassica rapa*. Plant Physiol. 118, 183–190.

Barret, P., Delourme, R., Renard, M., Domerque, F., Lessire, R., Delseny, M., Roscoe, T.J., 1998. A rapeseed *FAE1* gene is linked to the E1 locus associated with variation in the content of erucic acid. Theor. Appl. Genet. 96, 177–186.

Blacklock, B.J., Jaworski, J.G., 2002. Studies into factors contributing to substrate specificity of membrane-bound 3-ketoacyl-CoA synthases. Eur. J. Biochem. 269, 4789–4798.

Borisjuk, L., Neuberger, T., Schwender, J., Heinzel, N., Sunderharus, S., Fuchs, J., Hay, J.O., Tschiersch, H., Braun, H.-P., Denolf, P., Lambert, B., Jakob, P.M., Rolletschek, H., 2013. Seed architecture shapes embryo metabolism in oilseed rape. Plant Cell 25, 1625–1640.

Brandle, J.E., McVetty, P.B.E., 1989. Heterosis and combining ability in hybrids derived from oilseed rape cultivars and inbred lines. Can. J. Plant Sci. 69, 1191–1195.

Brandle, J.E., McVetty, P.B.E., 1990. Geographical diversity, parental section and heterosis in oilseed rape. Can. J. Plant Sci. 70, 935–940.

Brough, C.L., Coventry, J.M., Christie, W.W., Kroon, J.T.M., Brown, A.P., Barsby, T.L., Slabas, A.R., 1996. Towards the genetic engineering of triacylglycerols of defind fatty acid composition: major changes in erucic acid content at the *sn*-2 position affected by the introduction of a 1-acyl-*sn*-glycerol-3-phosphate acyltransferase from *Limnanthes douglasii* into oil seed rape. Mol. Breed. 2, 133–142.

Brown, A.P., Brough, C.L., Kroon, J.T.M., Slabas, A.R., 1995. Identificationof a cDNA that encodes a 1-acyl-*sn*-glycerol-3-phosphate acyltransferase from *Limnnanthes douglusii*. Plant Mol. Biol. 29, 267–278.

Buczek, B., Czepirski, L., 2004. Applicability of used rapeseed oil for production of biodiesel. INFORM 15, 186–188.

Buhr, T., Sato, S., Ebrahim, F., Xing, A., Zhou, Y., Mathiesen, M., Schweiger, B., Kinney, A.J., Staswick, P., Clemente, T., 2002. Ribozyme termination of RNA transcripts down-regulate seed fatty acid genes in transgenic soybean. Plant J. 30, 155–163.

Burns, D., Scarth, R., McVetty, P.B.E., 1991. Temperature and genotypic effects of pol cytoplasmic male sterility in summer rape. Can. J. Plant Sci. 71, 655–662.

Canola Council of Canada, 2010. Canola Oil Properties and Uses. http://www.canolacouncil.org/canola_oil_properties_and_uses.aspx.

Canvin, D.T., 1965. Effect of temperature on oil content and fatty acid composition of oils from several oil seed crops. Can. J. Bot. 43, 63–69.

Carlsson, A.S., 2009. Plant oils as feedstock alternatives to petroleum – a short survey of potential oil crop platforms. Biochemie 91, 665–670.

Cohen, D.B., October–November, 2012. Oilseed Crops & Enhanced Oil Recovery: Can Oilseeds Produced in Saskatchewan Be Used Effectively in EOR and What Advantages, If Any, Might They Have over Existing Solutions? Prepared for Innovation Saskatchewan. Maccabee Seed Company, Davis, California. 123 pp.

Curtis, J.M., Liu, G., Omonov, T.S., Kharraz, E., January 26, 2012. Polyol Synthesis from Fatty Acids and Oils. WIPO International Publication Number: WO 2012/009801/A1.

Cuthbert, J.L., McVetty, P.B.E., 2001. Plot to plot, row to row and plant to plant out crossing studies in oilseed rape. Can. J. Plant Sci. 81, 657–664.

Cuthbert, R.D., Crow, G., McVetty, P.B.E., 2009. Assessment of agronomic performance and heterosis for agronomic traits in hybrid high erucic acid rapeseed (HEAR). Can. J. Plant Sci. 89, 227–237.

Damude, H.G., Kinney, A.F., 2008. Enhancing plant seed oils for human nutrition. Plant Physiol. 147, 962–968.

Delourme, R., Falentin, C., Huteau, V., Clouet, V., Horvais, R., Gandon, B., Specel, S., Hanneton, L., Dheu, J.E., et al., 2006. Genetic control of oil content in oilseed rape (*Brassica napus* L.). Theor. Appl. Genet. 113, 1331–1345.

Doll, K.M., Erhan, S.Z., 2006. Synthesis and performance of surfactants based on epoxidized methyl oleate and glycerol. J. Surfact. Deterg. 9, 377–383.

Doll, K.M., Erhan, S.Z., 2009. Polyol and amino acid-based biosurfactants, builders, and hydrogels. In: Hayes, D.G., Kitamoto, D., Solaiman, D.K.Y., Ashby, R.D. (Eds.), Biobased Surfactants and Detergents. Synthesis, Properties, and Applications. AOCS Press, Urbana, Illinois, pp. 425–448.

Domergue, F., Chevalier, S., Santarelli, X., Cassagne, C., Lessire, R., 1999. Evidence that oleoyl-CoA and ATP-dependent elongations coexist in rapeseed (*Brassica napus* L.). Eur. J. Biochem. 263, 464–470.

Downey, R.K., Röbbelen, G., 1989. Brassica species. In: Röbbelen, G., Downey, R.K., Ashri, A. (Eds.), Oil Crops of the World. McGraw-Hill, New York, pp. 339–362.

Drexler, H., Spiekermann, P., Meyer, A., Domergue, F., Zank, T., Sperling, P., Abbadi, A., Heinz, E., 2003. Metabolic engineering of fatty acids for breeding of new oilseed crops: strategies, problems and first results. J. Plant Physiol. 160, 779–802.

Durrett, T.P., Benning, C., Ohlrogge, J., 2008. Plant triacylglycerols as feedstocks for the production of biofuels. Plant J. 54, 593–607.

Dyer, J.M., Stymne, S., Green, A.G., Carlsson, A.S., 2008. High-value oils from plants. Plant J. 54, 640–655.

Earls, J.D., White, J.E., Lopez, L.C., Lysenko, Z., Dettloff, M.L., Null, M.J., 2007. Amine-cured omega-epoxy fatty acid triglycerides: fundamental structure-property relationships. Polymer 48, 712–719.

Erhan, S.Z., 2005. Industrial Uses of Vegetable Oils. AOCS Press, Champaign, Illinois.

Erhan, S.Z., Perez, J.M. (Eds.), 2002. Biobased Industrial Fluids and Lubricants. AOCS Press, Champaign, Illinois.

Falk, K.C., 1999. Development of early maturing *Brassica carinata* for western Canada. In: 10th International Rapeseed Congress, Canberra, Australia, September 26–29. http://www.regional.org.au/au/gcirc/4/100.htm.

Fan, Z., Stefansson, B.R., 1986. Influence of temperature on sterility of two cytoplasmic male-sterility-inducing cytoplasms in rape (*Brassica napus* L.). Can. J. Plant Sci. 66, 229–234.

Fitchet, C.S., Laughton, N.G., Chappell, C.G., Khan, M.L., Tverezovskiy, V., Tomkinson, J., Fowler, P., 2005. Oil Ozonolysis. U.S. Patent Number 2005/0010,069 A1.

Fobert, P.R., Smith, M.A., Zou, J., Mietkiewska, E., Keller, W.A., Taylor, D.C., 2008. Developing Canadian seed oils as industrial feedstocks. Biofuels Bioprod. Bioref. 2, 206–214.

Frauen, M., Paulmann, W., 1999. Breeding of hybrids varieties of winter oilseed rape based on the MSL-system. In: 10th International Rapeseed Congress, Canberra Australia, pp. 258–263.

Frentzen, M., 1998. Acyltransferases from basic science to modified seed oils. Fett/Lipid 100, 161–166.

Friedt, W., Snowdon, R., 2009. Oilseed rape. In: Vollmann, J., Rajcan, I. (Eds.), Handbook of Plant Breeding. Oil Crops, vol. 4. Springer-Verlag, New York, pp. 91–126.

Fu, T.D., 1981. Production and research of rapeseed in the People's Republic of China. Eucarpia Cruciferae Newsl. 6, 6–7.

Getinet, A., Rakow, G., Downey, R.K., 1996. Agronomic performance and seed quality of Ethiopian mustard in Saskatchewan. Can. J. Plant Sci. 76, 387–392.

Gold, M.V., 2007. Sustainable Agriculture: Definitions and Terms, by AFSIC. http://afsic.nal.usda.gov/sustainable-agriculture-definitions-and-terms-1.

Graboski, M.S., McCormick, R.L., 1998. Combustion of fat and vegetable oil derived fuels in diesel engines. Prog. Energy Combust. Sci. 24, 125–164.

Goering, K.J., Eslick, R., Brelsford, D.L., 1965. A search for high erucic containing oils in the Cruciferae. Econ. Bot. 19, 251–256.

Grant, I., Beversdorf, W.D., 1985. Heterosis and combining ability estimates in spring oilseed rape (*Brassica napus* L.). Can. J. Genet. Cytol. 27, 472–478.

Gupta, S.K., Pratap, A., 2007. History, origin and evolution. Rapeseed breeding. In: Advances in Botanical Research, vol. 45. Elsevier Ltd, New York, pp. 1–21.

Han, J., Lühs, W., Sonntag, K., Zähringer, U., Borchardt, D.S., Wolter, F.P., Heinz, E., Frentzen, M., 2001. Functional characterization of β-ketoacyl-CoA synthase genes from *Brassica napus* L. Plant Mol. Biol. 46, 229–239.

Hartley, M.R., Lord, J.M., 2004. Cytotoxic ribosome-inactivating lectins from plants. Biochim. Biophys. Acta 1701, 1–14.

Hedge, I.C., 1976. A systematic and geographical survey of the old world cruciferae. In: Vaughan, J.G., MacLeod, A.J., Jones, B.M.G. (Eds.), The Biology and Chemistry of the Cruciferae. Academic Press, New York, pp. 10–45.

Heyn, F.W., 1978. Introgression of restorer genes from *Raphanus sativus* into male sterile *Brassica napus* and the genetics of fertility restoration. In: Proceedings of the 5th International Rapeseed Conference, Malmo, Sweden, vol. 1, pp. 82–83.

Hinata, K., Konno, N., 1979. Studies on a male sterile system having *B. campestris* nucleus and *D. muralis* cytoplasm. Breeding and some characteristics of this strain. Jpn. J. Breed. 29, 305–311.

Hirasaki, G.J., Miller, C.A., Puerto, M., 2011. Recent advances in surfactant EOR. SPE J. 16, 889–907.

Hu, Z., Wang, X., Zhan, G., Liu, G., Hua, W., Wang, H., 2009. Unusually large oil bodies are highly correlated with lower oil content in *Brassica napus*. Plant Cell Rep. 28, 541–549.

Hu, Z.-H., Hua, W., Zhang, L., Deng, L.-B., Liu, G.-H., Hao, W.-J., Wang, H.-Z., 2013. Seed structure characteristics to form ultrahigh oil content in rapeseed. PLoS One 8, e62099.

Hua, W., Li, R.-J., Zhan, G.-M., Liu, J., Li, J., Wang, X.-F., Liu, G.-H., Wang, H.-Z., 2012. Maternal control of seed oil content in *Brassica napus*: the role of silique wall photosynthesis. Plant J. 69, 432–444.

Irtelli, B., Navari-Izzo, F., 2008. Uptake kinetics of different arsenic species by *Brassica carinata*. Plant Soil 303, 105–113.

Irtelli, B., Peprucci, W.A., Navari-Izzo, F., 2009. Nicotianamine and histidine/proline are, respectively, the most important copper chelators in xylem sap of *Brassica carinata* under conditions of copper deficiency and excess. J. Exp. Bot. 60, 269–277.

Jadhav, A., Katavic, V., Marillia, E.-F., Giblin, E.M., Barton, D.L., Kumar, A., Sonntag, C., Babic, V., Keller, W.A., Taylor, D.C., 2005. Increased levels of erucic acid in *Brassica carinata* by co-suppression and antisense repression of the endogenous *FAD2* gene. Metab. Eng. 7, 215–220.

Jaworski, J., Cahoon, E.B., 2003. Industrial oils from transgenic plants. Curr. Opin. Plant Biol. 6, 178–184.

Johnson, D.L., Rhodes, B., Allen, R., 2002. Canola-based motor oils. In: Janick, J., Whipkey, A. (Eds.), Trends in New Crops and New Uses. ASHS Press, Alexandria, Virginia, pp. 29–33.

Jones, C.W., 1999. Application of hydrogen peroxide for the synthesis of fine chemicals. In: Applications of Hydrogen Peroxide and Derivatives (RSC Clean Technology Monographs). The Royal Society of Chemistry, Cambridge, United Kingdom, pp. 79–177.

Karmakar, A., Karmakar, S., Mukherjee, S., 2010. Properties of various plants and animals feedstocks for biodiesel production. Bioresour. Technol. 101, 7201–7210.

Katavic, V., Friesen, W., Barton, D.L., Gossen, K.K., Giblin, E.M., Luciw, L., An, J., Zou, J., MacKenzie, S.L., Keller, W.A., et al., 2000. Utility of the *Arabidopsis FAE1* and yeast *SLC1-1* genes for improvements in erucic acid and oil content in rapeseed. Biochem. Soc. Trans. 28, 935–937.

Katavic, V., Friesen, W., Barton, D.L., Gossen, K.K., Giblin, E.M., Luciw, T., An, J., Zou, J.-T., MacKenzie, S.L., Keller, W.A., et al., 2001. Improving erucic acid content in rapeseed through biotechnology: what can the Arabidopsis *FAE1* and the yeast *SLC1-1* genes contribute. Crop Sci. 41, 739–747.

Keller, W.A., Armstrong, K.C., 1983. Production of *Brassica napus* haploids through anther and microspore culture. In: Proc., 6th Int. Rapeseed Conference, Paris, France, pp. 239–245.

Kelly, A.A., Shaw, E., Powers, S.J., Kurup, S., Eastmond, P.J., 2013. Suppression of the *SUGAR-DEPENDENT1* triacylglycerol lipase family during seed development enhances oil yield in oilseed rape (*Brassica napus* L.). Plant Biotech. J. 11, 355–361.

Khan, M.L., Tomkinson, J., Fitchett, C., December 15, 2004. Oxidative Cleavege of Unsaturated Oilsand Products Obtained Therefrom. European Patent EP 1198443 B1.

Kinney, A.J., 1994. Genetic modification of the storage lipids of plants. Curr. Opin. Biotechnol. 5, 144–151.

Knothe, G., Steidley, K.R., 2005. Kinematic viscosity of biodiesel fuel components and related compounds. Influence of compound structure and comparison to petrodiesel fuel components. Fuel 84, 1059–1065.

Knutzon, D.S., Hayes, T.R., Wyrick, A., Xiong, H., Davies, H.M., Voelker, T.A., 1999. Lysophosphatidic acid acyltransferase from coconut endosperm mediates the insertion of laurate at the *sn*-2 position of triacylglycerols in lauric rapeseed oil and can increase total laurate levels. Plant Physiol. 120, 739–746.

Körbitz, W., 1995. Utilization of oil as a biodiesel fuel. In: Kimber, D., McGregor, D.I. (Eds.), Brassica Oilseeds. CAB International, Wallingford, Oxon, United Kingdom, pp. 339–352.

Landry, B.S., Hubert, N., Etoh, T., Harada, J.J., Lincoln, S.E., 1991. A genetic map for *Brassica napus* based on restriction fragment length polymorphisms detected with expressed DNA sequences. Genome 34, 543–552.

Lardizabal, K.D., Metz, J.G., Sakamoto, T., Hutton, W.C., Pollard, M.R., Lassner, M.W., 2000. Purification of a jojoba embryo wax synthase, cloning of its cDNA and production of high levels of wax in seeds of transgenic *Arabidopsis*. Plant Physiol. 122, 645–655.

La Scala, J., Wool, R.P., 2005a. Rheology of chemically modified triglycerides. J. App. Polym. Sci. 95, 774–783.

La Scala, J., Wool, R.P., 2005b. Property analysis of triglyceride-based thermosets. Polymer 46, 61–69.

Lassner, M.W., Levering, C.K., Davies, H.M., Knutzon, D.S., 1995. Lysophosphatidic acid acyltransferase from meadowfoam mediates insertion of erucic acid at the *sn*-2 position of triacylglycerol in transgenic rapeseed oil. Plant Physiol. 109, 1389–1394.

Leonard, C., 1994. Sources and commercial applications of high erucic vegetable oils. Lipid Technol. 4, 79–83.

Li, C.S., 1980. Classification and evolution of mustard crops (*Brassica juncea*) in China. Cruciferae Newsl. 5, 33–36.

Li, G., Quiros, C.F., 2001. Sequence-related amplified polymorphism (SRAP), a new marker system based on a simple PCR reaction: its application to mapping and gene tagging in Brassica. Theor. Appl. Genet. 103, 455–461.

Liu, Q., Singh, S.P., Green, A.G., 2002. High-stearic and high-oleic cottonseed oils produced by hairpin RNA-mediated post transcriptional gene silencing. Plant Physiol. 129, 1732–1743.

Liu, J., Hua, W., Zhan, G., Wei, F., Wang, X., Liu, G., Wang, H., 2010. Increasing seed mass and oil content in transgenic *Arabidopsis* by overexpression of *wri1-like* gene from *Brassica napus*. Plant Physiol. Biochem. 48, 9–15.

Liu, Q., Siloto, R.M.P., Lehner, R., Stone, S.J., Weselake, R.J., 2012a. Acyl-CoA:diacylglycerol acyltransferase: molecular biology, biochemistry and biotechnology. Prog. Lipid Res. 51, 350–377.

Liu, J., Hua, W., Yang, H.-L., Zhan, G.-M., Li, R.-J., Deng, L.-B., Wang, X.-F., Liu, G.-H., Wang, H.-Z., 2012b. The *BnDRF2* gene (*GRF2-like* gene from *Brassica napus*) enhances seed oil production through regulating cell number and plant photosynthesis. J. Exp. Bot. 63, 3727–3740.

Lu, J., Khot, S., Wool, R.P., 2005. New sheet molding compound resins from soybean oil. I. Synthesis and characterization. Polymer 46, 71–80.

Lu, C., Xin, Z., Ren, Z., Miquel, M., Browse, J., 2009. An enzyme regulating triacylglycerol composition is encoded by the *ROD1* gene of Arabidopsis. Proc. Natl. Acad. Sci. 106, 18837–18842.

Lois, E., 2007. Definition of biodiesel. Fuel 86, 1212–1213.

Lühs, W., Friedt, W., 1994. Non-food uses of vegetable oils and fatty acids. In: Murphy, D.J. (Ed.), Designer Oil Crops. VCH, Wienheim, Germany, pp. 73–130.

Lühs, W.W., Voss, A., Seyis, F., Friedt, W., 1999. Molecular genetics of erucic acid content in the genus *Brassica*. In: Proceedings of the the 10th International Rapeseed Congress: New Horizons for an Old Crop, Canberra, AustraliaAvailable at: http://www.regional.org.au/au/gcirc/4/442.htm (verified 23.01.06.).

Márquez-Lema, A., Fernández-Martínez, J.M., Pérez-Vich, B., Velasco, L., 2008. Development and characterisation of a *Brassica carinata* inbred line incorporating genes for low glucosinolate content from *B. juncea*. Euphytica 164, 365–375.

McCubbin, A., Dickinson, H.G., 1997. Self-incompatibility. In: Shivanna, K.R., Sawhney, V.K. (Eds.), Pollen Biotechnology for Crop Production and Improvement. Cambridge University Press, Cambridge, United Kindgom, pp. 199–217.

McGuire, J., Haslam, P.J., Bond, G., 2006. Ozonolysis in the production of chiral fine chemicals. In: Ager, D. (Ed.), Handbook of Chiral Chemicals, second ed. CRC Press, Boca Raton, Florida, pp. 165–184.

McVetty, P.B.E., Edie, S.A., Scarth, R., 1990. Comparison of the effect of *nap* and *pol* cytoplasms on the performance of intercultivar summer oilseed rape hybrids. Can. J. Plant Sci. 70, 117–126.

McVetty, P.B.E., Scarth, R., Rimmer, S.R., Van Den Berg, C.G.J., 1996a. Neptune high erucic low glucosinolate summer rape. Can. J. Plant Sci. 76, 343–344.

McVetty, P.B.E., Scarth, R., Rimmer, S.R., Van Den Berg, C.G.J., 1996b. Venus high erucic low glucosinolate summer rape. Can. J. Plant Sci. 76, 341–342.

McVetty, P.B.E., 1997. Cytoplasmic male sterility. In: Shivanna, K.R., Sawhney, V.K. (Eds.), Pollen Biotechnology for Crop Production and Improvement. Cambridge University Press, Cambridge, United Kingodm, pp. 155–182.

McVetty, P.B.E., Rimmer, S.R., Scarth, R., 1998. Castor high erucic low glucosinolate summer rape. Can. J. Plant Sci. 78, 305–306.

McVetty, P.B.E., Scarth, R., Rimmer, S.R., 1999. MillenniUM 01 high erucic low glucosinolate summer rape. Can. J. Plant Sci. 79, 251–252.

McVetty, P.B.E., Scarth, R., Rimmer, S.R., 2000a. MillenniUM 02 high erucic low glucosinolate summer rape. Can. J. Plant Sci. 80, 609–610.

McVetty, P.B.E., Scarth, R., Rimmer, S.R., 2000b. MillenniUM 03 high erucic low glucosinolate summer rape. Can. J. Plant Sci. 80, 611–612.

McVetty, P.B.E., Scarth, R., 2002. Breeding for improved oil quality in Brassica oilseed species. J. Crop Prod. 5, 345–369.

McVetty, P.B.E., Fernando, W.G.D., Scarth, R., Li, G., 2006a. Red River 1826 Roundup Ready high erucic acid low glucosinolate summer rape. Can. J. Plant Sci. 86, 1179–1180.

McVetty, P.B.E., Fernando, W.G.D., Scarth, R., Li, G., 2006b. Red River 1852 Roundup Ready high erucic acid low glucosinolate summer rape. Can. J. Plant Sci. 86, 1181–1182.

McVetty, P.B.E., Zelmer, C.D., 2007. Breeding herbicide tolerant oilseed rape cultivars. In: Gupta, S.K. (Ed.), Rapeseed Breeding. Advances in Botanical Research, vol. 45. Elsevier, San Diego, California, pp. 234–270.

McVetty, P.B.E., Fernando, W.G.D., Tahir, M., Zelmer, C., 2009. High erucic acid, low glucosinolate rapeseed (HEAR) cultivar development in Canada. In: Hou, C.T., Shaw, J.-F. (Eds.), Biocatalysis and Agricultural Biotechnology. Taylor and Francis, Boca Raton, Florida, pp. 43–62.

McVetty, P.B.E., Fernando, W.G.D., Li, G., Tahir, M., Zelmer, C.D., 2010. Red River 1997 Roundup Ready high erucic acid low glucosinolate summer rape. Can. J. Plant Sci. 90, 711–713.

Mietkiewska, E., Giblin, E.M., Wang, S., Barton, D.L., Dirpaul, J., Brost, J.M., Katavic, V., Taylor, D.C., 2004. Seed-specific heterologous expression of a nasturtium *FAE* gene in *Arabidopsis* results in a dramatic increase in the proportion of erucic acid. Plant Physiol. 136, 2665–2675.

Mietkiewska, E., Brost, J.M., Giblin, E.M., Barton, D.L., Taylor, D.C., 2007. Cloning and functional characterization of the *Fatty Acid Elongase 1* (*FAE1*) gene from high erucic *Crambe abyssinica* cv. Prophet. Plant Biotechnol. J. 5, 636–645.

Mietkiewska, E., Hoffman, T.L., Brost, J.M., Giblin, E.M., Barton, D.L., Francis, T., Zhang, Y., Taylor, D.C., 2008. Hairpin-RNA mediated silencing of endogenous *FAD2* gene combined with heterologous expression of *Crambe abyssinica FAE* gene causes a dramatic increase in the level of erucic acid in transgenic *Brassica carinata* seeds. Mol. Breed. 22, 619–627.

Millar, A.A., Kunst, L., 1997. Very-long-chain fatty acid biosynthesis is controlled through the expression and specificity of the condensing enzyme. Plant J. 12, 101–111.

Mittelbach, M., 2010. Meeting report from 2nd International Congress on biodiesel: the science and the technologies 15–17 November 2009, Munich, Germany. Eur. J. Lipid Sci. Technol. 112, 288–292.

Miquel, M., Browse, J., 1992. *Arabidopsis* mutants deficient in polyunsaturated fatty acid synthesis. Biochemical and genetic characterization of a plant oleoyl-phosphatidylcholine desaturase. J. Biol. Chem. 267, 1502–1509.

Murugesan, A., Umarani, C., Chinnusamy, T.R., Krishnan, M., Subramanian, R., Neduzchezhain, N., 2009. Production and analysis of bio-diesel from non-edible oils – a review. Renew. Sustain. Energy Rev. 13, 825–834.

Nath, U.K., Wilmer, J.A., Wallington, E.J., Becker, H.C., Möllers, C., 2009. Increasing erucic acid content through combination of endogenous low polyunsaturated fatty acids alleles with *Ld*-LPAZT + *Bn-fae1* transgenes in rapeseed. Theor. Appl. Genet. 118, 765–773.

Nagaharu, U., 1935. Genome analysis in *Brassica* with special reference to the experimental formation of *Brassica napus* and peculiar mode of fertilization. Jpn. J. Bot. 7, 389–452.

Ogura, H., 1968. Studies on a new male sterility system in Japanese radish with special reference to utilization of the sterility toward the practical raising of hybrid seed. Mem. Fac. Agric. Kagoshima Univ. 6, 39–78.

Omonov, T.S., Kharraz, E., Curtis, J.M., 2011. Ozonolysis of canola oil: a study of product yields and ozonolysis kinetics in different solvent systems. JAOCS 88, 689–705.

Palmer, C.E., Keller, W.F., 1997. Pollen embryos. In: Shivanna, K.R., Sawhney, V.K. (Eds.), Pollen Biotechnology for Crop Production and Improvement. Cambridge University Press, Cambridge, United Kingdom, pp. 392–422.

Pandian, A., Liu, Q., Hurlestone, C., Singh, S., Salisbury, P., Green, A., 2004. Development of nutritionally superior *Brassica napus* and *B. juncea* oils using RNAi-mediated gene silencing. In: 4th International Crop Science Congress Available at: http://www.cropscience.org.au/icsc2004/poster/5/1/3/703_pandiana.htm (verified 23.06.06.).

Pedroche, J., Yust, M.M., Lqari, H., Megias, C., Giron-Calle, J., Alaiz Vioque, J., Millan, F., 2007. Obtaining of *Brassica carinata* protein hydrolysates enriched in bioactive peptides using immobilized digestive proteases. Food Res. Int. 40, 931–938.

Pellan-Delourme, R., Eber, F., Renard, M., 1987. Male fertility restoration in *Brassica napus* with radish cytoplasmic male sterility. In: Proceedings 7th International Rapeseed Congress, Poznan Poland, pp. 199–203.

Pelletier, G.C., Primard, C., Vedel, F., Chetrit, P., Rouselle, R., Renard, M., 1983. Intergeneric cytoplasmic hybridization in Cruciferae by protoplast fusion. Mol. Gen. Genet. 191, 244–250.

Pelletier, G., Primard, C., Vedel, F., Chetrit, P., Renard, M., Pellan-Delourme, R., Mequida, J., 1987. In: Proceedings of the 7th International Rapeseed Congress, Pozan, Poland, pp. 113–118.

Peng, Q., Hu, Y., Wei, R., Zhang, Y., Guan, C., Ruan, Y., Liu, C., 2010. Simultaneous silencing of *FAD2* and *FAE1* genes affects both oleic acid and erucic acid contents in *Brassica napus* seeds. Plant Cell Rep. 29, 317–325.

Petrovic, Z.S., Zhang, W., Javni, I., 2005. Structure and properties of polyurethanes prepared from triglyceride polyols by ozonolysis. Biomacromolecules 6, 713–719.

Piazza, G.J., Foglia, T.A., 2001. Rapeseed oil for oleochemical usage. Eur. J. Lipid Sci. Technol. 103, 450–454.

Pilalis, E., Chatziioannour, A., Thomasset, B., Kolisis, F., 2011. An in silico compartmentalized metabolic model of *Brassica napus* enables the systemic study of regulatory aspects of plant central metabolism. Biotech. Bioeng. 108, 1673–1682.

Potter, T.D., Salisbury, P.A., Ballinger, D.J., Wratten, N., Mailer, R.J., 1995. Comparison of historical varieties of rapeseed and canola in Australia. In: Proceedings of the 9th International Rapeseed Congress. Canberra, Australia, pp. 365–367.

Potts, D.A., Rakow, G.W., Males, D.R., Woods, D.L., 2003. The development of canola-quality *Brassica juncea*. Can. J. Plant Sci. 83, 117–118.

Prakash, S., 1980. Cruciferous oilseeds in India. In: Tsunoda, S., Hinata, K., Gomez-Campo, C. (Eds.), *Brassica* and Wild Allies. Japan Scientific Press, Tokyo, Japan, pp. 151–163.

Prakash, S., Hinata, K., 1980. Taxonomy, cytogenetics and origin of crop *Brassica*, a review. Opera Bot. 55, 11–57.

Prilezhaev, N., 1909. Oxydation ungesättigter verbindungen mittels organischer superoxyde/oxidation of unsaturated compounds by means of organic peroxides. Ber. Dtsch. Chem. Ges. 42, 4811–4815.

Pryde, E.H., Anders, D.E., Teeter, H.M., Cowan, J.C., 1960. The ozonization of methyl oleate. J. Org. Chem. 25, 618–621.

Pryde, E.H., Anders, D.E., Teeter, H.M., Cowan, J.C., 1961. Ozonization of soybean oil – the preparation and some properties of aldehyde oils. JAOCS 38, 375–379.

Puyaubert, J., Garcia, C., Chevalier, S., Lessire, R., 2005. Acyl-CoA elongase, a key enzyme in the development of high-erucic acid rapeseed? Eur. J. Lipid Sci. Technol. 107, 263–267.

Rahman, H., Singer, S.D., Weselake, R.J., 2013. Development of low-linolenic acid *Brassica oleracea* lines through seed mutagenesis and molecular characterization of mutants. Theor. Appl. Genet. 126, 1587–1598.

Rahman, M., Sun, Z., McVetty, P.B.E., Li, G., 2008. High throughput genome specific molecular markers for erucic acid content gens in *Brassica napus* (L.). Theor. Appl. Genet. 117, 895–904.

Rakow, G., Woods, D.L., 1987. Out crossing in rape and mustard under Saskatchewan prairie conditions. Can. J. Plant Sci. 67, 147–151.

Ramos, M.J., Fernández, C.M., Casas, A., Rodríguez, L., Pérez, A., 2009. Influence of fatty acid composition of raw materials on biodiesel properties. Bioresour. Technol. 100, 261–268.

Ripley, V.L., Falk, K., Roslinski, V., 2006. Genetic diversity of *Brassica carinata* populations selected for improved seed quality. Acta Hortic. 706, 135–140.

Riungu, T.C., McVetty, P.B.E., 2004. Comparison of the effect of *mur* and *nap* cytoplasms on the performance of intercultivar summer rape hybrids. Can. J. Plant Sci. 84, 731–738.

Röbbelen, G., 1991. Rapeseed in a changing world: plant production potential. In: Proceedings of the 8th International Rapeseed Congress, Saskatoon, Canada, July 9–11, pp. 29–38.

Sawhney, V.K., 1997. Genetic male sterility. In: Shivanna, K.R., Sawhney, V.K. (Eds.), Pollen Biotechnology for Crop Production and Improvement. Cambridge University Press, Cambridge, United Kingdom, pp. 183–188.

Scarth, R., McVetty, P.B.E., Rimmer, S.R., Stefansson, B.R., 1991. Hero summer rape. Can. J. Plant Sci. 71, 865–866.

Scarth, R., McVetty, P.B.E., Rimmer, S.R., Daun, J.K., 1992. Breeding for specialty oil quality in canola rapeseed: the University of Manitoba program. In: MacKenzie, S.L., Taylor, D.C. (Eds.), Seed Oils for the Future. AOCS Press, Champaign, Illinois, pp. 171–176.

Scarth, R., McVetty, P.B.E., Rimmer, S.R., 1994. Mercury high erucic low glucosinolate summer rape. Can. J. Plant Sci. 74, 205–207.

Scarth, R., Tang, J., 2006. Modification of *Brassica* oil using conventional and transgenic approaches. Crop Sci. 46, 1225–1236.

Schwender, J., Goffman, F., Ohlrogge, J.B., Shachar-Hill, Y., 2004. Rubisco without the Calvin cycle improves the carbon efficiency of developing green seeds. Nature 432, 779–782.

Schwender, J., Hay, J.O., 2012. Predictive modeling of biomass component tradeoffs in *Brassica napus* developing oilseeds based on in silico manipulation of storage metabolism. Plant Physiol. 160, 1218–1236.

Sernyk, J.L., Stefansson, B.R., 1982. Heterosis in summer rape (*Brassica napus* L.). Can. J. Plant Sci. 63, 407–413.

Shahidi, F., 1990. Rapeseed and canola: global production and distribution. In: Shahidi, F. (Ed.), Canola and Rapeseed Production, Chemistry, Nutrition and Processing Technology. Van Nostrand Reinhold, New York, pp. 3–14.

Shukla, V.K.S., Blicher-Mathiesen, U., 1993. Studies in evaluation of unconventional oils from Southeast Asia. Fat Sci. Technol. 95, 367–369.

Siloto, R.M.P., Truksa, M., Brownfield, D., Good, A.G., Weselake, R.J., 2009a. Directed evolution of acyl-CoA:diacylglycerol acyltransferase: development and characterization of *Brassica napus DGAT1* mutagenized libraries. Plant Physiol. Biochem. 47, 456–461.

Siloto, R.M.P., Truksa, M., He, X., McKeon, T., Weselake, R.J., 2009b. Simple methods to detect triacylglycerol biosynthesis in a yeast-based recombinant system. Lipids 44, 963–973.

Singer, S.D., Greer, M.S., Mietkiewska, E., Pan, X., Weselake, R.J., 2013. Genetic engineering of lipid synthesis in seeds. In: Gupta, S.K. (Ed.), Biotechnology of Crucifers. Springer Science + Business Media, Berlin, pp. 111–149.

Sivaraman, I., Arumugam, N., Sodhi, Y.S., Gupta, V., Mukhopadhyay, A., Pradhan, A.K., Burma, P.K., Pental, D., 2004. Development of high oleic and low linoleic acid transgenics in a zero erucic acid. *Brassica juncea* L. (Indian mustard) line by antisense suppression of the *fad2* gene. Mol. Breed. 13, 365–375.

Snowdon, R.J., Luhs, W., Freidt, W., 2006. Oilseed rape. In: Cole, C. (Ed.), Genome Mapping and Molecular Breeding. vol. 2. Springer, Heidelberg, pp. 55–114.

Snyder, C.L., Yurchenko, O.P., Siloto, R.M.P., Chen, X., Liu, Q., Mietkiewska, E., Weselake, R.J., 2009. Acyltransferase action in the modification of seed oil biosynthesis. New Biotechnol. 26, 11–16.

Snyder, C.L., Weselake, R.J., 2012. Modification of seed oil formation in *Brassica* oilseed species. In: Thiyam-Holländer, U., Eskin, N.A.M., Matthäus, B. (Eds.), Canola and Rapeseed: Production, Processing, Food Quality, and Nutrition. CRC Press, Taylor & Francis Group, Boca Raton, Florida, pp. 101–124.

Soi, H.S., Mazalan, N.M., Satter, M.N., Ismail, T.N.M.T., Hassan, H.A., Lye, O.T., Ahmad, S., Wahid, M.B., Basiron, T.S.D.Y., December 8, 2009. Process for the Production of a Polyol Monomer. U.S. Patent 7629478 B2.

Stefansson, B.R., Downey, R.K., 1995. Rapeseed. In: Slinkard, A.E., Knott, D.R. (Eds.), Harvest of Gold. University Extension Press, University of Saskatchewan, Saskatoon, Saskatchewan, Canada, pp. 140–152.

Stefansson, B.R., Kondra, Z.P., 1975. Tower summer rape. Can. J. Plant Sci. 55, 343–344.

Stoutjesdijk, P.A., Hurlstone, C.J., Singh, S.P., Green, A.G., 2000. High-oleic acid Australian *Brassica napus* and *B. juncea* varieties produced by co-suppression of endogenous Δ12-desaturases. Biochem. Soc. Trans. 28, 938–940.

Stoutjesdijk, P.A., Singh, S.P., Liu, Q., Hurlstone, C.J., Waterhouse, P.A., Green, A.G., 2002. hpRNA-mediated targeting of the Arabidopsis *FAD2* gene gives highly efficient and stable silencing. Plant Physiol. 129, 1723–1731.

Sun, M., Hua, W., Liu, J., Huang, S., Wang, X., Liu, G., Wang, H., 2012. Design of new genome- and gene-sourced primers and identification of QTL for seed oil content in a specially high-oil *Brassica napus* cultivar. PLOS One 7, e47037.

Sun, Z., Wang, Z., Tu, J., Zhang, J., Yu, F., McVetty, P.B.E., Li, G., 2007. An ultra dense genetic recombination map for *Brassica napus*, consisting of 13551 SRAP markers. Theor. Appl. Genet. 114, 1305–1317.

Swern, D., Billen, G.N., Findley, T.W., Scanlan, J.T., 1945. Hydroxylation of monounsaturated fatty materials with hydrogen peroxide. J. Am. Chem. Soc. 67, 1786–1789.

Taylor, D.C., Barton, D.L., Giblin, E.M., MacKenzie, S.L., Van Den Berg, C.G.J., McVetty, P.B.E., 1995. Microsomal lyso-phosphatidic acid transferase from a *Brassica oleracea* cultivar incorporates erucic acid into the *sn*-2 position of seed triacylglycerols. Plant Physiol. 109, 409–420.

Taylor, D.C., Zhang, Y., Kumar, A., Francis, T., Giblin, E.M., Barton, D.L., Ferrie, J.R., Laroche, A., Shah, S., Zhu, W., Snyder, C.L., Hall, L., Rakow, G., Harwood, J.L., Weselake, R.J., 2009a. Molecular modification of triacylglycerol accumulation by over-expression of *DGAT1* to produce canola with increased seed oil content under field conditions. Botany 87, 533–543.

Taylor, D.C., Francis, T., Guo, Y., Brost, J.M., Katavic, V., Mietkiewska, E., Giblin, E.M., Lozinsky, S., Hoffman, T., 2009b. Molecular cloning and characterization of a *KCS* gene from *Cardamine graeca* and its heterologous expression in *Brassica* oilseeds to engineer high nervonic acid oils for potential medical and industrial use. Plant Biotechnol. J. 7, 925–938.

Taylor, D.C., Guo, Y., Katavic, V., Mietkiewska, E., Francis, T., Bettger, W., 2009c. New seed oils for improved human and animal health and as industrial feedstocks: genetic manipulation of the *Brassicaceae* to produce oils enriched in nervonic acid. In: Krishnan, H.B. (Ed.), Modification of Seed Composition to Promote Health and Nutrition. ASA-CSSA-SSSA Publishing, Madison, Wisconsin, pp. 219–233.

Taylor, D.C., Falk, K.C., Palmer, C.D., Hammerlindl, J., Babic, V., Mietkiewska, E., Jadhav, A., Marillia, E.-F., Francis, T., Hoffman, T., Giblin, E.M., Katavic, V., Keller, W.A., 2010a. *Brassica carinata* – a new molecular farming platform for delivering bio–industrial oil feedstocks: case studies of genetic modifications to improve very long–chain fatty acid and oil content in seeds. Biofuels Bioprod. Biorefin. 4, 538–561.

Taylor, D.C., 2010. New very long-chain fatty acid seed oils produced through introduction of strategic genes into *Brassica carinata*. INFORM 21, 602–605.

Taylor, D.C., Francis, T., Lozinsky, S., Hoffman, T., Giblin, M., Marillia, E.F., 2010b. Cloning and characterization of a *lysophosphatidic acid acyltransferase 2* (*LPAT2*) gene from *Tropaeolum majus* L. Open Plant Sci. J 4, 7–17.

Taylor, D.C., Smith, M., Fobert, P., Mietkiewska, E., Weselake, R.J., 2011. Metabolic engineering of higher plants to produce bio-industrial oils. In: Moo-Young, M. (Ed.), Comprehensive Biotechnology 2nd Edition, vol. 4. Elsevier, pp. 67–85.

Thompson, K.F., 1972. Cytoplasmic male sterility in oilseed rape. Heredity 29, 253–257.

Throckmorton, P.E., Pryde, E.H., 1972. Pilot run, plant design and cost-analysis for reductive ozonolysis of methyl soyate. JAOCS 49, 643–648.

Töpfer, R., Martini, N., Schell, J., 1995. Modification of plant lipid synthesis. Science 268, 681–686.

Tran, P., Graiver, D., Narayan, R., 2005. Ozone-mediated polyol synthesis from soybean oil. JAOCS 82, 653–659.

Van Dyne, D.L., Blasé, M.G., 1991. Commercialization of crops with high erucic acid for industrial uses. In: Rattray, J. (Ed.), Biotechnology of Plant Fats and Oils. AOCS, Champaign, Illinois, pp. 151–161.

Velasco, L., Goffman, F.D., Becker, H.C., 1998. Variability for the fatty acid composition of the seed oil in a germplasm collection of the genus Brassica. Genet. Resour. Crop Evol. 45, 371–382.

Voelker, T.A., Hayes, T.R., Cranmer, A.M., Turner, J.C., Davies, H.M., 1996. Genetic engineering of a quantitative trait: metabolic and genetic parameters influencing the accumulation of laurate in rapeseed. Plant J. 9, 229–241.

Warwick, S.I., Gugel, R.K., McDonald, B., Falk, K.C., 2006. Genetic variation and agronomic potential of Ethiopian mustard (*Brassica carinata* A. Braun) in western Canada. Genet. Resour. Crop Evol. 53, 297–312.

Weselake, R.J., Taylor, D.C., 1999. The study of storage lipid biosynthesis using microspore-derived cultures of oilseed rape. Prog. Lipid. Res. 38, 1–60.

Weselake, R.J., Shah, S., Tang, M., Quant, P.A., Snyder, C.L., Furukawa-Stoffer, T.L., Zhu, W., Taylor, D.C., Zou, J., Kumar, A., Hall, L., Laroche, A., Rakow, G., Raney, P., Moloney, M.M., Harwood, J.L., 2008. Metabolic control analysis is helpful for informed genetic manipulation of oilseed rape (*Brassica napus*) to increase seed oil content. J. Exp. Bot. 59, 3543–3549.

Weselake, R.J., Taylor, D.C., Rahman, M.H., Shah, S., Laroche, A., McVetty, P.B.E., Harwood, J.L., 2009. Increasing the flow of carbon into seed oil. Biotechnol. Adv. 27, 866–878.

Weselake, R.J., 2011. Genetic engineering approaches for trait development in Brassica oilseed species. In: Daun, J.K., Hickling, D., Eskin, N.A.M. (Eds.), Canola: Description, Variety Development, Agronomy, Composition, and Utilization. AOCS Press, Urbana, Illinois, pp. 57–91.

Williams, M.E., Leemans, J., Michaels, F., 1997. Male sterility through recombinant DNA technology. In: ;Shivanna, K.R., Sawhney, V.K. (Eds.), Pollen Biotechnology for Crop Production and Improvement. Cambridge University Press, Cambridge, United Kingdom, pp. 237–258.

Weier, D., Hanke, C., Eickelkamp, A., Lűhs, W., Dettendorfer, J., Schaffert, E., Mőllers, C., Friedt, W., Wolter, F.P., Frentzen, M., 1997. Trierucoylglycerol biosynthesis in transgenic plants of rapeseed (*Brassica napus* L.). FETT/Lipid 99, 160–165.

Wool, R.P., Sun, X.S., 2005. Bio-based Polymers and Composites. Elsevier, London, United Kingdom.

Xu, J., Francis, T., Mietkiewska, E., Giblin, E.M., Barton, D.L., Zhang, Y., Zhang, M., Taylor, D.C., 2008. Cloning and characterization of an Acyl-CoA-dependent Diacylglycerol Acyltransferase 1 (*DGAT1*) gene from *Tropaeolum majus* and a study of the functional motifs of the DGAT protein using site-directed mutagenesis to modify enzyme activity and oil content. Plant Biotechnol. J. 6, 799–818.

Yue, J., Narine, S.S., 2007. Separation and quantification of vegetable oil based polyols by high performance liquid chromatography with evaporative light scattering detection. JAOCS 84, 803–807.

Zhu, Y., Cao, Z., Xu, F., Huang, Y., Chen, M., Guo, W., Zhou, W., Zhu, J., Meng, J., Zou, J., Jiang, L., 2012. Analysis of gene expression profiles of two near-isogenic lines differing at a QTL region affecting oil content at high temperatures during seed maturation in oilseed rape (*Brassica napus* L.). Theor. Appl. Genet. 124, 515–531.

Zou, J.-T., Katavic, V., Giblin, E.M., Barton, D.L., MacKenzie, S.L., Keller, W.A., Hu, X., Taylor, D.C., 1997. Modification of seed oil content and acyl composition in Brassicaceae by expression of a yeast *sn*-2 acyltransferase gene. Plant Cell 9, 909–923.

Zou, J., Jiang, C., Cao, Z., Li, R., Long, Y., Chen, S., Meng, J., 2010. Association mapping of seed oil content in *Brassica napus* and comparison with quantitative trait loci identified from linkage mapping. Genome 53, 908–916.

Chapter 6

Flax (*Linum usitatissimum* L.)

Linda M. Hall
Department of Agricultural, Food and Nutritional Science, University of Alberta, Edmonton, AB, Canada

Helen Booker
Crop Development Centre, Department of Plant Sciences, University of Saskatchewan, Saskatoon, SK, Canada

Rodrigo M.P. Siloto
Department of Agricultural, Food and Nutritional Science, University of Alberta, Edmonton, AB, Canada

Amit J. Jhala
Department of Agricultural, Food and Nutritional Science, University of Alberta, Edmonton, AB, Canada; Department of Agronomy and Horticulture, University of Nebraska-Lincoln, Lincoln, NE, United States

Randall J. Weselake
Department of Agricultural, Food and Nutritional Sciences, University of Alberta, Edmonton, AB, Canada

INTRODUCTION

Flax (or linseed) is one of the predominant industrial oils seed crops grown in temperate climates. The seed oil of this oilseed crop is enriched in α-linolenic acid (ALA) ($18:3^{cis\Delta9,12,15}$). Because of this, flax oil readily polymerizes on exposure to oxygen, making it useful for a variety of industrial products, including varnish and linoleum, while meal from pressed seed is useful as animal feed. Industrial demand declined after the 1960s due to a shift to acrylic paints and vinyl floor coverings, but now there is a renewed interest in using biomass-derived feedstocks. In addition to industrial applications, ω-3-enriched flaxseed oil is gaining importance in livestock feed and aqua-feed applications. The oil is also recognized as a good source of ALA for the human diet. Additional flax constituents, including fiber and lignans (Czemplik and Szopa, 2009; Vaisey-Genser and Morris, 2003), may provide human health benefits. "Solin"-type flax, which is enriched in linoleic acid (LA; $18:2^{cis\Delta9,12}$) and low in ALA content (<5%), has greater oxidative stability than regular flax. Solin was developed

Industrial Oil Crops. http://dx.doi.org/10.1016/B978-1-893997-98-1.00006-3

for widespread food use in place of sunflower and safflower seed oils (Vrinten et al., 2005; Ntiamoah and Rowland, 1997; Rowland, 1991). This chapter discusses the origin, production, agronomy, genetics and breeding, oil synthesis biochemistry, and biotechnology of flax, with a focus on industrial applications of the crop.

DIFFERENTIATION OF FIBER FLAX AND OILSEED FLAX

Flax cultivars have been selected for production of either fiber (fiber flax) or oil (oilseed flax). Location of production, climatic adaptation, and morphology of these types now differ considerably. Oilseed-type plants are usually shorter, have more branches, and produce more seeds, while fiber flax types are generally taller, have few branches, and have been selected for fiber (Gill, 1987). Bast fibers from flax, derived as part of the phloem, are long (4 cm), have high tensile strength, and have a high quality of cellulose (Deyholos, 2006). Fine flax fibers are used for linens and textiles, while coarser fibers are used for nonwoven textiles and twine. Both flax types have a short tap root system with fibrous branches. Flax is relatively shallow rooted, with only 4–7% of root mass deeper than 60 cm (Gan et al., 2009).

Production of both linen and oil is generally not compatible in the same crop. Bast fibers become more lignified and less flexible during flower and seed set, so high-quality linen fiber is harvested prior to seed ripening and harvest (Deyholos, 2006). Additionally, in many regions where flax is grown for oil, weather conditions do not permit postharvest fiber processing (retting), which requires a relatively wet fall climate. The 202,000 ha of fiber flax are mainly located in China, Russia, Egypt, and near the northwestern European coast (Vromans, 2006; Green et al., 2008). For oilseed production, the straw is considered an impediment during harvest and for subsequent seeding operations and until recently has been burned. However, the potential for alternative fiber uses such as industrial fibers for composites, paper, and nonwoven fiber increases the potential for whole plant utilization (Irvine et al., 2010). The majority of flax in North America is grown for the oil.

Flax Area and Production

Currently, oilseed flax is primarily grown in Canada, China, USA, India, and Russia (FAOSTAT, 2013). Canada is the world's largest producer and exporter of flaxseed, a position it has held since 1994 (Fig. 6.1). Canada's and Russia's role as a flaxseed producer has increased while that of India, Argentina, Europe, and the USA decreased since 1995 (Agriculture and Agri-Food Canada, 2010; FAOSTAT, 2013). In Canada, the annual production of flaxseed averaged 706,000 tons from 1999 to 2009. In 2009, flax was seeded on ~631,000 ha in Canada with production of ~861,000 ton (Agriculture and Agri-Food Canada, 2010). Canada exports 80–90% of its total flaxseed produced, mainly to Europe,

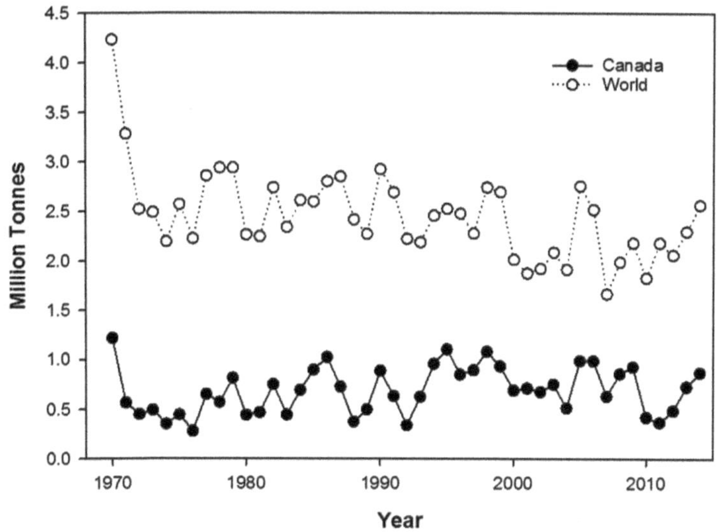

FIGURE 6.1 Global flaxseed production in million metric ton and the Canadian share of global production from 1997 to 2007. *Data adapted from the Food and Agriculture Organization (FAOSTAT data 2010).*

the USA, Japan, and South Korea (Flax Council of Canada, 2014a). The presence of transgene in Canadian shipments relates back to the mid-1990s introduction of GM flax variety CDC Triffid (McHughen et al., 1997). This variety was deregistered in 2001 due to concerns about the effect of its production on offshore markets. A decade after removal of transgenic flax from the commercial system in Canada, it was detected in grain shipments to Europe. This event resulted in disruption in trade between Canada and the European Union (EU). To demonstrate compliance the industry adopted a protocol testing grain samples (postharvest) using a real-time polymerase chain reaction (PCR) assay for the construct found in CDC Triffid. Since the inception of the testing protocol, the Flax Council of Canada has maintained a database of test results. Booker and Lamb (2012) evaluated GM presence in Canadian grain stocks and found the positive tests showed a downward trend, indicating removal of transgenic flax from the commercial system. However, low-level GM presence persists in Canadian grain stocks. A way forward for the industry in Canada includes renewal of seed stocks with reconstituted GM-free varieties (SaskFlax, 2013).

Origin and Domestication

The genus *Linum* is a large group with ~230 species (Heywood, 1993). The genus is divided into five sections, *Linum*, *Linastrum*, *Cathartolinum*, *Dasylinum*, and *Syllinum*, based on chromosome number, floral morphology, and interspecific compatibility (Gill, 1987). Cultivated flax, *Linum usitatissimum*, is placed in the section *Linum* and has 30 diploid chromosomes (Tammes, 1928;

Tutin et al., 1980) along with the proposed progenitors *L. angustifolium* and *L. bienne*, which share homostylous rather than heterosylous flowers.

The origin of flax lies in southern Europe, the Near East, or Central Asia (Tammes, 1925; Helbaek, 1959; Zeven, 1982). There is evidence of flax use in Neolithic cultures as a source of fiber. Flax fiber was identified in prehistoric sites in Israel and Syria (van Zeist and Bakker-Heeres, 1975). Flax was grown in Egypt between 4500 and 4000 BC and reached Switzerland by around 3000 BC; it was found somewhat later at Willium Hill in England (Smith, 1969). It is believed that Lois Hebert, the first farmer in Canada, introduced flax to Canada almost 400 years ago (Prairie Flax Products, 2007). By 1875, European settlers were growing flax on the Canadian prairies.

L. angustifolium, also known as pale flax, a closely related species of flax, is indigenous to the territory bordering the Mediterranean Sea, Iran, and the Canary Islands, and has a similarity to cultivated flax (Diederichsen and Hammer, 1995). Cultivated flax, *Linum usitatissimum*, and pale flax, *L. angustifolium*, are homostylous, inbreeding species and both species have similarity in chromosome number (Gill, 1987; Tammes, 1925). The genetically similar behavior of *L. angustifolium* and *L. usitatissimum* and the fact that they can easily hybridize with each other in any direction (male or female) and yield fertile hybrids (Gill, 1966) suggests that *L. angustifolium* is the wild progenitor of flax (Dillman and Goar, 1937). *L. usitatissimum* is differentiated from *L. angustifolium* by one translocation, while *L. africanum, L. corymbiferum*, and *L. decumbens* have departed from *L. angustifolium* in two translocations (Gill, 1966).

A molecular study indicates that *L. bienne* is the sister species of *Linum usitatissimum* (Vromans, 2006), but some researchers consider *L. angustifolium* and *L. bienne* to be the same species (Tutin et al., 1980; Zohary and Hopf, 2000). A more detailed molecular study, including genome comparisons with molecular markers of these three species (*L. angustifolium, L. bienne*, and *L. usitatissimum*), has confirmed that they are very closely related genetically (Muravenko et al., 2003). *L. bienne* can be considered as a subspecies of *L. usitatissimum* rather than a separate species, and *L. angustifolium* can be considered as a wild progenitor of cultivated flax. According to Richharia (1962), evolution of flax might be a case of convergent evolution. The interspecific hybridization of flax with several other wild relatives with $n = 15$ reported to occur in North America has not been studied or reported yet (Jhala et al., 2008). The history of domestication of flax has not been clearly delineated and may have occurred at several places simultaneously (Harlan, 1965). It is believed that the usage as a fiber plant has provided the impetus for its domestication (Dillman, 1936). Various characters of the cultivated flax including annual habit, nonshattering capsules, self-fertilization, branching habits, and variability in fatty acid profile are some of the examples of the breeding efforts and result of interaction of many inherited and environmental factors. The early and rapid distribution of flax cultivation in the Old World resulted in a wide range of flax landraces adapted to different use and environmental conditions (Vavilov, 1926).

Agronomy of Flax Production

Climatic Adaptation

Flax thrives best in regions with a moist, moderately warm climate—free of late frosts in spring, with sufficient moisture during growing period and where long, continued rainy spells do not alternate with long, continued dryness (Dillman and Goar, 1937). Flax is well adapted to western Canada where temperatures range from 10 to 25°C during growing season (Braidek, 1975). High temperatures in the absence of drought decreases seed set and reduces yield (Cross et al., 2003). In the United States, flax is primarily grown in north central states including North Dakota, South Dakota, Montana, and Minnesota.

Soil and Fertility Requirements

Flax production is optimal on well-drained, medium heavy soils, especially silty loam, clay loam, and silty clays (Braidek, 1975). In Canada, flax is adapted to the mixed grasslands and moist mixed grassland agro-ecological regions, except on poorly drained or sandy soils. Because of its shallow rooting character, flax extracts 95% or more water from the top 71 cm of soil. Water stagnant conditions may result in chlorosis and stunting of flax. Though flax plants are tolerant to a wide range of pH, the best flax development was recorded at pH 6.0 in sand culture (Sinha and Saxena, 1965).

Flax responds to increased nitrogen availability, but less so than cereals or canola (Grant et al., 1999). The optimum nitrogen requirements for flax are 100–120 kg (soil residual nitrate-N and inorganic fertilizer) (Lafond et al., 2008), depending on available moisture. Increasing nitrogen fertilizer from 67% to 100% increased seed yield, bolls/plant, seed weight, and seed oil content. Flax had higher seed nitrogen concentrations than crops such as cereals, but because of the reduced yield, nitrogen removal from soil tends to be lower. Seed-placed fertilizers may injure flax seedlings. A low rate of phosphate, <20 kg/ha of P_2O_5, may be seed-placed, but alternatively it may be placed to the side at a depth of 25 mm to improve phosphorus nutrition in the flax plant (Flax Council of Canada, 2014c).

Seed Selection, Time of Sowing, and Seeding Rates

Flax cultivar choices are limited compared to other crops (<15 varieties in Canada in 2010). Cultivar choices are made on the basis of yield, rust resistance, maturity, and lodging. Most varieties are resistant to rust and Fusarium wilt, and cultivar selection, in combination with crop rotation, has reduced the impact of diseases. Canadian growers predominantly use farm grown seeds, but following the recent contamination of flaxseed with a genetically engineered (GE) trait unapproved in Europe, the use of certified seed is now recommended.

Because flax is sensitive to spring frosts, seeding must be delayed until the risk of frost is reduced. Later planting was favored at most northerly Canadian sites and early planting at the most southerly sites (Lafond et al., 2008).

A seeding rate of ~35–45 kg/ha was recommended, which resulted in a crop stand of 300–400 plants/m^2.

Crop Rotation

In western Canada, the majority of flax is grown in rotation with cereals to prevent disease; at least 3 years is recommended between flax crops (Flax Council of Canada, 2014c). Yields of spring wheat were higher following flax than following canola or field pea (Bourgeois and Entz, 1996) and flax yields were higher after mycorrhizal wheat than nonmycorrhizal canola (Khakbazan et al., 2009). Flax should not be grown after legumes or potato because of the potential for infection from *Rhizoctonia* spp. (Johnston et al., 2002).

Crop Protection

Flax is less competitive than most western Canadian crops with respect to weeds, and weeds result in significant yield decreases (Lafond et al., 2008). Grass weeds can be relatively easily controlled (Wall, 1994a) but a limited number of herbicides are available for broadleaf weed control (Wall and Kenaschuk, 1996; Wall, 1994b). The lack of competition and herbicide tolerance are the reasons herbicide resistance was the first trait chosen for genetic engineering of flax (see Market Constraints to Genetically Engineered Flax). Insect pests of flax in western Canada include flax bollworm, *Heliothis ononis* (Denis and Schiffermiller); several species of cutworms including redbacked cutworm, *Euxoa ochrogaster* (Guenee), pale western cutworm, *Agrotis orthogonia* (Morrison), early cutworm, *Euxoa tristicula* (Morrison) and army cutworm, *Euxoa auxiliaris* (Grote); the bertha armyworm, *Mamestra configurata* (Walker), the beet webworm, *Loxostege sticticalis* (L.); the potato aphid, *Macrosiphum euphorbiae* (Thomas); the aster leafhopper, *Macrosteles quadrilineatus* (Forbes); and the tarnished plant bug, *Lygus lineolaris* (Palisot de Beauvois). With the exception of the flax bollworm, flax insect pests are polyphagous.

Fungal diseases of flax include rust caused by *Melampsora lini* (Ehrenb.) Desmaz.; fusarium wilt caused by *Fusarium oxysporum* Schlechtend.: Fr. f.sp. lini (Bolley) Snyder and Hansen; gray mold caused by *Botrytis cinerea* Fr.; alternaria blight caused by *Alternaria* spp.; powdery mildew caused by *Erysiphe polygoni* DC, *E. lini, E. cichracearum* DC, *Sphacelotheca lini, Leveillula taurica*, and *Oidum lini* Skoric; pasmo disease caused by *Septoria linicola* (Speg.) Garassini (telomorph). Fusarium wilt caused by *Fusarium oxysporum* f.sp. *lini* (Bolley) Snyder and Hansen continues to be a problem in Canadian flax fields (Mpofu and Rashid, 2001), but several varieties are resistant to rust. Cultivar selection for resistance to pasmo, caused by *Septoria linicola*, has been less successful. *Septoria linicola*, attacks above-ground parts of flax and overwinters in the soil on infected flax stubble. Flax is most susceptible to pasmo in the ripening stage under moist conditions. Prevention may be achieved by early seeding at the recommended rates to avoid high moisture conditions in the fall, following a rotation of at least 3 years between flax crops.

Flax is also susceptible to several viral diseases (Rashid, 2003). Viral diseases include crinkle, caused by oat blue dwarf virus (OBVD), and curly top, caused by the beet curly top virus (BCTV). Crinkle is transmitted by the six-spotted leafhopper, *Macrosteles fascifrons* (Uhl.), and results in stunting and reduced tillering. Curly top is transmitted by the leafhopper *Eutettix tenellus* (Baker) and results in general chlorosis and irregular leaf development where terminal leaves are grouped together.

Aster yellows is a disease caused by a phytoplasma, which is transmitted by the six-spotted leaf-hopper, *Macrosteles fascifrons* (Uhl.). Symptoms include stunted growth and abnormal flower development where floral parts are transformed into small leaves. It is beyond the scope of this chapter to discuss the control of insects and other pests of flax, and, thus, readers are directed to the Flax Council of Canada (2014c) and Rashid (2003).

Volunteer Flax

The small size of flaxseeds and difficult harvest conditions can result in high seed losses, as high as 6822 seeds/m^2 with the maximum flax yield loss of 331 kg/ha (Dexter et al., 2011). Volunteers emerge sooner following spring tillage than in direct seeding and are more numerous. Volunteer flax can persist and volunteers emerge for 2 years following flax harvest in western Canada (Dexter et al., 2010b).

Volunteer flax is considered a poor competitor, but it can interfere with harvesting operations, especially at high densities and can reduce seed quality (Knott, 1993); it may also contribute to the seed- and pollen-mediated gene flow (Dexter et al., 2010a,b; Jhala et al., 2008, 2011). Although the growing area of flax has not changed significantly over the past few years, the relative abundance of volunteer flax has increased from 2.0 to 15.3 m^{-2} in western Canada (Thomas et al., 1997). Volunteer flax was ranked as the 32nd most abundant weed on Canadian Prairies in the 1970s, but near the beginning of the 21st century it was the 26th (Leeson et al., 2005), indicating that it is becoming a problem in subsequent crops. In the 1997 Manitoba survey, volunteer flax was present in twice as many zero tillage fields, but at much lower average densities, compared to conventional tillage (Thomas et al., 1997), consistent with a recent small survey in Alberta (Dexter et al., 2010a). When volunteer flax is left uncontrolled, wheat yield may be reduced by 27% (Wall and Smith, 1999). In western Canada, a number of herbicides have been registered for volunteer flax control. Quinclorac provided good control of volunteer flax. Pre-emergence application of glyphosate reduced volunteer flax densities from 39 plants to <4 plants/m^2, while postemergence application of fluroxypyr plus MCPA or 2,4-D ester reduced volunteer flax densities to 2 plants/m^2 (Dexter et al., 2010a). In glufosinate-resistant canola, postemergence application of glufosinate effectively controlled volunteer flax (Jhala et al., 2010).

FLAX GENETICS AND BREEDING

Flax is a diploid ($2n=30$) autogamous (primarily self-pollinating) crop plant. Improvement in flax has lagged behind other oilseeds, including canola and soybean, in part because flax occupies a smaller niche as an oilseed and, consequently, received reduced resources for development. Genetic diversity within the crop is low, and cannot be readily supplemented by intraspecific hybridization. Finally, methods of hybrid seed production have not been developed.

Inheritance of crop characteristics can range for one major gene not influenced by environment (qualitative traits) to control by many genes, much influenced by the environment (quantitative traits). Many important traits in flax including days to flowering, days to maturity, plant height, plant branching and seed yield are quantitative, and demonstrate additive gene effects that are predominant, as in other autogamous crops (Patel and Chopde, 1981; Salej et al., 2007; Bhateria et al., 2006; Mohammadi et al., 2010). Selection of parents and in early generations is therefore critical in flax breeding programs.

Breeding Objectives

Conventional breeding of flax has focused on stabilizing yield across diverse environments, enhancing oil production and quality, discovering durable resistance to major wilt and rust diseases (Kenaschuk and Rowland, 1993; Mpofu and Rashid, 2001), improving lodging resistance, and adopting crop phenology for regional climatic limitations.

Seed Yield

Seed yield is a quantitative trait that is the most important in an oilseed flax breeding program. Numerous crop characteristics and environmental factors may influence seed yield, but little is published on basic crop characteristics of flax that affect yield, such as canopy expansion and light interception, dry matter production, and partitioning. During the reproductive phase, light use efficiency and harvest index are correlated with grain production under favorable growing conditions (D'Antuono and Rossini, 1995). Selection for seed yield is normally delayed until later generations when selection for complex traits is more effective.

Maturity

Early maturity is an important trait that protects crops from stresses such as disease, heat, drought, and frost. Early maturity aids harvest as immature flax remains green, which entangles machinery and reduces grain quality. The time from flax seeding to harvest can vary between 90 and 150 days (Diederichsen and Richards, 2003). Flax is considered a long day plant, where lengthening days hasten the reproductive phase of development, but the degree of photosensitivity in flax varies greatly.

Maturity is quantitatively inherited and little is known about the genetic basis of early maturity in flax. There is evidence, however, that major genes under epigenetic control contribute to the expression of early flowering. Gene methylation silences gene expression (reviewed in Suzuki and Bird, 2008 and Gehring and Henikoff, 2007) so that methlyated genes are transcriptionally inactive, while genes that are not methylated remain active. Early flowering flax genotypes from the *Royal* (*R*) genotype (an oil seed cultivar) and *Large* (*L*) genotroph derived from Stormont cirrus, a fiber cultivar (Durrant, 1971), were generated when germinating seedlings were treated with 5-azacytidine (Fieldes, 1994). The stable, heritable changes in early flowering lines resulted in shorter plants with fewer leaves that flower 7–13 days earlier than untreated control lines. Early flowering lines have a reduced late-vegetative phase (Fieldes and Harvey, 2004) and their DNA is hypomethylated compared to the untreated control lines (Fieldes et al., 2005; Brown et al., 2008). Fieldes and Amyot (1999) proposed a three loci genetic model to explain the phenotypic ratios observed in segregating generations of crosses between early flowering lines RE1 and LE1 and their corresponding control lines (RC and LC). For the early flowering phenotype to be observed, they proposed that all three loci must be demethylated. Furthermore, because all three loci are demethylated as a result of heritable effects of 5-azacytidine, only a critical threshold of floral stimulus is required to induce flowering.

Disease Resistance

Most early efforts in flax improvement in North America were aimed at breeding disease resistance (Kenaschuk and Rowland, 1993), and the most practical method of controlling disease remains the development and use of resistant varieties, combined with effective crop rotations.

The gene-for-gene theory of host–pathogen interactions was formulated by Harold Flor (1942) in flax using flax rust, *Melampsora lini*, as a model. For each host gene conferring resistance, he demonstrated the presence of a complementary gene in the pathogen determining avirulence. The disadvantage of this type of specific resistance/immunity is that the new pathogen races may evolve to overcome host resistance. However, there are a number of genes (*K*, *L*, *M*, and *P*) that may confer rust resistance in flax. Canadian varieties have alleles of rust resistance genes singly or in combination (S. Duguid, K. Rashid, personal communication). Hausner et al. (1999a,b) identified DNA molecular markers for specific alleles. All Canadian flax varieties are immune to local rust races and immunity to race specific rust has held up since the 1970s, and there has been no reoccurrence of flax rust in Canadian flax fields.

Fusarium wilt caused by *Fusarium oxysporum* f.sp. lini (Bolley) Snyder and Hansen is a limiting factor in flax production. Seedlings may be killed by the disease shortly after emergence, while delayed infections causes yellowing and wilting of leaves (Bailey et al., 2009). The fungus persists in the soil and crop debris for many years and wind-blown or run-off soil, or spores on seed may spread the fungus from one field to another. Canadian flax varieties for release in Canada must be resistant/moderately resistant to fusarium wilt; however, it has been very difficult to breed for in yellow seeded flax varieties (P. Dribnenki, personal communication).

Further characterization of flax (host) wilt (pathogen) interactions with a wide diversity of flax lines and fungal isolates and development of DNA markers for host resistance genes would speed progress for breeding for resistance to fusarium wilt in flax. However, screening varieties for fusarium wilt resistance in flax wilt nurseries is problematic. There are substantial differences in disease development and severity between years and fields due to differences in pathogen population structure, resulting variable results (Mpofu and Rashid, 2001). Diversity in tolerance to fusarium wilt has been reported (Diederichsen et al., 2008) with higher than average resistance for in flax accessions from East Asia, average resistance in those from North America and South America, and lower than average resistance in those from Europe (mostly fiber flax) and the Indian subcontinent. Spielmeyer et al. (1998a) demonstrated that two major quantitative trait loci (QTL) with additive effects accounted for 38% and 26% of the genetic variability for fusarium wilt resistance. Using amplified fragment length polymorphism (AFLP), they mapped the QTL to two linkage groups (Spielmeyer et al., 1998b). Early generation screening is most effective to incorporate major wilt resistance.

Pasmo disease caused by *Septoria linicola* (Speg.) Garassini is prevalent and widespread on flax in Western Canada (Rashid and Duguid, 2010). Most Canadian flax cultivars are susceptible and little is known about the pathogenicity of different *S. linicola* isolates. Rapid disease development is exhibited in areas with high humidity and temperature and yield of susceptible varieties infected during flowering can be reduced by 75% (Saskston, 1959). *S. linicola* survives from one season to the next on infected crop residues left in the field (Bailey et al., 2009). Field sanitation methods, such as tillage, are not advisable due to conflict with soil conservation (no till) practices.

Powdery mildew was first observed in Western Canada in 1997 (Bailey et al., 2009). Powdery mildew can spread quickly and its incidence and severity have increased sharply in Manitoba and Saskatchewan. The causal agent is the fungus *Oidium lini*. Little is known about the overwintering and host range of this fungus in western Canada. Early infections may cause severe defoliation of the flax plant and reduce the yield and quality of seed (Bailey et al., 2009). Certain Canadian flax varieties have demonstrated a good level of resistance to this disease and flax breeders are encouraged to incorporate and strive to maintain resistance to powdery mildew in new flax varieties.

Agronomic Traits

Lodging resistance and plant height are important traits in cultivar development. Lodging reduces seed yield in flax and may hinder harvesting of the crop.

Lodging in flax may be associated with incidence of pasmo (Vera et al., 2011). Plant height at maturity and lodging resistance are considered quantitative characters and are selected for using field trials in the target environment.

Seed Size

Seed size is an important trait in flax. Large seeds offer advantages such as superior vigor in germination and seedling establishment that translate into greater capacity for yield. Larger seeded varieties are less prone to losses during harvest and seed cleaning. There is potential to accumulate more oil in a larger seed as the embryo is relatively larger. This is a particularly important characteristic for flax, where the seed coat constitutes nearly 40% of the seed mass, but contributes very little to the oil content. Global flax collections (eg, Plant Gene Resources Center) (PGRC, 2011) contain flax lines with large seed size, and genetic crossing results in large seed progeny. Inheritance of seed size is quantitative and selecting from the extremes of a population distribution may result in lines with the targeted seed size (Mohammadi et al., 2010). However, producing a large seed cultivar that retains the field performance of elite varieties is challenging, as the genetic basis that underpins the yield is multifaceted. Until DNA markers for seed size are available for introgressing large seed loci into elite varieties, the only feasible way to select for large seed size together with yield is in later generations using field trials.

Seed Constituents

Seed oil and protein are highly heritable traits in flax and can be selected for in early generations (see Oil Composition and Properties). Canadian flax breeders aim for 50% oil and 25% protein content on a dry matter basis.

Seed color in flax was used in Canada to identify traditional high-ALA acid flax (brown seed) and low-ALA flax (yellow seed) Solin market types (see Mutation Breeding). The Canadian Grain Commission (2013) has removed the requirement for yellow seed coat to be a phenotypic marker for the Solin oil profile. This requirement can be removed because the two traits are not linked and randomly assort, allowing the yellow seed trait to be used for other oil profiles in flax. Mittapalli and Rowland (2003) elucidated the allelic–gene relationship of the dominant yellow gene, the variegated recessive gene, and unknown spontaneous recessive yellow genes in flax. Segregation analysis indicated seed color is governed by four independently inherited loci, designated as the dominant *Y1* allele and recessive *g*, *d*, and *b1* alleles. The variegated seed color is conditioned by a second recessive allele of the *b1* locus, designated as *b1vg*. The recessive genes (*g*, *d*, and *b1vg*), when homozygous recessive, are epistatic to the other loci carrying dominant alleles. Palmitic and variegated seed loci segregated independently (Saeidi and Rowland, 1997) and thus Solin varieties could be developed for both high palmitic acid and variegated seed color.

Conventional Breeding Methods

Breeding in autogamous crop plants involves creating genetic variability, selecting the best recombinants and fixing the genes by making them homozygous. Homozygous varieties are called pure-line varieties and are true breeding. They are productive in the environments in which they were bred, but may not perform well outside these environments. Breeding in flax predominantly involves producing pure-line varieties through pedigree breeding.

Selection of Parents

Selection of parental material is of paramount importance in a flax breeding program. Selection of parental germplasm begins with an understanding of the goal of the breeding program (see Breeding Objectives) and the plant breeder must consider many characters for improvement and assign priorities to the traits considered for improvement. The way parents are selected depends on the trait(s) of interest, the purpose of the cross, the relative importance of characters other than yield, the pedigree of the pure-lines, and the resources and time available. Greater genetic diversity between the parents, results in greater genetic variability and likelihood of transgressive segregants, progeny that combine the genes of parental genotypes and have a phenotypic expression level superior to either parent.

Genetically diverse parents with complementary characteristics are chosen to create a segregating population. The parental lines can be from many different sources including existing cultivars, adapted elite breeding lines, or new introductions from global flax collections. The Plant Gene Resources Canada (PGRC, 2011), Saskatchewan, provides access to more than 3300 accessions of *Linum* L. germplasm, one of the larger collections of flax in the world (Diederichsen and Fu, 2008). A core world collection of 500 distinct flax accessions has been selected, based on genetic diversity studies (Fu, 2006), phenology, and morphological and agronomic traits (Diederichsen and Raney, 2006; Diederichsen et al., 2006, 2008; Diederichsen and Ulrich, 2009), as well as importance to Canadian agriculture (Kenaschuk and Rowland, 1993).

Inbreeding Selection and Line Evaluation

Pure-Line Method

Commercial flax cultivars are pure-lines developed via hybridization of inbred lines with complementary traits followed by pedigree selection of recombinant lines from segregating progeny (Culbertson, 1954; Kenaschuk, 1975). The aim is to obtain a pure-line cultivar that combines all the genes and meets the breeding objectives.

Pedigree Method of Breeding

Pedigree selection method begins with the intercrossing of inbred lines, followed by selfing of the F_1 hybrid plants to produce a large F_2 seed population.

F_2 seed from individual crosses is sown the following growing season and single plants are harvested from F_2 plots. Seed harvested from single plants planted in F_3 progeny rows. Progeny row selections within F_3 families based on stand maturity and vigor are made and single plants are harvested, along with bulk seed from the row. Seed quality traits such as oil quality, oil content, and protein content are determined using near infrared resonance or gas chromatography from samples taken from the bulk seed lots, and data used to further scale down the population of desirable lines. Seed harvested from F_3 single plants is used to establish F_4 rows and the selection cycle of between sister-lines and within-family selection continues and is repeated in the F_5 generation. In the F_5 generation, the bulk seed is used to establish replicated small yield plots to facilitate between line selections for seed yield, and the single plant selections are grown as progeny rows. Superior F_5 lines are identified on the basis of yield and seed quality attributes in the replicated plots, and sub-lines of these are used to establish the subsequent F_6 generation. Selected pure-lines are advanced into multilocation yield trials to compare their performance and adaptability with that of existing cultivars. In the F_6 generation, the bulk seed from F_6 rows is used to establish small increase plots for the development of breeder seed.

Green et al. (2008) describe a disease resistance screening program for flax wilt where large populations of F_2 plants are grown in a field on soil that is heavily infected with *Fusarium oxysporum*, the causal organism of flax wilt. The selected F_2 plants are then retested as F_3 progeny rows to identify homozygous wilt-resistant lines. In parallel, F_3 plants may be inoculated with specific strains of flax rust (*Melamspora lini*) under glasshouse conditions to identify rust-resistant lines. This method is limited by available space and time, and therefore disease-resistance screening generally consists of F_6 rows planted in wilt nurseries. Later-generation pure-line plants under glasshouse conditions may also be inoculated with specific races of flax rust (*Melampsora lini*) to identify and eliminate compatible lines. Disease-resistance screening in later generations is less than ideal as segregants with major resistance genes may not be carried forward in the breeding program, resulting in advanced breeding lines without the major wilt and rust resistance genes. The further development of DNA markers for major wilt and rust resistance genes will allow breeders to screen in the F_2 and F_3 generation for homozygous wilt- and rust-resistant lines, thus circumventing the reliance on disease-resistance phenotyping in the fieldhouse or glasshouse.

Bulk Population Method of Breeding

The bulk population method of selection is amenable to small grain crops that cannot be harvested singly, which is a necessary requirement of the pedigree selection method. Selection is delayed until later generations when selection for more complex characters (eg, seed yield) is more effective. In the bulk method, a segregating population is harvested together and a sample of the seed is used to plant the next generation. For example, the F_2 population is planted and seed harvested from the entire population and bulked. Seed is randomly obtained from the bulked seed population and a plot of F_3 is seeded. The F_3 plot is similarly harvested and a plot of F_4 planted. This is continued until the F_6 generation. Selection begins in the F_6 generation, which is largely homozygous. Large populations of F_6 need to be planted for selection to be effective at this stage. Two methods of selection can be practiced at this stage: either pure-line selection (single plant selections, plant to row progenies, selections among homozygous lines) or mass selection (large number of individual plants selected and bulked and carried forward). The bulk method can produce lines with the desired traits if the environment where the method is conducted favors those characteristics (Orf, 2008).

Single-Seed Descent

Goulden (1939) first proposed single-seed descent as an alternative to pedigree and bulk breeding methods. He noted that breeding in self-pollinated crops involves two processes: achieving homozygosity and selection of traits with high and low heritability. He noted that in the pedigree method both processes are conducted simultaneously, resulting in characters with low heritability (such as seed yield) not selected for efficiently. Single plant selection (plant to row progenies), however, is an ideal way to identify transgressive segregants and achieve homozygosity, whereas in the bulk breeding method, selecting for rare transgressive segregants is not possible and bulking seeds is not an efficient way of achieving homozygosity.

In single-seed descent, attaining homozygosity and selection are separated into two different phases. Brim (1966) described single-seed descent as a modification of the pedigree selection method that allows accelerated development of recombinant inbred lines (RILs). RIL populations are derived from single-seed descent of large F_2 populations involving crosses of diverse flax lines to a high yielding and adapted cultivar. Single bolls/seeds are retained from each plant to produce subsequent generations, which were similarly advanced by single-seed descent to an advanced generation (eg, F_6). The generation cycle time can be reduced by growing the single-seed descent generation in the growth chamber under conditions of high plant density and restricted nutrition, which allows for three generations per year. This selection technique allows homozygosity to be achieved in a short time and the fixing of traits within the RILs. No selection is conducted within a population, so the RIL populations segregate (differ) for the traits in the contrasting parental lines. Selection can begin in the F_7 generation when homozygousity is achieved and additive gene effects are high. Furthermore, the variation in the F_2 is represented in the F_7 through single-seed descent. The F_7 can be grown as plant to row progenies. Undesirable lines can be removed. Selected F_8 lines can be grown in multirow plots and preliminary yield tests conducted.

RIL populations are ideal for the genetic study of traits and the genetic mapping of DNA markers of differing traits in the RIL population. Estimation of the genetic and environmental components of traits is possible when the RIL population is grown in different environments, across years, when genotype by environment interactions are integrated in QTL analysis, thus resulting in DNA markers reflecting the amount and direction of reaction of the plant to environment (Van Bueren Lammerts et al., 2010).

Backcross Breeding

Backcross breeding is used to incorporate simply inherited traits from unadapted donor parents into recipient lines and involves repeated cycles of crossing to the recipient line (recurrent parent), followed by selection of the trait being transferred (Kenaschuk, 1975). Backcross breeding was first used in flax to develop a set of rust differentials by transferring individual rust resistance alleles into the cv. Bison (Flor, 1955). These rust differentials have been extensively utilized in flax rust research and as a source of resistance genes for backcrossing into flax varieties to overcome their susceptibility to new races of flax rust (Green et al., 2008). Backcross breeding plays an important role in transfer of novel alleles from mutant populations into elite recipient lines. Backcross breeding was used to intogress the (*ln1* and *ln2*) and a yellow seed color marker gene into Canadian flax cultivars McGregor and Norlin, to develop Linola cultivars (Dribnenki and Green, 1995; Dribnenki et al., 1996) and CDC Gold.

Recurrent Selection

Recurrent selection is a variation of backcross breeding, where selection for performance is practiced within consecutive segregating progeny generations after the population has undergone selection for the major trait being transferred (Green et al., 2008). A number of high-performing lines are crossed back to the recurrent parent to initiate another cycle with up to five cycles of hybridization back to the recurrent parent. This approach allows for the transmission of the major trait, while providing the possibility of increasing overall performance above that of the recurrent parent through contribution of favorable minor genes from the donor parent.

Early-Generation Testing

Early-generation testing is designed to identify bulk hybrid populations so that only families that have the greatest potential are carried forward in the breeding program. For the pedigree method to be effective, it is necessary that selection be confined to visible characteristics with high heritability in early generations. A large number of F_2 progeny rows must be carried forward to the F_6 generation, so that poorly heritable characters, for example, seed yield, can be selected at a later generation. Early-generation testing and a modified form of recurrent selection are being tested to improve seed yield in flax. Using field trials, bulk F_3 populations are tested. $F_{3:4}$ plants from the highest yielding families are selected from increase plots and grown in an off-season nursery as single-plant progeny rows. F_5 seed from harvested rows is used to grow preliminary yield trials. The superior lines are further evaluated using field trials and the best lines are intercrossed to begin another breeding cycle. This method of selection is designed to improve the frequency of favorable alleles (for the trait in this example, seed yield undergoing selection) in a population.

Pollen-Mediated Gene Flow

While flax is considered primarily autogamous, pollen-mediated gene flow may occur and allow gene movement between pure breeding lines (Jhala et al., 2009). Pollen-mediated gene flow was quantified between two cultivars of flax in western Canada using the xenia effect of dominant alleles of high-ALA to the low-ALA trait as a marker (Jhala et al., 2011). Frequency of gene flow was highest near the source: 0.0185 at 0.1 m but declined rapidly with distance. Outcrossing was reduced by 50% (O_{50}) and 90% (O_{90}) between 0.85 and 2.64 m and 5.68–17.56 m, respectively. No gene flow was detected at any site or year >35 m distance from the pollen source. Low levels of gene flow are unlikely to be problematic for seed growers but must be considered when isolation distances are established within breeding programs.

Male Sterility

Male sterility is the inability of the plant to produce functional anthers or pollen, through chromosomal aberrations, gene action, or cytoplasmic influences. Plant breeders make use of cytoplasmic male sterility and pollen fertility restorer genes to take advantage of hybrid vigor and superior performance of the hybrid over the parental lines due to heterosis. Many components critical for commercial hybrid production (Palmer et al., 2001) are lacking for flax: a stable male-sterile, female-fertile sterility system; a selection system to obtain 100% female plants that set seed normally and can be mechanically harvested; and an efficient pollen transfer mechanism from pollen parent to female parent. Considerable research is needed before an efficient, productive, and cost-effective system of hybrid seed production could be available in flax (Green et al., 2008).

Mutation Breeding

The primary strategy in mutation-based breeding has been to upgrade elite lines by altering one or two production-limiting or quality traits (Ahloowalia et al., 2004; Rowland and Bhatty, 1990) using radiation or chemical mutagenesis. Mutation breeding in flax led to the development of a new type of edible flaxseed oil that has nearly eliminated ALA (Green and Marshall, 1984; Rowland, 1991). The deficient ALA trait is known to be controlled by two recessive genes (*ln1* and *ln2*) at independent loci (Green et al., 1984; Green, 1986c; Rowland, 1991; Ntiamoah and Rowland, 1997). Vrinten et al. (2005) reported the cloning of two members of the *FATTY ACID DESATURASE* (*FAD3*) family of microsomal Δ15-FADs, designated *LuFAD3A* and *LuFAD3B* (to be discussed further under Flaxseed Oil). Mutation breeding has a much greater potential to enhance the available traits in flax, including further modification to fatty acid profiles and the reduction of bast fiber content (McKenzie, 2011). Mutation breeding may have potential in the future to remove undesirable compounds from flaxseed, such as cyanogenic glycosides and linatine (Green et al., 2008).

Use of Genetic Markers in Flax Breeding

The development of genetic markers in flax has not kept pace with other oil seed crops. Rapid amplification of polymorphic DNA (RAPD) markers has been used to assess the genetic diversity of North American flax lines (Fu et al., 2003a) and to test for the presence of off-types in breeding lines (Fu et al., 2003b; McKenzie, 2011). The number of available RAPD was limited and their usefulness was limited. Recently expressed sequence tags-single sequence repeats (EST-SSR) markers for flax have been developed (Roose-Amsaleg et al., 2006). Cloutier et al. (2009) reported the development of 275 polymorphic EST-SSRs which can be used to develop physical maps, QTL mapping, and potential to decrease the time for cultivar development. Following development of these markers, a consensus genetic and physical map using DNA markers in combination with the recent sequencing of the flax genome (http://www.phytozome.org/flax.php; Wang et al., 2012) was reported (Cloutier et al., 2012). This map was generated using three different mapping populations (CDC Bethune/MacBeth, E1474/Viking, and SP2047/UGG5-5). The consensus map has 670 DNA markers anchored to 204 of the 416 fingerprinted contigs of the physical map and the overall map density is 2.0 cM for markers arranged on 15 linkage groups. Genome-wide SNP discovery in flax is now possible using genotyping-by-sequencing approach that can be used to identify QTLs/regions of the genome linked to novel genes in this crop plant for disease resistance, seed and fiber yield, and traits of importance to human and animal health (Kumar et al., 2012). These genomic resources will be available for research purposes.

Market Constraints to Genetically Engineered Flax

Flax was among the first crop species to be genetically engineered to insert genes of potential agronomic value. Herbicide resistant traits have been expressed in flax including chlorsulfuron and metsulfuron methyl resistance (McSheffrey et al., 1992), glufosinate-ammonium (McHughen and Holm, 1995), and glyphosate resistance (Jordan and McHughen, 1988). The first GE flax cultivar "CDC Triffid" was registered in Canada in 1997–98 and was approved for food and feed and environmental release (McHughen et al., 1997). In 2001, prior to broad-scale release, this cultivar was deregistered at the request of the flax industry to protect the Canadian flaxseed export market to European countries (McSheffrey et al., 1992). In 2009, traces of CDC Triffid were found in Germany (Bedard, 2009). The Flax Council of Canada (2014b) has detected widespread, low-level presence of GE flax in commercial flax stocks and by 2014 industry stewardships programs were successful in clearing most of the CDC Triffid from the Canadian flax system. (Reference Flax Council of Canada: http://flaxcouncil.ca/flax-council-of-canada-stewardship-program-update/). The source(s) of contamination have not yet been determined but are not associated with a single point source. Consumer and political concerns about GE crops for food and feed continue to be pervasive in the EU, and this incident temporarily halted exports of Canadian flaxseeds to key markets including the EU and Brazil (Flax Council of Canada, 2010). Since identification of CDC Triffid, extensive flaxseed testing has been instituted in Canada: prior to planting, post-harvest, at initial receptor sites (elevators), and at grain terminals prior to export. The majority (>80%) of grain exports to the EU take place from Thunder Bay, and in 2011, 5% of flax samples were positive for the presence of CDC Triffid. Testing of adventitious presence, a critical element of regulatory compliance, is confounded by the practical level of detection of PCR assays (0.01% or 1 GE seed in 9999 conventional seeds) and the large sources of error inherent in taking representative and random samples in large seed lots. The testing protocol implemented requires collection of a 2-kg sample of any flax entering the handling system, and testing of four 60-g subsamples for the presence of GE flax (Canadian Grain Commission, 2010). The sampling protocol presumably gives a 95% probability (or 5% error) of detecting 1 GE seed in 9999 non-GE flaxseeds (Remund et al., 2001; Whitaker et al., 2001). Lamb and Booker (2011) demonstrated, however, that low levels of presumed contamination are indistinguishable from the number of positive tests expected from a clean seed lot given the potential rates of false positives. This finding has significant implications for the testing of flaxseed lots for the presence of GE seed. While testing will be required for the foreseeable future to reduce the risk of product rejection, zero tolerance in the bulk grain handling system is neither verifiable nor achievable technically. The reader is referred to Jhala et al. (2009) for a more in-depth discussion of the potential benefits and risks, regulatory aspects, and gene-flow issues associated with GE flax.

FLAXSEED OIL

The vegetable oil obtained from flaxseed has been used for many centuries, and flax cultivation dates back to a few millennia (Henriksen and Robinson, 1996). Phylogenetic analysis has indicated that flax was initially domesticated for oil instead of fiber (Allaby et al., 2005). This ALA-rich oil has unique properties that are useful for food and non-food applications. Food-grade oil can be obtained by cold-pressing and commercialized as "flaxseed oil." Alternatively, extraction with solvents can be performed when the oil is used for nonfood industrial applications. In this case the oil is often referred to as "linseed oil" and can be used in the production of paints, varnishes, and linoleum flooring. Recently, considerable attention has also been devoted to flaxseed as a functional food ingredient due to reported health benefits (Goyal et al., 2014). The specific components, which contribute mostly to these benefits, have not been completely elucidated, but most of the research has focused on oil components, lignans, and soluble polysaccharides.

Oil Composition and Properties

The oil content of flaxseed ranges between 35% and 45% (w/w), although higher values have been reported (Western Canadian flaxseed, 2010). The main component of most vegetable oils is triacylglycerol (TAG), which accumulates during the maturation stage of seed development and is used by the plant as a source of reduced carbon to support seedling growth after germination. The chemical composition of TAGs from flaxseed oil is well characterized (Vereshchagin and Novitskaya, 1965). The proportions of the five major fatty acids, namely ALA, palmitic, stearic, oleic, and LA, found in flaxseed oil are shown in Table 6.1. ALA can accumulate to levels as high as 64% depending on the cultivar and growth conditions (Green and Marshall, 1981). It has been well documented that environmental factors such as soil and climate can influence fatty acid composition of seed oils. Oleaginous crops cultivated at lower temperatures usually accumulate higher levels of polyunsaturated fatty acids (PUFAs) at the expense of saturated and monounsaturated fatty acids (Green, 1986b; Deng and Scarth, 1998). This effect is evident when comparing flaxseed cultivated in North Dakota with similar varieties grown in western Canada (Hettiarachy et al., 1990).

The high content of ALA influences some physicochemical properties of vegetable oils. For example, flaxseed oil with high ALA content has a higher specific gravity and lower melting and flash points than the oil of a determined flax cultivar with low-ALA content (Przybylski, 2005). The unsaponifiable matter composes a minor fraction of vegetable oils. However, it contains compounds that play an important role in preventing lipid oxidation, which affects the flavor and nutritional quality of the oil (Rathjen and Steinhart, 1997). Among such compounds are phytosterols and antioxidants (tocopherols, tocotrienols, and plastochromanol-8) (Table 6.1). Flaxseed oil contains a similar amount of unsaponifiable components compared to other vegetable oils. It is characterized, however, by relatively low amounts of total tocopherols and high contents of $\Delta5$-avenasterol, tocotrienols, and plastochromanol-8 (Velasco and Goffman, 2000; Ahmed et al., 2005). Plastochromanol-8 is a potent antioxidant (Olejnik et al., 1997) and might play a role in the stability of flaxseed oil.

Processing of Flax Oil for Industrial Applications

A number of vegetable oils have been traditionally used in industrial coating formulations for their unique fatty acid compositions. These include oils from

TABLE 6.1 Properties of Flax and Linola™ Oils

Parameter	Flax Oil	Linola™ Oil[a]
Relative density (20°C/water at 4°C)	0.93–0.94	0.92
Refractive index (n_D 20°C)	1.48	1.47
Viscosity (cp)	–	46.4
Iodine value	182–203	144
Unsaponifiable matter (%)	0.1–1.7	0.6
Saponification value (mg HOH/g)	187–195	185
Free fatty acids (% oleic)	0.1–2.0	<0.02
Triacylglycerol (%)	94–98	96–98
Fatty Acid	**%, wt**	
14:0	Trace	Trace
16:0	5.0–7.0	5.6–6.1
16:1	Trace-0.2	–
18:0	2.0–6.0	3.8–4.0
18:1 (*cis*Δ9)	17.9–19.0	15.5–15.9
18:2 (*cis*Δ9,12)	14.7–24.1	71.3–71.8
α-18:3 (*cis*Δ9,12,15)	47.4–58.7	2.0
20:0	Trace-0.4	–
Sterols	**%, wt**	
Brassicasterol	1	1
Campesterol	27	23
Stigmasterol	8	4
β-Sitosterol	50	54
Δ^5-Avenasterol	10	18
Total sterols (g/kg)	2.3	2.2
Tocopherols	**ppm**	
α-Tocopherol	20	15
γ-Tocopherol	200	200
δ-Tocopherol	7	5
Plastochromanol-8	120	110
Total tocopherols	347	330

[a]*Refined, bleached and deodorized.*
Adapted from Przybylski, R., 2005 Flax oil and high linolenic oils. In: Shahidi, F. (Ed.), Bailey's Industrial Oil and Fat Products. John Wiley & Sons, Inc., Hoboken, NJ, pp. 281–298.

tung, perilla, oiticica, castor, and flaxseed (Derksen et al., 1995). The high degree of desaturation of flax oil makes it highly susceptible for oxidation at high temperatures, leading to cross-linking and polymerization reactions. As a result, it forms a soft and durable film upon air exposure, a property that is advantageous for paints and varnishes. In addition, flax oil turn paints more fluid, transparent, and glossy. To increase its binder properties, flax oil is refined to remove lecithins and free fatty acids. The refining process provides partial oxidation of double bonds and initiates polymerization, resulting in high viscosity and decreasing the time required for drying. Depending on the application, different refining processes can be used to achieve the desired balance of viscosity and drying time. For example, to improve flow of oil based paints, flax oil is "boiled" at temperature of 137.5°C in the presence of zirconium, manganese, or cobalt until the specific gravity reaches 0.942 at 15.6°C. Using this process, flax oil viscosity increases from 40 cP to 80–120 cP, while the drying time decreases from few days to 12–20 h. To improve some physicochemical properties of highly pigmented paints (eg, for undercoats and wall paints), "blown oil" is prepared by increasing the degree of polymerization, removing the metal components previously mentioned, thus increasing the viscosity to >3 P and reducing the drying time to 24–36 h. Processed flax oil can be also be used as an ingredient of ink formulations. In this case, "stand oil" is prepared by increasing the temperature to 290°C under air exposure until the viscosity reaches ~200 P. Flax oil is also used for the production of linoleum flooring, leather-imitation, among other products (Jhala and Hall, 2010; Rowland et al., 1995).

Development and Uses of Low–α-Linolenic Flaxseed Varieties

Despite of many uses of flaxseed oil, the high ALA content renders low oxidative stability at high temperatures, making this oil unsuitable for cooking. To expand the potential markets for flaxseed, low-ALA flax varieties were developed by the Commonwealth Scientific and Industrial Research Organization (CSIRO) of Australia in partnership with the United Grain Growers Ltd. (currently Viterra, Canada) using traditional breeding techniques (Green, 1986a). These lines, which are known as Linola™, or Solin in Canada, have a yellow seed color that helps distinguish them from the traditionally brown flaxseed. Low-ALA varieties are deficient in the enzyme that converts LA to ALA (to be discussed in the next section), with ALA composing not more than 2% of total seed fatty acids. Consequently, LA accumulates at high levels of ~70% (Table 6.1), which is comparable to other high-LA oils such as sunflower and safflower. This vegetable oil has increased oxidative stability and low-ALA varieties can be used in conventional oilseed processing for seed crushing and oil refinement (Green and Dribnenki, 1994). Deodorized Linola™ oil has comparable flavor quality to canola oil and has been approved by the Food and Drug Administration for use as cooking and salad oil and as an ingredient in food products. Moreover, Linola™ can be cultivated in the same regions as traditional flaxseed

(Weiss, 1993), representing an alternative for a high-LA oil seed production in climate conditions from the Canadian Prairies and northern Europe. Further improvement of low-ALA flax varieties include enhancing the feeding quality of the meal and reducing saturated fatty acid content.

Potential for Seed Oil Modification

The value and industrial usefulness of an oil increase if it is highly enriched in a particular fatty acid (Taylor et al., 2011). For example, castor bean contains ~90% ricinoleic acid, which has contributed to its use numerous industrial applications including the production polymers such as nylon 6, 10 (Somerville and Bonetta, 2001). Although flax is highly enriched in ALA, increasing the content of this PUFA further could increase the industrial usefulness of flax oil and also render it more useful for genetic engineering to produce other industrial fatty acids from ALA. The same may apply to Solin that contains high levels of LA.

Seed Oil Formation

The process of seed oil formation has been reviewed in considerable detail elsewhere (eg, Weselake et al., 2009; Taylor et al., 2011; Bates and Browse, 2012). Here, we present a brief overview of fatty acid and TAG biosynthesis with a focus on aspects relevant to flax. A generalized schematic representation of this process in developing oleaginous seeds is depicted in Fig. 6.2. After de novo synthesis in plastids, fatty acids are exported as saturated (without double bonds) or monounsaturated (single double bond) entities to the cytoplasm. Several enzymes involved in the biosynthesis of fatty acids including β-ketoacyl CoA synthase, fatty acid elongase, stearoyl-ACP desaturase (SAD), and FAD have been previously identified in flaxseed (Fofana et al., 2004). SAD is a soluble enzyme that catalyzes the production of monounsaturated fatty acids, while attached to acyl carrier protein, in plastid. When fatty acids are released in the cytoplasm, they become esterified with coenzyme A (CoA), forming the main substrate for the acyl-CoA–dependent assembly of phospholipids and TAGs.

In the endoplasmic reticulum (ER), additional double bonds are produced by the catalytic action of membrane-bound desaturases. The FAD2 catalyzes the conversion of oleic acid ($18{:}1^{cis\Delta 9}$) to LA ($18{:}2^{cis\Delta 9,12}$), while the fatty acid Δ15 desaturase (FAD3) catalyzes the conversion of LA to ALA ($18{:}3^{cis\Delta 9,12,15}$). These reactions utilize the acyl chains at the *sn*-2 position of phosphatidylcholine (PC) as substrates in a catalytic process that requires an associated electron transport chain containing cytochrome b5 (Smith et al., 1990). In flaxseed, two genes encoding FAD3 have been characterized (*LuFAD3A* and *LuFAD3B*) (Vrinten et al., 2005). Point mutations in these genes lead to inactive polypeptides, co-segregating with low levels of unsaturation in the low-ALA flaxseed varieties. This indicates that *LuFAD3A* and *LuFAD3B* encode the main enzymes responsible for FAD3 enzyme activity in flaxseed, accounting for the key genetic factors responsible for the uniqueness of this vegetable oil. Banik et al. (2011)

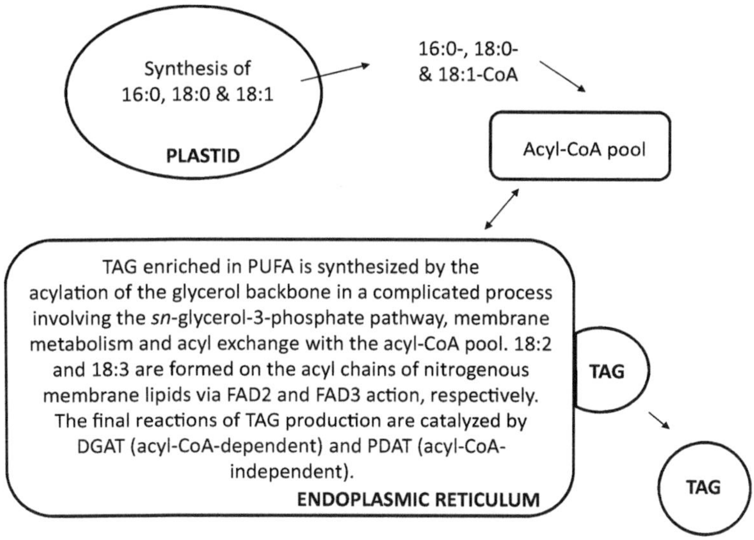

FIGURE 6.2 Simplification of triacylglycerol (TAG) synthesis in a developing seeds of oleaginous plants. *CoA*, coenzyme A; *FAD*, fatty acid desaturase; *PUFA*, polyunsaturated fatty acids; 16:0, palmitic acid; 18:0, stearic acid; 18:1, oleic acid (*cis*Δ9); 18:2, linoleic acid (*cis*Δ9,12); 18:3, α-linolenic acid (*cis*Δ9,12,15). *Adapted from Weselake, R.J., Shah, S., Taylor, D.C., Rahman, H., Shah, S., Laroche, A., McVetty, P.B.E., Harwood, J.L., 2009. Increasing the flow of carbon into seed oil. Biotechnol. Adv. 27, 866–878 and Bates, P.D., Browse, J. 2012. The significance of different diacylglycerol synthesis pathways on plant oil composition and bioengineering. Front. Plant Sci. 3, 147.*

investigated the expression of *FAD3A*, *FAD3B*, and a novel *FAD3C* during seed development in high-, moderate-, and low-ALA genotypes of flax. Their results demonstrated that FAD3A and FAD3B were responsible for ALA accumulation in flaxseeds but the novel FAD3C did not appear to have a major role in this process. Recently, Khadake et al. (2011) reported that variants of FAD3 in flax differed in the rate at which they could catalyze the conversion of LA to ALA. Subtle changes in amino acid sequence appeared to account for the variation in enzyme activity. Expression of *SAD* and *FAD2* has been shown to be differentially modulated during flaxseed development (Fofana et al., 2006). *SAD* was maximally expressed at very early stages of seed development whereas *FAD2* showed maximal expression 16 days after anthesis. Production of LA did not positively correlate with *FAD2* expression. Instead, ALA levels steadily rose, which suggested that LA was a transient substrate that was rapidly utilized by FAD3. Khadake et al. (2009) have cloned a second *FAD2* from flax genomic DNA, which was designated *LuFAD2-2*. The enzyme was encoded by an intronless gene and exhibited 85% similarity in amino acid sequence to the previously reported FAD2. Functional expression of LuFAD2-2 in yeast suggested that the recombinant enzyme had a similar substrate specificity to LuFAD2.

The process of TAG formation is intricately associated with membrane metabolism. Various mechanisms are operative in moving PUFA from

PUFA-enriched PC into TAG (Weselake et al., 2009; Bates and Browse, 2012). Although TAG can be produced via the *sn*-glycerol-3-phosphate pathway, most of the TAG formation in PUFA-enriched crops involves modifications at the level of PC. Diacylglycerol acyltransferase (DGAT) catalyzes the acyl-CoA–dependent acylation of *sn*-1,2 diacylglycerol (DAG) to produce TAG (Lung and Weselake, 2006; Liu et al., 2012). At least two distinct membrane-bound polypeptides, referred to as DGAT1 and DGAT2, have been identified in plants (Hobbs et al., 1999; Shockey et al., 2006). A soluble polypeptide displaying DGAT activity has been reported although its role for seed oil formation is yet to be determined (Saha et al., 2006). In some oilseed species, the level of DGAT activity during seed development has a major influence in the flow of carbon into TAG (Perry and Harwood, 1993; Zou et al., 1999; Cahoon et al., 2007; Zheng et al., 2008). It has been previously reported that DGAT activity from microsomal membranes of flaxseed displays high specificity for ALA (Sorensen et al., 2005). Another pathway leading to TAG formation is catalyzed by phospholipid:diacylglycerol transferase (PDAT), which transfers the acyl moiety from the *sn*-2 position of PC to *sn*-1,2-DAG (Dahlqvist et al., 2000). This enzyme might have a substantial influence on the incorporation of ALA into TAG since it directly utilizes the product of FAD3 as a substrate. It has been demonstrated that PDAT and DGAT reactions have overlapping functions in oil formation in Arabidopsis (Zhang et al., 2009). Rao et al. (2008) investigated the fatty acid composition of PC and TAG during seed development in flax in relation to *FAD3* expression. Although FAD3 appeared to have a major role in producing ALA-enriched TAG, there also appeared be selective transfer of ALA into TAG from PC presumably catalyzed by acyltransferase(s). Recently, several *DGAT* and *PDAT* genes were cloned from developing flaxseed and functionally characterized in yeast and Arabidopsis (Pan et al., 2013). This study indicated that developing flaxseed produced a pair of PDAT homologs exhibiting a preference for substrates containing ALA. Thus, the encoding genes may be useful in genetic engineering or molecular breeding strategies aimed at further increasing the ALA content of flax oil. It is anticipated that the recently developed genomic resources for flax (Venglat et al., 2011; Wang et al., 2012) will further accelerate research on deciphering the details of oil biosynthesis in this crop.

TAG is stored in organelles known as oilbodies, which are a core of neutral lipids surrounded by a monolayer of phospholipids and coated by proteins known as oleosins (Huang, 1996). Oleosins function as structural proteins, preventing the coalescence of oilbodies during seed maturation, and preserving these organelles as small entities, which may facilitate TAG mobilization during seedling growth (Siloto et al., 2006). Flax oil bodies have an average diameter of 1.34 μm, corresponding to twice the average size of canola oilbodies as a result of lower oleosin to TAG ratio (Tzen et al., 1993). The *OLEOSIN* gene family in flaxseed is composed of two genes encoding a high molecular mass isoform (18.6 kDa) and four genes encoding a low molecular mass isoform (16.03 kDa) (Chaudhary, 1975).

High-Palmitic Acid Flaxseed

While low levels of saturated fatty acids are generally desired for cooking oils, small-chain saturated fatty acids such as palmitic acid are used for the production of margarines, shortening and soaps. A flaxseed cultivar with double content of palmitic acid was developed by Rowland et al. (2002) to target this market niche. Heterologous expression of a medium-chain thioestearase from *Cuphea wrightii* in this variety further increased the level of palmitic acid to 39%, resulting also in the production of lauric (C12:0) and myristic (C14:0) acids that have other industrial applications (Rowland et al., 2002). For example, lauric acid–enriched oils are useful for production of detergents (Taylor et al., 2011).

Potential for Producing Oils Containing Conjugated Octadecatrienoic Acids

PUFAs are usually synthesized having two single bonds separating each double bond. Some plant species deviate from this, producing linolenic acids with at least one pair of double bonds separated by one single bond. Plant oils enriched in conjugated octadecatrienoic acids are useful raw materials for the production of organic coatings and polymers (Derksen et al., 1995; www.cyberlipid.org/fa/acid0003.htm). For example, the oil from tung tree (*Aleurites fordii*) seeds, which is rich in α-eleostearic acid ($18{:}3^{\Delta cis9,trans11,13}$), is a very effective drying oil, requiring lower levels of oxygen for hardening and drying more quickly than flax oil (Gunstone and Harwood, 2007). Other industrially useful octadecatrienoic acids include calendic acid ($18{:}3^{\Delta trans8,10,cis12}$) from marigold (*Calendula officinalis*) seed, licanic acid ($18{:}3^{4\text{-oxo}\Delta cis9,trans11,13}$) from the oiticica tree (*Lieania rigida*), and punicic acid ($18{:}3^{\Delta cis9,trans11,cis13}$) from pomegranate (*Punica granatum*) seed (www.lipidlibrary.aocs.org/plantbio/unusualfa/index.htm).

Conjugated double bonds can be introduced through the catalytic action of a homolog of FAD2 known as fatty acid conjugase (FADX), which utilizes LA as substrate (Rawat et al., 2012). Examples of linolenic acids that are derived from LA, including some conjugated trienes, are shown in Fig. 6.3. An FADX from pomegranate catalyzes the conversion of the Δ12 double bond of LA into two conjugated double bonds (*cis*Δ11 and *cis*Δ13), forming punicic acid (Hornung et al., 2002; Iwabuchi et al., 2003). In the case of calendic acid, an FADX from marigold catalyzes the modification of the Δ9 double bond of LA, producing conjugated *trans*Δ8 and *trans*Δ10 double bonds (Cahoon et al., 2001). Formation of α-eleostearic acid in tung tree seed is catalyzed by an FADX that catalyzes the conversion of the Δ12 double bond of LA into conjugated *trans*Δ11 and *trans*Δ13 double bonds (Dyer et al., 2002). Expression of FADX from these species in oilseed crops has the potential to be used as a biotechnological tool for large-scale production of seed oil containing conjugated fatty acids. High-LA flaxseed (ie, Solin or Linola™) logically represents an attractive target for production of oil containing conjugated octadecatrienoic acids. A combination of ALA, α-eleostearic, and other conjugated octadecatrienoic acids could potentially produce unique oils with novel properties and industrial applications.

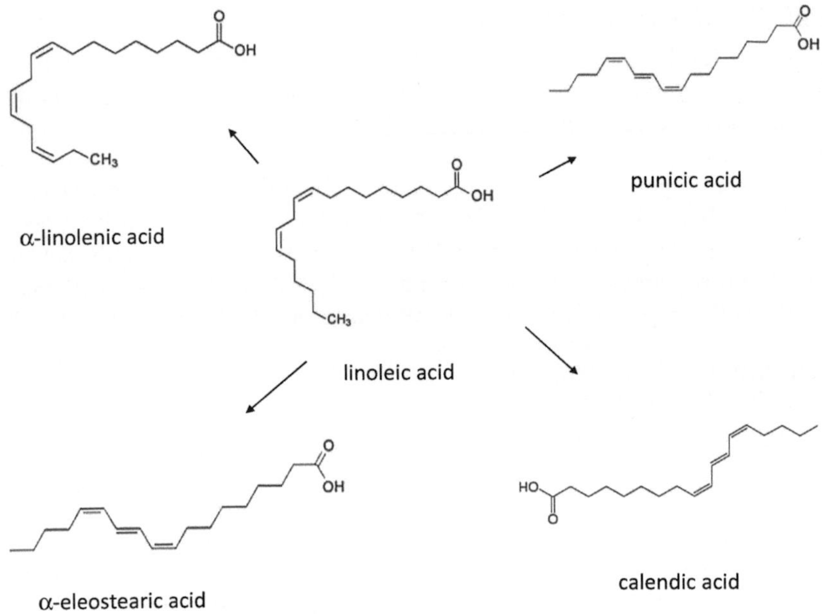

FIGURE 6.3 Examples of linolenic acids derived from linoleic acid. The formation of α-linolenic acid, which contains methylene-interrupted double bonds, is catalyzed by fatty acid desaturase (FAD) 2. The formation of punicic acid, α-eleostearic acid and calendic acid, which contain conjugated or non–methylene-interrupted double bonds, are catalyzed by divergent forms of FAD2 (FADX). The substrates of these FAD2 or FADX are nitrogenous phospholipids (eg, phosphatidylcholine). *See Cahoon, E.B., Dietrich, C.R., Meyer, K., Damude, H.G., Dyer, J.M., Kinney, A.J., 2006. Conjugated fatty acids accumulate to high levels in phospholipids of metabolically engineered soybean and Arabidopsis seeds. Phytochemistry, 67, 1166–1176 and* www.lipidlibrary.aocs.org/plantbio/unusualfa/index.htm.

Engineering the Antioxidant Capacity of Flax

Although Europe has traditionally shown strong resistance to the development of GE crops, in the past decade there has been intensive research on transgenic flax conducted by investigators from Wroclaw University in Poland. One area of their investigations has focused on increasing the antioxidant capacity of flax, which has the potential to protect PUFA from oxidation (Lorenc-Kukula et al., 2005; Żuk et al., 2011, 2012). In addition to decreasing the potential for PUFA rancidity, the antioxidants themselves may also provide health benefits in the human diet. In a recent study, expression of the gene encoding chalcone synthase was co-suppressed in developing Linola-type flax in order to partially redirect substrates from flavonoid biosynthesis to other routes in the phenylpropanoid pathway (Zuk et al., 2012). This modification resulted in overproduction of hydrolyzable tannins in the transgenic plants, which changed seeds from light yellow to brown. The antioxidant capacity was significantly increased by 40–50% in extracts of seeds from transgenic lines and the stability of the oil was increased as indicated by a decrease in peroxide value by ~72%. Interestingly, the proportion of ALA increased by ~25-fold in the seed oil transgenic plants compared to the control. The high accumulation of ALA was difficult to explain, but the authors speculated that FAD3 may have somehow been activated. Transgenic plants also showed increased resistance to *Fusarium* spp. infection. In related work, Fujisawa et al. (2008) expressed a bacterial phytoene synthase gene, *crtB*, in developing flaxseed which resulted in a 7.8- to 18.6-fold increase in total carotenoids in seeds. Lutein was the only carotenoid that was detected in control seeds. β-Carotene and α-carotene were present in a ratio of ~3:1. β-Carotene is the main source of dietary vitamin A. The various approaches to increase the antioxidant capacity of flaxseed could prove useful in the context of further increasing the ALA content of flax oil or in engineering the production of oils containing conjuagated ocatdecatrienoic acids, which are even more prone to oxidation than ALA.

Engineering Flax Fiber

Flax fibers can be separated from other parts of the stem by the process of retting providing raw material for the textile industry (Easson et al., 1996). Overproduction of flavonoids in flaxseed through genetic engineering has been shown to increase the antioxidative capacity of flax fiber, suggesting the modified fibers may be useful for biomedical applications such as wound dressings (Żuk et al., 2011). FT-IR analysis of fibers from transgenic plants indicated a change in the arrangement of cellulose polymer in the fibers and there was also a significant decrease in the quantity of hydrogen bonds. Recently, the introduction into flax of a cDNA encoding a potato β-1,3-glucanse resulted in fibers with increased levels of pectins and phenolics and decreased levels of lignin (Wojtasik et al., 2013). The mechanical properties of the resulting fibers were improved by their biochemical composition and the increased antioxidant capacity of the fibers suggested they would be useful in biomedical applications.

Researchers at the University of Wroclaw have also engineered flax to accumulate bioplastic in the growing fibers (Wróbel et al., 2004). Three bacterial genes were expressed in developing flax stems so as to produce poly-β-hydroxybutyrate, a biodegradable plastic. There was a substantial improvement in the elastic properties of the fibers from the GE plants and there was a decrease in lignin, pectin, and hemicellulose content, which improved retting efficiency (Wróbel-Kwiatkowska et al., 2007). In addition, there was an increased resistance to pathogen infection in transgenic plants which was attributable to an increase in phenolic acid content. In more recent work, the investigators have explored the properties of composites prepared by combining bioplastic flax fibers with polylactic acid or poly-ε-caprolactone, which served as the matrix while the bioplastic flax fibers served as reinforcement (Wróbel-Kwiatkowska et al., 2012). Composites of polylactic acid and bioplastic flax fibers appeared to display bacteriostatic, platelet antiaggregation and noncytotoxic effects. The investigators suggested that the composites might find applications as implants to replace damaged human tissues.

CLOSING COMMENTS

In addition to its traditional industrial uses in the production of linoleum and paints and varnishes, flax has the potential to become a platform crop for even more powerful drying agents, antioxidants, and high performance fibers. Flax oil has the potential to be part of a growing specialty oils market that could drive a substantial increase in production of the crop. To compete for the potentially large modified oils markets, however, flax must overcome major challenges. Although the crop is amenable to genetic transformation, it has yet to benefit, like other oilseeds crops have, from enhanced agronomic traits such as herbicide-resistance, enhanced yield from hybrid breeding and increased breeding efficiency from the routine use of marker-assisted breeding for trait development. Flax has not yet received major investments in trait and cultivar development. Seed size, yield, yield stability, disease resistance, and weed control remain significant barriers for flax production. The recent sequencing of the flax genome (/www.phytozome.org/flax.php; Wang et al., 2012), however, represents a major milestone in advancement of this crop and is sure to provide valuable tools for genomics-assisted breeding of this crop in addition to identifying targets for biotechnological modification of the crop.

ACKNOWLEDGMENTS

R.J.W. thanks Alberta Enterprise and Advanced Education, AVAC Ltd., Genome Canada, Genome Prairie, the Canada Research Chairs Program, the Canada Foundation for Innovation, and Genome Canada and Genome Prairie for supporting his research on flax. H.B. would like to thank University of Saskatchewan, Saskatchewan Ministry of Agriculture, SeCan, SaskFlax Development Commission, and Genome Canada and Genome Prairie for research support. L.M.H. would like to acknowledge the contribution of Alberta Ingenuity, Alberta Advanced Education, the University of Alberta, Alberta Agriculture and Rural Development, Vitera, Alberta Crop Industrial Development Fund, and Natural Science and Engineering Research Council. The authors thank Crystal Snyder for her critical assessment of the manuscript.

REFERENCES

Agriculture and Agri-Food Canada, 2010. Canada: Grains and Oilseeds Outlook. http://www.agr.gc.ca/pol/mad-dam/index_e.php?s1=pubs&s2=go-co&s3=php&page=go-co_2011-06-07 (accessed 30.01.10.).

Ahloowalia, B.S., Nichterlein, K., Maluszynski, M., 2004. Global impact of mutation-derived varieties. Euphytica: Int. J. Plant Breed. 135, 187–204.

Ahmed, M.K., Daun, J.K., Przybylski, R., 2005. FT-IR based methodology for quantitation of total tocopherols, tocotrienols and plastochromanol-8 in vegetable oils. J. Food Compos. Anal. 18, 359–364.

Allaby, R.G., Peterson, G.W., Merriwether, D.A., Fu, Y.B., 2005. Evidence of the domestication history of flax (*Linum usitatissium* L.) from genetic diversity of the *sad2* locus. Theor. Appl. Genet. 112, 5865.

Bailey, K.L., Gossen, B.D., Gugel, R.K., Morrall, R.A.A., 2009. Diseases of flax. In: Bailey, K.L., Gossen, B.D., Gugel, R.K., Morall, R.A.A. (Eds.), Diseases of Field Crops of Canada. The Canadian Phytopathological Society, Saskatoon, Saskatchewan, pp. 147–154.

Banik, M., Duguid, S., Cloutier, S., 2011. Transcript profiling and gene characterization of three fatty acid desaturase genes in high, moderate, and low linolenic acid genotypes of flax (*Linum usitatissium* L.) and their role in linolenic acid accumulation. Genome 54, 471–483.

Bates, P.D., Browse, J., 2012. The significance of different diacylglycerol synthesis pathways on plant oil composition and bioengineering. Front. Plant Sci. 3, 147.

Bedard, D., 2009. Deregistered GM Flax Pops up in Europe. http://www.albertafarmexpress.ca/news/deregistered-canadian-gm-flax-pops-up-in-europe/1000340642/ (accessed 18.09.09.).

Bhateria, S., Sood, S.P., Pathania, A., 2006. Genetic analysis of quantitative traits across environments in flax (*Linum usitatissimum* L.). Euphytica 150, 185–194.

Booker, H.M., Lamb, E.G., 2012. Quantification of low-level GM seed presence in Canadian commercial flax stocks. AgBioForum 15, 31–35.

Bourgeois, L., Entz, M.H., 1996. Influence of previous crop type on yield of spring wheat: analysis of commercial field data. Can. J. Plant Sci. 76, 457–459.

Braidek, J.G., 1975. Agronomic considerations in flax production. In: Harapiak, J.T. (Ed.), Oil Seed and Pulse Crops in Western Canada: A Symposium. Western Co-operative Fertilizers, Calgary, Alberta, pp. 279–285.

Brim, C.A., 1966. A modified pedigree method of selection in soybeans. Crop Sci. 6, 220.

Brown, J.C.L., De Decker, M.M., Fieldes, M.A., 2008. A comparative analysis of developmental profiles for DNA methylation in 5-azacytidine-induced early-flowering flax lines and their control. Plant Sci. 175, 217–225.

Cahoon, E.B., Ripp, K.G., Hall, S.E., Kinney, A.J., 2001. Formation of conjugated delta(8), delta(10)-double bonds by delta(12)-oleic-acid desaturase-related enzymes – biosynthetic origin of calendic acid. J. Biol. Chem. 276, 2637–2643.

Cahoon, E.B., Dietrich, C.R., Meyer, K., Damude, H.G., Dyer, J.M., Kinney, A.J., 2006. Conjugated fatty acids accumulate to high levels in phospholipids of metabolically engineered soybean and *Arabidopsis* seeds. Phytochemistry 67, 1166–1176.

Cahoon, E.B., Shockey, J.M., Dietrich, C.R., Gidda, S.K., Mullen, R.T., Dyer, J.M., 2007. Engineering oilseeds for sustainable production of industrial and nutritional feedstocks: solving bottlenecks in fatty acid flux. Curr. Opin. Plant Biol. 10, 236–244.

Canadian Grain Commission, 2010. Sampling and Testing Protocol for Canadian Flaxseed Exported to the European Union. http://www.grainscanada.gc.ca/gmflax-lingm/stpf-peevl-eng.htm (accessed 29.12.15.).

Canadian Grain Commission, 2013. Canada Western Solin No Longer Regulated as of August 1 2013. http://www.grainscanada.gc.ca/media-medias/press-presse/2012/2012-12-12-eng.htm (accessed 15.07.13.).

Chaudhary, S., 1975. Molecular Biology of Flax (*Linum usitatissimum*) Seed Oleosin Genes (Ph.D. thesis). University of Calgary, Calgary, Alberta, pp. 1–290.

Cloutier, S., Duguid, S., Datla, R., Niu, Z., 2009. Development and analysis of EST-SSRs for flax (*Linum usitatissimum L.*). T.A.G. 119, 53–63.

Cloutier, S., Ragupathy, R., Miranda, E., Radovanovic, N., Reimer, E., Walichnowski, A., Ward, K., Rowland, G., Duguid, S., Banik, M., 2012. Integrated consensus genetic and physical maps of flax (*Linum usitatissimum* L.). Theor. Appl. Genet. 125 (8), 1783–1795.

Cross, R.H., Mckay, S.A.B., McHughen, A.G., Bonham-Smith, P.C., 2003. Heat stress effects on reproduction and seed set in *Linum usitatissimum* L. (flax). Plant Cell Environ. 26, 1013–1020.

Culbertson, J.O., 1954. Breeding flax. Adv. Agron. 6, 174–178.

Czemplik, M., Szopa, J., 2009. Optimizing biomedical and industrial products development based on flax. CAB Rev. Perspect. Agric. Vet. Sci. Nutr. Nat. Resour. 4 (062), 1–10. www.cyberlipid.org/fa/acid0003.htm.

D'Antuono, L.F., Rossini, F., 1995. Experimental estimation of flax (*Linum usitatissimum* L.) crop parameters. Ind. Crops Prod. 3, 261–271.

Dahlqvist, A., Stahl, U., Lenman, M., Banas, A., Lee, M., Sandager, L., Ronne, H., Stymne, S., 2000. Phospholipid:diacylglycerol acyltransferase: an enzyme that catalyzes the acyl-CoA-independent formation of triacylglycerol in yeast and plants. Proc. Natl. Acad. Sci. U.S.A. 97, 6487–6492.

Deng, X., Scarth, R., 1998. Temperature effects on fatty acid composition development of low-linolenic oilseed rape (*Brassica napus* L.). J. Am. Oil Chem. Soc. 75, 759–766.

Derksen, J.T.P., Cuperus, P.F., Kolster, P., 1995. Paints and coatings from renewable resources. Ind. Crops Prod. 3, 225–236.

Dexter, J.E., Jhala, A.J., Hills, M.J., Yang, R.-C., Topinka, A.K., Weselake, R.J., Hall, L.M., 2010a. Emergence and persistence of volunteer flax (*Linum usitatissimum* L.) in western Canadian cropping systems. Agron. J. 102, 1321–1328.

Dexter, J.E., Jhala, A.J., Hills, M.J., Yang, R.-C., Topinka, A.K., Weselake, R.J., Hall, L.M., 2010b. Quantification and mitigation of adventitious presence of volunteer flax (*Linum usitatissimum* L.) in wheat. Weed Sci. 58, 80–88.

Dexter, J.E., Jhala, A.J., Yang, R.-C., Hills, M.J., Weselake, R.J., Hall, L.M., 2011. Harvest loss and seed bank longevity of flax (*Linum usitatissimum* L.): Implications for seed-mediated gene flow. Weed Sci. 59, 61–67.

Deyholos, M.K., 2006. Bast fiber of flax (*Linum usitatissimum* L.): biological foundations of its ancient and modern uses. Isr. J. Plant Sci 54, 273–280.

Diederichsen, A., Fu, Y.B., 2008. Flax genetic diversity as the raw material for future success. Abstract No:51. In: FAO/ESCORENA International Conference, Canada, July 21–23, 2008.

Diederichsen, A., Ulrich, A., 2009. Variability in stem fibre content and its association with other characteristics in 1177 flax (*Linum usitatissimum* L.) genebank accessions. Ind. Crops Prod. 30, 33–39.

Diederichsen, A., Raney, J.P., 2006. Seed colour, seed weight and seed oil content in flax (*Linum usitatissimum* L.) accessions held by Plant Gene Resources of Canada. Plant Breed. 125, 372–377.

Diederichsen, A., Richards, K., 2003. Cultivated flax and the genus *Linum* L.: taxonomy and germplasm conservation. In: Muir, A.D., Westcott, N.D. (Eds.), Flax: The Genus *Linum*. Taylor and Francis Ltd, New York, NY, pp. 22–54.

Diederichsen, A., Rozhimina, R.A., Kudrjavceva, L.P., 2008. Variation patterns within 153 flax (*Linum usitatissimum* L.) genebank accessions based on evaluation for resistance to fusarium wilt, anthracnose and pasmo. Plant Genet. Resour. Charact. Utiliz. 6, 22–32.

Diederichsen, A., Rozhmina, T.A., Zhuchenko, A.A., Richards, K.W., 2006. Screening for broad adaptation in 96 flax (*Linum usitatissimum* L.) accessions under dry and warm conditions in Canada and Russia. FAO/IPGRI Plant Genet. Resour. News 146, 7–14.

Diederichsen, A., Hammer, K., 1995. Variation of cultivated flax (*Linum usitatissimum* L. subsp. *usitatissimum*) and its wild progenitor pale flax (subsp. *angustifolium* (Huds) Thell). Genet. Resour. Crop Evol. 42, 263–272.

Dillman, A.C., 1936. Improvement in Flax. Year Book of Agriculture. USDA, Washington, DC. 784 p.

Dillman, A.C., Goar, L.G., 1937. Flaxseed Production in the Far Western States. Farmer's Bulletin no. 1792. 1 p.

Dribnenki, J.C.P., Green, A.G., Atlin, G.N., 1996. Linola '989' low linolenic flax. C.J.P.S. 76, 329–331.

Dribnenki, J.C.P., Green, A.G., 1995. Linola '947' low linolenic flax. Can. J. Plant Sci. 75, 201–202.

Durrant, A., 1971. Induction and growth of flax genotrophs. Heredity 27, 277–298.

Dyer, J.M., Chapital, D.C., Kuan, J.C.W., Mullen, R.T., Turner, C., Mckeon, T.A., Pepperman, A.B., 2002. Molecular analysis of a bifunctional fatty acid conjugase/desaturase from tung. Implications for the evolution of plant fatty acid diversity. Plant Physiol. 130, 2027–2038.

Easson, L.D., Molloy, R., 1996. Retting-a key process in the production of high value fibre from flax. Outlook Agric. 25, 235–242.

FAOSTAT, 2013. Food and Agriculture Organization. http://faostat3.fao.org/home/index.html (accessed 15.06.13.).

Fieldes, M.A., Harvey, C.G., 2004. Differences in developmental programming and node number at flowering in the 5-azacytidine-induced early flowering flax lines and their controls. Int. J. Plant Sci. 165, 695–706.

Fieldes, M.A., 1994. Heritable effects of 5-azacytidine treatments on the growth and development of flax (*Linum usitatissimum*) genotrophs and genotypes. Genome 37, 1–11.

Fieldes, M.A., Amyot, L.M., 1999. Epigenetic control of early flowering in flax lines induced by 5-azacytidine applied to germinating seed. J. Hered. 90, 199–206.

Fieldes, M.A., Schaeffer, S.F., Krech, M.J., Brown, J.C.L., 2005. DNA hypomethylation in 5-azacytidine induced, early-flowering lines of flax. Theor. Appl. Genet. 111, 126–149.

Flax Council of Canada, 2014a. Canada – A Flax Leader. http://www.flaxcouncil.ca/english/index.jsp?p=what2&mp=what (accessed 28.01.14a.).

Flax Council of Canada, 2014b. GMO Flax Update 20 January 2010. http://www.flaxcouncil.ca/files/web/GMO%20Flax%20Update%20No%201%20REVISED%2028%20September%202009.pdf (accessed 28.01.14b.).

Flax Council of Canada, 2014c. Growing Flax. http://www.flaxcouncil.ca/english/index.jsp?p=growing (accessed 28.01.14c.).

Flor, H.H., 1955. Host-parasite interaction in flax rust—its genetics and other implications. Phytopathology 45, 680–685.

Flor, H.H., 1942. Inheritance of pathogenicity in *Melampsora lini*. Phytopathology 32, 653–669.

Fofana, B., Duguid, S., Cloutier, S., 2004. Cloning of fatty acid biosynthetic genes beta-ketoacyl CoA synthase, fatty acid elongase, stearoyl-ACP desaturase, and fatty acid desaturase and analysis of expression in the early developmental stages of flax (*Linum usitatissimum* L.) seeds. Plant Sci. 166, 1487–1496.

Fofana, B., Cloutier, S., Duguid, S., Ching, J., Rampitsch, C., 2006. Gene expression of stearoyl-ACP desaturase and delta 12 fatty acid desaturase 2 is modulated during seed development of flax (*Linum usitatissium*). Lipids 41, 705–712.

Fu, Y.B., Richards, K.W., Duguid, S.D., Rowland, G.G., 2003a. RAPD analysis of 54 North American flax cultivars. Crop Sci. 43, 1510–1515.

Fu, Y.B., Guerin, S., Peterson, G.W., Diederichsen, A., Rowland, G.G., Richards, K.W., 2003b. RAPD analysis of genetic variability of regenerated seeds in the Canadian flax cultivar CDC Normandy. Seed Sci. Technol. 31, 207–211.

Fu, Y.B., 2006. Redundancy and distinctness in flax germplasm as revealed by RAPD dissimilarity. Plant Genet. Resour. 4, 117–124.

Fujisawa, M., Watanabe, M., Choi, S.-K., Teramoto, M., Ohyama, K., Misawa, N., 2008. Enrichment of carotenoids in flaxseed (*Linum usitatissiumum*) by metabolic engineering with introduction of bacterial phytoene synthase gene *crtB*. J. Biosci. Bioeng. 105, 636–641.

Gan, Y., Basnyat, P., McDonald, C.L., Campbell, C.A., Liu, L., 2009. Water use and distribution profile under pulse and oilseed crops in semiarid northern high latitude areas. Agric. Water Manage. 96, 337–348.

Gehring, M., Henikoff, S., 2007. DNA methylation dynamics in plant genomes. Biochim. Biophys. Acta 1769, 276–286.

Gill, K.S., 1966. Evolutionary Relationship Among *Linum* Species (Ph.D.). University of California, Riverside, CA, USA. 260 pp.

Gill, K.S., 1987. Flax. Indian Council of Agricultural Research, New Delhi, India. 386 p.

Goulden, C.H., 1939. Problems in plant selection. Conference information. In: Proc. 7th Int. Genet. Congr. August 23–30, 1939, pp. 132–133.

Grant, C.A., Dribnenki, J.C.P., Bailey, L.D., 1999. A comparison of yield response of solin (cv. Linola 947) and flax (cv. McGregor and Vimy) to the application of nitrogen, phosphorus and provide (*Penicillum bilaji*). Can. J. Plant Sci. 79, 527–533.

Green, A.G., 1986a. A mutant genotype of flax (*Linum usitatissimum* L.) containing very low levels of linolenic acid in its seed oil. Can. J. Plant Sci. 6, 499–503.

Green, A.G., Marshall, D.R., 1984. Isolation of induced mutants in flax (*Linum usitatissimum*) having reduced linolenic acid content. Euphytica 33, 321–328.

Green, A.G., Chen, Y., Singh, S.P., Dribnenki, 2008. Flax. In: Chittaranjan, K., Hall, T.C. (Eds.), Compendium of Transgenic Crop Plants: Transgenic Oilseed Crops. Blackwell Publishing Ltd, Chicester, UK, pp. 200–226.

Green, A.G., 1986b. Effect of temperature during seed maturation on the oil composition of low-linolenic genotypes of flax. Crop Sci. 26, 961–965.

Green, A.G., Dribnenki, J.C.P., 1994. Linola – a new premium polyunsaturated oil. Lipid Tech. 6, 29–33.

Green, A.G., Marshall, D.R., 1981. Variation for oil quantity and quality in flax (*Linum usitatissimum*). Aust. J. Agric. Res. 32, 599–607.

Green, A.G., 1986c. Genetic control of polyunsaturated fatty acid biosynthesis in flax (*Linum usitatissimum)* seed oil. Theor. Appl. Genet. 72, 654–661.

Goyal, A., Sharma, V., Upadhyay, N., Gill, S., Sihag, M., 2014. Flax and flaxseed oil: an ancient medicine & modern functional food. J. Food Sci. Tech 1–21.

Gunstone, F.D., Harwood, J.L., 2007. In: Dijkstra, A.J. (Ed.), The Lipid Handbook. CRC Press, Boca Raton, FL, pp. 37–142.

Harlan, J.R., 1965. Possible role of weed races in evolution of cultivated plants. Euphytica 14, 173–176.

Hausner, G., Rashid, K.Y., Kenaschuk, E.O., Procunier, J.D., 1999a. The development of codominant PCR/RFLP based markers for the flax rust-resistance alleles at the L locus. Genome 42, 1–8.

Hausner, G., Rashid, K.Y., Kenaschuk, E.O., Procunier, J.D., 1999b. The identification of a cleaved amplified polymorphic sequence (CAPS) marker for the flax rust resistance gene M3. Can. J. Plant Pathol. 21, 187–192.

Helbaek, H., 1959. Domestication of food plants in the old world. Science 130, 365–372.

Henriksen, P.S., Robinson, D., 1996. Early iron age agriculture: archaeobotanical evidence from an underground granary at overbygard in northern Jutland, Denmark. Veg. Hist. Archaeobot. 5, 1–11.

Hettiarachy, N.S., Hareland, G.A., Ostenson, A., Baldner-Shank, G., 1990. Chemical composition of eleven flaxseed varieties grown in North Dakota. In: Proceedings of the 53rd Annual Flax Institute of United States, North Dakota, January, 25–26th, 1990, Fargo, North Dakota, U.S.A., pp. 36–40.

Heywood, V.H., 1993. Flowering Plants of the World. Oxford University Press, Oxford, UK, pp. 207–208.

Hobbs, D.H., Lu, C.F., Hills, M.J., 1999. Cloning of a cDNA encoding diacylglycerol acyltransferase from *Arabidopsis thaliana* and its functional expression. FEBS Lett. 452, 145–149.

Hornung, E., Pernstich, C., Feussner, I., 2002. Formation of conjugated delta(11) delta(13)-double bonds by delta(12)-linoleic acid (1,4)-acyl-lipid-desaturase in pomegranate seeds. Eur. J. Biochem. 269, 4852–4859.

Huang, A.H.C., 1996. Oleosins and oil bodies in seeds and other organs. Plant Physiol. 110, 1055–1061.

Irvine, R.B., McConnell, J., Lafond, G.P., May, W.E., Hultgreen, G., Ulrich, A., Stonehouse, K., Chalmers, S., Stevenson, F.C., 2010. Impact of production practices on fiber yield of oilseed flax under Canadian prairie conditions. Can. J. Plant Sci. 90, 61–70.

Iwabuchi, M., Kohno-Murase, J., Imamura, J., 2003. Delta 12-oleate desaturase-related enzymes associated with formation of conjugated trans-delta 11, cis-delta 13 double bonds. J. Biol. Chem. 278, 4603–4610.

Jhala, A.J., Bhatt, H., Topinka, A.K., Hall, L.M., 2011. Pollen mediated gene flow in flax: can genetically engineered and organic flax co-exist? Heredity 106, 557–566.

Jhala, A.J., Hall, L.M., 2010. Flax (*Linum usitatissimum* L.): current uses and future applications. Aust. J. Appl. Sci. 4, 4304–4312.

Jhala, A.J., Hall, L.M., Hall, J.C., 2008. Potential hybridization of flax (*Linum usitatissimum* L.) with weedy and wild relatives: an avenue for movement of engineered genes? Crop Sci. 48, 825–840.

Jhala, A.J., Weselake, R.J., Hall, L.M., 2009. Genetically engineered flax: potential benefits, risks, regulations, and mitigation of transgene movement. Crop Sci. 49, 1943–1954.

Jhala, A.J., Raatz, L., Dexter, J.E., Hall, L.M., 2010. Adventitious presence: volunteer flax (*Linum usitatissimum* L.) in herbicide resistant canola (*Brassica napus* L.). Weed Technol. 24, 244–252.

Johnston, A.M., Tanaka, D.L., Miller, P.R., Brandt, S.A., Nielsen, D.C., Lafond, G.P., Riveland, N.R., 2002. Oilseed crops for semiarid cropping systems in the Northern Great Plains. Agron. J. 94, 231–240.

Jordan, M.C., McHughen, A., 1988. Glyphosate tolerant flax plants from Agrobacterium mediated gene-transfer. Plant Cell Rep. 7, 281–284.

Khadake, R.M., Ranjekar, P.K., Harsulkar, A.M., 2009. Cloning of a novel omega-6 desaturase from flax (*Linum usitatissimum* L.) and its functional analysis in *Sacchromyces cerevisiae*. Mol. Biotechnol. 42, 168–174.

Khadake, R., Khonde, V., Mhaske, V., Ranjekar, P.A., 2011. Harsulkar. Functional and bioinformatics characterization of sequence variants of *Fad3* gene from flax. J. Sci. Food Agric. 91, 2689–2696.

Kenaschuk, E.O., Rowland, G.G., 1993. Flax. In: Slinkard, A.E., Knott, D.R. (Eds.), Harvest of Gold: The History of Field Crop Breeding in Canada. University of Saskatchewan, Saskatoon, Saskatchewan, pp. 173–176.

Kenaschuk, E.O., 1975. In: Harpiak, J.T. (Ed.), Oilseed and Pulse Crops in Western Canada—A Symposium. Western Co-operative Fertilizers, Calgary, Alberta, pp. 203–221.

Khakbazan, M., McLaren, D.L., Monreal, M., Mohr, R.M., Grant, C.A., Irvine, R.B., 2009. Influence of alternative management methods on the economics of flax production in the Black Soil Zone. Can. J. Plant Sci. 89, 903–913.

Knott, C.M., 1993. Volunteer flax control in vining peas. Aspects Appl. Biol. 35, 179–182.

Kumar, S., You, F.M., Cloutier, S., 2012. Genome wide SNP discovery in flax through next generation sequencing of reduced representation libraries. BMC Genomics 13, 684–695.

Lafond, G.P., Irvine, B., Johnston, A.M., May, W.E., McAndrew, D.W., Shirtliff, S.J., Stevenson, F.C., 2008. Impact of agronomic factors on seed yield formation and quality in flax. Can. J. Plant Sci. 88, 485–500.

Lamb, E.G., Booker, H.M., 2011. Quantification of low level genetically modified (GM) seed presence in large seed lots: a case study of GM seed in Canadian flax breeder seed lots. Seed Sci. Res. 21, 315–321.

Leeson, J.Y., Thomas, A.G., Hall, L.M., Brenzil, C.A., Andrews, T., Brown, K.R., Van Acker, R.C., 2005. Prairie Weed Surveys of Cereal, Oilseed and Pulse Crops from the 1970s to the 2000s. Weed Survey Series Publication 05-1CD. 395 pp.

Liu, Q., Siloto, R.M.P., Lehner, R., Stone, S.J., Weselake, R.J., 2012. Acyl-CoA:diacylglycerol acyltransferase: molecular biology, biochemistry and biotechnology. Prog. Lipid. Res. 51, 350–377.

Lorenc-Kukula, K., Amarowicz, R., Oszmiański, J., Doermann, P., Starzycki, M., Skala, J., Żuk, M., Kulma, A., Szopa, J., 2005. Pleiotropic effect of phenolic compounds content increases in transgenic flax plant. J. Agric. Food Chem. 53, 3685–3692.

Lung, S.C., Weselake, R.J., 2006. Diacylglycerol acyltransferase: a key mediator of plant triacylglycerol synthesis. Lipids 41, 1073–1088.

McHughen, A., Holm, F.A., 1995. Transgenic flax with environmentally and agronomically sustainable attributes. Transgenic Res. 4, 3–11.

McHughen, A., Rowland, G.G., Holm, F.A., Bhatty, R.S., Kenaschuk, E.O., 1997. CDC Triffid transgenic flax. Can. J. Plant Sci. 77, 641–643.

McKenzie, R.R., 2011. Genetic and Hormonal Regulation of Stem Vascular Tissue Development in Flax (*Linum usitatissimum* L.) (M.Sc. thesis). University of Alberta, Edmonton, Alberta, pp. 1–220.

McSheffrey, S.A., McHughen, A., Devine, M.D., 1992. Characterization of transgenic sulfonylurea-resistant flax (*Linum usitatissimum*). Theor. Appl. Genet. 84, 480–486.

Mittapalli, O., Rowland, G.G., 2003. Inheritance of seed color in flax. Crop Sci. 43, 1945–1951.

Mohammadi, A.A., Saeidi, G., Arzani, A., 2010. Genetic analysis of some agronomic traits in flax (*Linum usitatissimum* L.). Aust. J. Crop Sci. 4, 343–352.

Mpofu, S.I., Rashid, K.Y., 2001. Vegetative compatibility groups within *Fusarium oxysporum* f.sp. *lini* from *Linum usitatissimum* (flax) wilt nurseries in western Canada. Can. J. Bot. 79, 836–843.

Muravenko, O.V., Lemesh, V.A., Samatadze, T.E., Amosova, A.V., Grushetskaya, Z.E., Popov, V.K., Semenova, O.Y., Khotyuleva, V.L., Zelenin, A.V., 2003. Genome comparisons with chromosomal and molecular markers for three closely related flax species and their hybrids. Russ. J. Genet. 39, 414–421.

Ntiamoah, C., Rowland, G.G., 1997. Inheritance and characterization of two low linolenic acid EMS-induced McGregor flax (*Linum usitatissimum*). Can J. Plant Sci. 77, 353–358.

Olejnik, D., Gogolewski, M., NogalaKalucka, M., 1997. Isolation and some properties of plastochromanol-8. Nahrung-Food 41, 101–104.

Orf, J.H., 2008. Breeding, genetics and production of soybeans. In: Johnson, L., Galloway, R., White, P.J. (Eds.), Soybeans: Chemistry, Production, Processing and Utilization. American Oil Chemists' Society, Urbana, IL, pp. 87–119.

Palmer, R.G., Gai, J., Sun, H., Burton, J.W., 2001. Production and evaluation of hybrid soybean. Plant Breed. Rev. 21, 263–307.

Pan, X., Siloto, R.M.P., Wickramarathna, A.D., Mietkiewska, E., Weselake, R.J., 2013. Identification of a pair of phospholipid:diacylglycerol acyltransferases from developing flax (*Linum usitatissimum* L.) seed catalyzing the selective production of trilinolenin. J. Biol. Chem. 288, 24173–24188.

Patel, V.D., Chopde, P.R., 1981. Combining ability analysis over environments in diallel crosses of flax (*Linum usitatissimum* L.). Theor. Appl. Genet. 60, 339–343.

Perry, H.J., Harwood, J.L., 1993. Changes in the lipid-content of developing seeds of *Brassica napus*. Phytochemistry 32, 1411–1415.

PGRC, 2011. Plant Gene Resources of Canada. http://pgrc3.agr.ca/about-propos_e.html (accessed 06.11.11.).

www.phytozome.org/flax.php.

Prairie Flax Products, 2007. Flax History. http://www.prairieflax.com/ (accessed 07.02.11.).

Przybylski, R., 2005. Flax oil and high linolenic oils. In: Shahidi, F. (Ed.), Bailey's Industrial Oil and Fat Products. John Wiley & Sons, Inc., Hoboken, NJ, pp. 281–298.

Rao, S., Abdel-Reheem, M., Bhella, R., McCracken, C., Hildebrand, D., 2008. Characteristics of high alpha-linolenic acid accumulation in seed oils. Lipids 43, 749–755.

Rashid, K.Y., 2003. Principal diseases of flax. In: Muir, A.D., Westcott, N.D. (Eds.), Flax: The Genus *Linum*. Taylor and Francis Ltd, New York, NY, pp. 92–123.

Rashid, K.Y., Duguid, S., 2010. Diseases of flax in Manitoba and Saskatchewan in 2009. Can. Plant Dis. Surv. 90, 136–137.

Rathjen, T., Steinhart, H., 1997. Natural antioxidants in lipids. In: Mossoba, M.M. (Ed.), New Techniques and Applications in Lipid Analysis. AOCS Press, Champaign, IL, pp. 341–355.

Rawat, R., Xiao-Hong, Y., Sweet, M., Shanklin, J., 2012. Conjugated fatty acid synthesis: residues 111 and 115 influence product partitioning of the *Momordica charantia* conjugase. J. Biol. Chem. 287, 16230–16237.

Remund, K.M., Dixon, D.A., Wright, D.L., Holden, L.R., 2001. Statistical considerations in seed purity testing for transgenic traits. Seed Sci. Res. 11, 101–120.

Richharia, R.H., 1962. Flax. The Indian Central Oilseeds Committee, Hyderabad, India. 155 p.

Roose-Amsaleg, C., Cariou-Phad, E., Vautrin, D., Taverneer, R., Solignac, M., 2006. Polymorphic microsatellite loci in *Linum usitatissimum*. Mol. Ecol. Notes 6, 796–799.

Rowland, G.G., 1991. An EMS-induced low-linolenic-acid mutant in McGregor flax (*Linum usitatissimum* L.). Can. J. Plant Sci. 71, 393–396.

Rowland, G., Wilen, R., Hormis, Y., Saeidi, G., Ntiamoah, C., 2002. Modification of flax (*Linum usitatissimum*) by induced mutagenesis and transformation. In: International Atomic Energy Agency, Plant Breeding and Genetics Section, August 26–30th, 2002, Vienna, Austria, pp. 143–150.

Rowland, G.G., McHughen, A., Bhatty, R.S., Mackenzie, S.L., Taylor, D.C., 1995. The application of chemical mutagenesis and biotechnology to the modification of flax (*Linum usitatissimum* L.). Euphytica 85, 317–321.

Rowland, G.G., Bhatty, R.S., 1990. Ethyl methanesulfonate induced fatty acid mutations in flax. JAOCS 67, 213–214.

Saeidi, G., Rowland, G.G., 1997. The inheritance of variegated seed color and palmitic acid in flax. J. Hered. 88, 466–468.

Saha, S., Enugutti, B., Rajakumari, S., Rajasekharan, R., 2006. Cytosolic triacylglycerol biosynthetic pathway in oilseeds. Molecular cloning and expression of peanut cytosolic diacylglycerol acyltransferase. Plant Physiol. 141, 1533–1543.

Salej, S., Kalia, N.R., Bhateria, S., Sanjeev, K., 2007. Detection of genetic components of variation for some biometrical traits in *Linum usitatissimum* L. in sub-mountain Himalayan region. Euphytica 155, 107–115.

SaskFlax, 2013. Rebooting Canada's Flax Industry. http://www.saskflax.com/rebootCanadasflaxseed.html (accessed 15.07.13.).

Saskston, W.E., 1959. Pasmo – past, present, and future. In: Proceedings of the Flax Institute of the United States, November 19th, 1959, Fargo, North Dakota, U.S.A., pp. 3–5.

Shockey, J.M., Gidda, S.K., Chapital, D.C., Kuan, J.C., Dhanoa, P.K., Bland, J.M., Rothstein, S.J., Mullen, R.T., Dyer, J.M., 2006. Tung tree DGAT1 and DGAT2 have nonredundant functions in triacylglycerol biosynthesis and are localized to different subdomains of the endoplasmic reticulum. Plant Cell 18, 2294–2313.

Siloto, R.M.P., Findlay, K., Lopez-Villalobos, A., Yeung, E.C., Nykiforuk, C.L., Moloney, M.M., 2006. The accumulation of oleosins determines the size of seed oilbodies in *Arabidopsis*. Plant Cell 18, 1961–1974.

Sinha, S.K., Saxena, S.S., 1965. Growth and yield of *Linum usitatissimum* L. at different levels of pH in sand culture. Indian J. Agron. 10, 283–288.

Smith, C.E., 1969. From Vavilov to the present – A review. Econ. Bot. 23, 2–19.

Smith, M.A., Cross, A.R., Jones, O.T.G., Griffiths, W.T., Stymne, S., Stobart, K., 1990. Electron-transport components of the 1-acyl-2-oleoyl-sn-glycero-3-phosphocholine delta-12-desaturase (delta-12-desaturase) in microsomal preparations from developing safflower (*Carthamus tinctorius* L.) cotyledons. Biochem. J. 272, 23–29.

Somerville, C.R., Bonetta, D., 2001. Plants as factories for technical materials. Plant Physiol. 125, 168–171.

Sorensen, B.M., Furukawa-Stoffer, T.L., Marshall, K.S., Page, E.K., Mir, Z., Forster, R.J., Weselake, R.J., 2005. Storage lipid accumulation and acyltransferase action in developing flaxseed. Lipids 40, 1043–1049.

Spielmeyer, W., Green, A.G., Lagudah, E.S., Mendham, N., 1998a. Inheritance of resistance to flax wilt (*Fusarium oxysporum* f.sp.lini) in a doubled haploid population of *Linum usitatissimum* L. Euphytica 101, 287–291.

Spielmeyer, W., Green, A.G., Bittisnich, D., Mendham, N., Lagudah, E.S., 1998b. Identification of quantitative trait loci contributing to Fusarium wilt resistance on an AFLP linkage map of flax (*Linum usitatissimum*). Theor. Appl. Genet. 97, 633–641.

Suzuki, M.M., Bird, A., 2008. DNA methylation landscapes: provocative insights for epigenomics. Nat. Rev. Genet. 9, 65–476.

Tammes, T., 1928. The genetics of the genus *Linum*. In: Lotsy, J.P., Goddijn, W.A. (Eds.), Bibliographia Genetica. Martinus Nijhoff, The Netherlands, pp. 1–34.

Tammes, T., 1925. Mutation and evolution. Z. Induct. Abstamm. u. VererbLehre 36, 417–426.

Taylor, D.C., Smith, M.A., Fobert, P., Mietkiewska, E., Weselake, R.J., 2011. Metabolic engineering of higher plants to produce bio-industrial oils. In: Moo-Young, M. (Ed.), Comprehensive Biotechnology, vol. 4. second ed. Elsevier B.V., pp. 67–85.

Thomas, A.G., Kelner, D.J., Wise, R.F., Frick, B.L., 1997. Manitoba Weed Survey: Comparing Zero and Conventional Tillage Crop Production Systems. Weed Survey Series Publication 97-1. Manitoba Agriculture, Soils and Crops Branch, Carman Manitoba, R0G 0J0. 130 p.

Tutin, T.G., Heywood, V.H., Burges, N.A., Valentine, D.H., 1980. Flora Europaea. Cambridge University Press, Cambridge, UK. 480 p.

Tzen, J.T.C., Cao, Y.Z., Laurent, P., Ratnayake, C., Huang, A.H.C., 1993. Lipids, proteins, and structure of seed oil bodies from diverse species. Plant Physiol. 101, 267–276.

Vaisey-Genser, M., Morris, D.H., 2003. History of the cultivation and uses of flaxseed. In: Muir, A.D., Westcott, N.D. (Eds.), Flax: The Genus *Linum*. Taylor and Francis Ltd, New York, NY, pp. 1–22.

van Bueren Lammerts, E.T., Backes, G., De Vriend, H., Ostergard, H., 2010. The role of molecular markers and marker assisted selection in breeding for organic agriculture. Euphytica 175, 51–64.

van Zeist, W., Bakker-Heeres, J.A.H., 1975. Evidence for flax cultivation before 600 BC. J. Archaeol. Sci. 2, 215–219.

Vavilov, N.I., 1926. Studies in the Origin of Cultivated Plants. Leningrad, Vsesoiuz. Inst. Priklad, Moscow, Russia. 248 p.

Velasco, L., Goffman, F.D., 2000. Tocopherol, plastochromanol and fatty acid patterns in the genus *Linum*. Plant Syst. Evol. 221, 77–88.

Venglat, P., Xiang, D., Qiu, S., Stone, S.L., Tibiche, C., Cram, D., Alting-Mees, M., Nowak, J., Cloutier, S., Deyholos, M., Bekkaoui, F., Sharpe, A., Wang, E., Rowland, G., Selvaraj, G., Datla, R., 2011. Gene expression analysis of flax seed development. BMC Plant Biol. 11, 74.

Vera, C.L., Duguid, S.D., Fox, S.L., Rashid, K.Y., Dribnenki, J.C.P., Clarke, F.R., 2011. Short Communication: comparative effect of lodging on seed yield of flax and wheat. Can. J. Plant Sci. 92, 39–43.

Vereshchagin, A.G., Novitskaya, G.V., 1965. Triglyceride composition of flax oil. J. Am. Oil Chem. Soc. 42, 970–975.

Vrinten, P., Hu, Z., Munchinsky, M.A., Rowland, G., Qiu, X., 2005. Two *FAD3* desaturase genes control the level of linolenic acid in flax seed. Plant Physiol. 139, 79–87.

Vromans, J., 2006. Molecular Genetic Studies in Flax (*Linum usitatissimum* L.) (Ph.D. thesis). Wageningen University, Wageningen, The Netherlands, 226 p.

Wall, D.A., 1994a. Fluazifop-p tank-mixtures with clethodim for annual grass control in flax (*Linum usitatissimum*). Weed Tech. 8, 673–678.

Wall, D.A., 1994b. Response of flax and lentil to seeding rates, depths and spring application of dinitroanaline herbicides. Can. J. Plant Sci. 74, 875–882.

Wall, D.A., Kenaschuk, E.O., 1996. Flax tolerance to thifensulfuron and tribenuron. Can. J. Plant Sci. 76, 899–905.

Wall, D.A., Smith, M.A.H., 1999. Control of volunteer flax in wheat. Can. J. Plant Sci. 79, 463–468.

Wang, Z., Hobson, N., Galindo, L., Zhu, S., Shi, D., McDill, J., Yang, L., Hawkins, S., Neutelings, G., Datla, R., Lambert, G., Galbraith, D.W., Grassa, C.J., Geraldes, A., Cronk, Q.C., Cullis, C., Dash, P.K., Kumar, P.A., Cloutier, S., Sharpe, A.G., Wong, G.-K.S., Wang, J., Deyholos, M.K., 2012. The genome of flax (*Linum sitatissiumum*) assembled *de novo* from short shotgun sequence reads. Plant J. 72, 461–473.

Weiss, E., 1993. Linola – a new edible oil. Oils Fats Int. 9, 23–24.

Weselake, R.J., Shah, S., Taylor, D.C., Rahman, H., Shah, S., Laroche, A., McVetty, P.B.E., Harwood, J.L., 2009. Increasing the flow of carbon into seed oil. Biotechnol. Adv. 27, 866–878.

Western Canadian flaxseed, 2010. Scientific Analysis of Harvest and Export Quality. http://www.grainscanada.gc.ca/flax-lin/hqfm-mqrl-eng.htm (accessed 23.11.10.).

Whitaker, T.B., Freese, L., Giesbrecht, F.G., Slate, A.B., 2001. Sampling grain shipments to detect genetically modified seed. J. AOAC Int. 84, 1941–1946.

Wotjasik, W., Kulma, A., Dymińska, L., Hanuza, J., Żebrowski, J., Szopa, J., 2013. Fibres from flax overproducing β-1,3-glucanase show increased accumulation of pectin and phenolics and thus higher antioxidant capacity. BMC Biotechnol. 13, 10.

Wróbel, M., Żebrowski, J., Szopa, J., 2004. Polyhydroxybutyrate synthesis in transgenic flax. J. Biotechnol. 8, 41–54.

Wróbel-Kwiatkowska, M., Czemplik, M., Kulma, A., Żuk, M., Kaczmar, J., Dymińska, L., Hanuza, J., Ptak, M., Szopa, J., 2012. New biocomposites based on bioplastic flax fibers and biodegradable polymers. Biotechnol. Prog. 28, 1336–1346.

Wróbel-Kwiatkowska, M., Żebrowski, J., Starzycki, M., Oszmiański, J., Szopa, J., 2007. Engineering of PHB synthesis causes improved elastic properties of flax fibers. Biotechnol. Prog. 23, 269–277.

Zeven, A.C., 1982. Dictionary of Cultivated Plants and Their Centers of Diversity, Excluding Ornamentals, Forest Tress and Lower Plants. Center for Agricultural Publishing and Documentation, Wageningen, The Netherlands. 263 p.

Zhang, M., Fan, J.L., Taylor, D.C., Ohlrogge, J.B., 2009. DGAT1 and PDAT1 acyltransferases have overlapping functions in *Arabidopsis* triacylglycerol biosynthesis and are essential for normal pollen and seed development. Plant Cell 21, 3885–3901.

Zheng, P., Allen, W.B., Roesler, K., Williams, M.E., Zhang, S., Li, J., Glassman, K., Ranch, J., Nubel, D., Solawetz, W., Bhattramakki, D., Llaca, V., Deschamps, S., Zhong, G.Y., Tarczynski, M.C., Shen, B., 2008. A phenylalanine in DGAT is a key determinant of oil content and composition in maize. Nat. Genet. 40, 367–372.

Zohary, D., Hopf, M., 2000. Domestication of Plants in the Old Word: The Origin and Spread of Cultivated Plants in West Asia, Europe and the Nile Valley. Oxford University Press, Oxford, UK. 316 p.

Zou, J.T., Wei, Y.D., Jako, C., Kumar, A., Selvaraj, G., Taylor, D.C., 1999. The *Arabidopsis thaliana* TAG1 mutant has a mutation in a diacylglycerol acyltransferase gene. Plant J. 19, 645–653.

Żuk, M., Kulma, A., Dymińska, L., Szoltysek, K., Prescha, A., Hanuza, J., Szopa, J., 2011. Flavonoid engineering of flax potentiate its biotechnological application. BMC Biotechnol. 11, 10.

Żuk, M., Prescha, A., Stryczewska, M., Szopa, J., 2012. Engineering flax plants to increase their antioxidant capacity and improve oil composition and stability. J. Agric. Food Chem. 60, 5003–5012.

Chapter 7

Crambe (*Crambe abyssinica*)

Li-Hua Zhu
Department of Plant Breeding, Swedish University of Agricultural Sciences, Alnarp, Sweden

ORIGIN AND HISTORY

Crambe (*Crambe abyssinica* Hochst) is an annual oilseed crop that is already in commercial production in some regions. It is originated in the Mediterranean region and parts of eastern Africa; however, it has become adapted to colder and drier regions (Weiss, 2000). It has been grown in tropical and subtropical Africa, the Near East, Central and West Asia, Europe, the United States, and South America. It was first used as a crop in 1933 at the Boronez Botanical Station, USSR, and was a part of a Swedish breeding program in 1949 (Oplinger et al., 1991). Intensive studies on crambe cultivation had been carried out in the 1960s. More recently, the interest in developing crambe as an industrial crop has been retained in an effort to find out renewable vegetative oils for replacing the fossil oil.

CRAMBE BOTANY

Crambe belongs to the family Brassicaceae. It has a number of branches and leaves are oval shaped (Fig. 7.1A). Plant height can reach over 1 m if growing conditions allow. Flowers are small and white (Fig. 7.1B), which are initially compressed, but later loosely distributed along the elongated flower stalk. Crambe is predominantly self-pollinating, but 9–14% cross-pollination has been reported (Vollmann and Ruckenbauer, 1991). Flowering is indeterminate and the flowering period can last for over 2 months with simultaneous branching under suitable growing conditions (Fig. 7.1C). Each fruit bears a single seed, which is enclosed in pod or silique when it matures. The pod shatter does not normally occur. Mature fruits are spherical with diameter of ~2.5 mm and color in tan (Fig. 7.1D). The 1000 seed weight is ~7.5 g including silique, which accounts for ~20% of the seed weight.

SEED COMPOSITION

Crambe seeds contain about 26–38% oil, 21.6–26% protein, 16.4–18% fiber, and 5.8% ash (Lazzeri et al., 1994; Kulig and Pisulewska, 2000; Lalas et al., 2012).

Industrial Oil Crops. http://dx.doi.org/10.1016/B978-1-893997-98-1.00007-5

FIGURE 7.1 Leaves (A), flowers (B), growing plants (C), and mature seeds (D) of *Crambe abyssinica*.

The K, P, Ca, and Mg content of the fruit is 5.8, 7.3, 4.3, and 2.2 g/kg on dry matter base, respectively (Kulig and Pisulewska, 2000). The seed water content is ~7–10% and higher than 11% may accelerate seed aging or compromise seed viability (Werner et al., 2013).

The fatty acid (FA) composition in the seed oil of crambe consists of 1–4% palmitic acid (C16:0), 0.1–0.5% palmitoleic acid (C16:1), 0.5–2.0 stearic acid (C18:0), 10.0–25.0% oleic acid (C18:1), 7.0–15.0% linoleic acid (C18:2), 2.0–5.0% linolenic acid (C18:3), 0.5–2.0% arachidic acid (C20:0), 2.0–6.0% eicosenoic acid (C20:1), 0.0–0.5% eicosadienoic acid (C20:2), 1.0–3.0% behenic acid (C22:0), 50.0–65.0% erucic acid (*cis*-13-docosenoic acid) (C22:1), and 0.0–1.0% lignoceric acid (C24:0) (Lazzeri et al., 1994; Wang et al., 1995; Finiquerra et al., 2001; Li et al., 2012). The oil contains high levels of β-sitosterol (51.93%), campestanol (21.98%), and brassicasterol (12.35%). Moreover, the seed oil also contains α-, β-, and δ-tocopherols at

7.67, 125.04, and 3.99 mg/kg, respectively (Lalas et al., 2012). However, it should be noted that the oil content and composition vary depending on growing conditions and management (Yuldasheva et al., 2011).

CRAMBE AS AN INDUSTRIAL OIL CROP

Owing to its high level of erucic acid (C22:1), crambe is considered as a dedicated oilseed crop for industrial oil production as erucic acid is an important feedstock in the oleochemical industry (Lazzeri et al., 1994; Leonard, 1994). Also due to its diverse potential industrial applications, erucic acid oil has attracted wide interests from many companies to invest more resources for optimizing oil extraction and for evaluating the potential uses of this oil as raw material for hydraulic fluids, oleochemicals, lubricants, additives, and as a starting material for new fibers, resins, plastics, lacquers, etc. (Carlsson et al., 2011). Currently, erucic acid oil is mainly produced in high-erucic acid rapeseed oil (Piazza and Foglia, 2001). As an industrial oil crop, crambe has clear advantage over the rapeseed as it does not cross over with other food oilseed crops including rapeseed, and it contains naturally much higher erucic acid in its seed oil. The oxidative stability of the biodiesel produced from the crambe oil has been shown to be more stable than soybean oil (Wazilewski et al., 2013), providing one more potential use of crambe as an oil crop.

Crambe seed cake contains high levels of proteins and fibers but also high levels of glucosinolates (3–6% (w/w)) that are toxic after hydroxylation and must be removed if the seed cake will be used as potential nutritional feeds for animals. The high protein and fiber levels make crambe seed cake a potential valuable byproduct in different nonfood fields. The crambe seed cake has been studied for protein-based plastic production (Newson, 2015). In addition, the crambe seed cake may be useful as an adsorbent material in removing the toxic metals such as cadmium, lead, and chromium from contaminated aqueous solutions (Artus, 2006; Oliveira et al., 2009; Gonçalves et al., 2013; Rubio et al., 2013).

CRAMBE CULTIVATION

Crambe cultivation was started already in the 1930s in the former USSR. Its agronomical characteristics and industrial uses had been extensively studied in the 1960s. Testing of the crop was extended to more regions in USSR, Sweden, and Poland. After the Second World War, the crop reached an area of 25,000 ha in Poland (Mastebroek et al., 1994). Its cultivation and commercial processing were expanding rapidly in both North America and Europe as well as Asia (Lazzeri et al., 1993; Warwick and Gugel, 2003) in the 1990s. For example, the varieties Belann and Belenzian, cultivated in the Po Valley or Padan Plain, northern Italy in the early 1990s, showed a good adaptability to this latitude, good yields (up to 3.2 ton/ha), short cultivation period (90–100 days), low seed dehiscence even at full maturity, optimal resistance to water shortage, and good resistance toward fungal, bacterial, and insect attacks (Fontana et al., 1998; Lazzeri et al., 1994). As a major source of erucic acid, the cultivation of crambe has recently been tested and recommended in southern Europe (Zanetti et al., 2012).

Crambe has wide climatic and agronomical adaptations. It is in general tolerant to drought and chill and moderate tolerant to saline soils but sensitive to waterlogging. It can be planted in semiarid and arid ecosystems where saline soils and saline irrigation water are common (Ionov et al., 2013). It can also be cultivated with good yields both in central Europe, as a spring-summer crop, and in southern Mediterranean countries as a winter crop. The yield can range from 179 to 3200 ton/ha depending on soil and climate conditions, sowing date, and management (Mastebroek et al., 1994; Johnson et al., 1995; Fontana et al., 1998; Ionov et al., 2013). However, for achieving high yields, well-drained and fertile soils of moderately coarse to fine texture (pH 6.0–7.0 or slightly higher) are required. The oil content decreased as the irrigation-water salinity increased (Francois and Kleiman, 1990; Yuldasheva et al., 2011). On the other hand, crambe cultivation was initiated in small field plots in the Aral Sea area where soil salinity ranged between 3% and 10% (Kamalov, 1995), suggesting its potential of salt tolerance. Water stress conditions induced by polyethylene glycol solutions significantly reduced seed germination and vigor in crambe (Toledo et al., 2011).

Fertilizers are also needed to achieve high yields. In general, phosphorus and potassium recommendations for small grains can be applied to crambe. It has been tested that ~50 kg/ha P_2O_5 and ~89 kg/ha K_2O are recommended. Crambe also responds to nitrogen fertilizer with ranging from ~89 to ~150 kg/ha (125 kg/ha in Sweden and 150 kg/ha in England) of actual N recommended. Leaf application of zinc did not clearly affect the vegetative and reproductive development of crambe; however, the phosphorus applied on seedling stage increased oil content and 1000 seed weight, and thus an increased seed yield (Benetoli da Silva et al., 2011). Credits for legume or manure N should be taken when appropriate. Moreover, crambe may respond to sulfur (22–28 kg/ha) on low sulfur soils (Oplinger et al., 1991).

Weed control is important for achieving high yields in crambe. In addition, weeds can also contribute to seed moisture (Stougaard and Moomaw, 1991), thus negatively affecting germination. Competition from common lambsquarters (*Chenopodium album* L.), pigweed (*Amaranthus* sp.), and foxtail (*Setaria* sp.) has been reported to hinder crambe production (Agricultural Research Service, 1966). Crambe has been shown to be tolerant to soil-applied trifluralin, pendimethalin, and cinmethylin, or DPX-A7881 applied postemergence. Postemergence applications of chlorimuron or imazethapyr were injurious to crambe (Stougaard and Moomaw, 1991).

Sowing distance ranges from 12.5 to 90 cm between rows. Narrow rows tend to reduce branching but provide more uniform maturity. Row widths of 15–30 cm can provide high yields if weeds are not a serious problem. In Europe, for instance, Sweden or England, the row width of 12.5 cm has been used for achieving high yields. Rows wider than 76 cm tend to lodge, making harvest difficult. Planting depth is a critical factor for crambe to obtain acceptable yields (Oplinger et al., 1991). It is recommended to have 0.6 cm in normal soils and up to 2.5 cm in drier regions. A seeding rate of 11–22 kg/ha is recommended (Carlson et al., 1996).

Compared with rapeseed, crambe is generally more resistant to diseases and insects. A study has shown that crambe is more resistant to the flea beetle, *Phyllotreta cruciferae* (Goeze), than rapeseed (Andersson et al., 1992). A field survey carried out in the west of Paraná, Brazil, has identified the following two diseases: black rot caused by *Xanthomona scampestris* pv. *Campestris* (Xcc), resulting in damping off in seedlings, and *Alternaria brassicae*, causing spots in adult plants (Moers et al., 2012). The bacterial blight caused by *Xanthomonas campestris*, closely related to Xcc, was first identified in the United States (Mihail et al., 1993). Seeds can be treated with a fungicide or soaked in water at 60°C for 20 min or at 54°C for 30 min to reduce the infection of *A. brassicae* (Oplinger et al., 1991).

Crambe could be cultivated as a cover crop. The decomposition and nutrient release from crop residues of crambe in the implementation of no-tillage system has been reported. Heinz et al. (2011) have reported that the kinetics of residue decomposition in crambe were similar to those of nutrient release with an initial rapid phase followed by a slower one. K, P, and Mg are released more quickly for subsequent crops, while the increased speed of nutrients release occurred around 15 days after the biomass management. Due to a high content of glucosinolates, crambe could also be used as a crop for biofumigation.

GENETIC RESOURCES AND CONVENTIONAL BREEDING

Crambe is hexaploid with the chromosome number of $2n = 6x = 90$. The conventional breeding by successive selection within *C. abyssinica* began already in some European countries in the 1950s and improved strains from Russia and Sweden were introduced into Canada and the United States in the late 1950s. Shortly later, three new varieties (Prophet, Indy, and Meyer) were bred and released in the 1970s by the Purdue University Agricultural Experimental Station, Maryland, USA. The former two came from selection of accessions and the last one through a cross of two accessions. Two new cultivars, "BelAnn" and "BelEnzian," bred by introgression and mass selection were released in 1985 by the USDA-ARS at Beltsville, Maryland, USA, as high yielding sources of erucic acid (Campbell et al., 1986; Mastebroek et al., 1994).

In the early 1990s, the DLO-Centre for Plant Breeding and Reproduction Research in the Netherlands started a breeding program in crambe. They evaluated agronomic traits of more than 20 accessions of crambe collected from cultivars or breeding lines from Sweden, USA, Germany, USSR, Poland, Bulgaria, and Denmark. Two years field evaluation showed a large variation in earliness, length of the top branch, 1000 seed weight and FA composition, depending on year. Earliness had a large impact on crop development and seed quality. Seed yield and oil content were considerably affected by environmental conditions. Little variation was found in the content of erucic acid and glucosinolates in the seeds (Mastebroek et al., 1994). Shortly after, Mastebroek and Lange (1997) reported that they made crosses between three accessions, two early flowering American breeding lines and a late flowering European land race. Estimation of broad-sense heritability on the F_3 population revealed high values for 1000 seed weight, moderate for plant height, earliness and seed oil content, and low for seed yield and number of seeds produced per square meter. It was concluded under Dutch environmental conditions, selection for earliness might be an effective tool to improve the total yield of seed oil.

Huang et al. (2013) have reported that the plant height is significantly correlated with the branching height and the number of first branches, while the branching height is significantly correlated with the pod number of primary inflorescence and the number of secondary branches but negatively correlated with the number of first branches. The 1000 seed weight is significantly correlated with seed yield per plant. Lara-Fioreze et al. (2013) have reported that the seed germination and vigor are positively correlated, but 1000 seed weight and oil content are not correlated with the seed germination or vigor.

C. abyssinica shows low genetic variability (Warwick and Gugel, 2003) on some important agronomic traits, such as hardiness, maturity time, and pest resistance (Mastebroek et al., 1994). This could be possibly improved by crossing the species with the related taxon, for instance, *C. hispanica* (Mulder and Mastebroek, 1996). On the other hand, crambe has a number of superior agronomic traits that may serve as a source of genes for *Brassica* crop improvement, particularly for flea beetle resistance (Anderson et al., 1992). With help of embryo/ovary recovery techniques, artificial crosses between *C. abyssinica* and *Brassica* can be carried out to overcome the crossing barriers between species (Wang and Luo, 1998; Warwick and Gugel, 2003). Wang et al. (2004) studied sexual progenies of asymmetric somatic hybrids between *B. napus* and *C. abyssinica* and demonstrated that a part of the *C. abyssinica* genome can be transferred into *B. napus* via asymmetric hybridization and maintained in sexual progenies of the hybrids.

CRAMBE IMPROVEMENT USING GENE TECHNOLOGY

The low genetic variability in *C. abyssinica* limits its genetic improvement through conventional breeding methods. Genetic modification is an important complementary method to conventional breeding for improving plant properties. Li et al. (2010, 2011) have recently developed a regeneration protocol with over 95% of regeneration frequency and a transformation protocol with 2.1% of transformation frequency using hypocotyles as explants and kanamycin as the selective agent. Chhikara et al. (2012) reported another protocol with 50–70%

regeneration and 6.7–8.3% transformation frequency using also hypocotyls as explants but hygromycin as the selective agent. Shortly after, Li et al. (2013) reported a modified transformation protocol with 9.9% of transformation frequency using hypocotyles as explants and hygromycin for selection. Qi et al. (2014) reported a crambe transformation protocol using cotyledonary nodes as explants. In addition, Głąb et al. (2013) have reported the successful infection of *Agrobacterium rhizogenes* of *C. abyssinica* to achieve hairy root formation. Using the *Agrobacterium tumefaciens*–mediated transformation protocol, Li et al. (2012) have successfully transformed three target genes into crambe for increasing erucic acid levels in the seed oil and obtained a transgenic line with 73% of erucic acid of the total seed oil compared with 60% in the wild type and transgenic lines with production of wax esters in the seed oil (Zhu et al., 2016) as well as production of high oleic acid oil (Li et al., 2015). The enzymatic studies on the high erucic acid line revealed the bottleneck in further increasing erucic acid level is at the early seed development stage (Guan et al., 2013). Further studies on downregulation of other fatty acid biosynthesis–related genes revealed a more complex feature of fatty acid synthesis (Guan et al., 2014). These achievements have demonstrated the feasibility of developing crambe into a bioplatform for industrial feedstock production. Some important genes involved in the oil biosynthesis have been characterized in crambe. Three alleles of *FAD2* genes (*FAD2-1*, *FAD2-2*, and *FAD2-3*) encoding for the FA desaturase enzyme have been identified very recently, and *FAD2-3* has shown to be highly expressed during the seed development (Cheng et al., 2013). It has also been reported that the differential expression of transcripts encoding glutathione-*S*-transferases, antioxidants, sulfur metabolism, heat-shock proteins, metal transporters, and enzymes in the ubiquitination pathway of protein degradation as well as several unknown novel proteins serve as molecular evidence for the physiological responses to arsenate stress in plants (Paulose et al., 2010).

Different molecular methods are required for evaluating transgenic lines or gene expression levels. Studies have shown that for Southern blot analysis on crambe, more than 50 μg DNA is required for a clear band. The choice of enzymes for DNA digestion was not rigid for confirmation of the T-DNA integration, while for determining the copy number of transgenes, suitable enzymes should be chosen. Increasing the enzyme concentration could improve the digestion and 20 μL enzyme was recommended for a complete digestion of up to 80 μg crambe DNA. For quantitative real-time polymerase chain reaction (qRT-PCR) analysis, sampling at ~20 days after flowering was observed to be the suitable time for quantitative real time PCR (qRT-PCR) analysis of the genes involved in the seed oil biosynthesis (Li et al., 2013).

FUTURE PERSPECTIVES

Crambe is not a new oilseed crop, but it has been seriously reconsidered in the recent years as a dedicated industrial oil crop. Yet, compared with other oilseed crops, research on crambe breeding and cultivation is relatively less intensive and needs to be enhanced. With help of modern breeding technologies, such as genetic transformation and site-directed mutagenesis, the potential in improving oil qualities and quantities as well as other important agronomic traits of crambe has become greater. With the recent establishment of efficient transformation protocols and other modern molecular tools, it is now possible to genetically modify crambe as needed. Since the genome sequence of *C. abyssinica* is not yet available and the molecular controls of many agronomic traits are unrivaled in crambe, the genetic improvement of these important traits is currently limited. However, the DNA sequence information of some genes obtained in crambe has shown a great degree of homology to *Arabidopsis* genes, which would facilitate the identification of target genes for a given trait for genetic improvement of crambe. In addition as an oilseed crop, the potential of using the seed cake as protein-based plastics and the whole aboveground biomass of crambe for diverse uses should be further studied in order to develop crambe into a multipurpose crop.

REFERENCES

Agricultural Research Service, 1966. Culture of Cambe – A New Industial Oilseed Crop. Prod. Res. Rep. No. 95. USDA-ARS, Washington, DC.

Anderson, M.D., Peng, C., Weiss, M.J., 1992. *Crambe*, *Crambe abyssinica* Hochst, as a flea beetle resistant crop (*Coleoptera: Chrysomelidae*). J. Econ. Entomol. 85, 594–600.

Artus, N.N., 2006. Arsenic and cadmium phytoextraction potential of crambe compared with Indian mustard. J. Plant Nutr. 29, 667–679.

Benetoli da Silva, T.R., Lavagnolli, R.F., Nolla, A., 2011. Zinc and phosphorus fertilization of crambe (*Crambe abyssinica* Hoechst). J. Food Agric. Environ. 9 (1), 264–267.

Campbell, T.A., Crock, J., Williams, J.H., Hang, A.N., Sigafus, R.E., Schneiter, A.A., McClain, C.R., Graves, C.R., Wooley, D.G., Kleiman, R., Adamson, W.C., 1986. Registration of 'Belann' and 'Belenzian' crambe. Crop Sci. 26 (5), 1082–1083.

Carlson, K.D., Gardner, J.C., Anderson, V.L., Hanzel, J.J., 1996. Crambe: new crop success. In: Janick, J. (Ed.), Progress in New Crops. ASHS Press, Alexandria, VA, pp. 306–322.

Carlsson, A.S., Lindberg Yilmaz, J., Green, A.G., Stymne, S., Hofvander, P., 2011. Replacing fossil oil with fresh oil – with what and for what? Eur. J. Lipid Sci. Tech. 113 (7), 812–831.

Cheng, J., Zhu, L.-H., Salentijn, E.M.J., Huang, B., Gruber, J., Dechesne, A.C., Krens, F.A., Qi, W., Visser, R.G.F., van Loo, E.N., 2013. Functional analysis of the omega-6 fatty acid desaturase (*CaFAD2*) gene family of the oil seed crop *Crambe abyssinica*. BMC Plant Biol. 13, 146.

Chhikara, S., Duttaa, I., Paulosea, B., Jaiwalb, P.K., Dhankhera, O.P., 2012. Development of an *Agrobacterium*-mediated stable transformation method for industrial oilseed crop *Crambe abyssinica* 'BelAnn'. Ind. Crops Prod. 37, 457–465.

Finiquerra, M.G., Lori, R., Palmieri, S., 2001. Soluble and total myrosinase activity in defatted *Crambe abyssinica* meal. J. Agric. Food Chem. 49, 840–845.

Fontana, F., Lazzeri, L., Malaguti, L., Galletti, S., 1998. Agronomic characterization of some *Crambe abyssinica* genotypes in a locality of the Po Valley. Eur. J. Agron. 9, 117–126.

Francois, L.E., Kleiman, R., 1990. Salinity effects on vegetative growth, seed yield, and fatty acid composition of crambe. Agron. J. 82, 1110–1114.

Głąb, B., Furmanek, T., Miklaszewska, M., Banas, A., Królicka, A., 2013. Lipids in hairy roots and non-Agrobacterium induced roots of *Crambe abyssinica*. Acta Physiol. Plant 35, 2137–2145.

Gonçalves Jr., A.C., Rubio, F., Meneghel, A.P., Coelho, G.F., Dragunski, D.C., Strey, L., 2013. The use of *Crambe abyssinica* seeds as adsorbent in the removal of metals from waters. Rev. Bras. Eng. Agríc. Ambiental 17 (3), 306–311.

Guan, R., Lager, I., Li, X.Y., Stymne, S., Zhu, L.-H., 2013. Bottlenecks in erucic acid accumulation in genetically engineered ultra-high erucic acid *Crambe abyssinica*. Plant Biotech. J. 12 (2), 193–203.

Guan, R., Li, X., Hofvander, P., Zhou, X.-R., Wang, D., Stymne, S., Zhu, L.-H., 2014. RNAi targeting putative genes in phosphatidylcholine turnover results in significant change in fatty acid composition in *Crambe abyssinica* seed oil. Lipids 50, 407–416.

Heinz, R., Garbiate, M.V., Neto, A.L.V., de Sousa Mota, L.H., Correia, A.M.P., Vitorino, A.C.T., 2011. Decomposition and nutrient release of crambe and fodder radish residues. Ciênc. Rural 41, 1549–1554.

Huang, B., Yang, Y., Luo, T., Wu, S., Du, X., Cai, D., van Loo, E.N., Huang, B., 2013. Components in *Crambe abyssinica*, a promising industrial oil crop. Am. J. Plant Sci. 4, 42–47.

Ionov, M., Yuldasheva, N., Ulchenko, N., Glushenkova, A.I., Heuer, B., 2013. Growth, development and yield of *Crambe abyssinica* under saline irrigation in the greenhouse. J. Agron. Crop Sci. 199, 331–339.

Johnson, B.L., McKay, K.R., Schneiter, A.A., Hanson, B.K., Schatz, B.G., 1995. Influence of planting date on canola and crambe production. J. Prod. Agric. 8, 594–599.

Kamalov, S.H., 1995. Working out of technology of clayey saline phyto-amelioration in the southern part of Aral sea. J. Arid Land Stud. 55, 311–314.

Kulig, B., Pisulewska, E., 2000. Remove from marked records fruit yield and chemical composition of crambe genotypes (*Crambe abyssinica* Hochst.) as affected by the growing seasons. Rośliny Oleiste 21, 631–639.

Lalas, S., Gortzi, O., Athanasiadis, V., Dourtoglou, E., Dourtoglou, V., 2012. Full characterisation of *Crambe abyssinica* hochst. Seed oil. J. Am. Oil Chem. Soc. (JAOCS) 89, 2253–2258.

Lara-Fioreze, A.C.C., Tomaz, C.A., Fioreze, S.L., Pilon, C., Zanotto, M.D., 2013. Genetic diversity among progenies of *Crambe abyssinica* Hochst for seed traits. Ind. Crops Prod. 50, 771–775.

Lazzeri, L., Conte, L.S., Palmieri, S., 1993. Agronomical and technological characterization of Italian-grown *Crambe abyssinica*. Agric. Med. 123, 55–64.

Lazzari, L., Leoni, O., Conte, L.S., Palmieri, S., 1994. Some technological and potential uses of *Crambe abyssinica* products. Ind. Crops Prod. 3, 103–112.

Leonard, C., 1994. Sources and commercial applications of high erucic vegetable oils. Lipid Tech. 6 (4), 79–83.

Li, X.Y., Ahlman, A., Yan, X.F., Lindgren, H., Zhu, L.H., 2010. Genetic transformation of the oilseed crop *Crambe abyssinica*. Plant Cell Tissue Organ Cult. 100, 149–156.

Li, X.Y., Ahlman, A., Lindgren, H., Zhu, L.H., 2011. Highly efficient in vitro regeneration of the industrial oilseed crop *Crambe abyssinica*. Ind. Crops Prod. 33, 170–175.

Li, X.Y., van Loo, E.N., Gruber, J., Fan, J., Guan, R., Frentzen, M., Stymne, S., Zhu, L.-H., 2012. Development of ultra-high erucic acid oil in the industrial oil crop *Crambe abyssinica*. Plant Biotech. J. 10, 862–870.

Li, X.Y., Fan, J., Gruber, J., Guan, R., Frentzen, M., Zhu, L.-H., 2013. Efficient selection and evaluation of transgenic lines of *Crambe abyssinica*. Front. Plant Sci. 4, 162. http://dx.doi.org/10.3389/fpls. 2013.00162.

Li, X., Mei, D., Liu, Q., Fan, J., Singh, S., Green, A., Zhou, X.-R., Zhu, L.-H., 2015. Down-regulation of crambe fatty acid desaturase and elongase in Arabidopsis and crambe resulted in significantly increased oleic acid content in seed oil. Plant Biotechnol. J. 1–9. http://dx.doi.org/10.1111/pbi.12386.

Mastebroek, H.D., Wallenburg, S.C., van Soest, L.J.M., 1994. Variation for agronomic characteristics in crambe (*Crambe abyssinica* Hochst. ex Fries). Ind. Crops Prod. 2, 129–136.

Mastebroek, H.D., Lange, W., 1997. Progress in a crambe cross breeding programme. Ind. Crops Prod. 6, 221–227.

Mulder, J.H., Mastebroek, H.D., 1996. Variation for agronomic characteristics in *Crambe hispanica*, a wild relative of *Crambe abyssinica*. Euphytica 89, 267–278.

Mihail, J.D., Taylor, S.J., Verslues, P.E., Hodge, N.C., 1993. Bacterial blight of *Crambe abyssinica* in Missouri caused by *Xanthomonas campestris*. Plant Dis. 77, 569–574.

Moers, E.M., Kuhn, O.J., Gonçalves Jr., A.C., Franzener, G., Stangarlin, J.R., 2012. Remove from marked records survey of diseases in the culture of crambe (*Crambe abyssinica* Hochst) in the west of Paraná. Sci. Agrar. Paran. 11, 35–48.

Newson, W.R., 2015. Bio-based materials from Crambe and Carinata industrial oilseed meals – compression moulded and extruded oilseed meal plastics, vol. 127 (Doctoral thesis). Swedish University of Agricultural Sciences, Alnarp. ISBN: 978-91-576-8352-9 (print version); ISBN: 978-91-576-8453-6 (electronic version).

Oliveira, L.S., Franca, A.S., Oliveira, V.F., Salles, A., 2009. Use of *Crambe abyssinica* press cake as a biosorbent for waste water treatment. In: International Conference on Chemical, Biological & Environmental Engineering, Singapore. Proceedings of the 2009 International Conference on Chemical, Biological & Environmental Engineering, Londres: Glamorgan University, 2009, pp. 25–28.

Oplinger, E.S., Oelke, E.A., Kaminski, A.R., Putnam, D.H., Teynor, T.M., Doll, J.D., Kelling, K.A., Durgan, B.R., Noetzel, D.M., 1991. Alternative Field Crops Manual, Crambe. University of Wisconsin-Extension, Cooperative Extension; University of Minnesota: Center for Alternative Plant & Animal Products and the Minnesota Extension Service. http://www.hort.purdue.edu/newcrop/afcm/crambe.html.

Piazza, G.J., Foglia, T.A., 2001. Rapeseed oil for oleo-chemical uses. Eur. J. Lipid Sci. Tech. 103 (7), 405–454.

Paulose, B., Kandasamy, S., Dhankher, O.P., 2010. Expression profiling of *Crambe abyssinica* under arsenate stress identifies genes and gene networks involved in arsenic metabolism and detoxification. BMC Plant Biol. 10 (108). http://www.biomedcentral.com/1471-2229/10/108.

Qi, W., Tinnenbroek-Cape, I.E.M., Schaart, J.G., Huang, B., Cheng, J., Visser, R.G.F., Van Loo, E.N., Krens, F.A., 2014. Regeneration and transformation of *Crambe abyssinica*. BMC Plant Biol. 14, 235.

Rubio, F., Gonc¸alves Jr., A.C., Dragunski, D.C., Tarley, C.R.T., Meneghel, A.P., Schwantes, D., 2013. A *Crambe abyssinica* seed by-product as biosorbent for lead(II) removal from water. Desalin. Water Treat. 1–10. http://dx.doi.org/10.1080/19443994.2013.836995. www.deswater.com.

Stougaard, R.N., Moomaw, R.S., 1991. Crambe (*Crambe abyssinica*) tolerance to herbicides. Weed Tech. 5 (3), 566–569.

Toledo, M.Z., Teixeira, R.N., Ferrari, T.B., Ferreira, G., Cavariani, C., Cataneo, A.C., 2011. Physiological quality and enzymatic activity of crambe seeds after the accelerated aging test. Acta Sci. Agron. 33 (4), 687–694.

Vollmann, J., Ruckenbauer, P., 1991. Estimation of outcrossing rates in Crambe (*Crambe abyssinica* Hochst. ex. Fries) using a dominant morphological marker gene. Bodenkultur 42, 361–366.

Warwick, S.I., Gugel, R.K., 2003. Genetic variation in the *Crambe abyssinica – C. hispanica – C. glabrata* complex. Gen. Res. Crop Evol. 50, 291–305.

Wang, Y.P., Luo, P., Li, X.F., 1995. Preliminary study on *Crambe abyssinica*. Acta Bot. Yunnanica 17, 169–174.

Wang, Y.P., Luo, P., 1998. Intergeneric hybridization between *Brassica* species and *Crambe abyssinica*. Euphytica 101, 1–7.

Wang, Y.P., Snowdon, R.J., Rudloff, E., Wehling, P., Friedt, W., Sonntag, K., 2004. Cytogenetic characterization and *fae1* gene variation in progenies from asymmetric somatic hybrids between *Brassica napus* and *Crambe abyssinica*. Genome 47 (4), 724–731.

Wazilewski, W.T., Bariccatti, R.A., Martins, G.I., Secco, D., Melegari de Souza, S.N., Rosa, H.A., Chaves, L.I., 2013. Study of the methyl crambe (*Crambe abyssin*ica Hochst) and soybean biodiesel oxidative stability. Ind. Crops Prod. 43, 207–212.

Weiss, E.A., 2000. Oilseed Crops, second ed. Blackwell Science.

Werner, E.T., Lopes, J.C., Junior, D.G., Luber, J., Teixeira do Amaral, J.A., 2013. Accelerated aging test to evaluate the quality of crambe (*Crambe abyssinica* Hochst – Brassicaceae) seed physiology. Idesia 31 (1), 35–43.

Yuldasheva, N.K., Ulchenko, N.T., Bekker, N.P., Chernenko, T.V., Skosyreva, O.V., Glushenkova, A.I., Mustaev, F.A., Ionov, M.V., Heuer, B., 2011. Influence of irrigation-water salinity on lipids of *Crambe abyssinica* seeds. Chem. Nat. Comp. 46 (6), 862–865.

Zanetti, F., Mosca, G., Rampin, E., Vamerali, T., 2012. Adaptability and sustainable management of high-erucic Brassicaceae in mediterranean environment. In: Akpan, U.G. (Ed.), Oilseeds. ISBN: 978-953-51-0665-4. InTech, Available from: http://www.intechopen.com/books/oilseeds/adaptabilityand-sustainable-management-of-high-erucic-brassicaceae-in-mediterranean-environment.

Zhu, L.H., Krens, F., Smith, M.A., Li, X., Qi, W., van Loo, E.N., Iven, T., Feussner, I., Nazarenus, T.J., Huai, D., Taylor, D.C., Zhou, X.R., Green, R.G., Shockey, J., Klasson, K.T., Mullen, R.T., Huang, B., Dyer, J.M., Cahoon, E.B., 2016. Dedicated industrial oilseed crops as metabolic engineering platforms for sustainable industrial feedstock production. Sci. Rep. (accepted).

Chapter 8

Camelina (*Camelina sativa*)

Eric J. Murphy
Department of Basic Sciences, School of Medicine and Health Sciences, University of North Dakota, Grand Forks, ND, United States; Agragen, LLC, Cincinnati, OH, United States; Agragen OY, Helsinki, Finland

INTRODUCTION

Camelina sativa (L.) Crantz, a member of the Brassicaceae family, has gained tremendous attention as a reemerging oil seed crop. Camelina has many positive attributes that make it unique among oil seed crops, which is the primary reason it has generated much excitement. One of the key issues is that Camelina is adaptable to many different environmental conditions and the only true limitation is heavy clay soils and organic soils (Gugel and Falk, 2006; Plessers et al., 1962; Zubr, 1997). Although seed yields and oil content are highly variable based on environment (Gugel and Falk, 2006; Plessers et al., 1962; Vollman et al., 1996; Zubr, 2003), it has outperformed rapeseed in trials under identical drought conditions (Gugel and Falk, 2006; Zubr, 1997). This combined with the relatively low inputs required (Urbaniak et al., 2008; Zubr, 1997) makes Camelina an ideal crop for use on less productive lands and in areas without sufficient rainfall to support other crops. When produced under these circumstances, Camelina would not be displacing crops used for food production and positively addresses the food for fuel debate that often plagues the use of crop oils for fuel production. Further, Camelina provides a high value meal with sufficient residual lipid content (5–10%) that has a protein profile similar to soy meal (Zubr, 1997). This makes Camelina meal an ideal meal to add to rations for poultry and other livestock (Frame et al., 2007; Plessers et al., 1962; Zubr, 1997).

Camelina is considered an ancient oil seed crop that has been used and cultivated since the bronze and iron ages (Bouby, 1998; Francis and Warwick, 2009; Plessers et al., 1962). Evidence suggests that cultivation of Camelina occurred in modern France during the early Iron Age or about 500 BC (Bouby, 1998). This is consistent with the reports of Camelina production throughout Europe during this same time period and reports that its meal was used for livestock rations and oil for lighting and making of soap (Zubr, 1997). Although Camelina was first noted in North American fields in the United States in 1838–40 (Torrey and Gray, 1838), the first intentional planting of Camelina in North America

Industrial Oil Crops. http://dx.doi.org/10.1016/B978-1-893997-98-1.00008-7

occurred in the early settlements in the Red River Valley of Manitoba in 1863 (Scoggan, 1957). Its planting continued to occur westward until it was abandoned as a crop a few years later (Scoggan, 1957). However, what has driven the reintroduction of Camelina as a commercial crop in North America and elsewhere?

The answer to this question is focused on the needs to find an adaptable, sustainable oil seed crop that produces oil in sufficient quantity and of sufficient quality to serve as a major feedstock for biofuel production. Initial excitement was generated by a report demonstrating that Camelina oil provided fatty acid methyl esters with a yield similar to rapeseed oil (Gugel and Falk, 2006). Despite the high iodine number associated with the relatively elevated amount of polyunsaturated fatty acids (PUFA) found in Camelina-derived oil, the engine performance of Camelina-derived biodiesel is on par with that produced from other oil seeds (Gugel and Falk, 2006). However, the inconsistent quality of the biodiesel produced by Camelina became an important concern, and we have seen a similar inconsistency with regard to Camelina oil–derived biodiesel meeting the boiling point specifications required for biodiesel. This suggests that Camelina-derived oil is not the best performing feedstock for first-generation biofuels such as biodiesel. On the contrary, Camelina-derived oil is an excellent feedstock for second-generation biofuels such as biojet due to the adaptability and sustainability of large-scale Camelina production. There are many reports in the popular press regarding the success of a number of refiners using Camelina-derived feedstock for production of second generation biofuels such as biojet. In this case, there are limited issues with regards to the high amount of PUFA found in Camelina oil as these fuels are all fully hydrogenated during production; hence, the major issue facing producers is to find a source of feedstock that is sustainable and environmentally friendly.

Despite the focus by many on the use of Camelina oil as a feedstock for biojet and other second-generation fuels, two characteristics of Camelina make it a desirable sustainable source of a feedstock for industrial oils. The recent demonstration of production of wax esters in Camelina (Iven et al., 2015), which are similar to that derived from sperm whale oil, demonstrates the usefulness of Camelina to serve as an environmentally friendly and sustainable source of these wax esters. Further recent modifications of Camelina oil to provide short-chain saturated fatty acids by others (Kim et al., 2015) and ourselves provide an oil that is more amenable to traditional biodiesel production, jet fuel production, and as a source of fatty acids for consumer products. Two groups have used a genetic engineering approach to produce high levels of docosahexaenoic acid (DHA, 22:6n-3) in Camelina (Mansour et al., 2014; Petrie et al., 2014; Ruiz-Lopez et al., 2014) as well as a high level of eicosapentaenoic acid (EPA, 20:5n-3) (Ruiz-Lopez et al., 2014). One of these oils is remarkably similar to fish oil (Ruiz-Lopez et al., 2014) and demonstrates the useful of bioengineering to produce a product to meet an ever-increasing world demand for fish oil. Engineering Camelina to produce a high level of vaccenic acid (18:1n-7), a fatty acid with pharmaceutical and industrial uses (Nguyen et al., 2015), further substantiates the usefulness of biotechnology in modifying Camelina oil to produce feedstock for fuel, industrial, and pharmaceutical uses in a manner that is sustainable and environmentally favorable.

The sustainable, adaptable nature of Camelina lends it to being amenable for planting on less than optimal land. As alluded to earlier, this does not necessitate the displacement of food crops for growing Camelina as it can be placed on less productive land. In addition, this limits the need for using common food oils, such as from canola or soy beans, in biofuel production, keeping these oils available for human consumption. This is an absolute critical consideration for plant-based biofuel feedstock production. It is estimated that in the United States alone there are 43 million hectares (Mha) of land that may be suitable for production of crops like Camelina that is not currently in production nor is it high-yield pasture land (Cai et al., 2011). Worldwide, the amount of abandoned or degraded land is 320–702 Mha, and it is estimated that with a crop similar to Camelina, this could provide 10–52% of the current worldwide liquid fuel consumption (Cai et al., 2011). This amount of land is even greater if marginally productive grassland is converted to land for growing biofuel feed stocks, raising the amount of potential land to over 1000 Mha worldwide. Hence, if a fraction of this land is shifted into production of an ideal crop like Camelina, there is addition of both feed stock for biofuel production as well as a large source of meal for animal feeds, freeing up other traditional meal sources for human consumption.

In summary, Camelina is well positioned to become a sustainable crop for use of its oil in biofuel production and the meal byproduct as a feed ration for livestock. However, the potential for Camelina as a source of oils for industrial use and for the chemical industry remain viable options as well as using genetically modified Camelina for production of healthy oils for human consumption. This chapter will address the use of Camelina for oil production, the meal for livestock rations, the Camelina genome, and the transformation of Camelina for genetically modified oil production and pharmaceutical production.

CAMELINA PRODUCTION AS AN OIL SEED CROP

The first set of field trials examining the potential of Camelina in North America noted that "Camelina is an oilseed plant which merits serious consideration as a potential crop in northern areas because of its earliness, high yield and oil content" (Plessers et al., 1962). These authors noted that over 6 years of cultivation in Ottawa, Ontario, the crop was resistant to damage by flea beetles, which has been reported in subsequent trials by others (Gugel and Falk, 2006). The 3-year trial at the Ottawa station demonstrates on average an 80-day period to maturity and yields of 1751 pounds per acre (1965 kg/ha) (Plessers et al., 1962). At a second trial location at Fort Vermillion, Alberta, the time to maturity is 98 days and the yield is 1473 pound per acre (1649 kg/ha)

(Plessers et al., 1962). In these trials, the yield for Camelina is better than that for either rapeseed or flaxseed grown in the same location. This first study demonstrated that Camelina is an excellent northern climate crop that is not susceptible to temperatures as low as −11°C but is highly susceptible to herbicides (Plessers et al., 1962). This is consistent with our observations that Camelina is susceptible to very low levels of residual class 2 herbicides in the soil. Furthermore, this first study reports that Camelina matures about 4 weeks more quickly than flax and that the protein yield is better than flax and has an average protein content of about 40%. Hence, this initial trial in 1958–60 demonstrates that Camelina has vast potential in the northern portions of the United States and in southern Canada as an alternative oil seed crop with a significant benefit to cold tolerance and quick maturity.

This initial trial was followed by additional trials for Camelina in Finland, Denmark, and England that show Camelina has on average oil content of 40%, while the locations in Denmark report an oil content approaching 44% (Zubr, 2003). A significant issue is the limited effort in breeding Camelina and the overall limited observed impact of accessions on yield. This is substantiated in this trial where three different cultivars had oil content varying from 41% to 43% (Zubr, 2003). Others have shown a similar limited impact of different accessions of Camelina on oil content (Gugel and Falk, 2006; Vollman et al., 1996; Zubr and Matthäus, 2002), while one trial indicates a variation in oil content from 30% to 38% in 13 different accessions planted in Rosemount, Minnesota (Budin et al., 1995).

Similar to other oil seed crops, weather has a key role in oil deposition and yield (Vollman et al., 1996; Zubr, 2003; Zubr and Matthäus, 2002), although while some accessions of Camelina demonstrate higher oil content (Budin et al., 1995), for the most part oil content and overall yield are influenced to a greater extent by location and weather. However, Camelina does have a greater degree of tolerance to drought than many other oil seed crops (Gugel and Falk, 2006; Vollman et al., 1996). When moisture is abundant, rapeseed has a greater yield in trials growing both plants in the same location. Under drought conditions, Camelina yield is significantly less impacted than rapeseed (Gugel and Falk, 2006), although its yield is reduced relative to nondrought conditions (Gugel and Falk, 2006; Vollman et al., 1996).

While Camelina is often marketed as a low-input crop that grows in marginal soils, the reality is that yield and oil deposition is influenced by the application of nitrogen and phosphorous. The application of phosphorous (30 kg/ha) and nitrogen (70–100 kg/ha) produces an ideal oil and seed yield (Zubr, 1997), while application of nitrogen above 60 kg/ha suppresses oil content (Urbaniak et al., 2008). Under these high-nitrogen conditions, the protein content of the seed is increased in lieu of oil production, and there is a reduction in monounsaturated fatty acids (18:1n-9 and 20:1n-9) (Urbaniak et al., 2008). Thus, application of nitrogen at rate of 50 kg/ha and phosphorous at 30 kg/ha is sufficient to produce a crop with a high yield and oil content.

Camelina is tolerant to many diseases and pest that commonly impact other oil seed crops. Further, Camelina is resistant to weed infiltration and demonstrates an ability to suppress key annual weeds (Budin et al., 1995; Saucke and Ackermann, 2006), suggesting that mixed cropping systems with Camelina might be desirable in organic farming operations. For instance, mixed cropping of field peas with Camelina increases the yield of field peas (Saucke and Ackermann, 2006). As noted earlier, Camelina is resistant to flea beetle infestations (Gugel and Falk, 2006; Plessers et al., 1962), as well as insect pressure in general (Budin et al., 1995). Further, it is resistant to blackleg disease but is susceptible to asters yellow disease (Gugel and Falk, 2006; Séguin-Swartz et al., 2009), as well as downy mildew and sclerotinia (Séguin-Swartz et al., 2009; Urbaniak et al., 2008). However, there are accessions resistant to sclerotinia and downy mildew (Séguin-Swartz et al., 2009), suggesting increased breeding efforts may produce more-resistant varieties for commercial production. We have noted that under greenhouse conditions, Camelina is susceptible to various fungal pressures that are easily treated. Overall, many of the common diseases associated with other oil seed crops are without effect on Camelina and, when combined with its resistance to insects, this provides additional rationale for use of this plant as a commercial oil seed crop.

CAMELINA OIL COMPOSITION AND STABILITY

Camelina oil is a rich source of alpha-linolenic acid (ALA, 18:3n-3) (Budin et al., 1995; Eidhin et al., 2003b; Murphy, 2011; Rodriquez-Rodriquez et al., 2013; Song and Thornally, 2007; Zubr, 1997; Zubr and Matthäus, 2002), but it is also unique among n-3 fatty acid–containing oil seeds in that the oil is more balanced in n-3 (18:3n-3) and in n-6 (linoleic acid, LNA, 18:2n-6) fatty acids (Budin et al., 1995; Murphy, 2011; Rodriquez-Rodriquez et al., 2013; Zubr, 1997; Zubr and Matthäus, 2002). The n-3/n-6 ratio for Camelina oil varies from around 1.3 to 2 (Budin et al., 1995; Murphy, 2011; Rodriquez-Rodriquez et al., 2013; Zubr, 1997; Zubr and Matthäus, 2002), whereas the value for the n-3/n-6 ratio is substantially lower in canola oil (~0.52), soybean oil (~0.15), safflower oil (~0.013), sunflower oil (~0.014), and corn oil (~0.018) and substantially higher in flax oil (~4.2) (calculated from oil composition data provided by Canola Council of Canada). Hence, unlike many commonly used oils for human and animal consumption, Camelina oil is much more balanced in the n-3 and n-6 fatty acid content, providing a dietary source for both of these essential fatty acids.

In our hands, we have found that 18:3n-3 content in field grown Camelina (38%) differs from that grown under greenhouse conditions (26–28%) (Murphy, 2011). This difference may be the result, in part, due to differences in the Camelina variety. However, others have shown that differences in oil composition between different accessions are relatively minor (Rodriquez-Rodriquez et al., 2013; Zubr and Matthäus, 2002), although similar to our greenhouse-grown Camelina, winter and spring varieties have a larger difference in 18:3n-3 content (Zubr, 1997). This difference is accounted for by an increase in 18:2n-6 content, suggesting a reduction in desaturation of 18:2n-6 into 18:3n-3 (Zubr, 1997). A recent study reports the kinetics for oil deposition during Camelina seed development, with oil content reaching a maximum 40 days after flowering (Rodriquez-Rodriquez et al., 2013). Another unique feature of Camelina oil is the high 20:1n-9 content (Budin et al., 1995; Murphy, 2011; Rodriquez-Rodriquez et al., 2013; Urbaniak et al., 2008; Zubr, 1997; Zubr and Matthäus, 2002), as well as a consistent content of 22:1n-9 between 2% and 3%. Unlike many oil seeds, the composition of fatty acids found in Camelina oil is consistent with a robust elongase activity as 20:1n-9 is an abundant fatty acid (~12%) in this plant and is derived from the elongation of 18:1n-9 and 22:1n-9 is less abundant but the elongation product of 20:1n-9 (Budin et al., 1995; Murphy, 2011; Rodriquez-Rodriquez et al., 2013; Urbaniak et al., 2008; Zubr, 1997; Zubr and Matthäus, 2002). Hence, Camelina oil is a rich source of 18:3n-3, while also providing a significant amount of 18:2n-6 as well as monounsaturated fatty acids.

However, because PUFA constitute about 50% of the total fatty acids found in Camelina oil (Budin et al., 1995; Eidhin et al., 2003b; Murphy, 2011; Rodriquez-Rodriquez et al., 2013; Song and Thornally, 2007; Zubr, 1997; Zubr and Matthäus, 2002), it suggests that the oil has low oxidative stability (Eidhin et al., 2003b). Like many other PUFA-containing oils, Camelina oil contains many different molecules with antioxidant capacity. Camelina oil is rich in tocopherols, with the total tocopherol content reported between 700 and 800 mg/kg of seed (Abramovic et al., 2007; Zubr, 1997; Zubr and Matthäus, 2002), with γ-tocopherol accounting for over 90% of the total (Abramovic et al., 2007; Zubr and Matthäus, 2002). However, Camelina is also high in total phenolic compounds and flavonoids, with a high flavonoid amount (143 mg/kg of seed) (Matthäus, 2002) and polar phenolic compounds as high as 439 mg/kg of seed (Abramovic et al., 2007). Because many of these antioxidant molecules are found in the oil, this increases the overall oxidative stability of the oil. Reports are varied but indicate that Camelina oil is more stable than flax oil and fish oil but less stable than sesame, corn, olive, and sunflower oil (Eidhin et al., 2003b). Of these oils, only olive oil and rapeseed oil have lower peroxide values after 16 days at 65°C, whereas under these conditions, conjugate triene formation in Camelina oil is much greater than in many other oils. However, another study reports that Camelina forms peroxides similar to the amount found in flax oil, although the rancimat test indicates greater stability than flax oil (Szterk et al., 2010). Thus, overall despite the high PUFA content, Camelina oil demonstrates oxidative stability similar to other rich n-3–containing oil seeds due to the level of antioxidant compounds found in the oil.

It is important to note that unlike many oil seeds, Camelina oil contains cholesterol (Shukla et al., 2002; Szterk et al., 2010). The major sterols in Camelina oil are sitosterol (1884 μg/g of oil), campesterol (893 μg/g of oil), cholesterol (188 μg/g of oil), brassicasterol (133 μg/g of oil), and stigmasterol

(103 μg/g of oil) (Shukla et al., 2002). Thus, although cholesterol content is not a major sterol in Camelina oil, it comprises a significant amount of the sterols reported.

CAMELINA MEAL QUALITY AND USES IN LIVESTOCK

The first scholarly publication (1962) detailing the potential for Camelina use in North America as an oil seed crop notes that the meal contains slightly more protein than meal from either rape or flax and that this additional protein content lends it to being used as a livestock feed (Plessers et al., 1962). This is important, as a major byproduct of growing Camelina for oil production for use in biofuels and industrial oils is a large amount of meal. Hence, the usefulness of the meal as a protein supplement for livestock rations provides additional revenue for those producing Camelina, as well as lessening the impact of shifting marginal lands into Camelina production. Clearly, this reduces the fuel for food debate often at the center of the controversy over use of oils commonly consumed by humans for biofuel production.

The protein in Camelina meal has a very good amino acid profile, with arginine, cysteine, glycine, lysine, methionine, and threonine content comparable to that of soy meal (Pekel et al., 2009; Zubr, 1997). This amino acid composition makes Camelina meal an attractive meal for use in poultry (Acamovic et al., 1999). We and others have used Camelina meal in poultry production as well as for a feed ration for laying hens. Use of 5% and 10% Camelina meal in the rations for broilers increases the n-3 fatty acid content of the meat without a change in the sensory quality of the meat (Ryhänen et al., 2007). Although there was a noted reduction in growth, consistent with the observed reduction in food intake and conversion ratio in the broilers consuming Camelina meal, there is no reported change in the thyroid gland (Ryhänen et al., 2007) (see section: Antinutritive Compounds in Camelina Meal). An important observation in our studies is that poultry are sensitive to the residual oil content in Camelina meal and residual oil content above 10% in the meal reduces overall feed consumption in poultry, reducing the feed-to-gain ratio. Again, this reduction in growth is observed using 10% Camelina meal in poultry starter diets, but a similar reduction is also observed using 10% flax meal in the same study (Pekel et al., 2009). Interestingly, in this study the addition of Cu^{2+} (150 mg/kg feed) in the diet in the form of copper sulfate normalized the impact of Camelina on body weight but not that of flax. Another important point is that the protein in flax has significantly less essential amino acids than those found in Camelina meal (Pekel et al., 2009).

In turkey production, the use of Camelina meal (5%, 15%, and 20%) in a starter diet is reported, and it is noted that Camelina meal is a good source of quality protein but that inclusion above 5% should be avoided in starter diets (Frame et al., 2007). We have noted no change in feed-to-gain or in body weight gain in broilers using up to 20% Camelina meal, but it is important to note that the meal had no more than 10% residual oil; thus, there was no impact of feed consumption (unpublished data). This is substantiated in the literature where the inclusion of 10% Camelina meal does not reduce feed efficiency or body weight gain (Azizal et al., 2010b). In this study, the total n-3 fatty acid content of the meat is increased twofold in white meat and 2.9-fold in dark meat in the 10% Camelina meal group, indicating that inclusion of Camelina meal provides a good source of protein, energy, and essential fatty acids for poultry (Azizal et al., 2010b). Overall, the inclusion of Camelina meal in poultry diets is a promising source of high-quality protein with a residual oil composition high in n-3 fatty acids, which offers a meal with an amino acid composition similar to soy but with the added benefit of essential PUFA that increase n-3 fatty acids in the meat.

Because of the high level of antioxidants in Camelina seed, when the seed is pressed there is an enhanced antioxidant capacity in the meal (Quezada and Cherian, 2012). In this study, the antioxidants found in low residual oil–containing meal (9.9%) and high residual oil–containing meal (24.6%) were assessed (Quezada and Cherian, 2012). The results indicate that the content of phenolic and flavonoid compounds are highest in the low residual oil–containing meal, whereas the high residual oil contains more α- and γ-tocopherols (Quezada and Cherian, 2012). However, for these two meals, the low residual oil–containing meal has an overall greater antioxidant capacity based on a variety of different tests.

This drives one to question whether these antioxidants provide any advantage when used in poultry diets? Indeed, inclusion of 5% or 10% Camelina meal in broiler rations reduces thiobarbituric acid reactive substance (TBARS) during short-term (2 days) and long-term (90 days) storage of thigh meat up to 49% and 36%, respectively (Azizal et al., 2010a). Interestingly, only thigh meat from the 10% Camelina meal ration has a reduction in cooking associated increase in TBARS, indicating that the enhanced antioxidant capacity of meat reduces formation of these lipid oxidation products during cooking. Further, only the group with enhanced antioxidant activity in breast meat is from the 10% Camelina meal group. Thus, the antioxidant compounds found in Camelina seed enhance the antioxidant capacity of the meal and increases these compounds, including γ-tocopherol, in the meat of broilers. The ideal inclusion rate for this enhanced antioxidant capacity in the meat is 10%, which enhances the long-term storage of the meat compared to inclusion of 5% Camelina meal in the rations or control diet (Azizal et al., 2010a).

Inclusion of Camelina meal in hen rations increases the n-3 fatty acid content of eggs in a dose-dependent manner (Kakani et al., 2012). In addition, Camelina oil (5% inclusion) itself increases the n-3 fatty acid content of the eggs about 3.5-fold without any change in egg production (Rokka et al., 2002). Further, the impact of Camelina oil inclusion on sensory qualities of the eggs is less than that with an equivalent amount of flax oil. In a study using Camelina meal with about 12% residual oil, there is a noted reduction in egg production in the group with inclusion of 15% Camelina meal in the ration as well as a marked increase in TBARS in the eggs, consistent with an eightfold increase in the proportion of 22:6n-3 (Cherian et al., 2009). On the other hand, in eggs from hens fed 10% Camelina meal containing 10% residual oil, total n-3 content is increased 2.5-fold and 22:6n-3 content is increased nearly 2.5-fold as well (Kakani et al., 2012). Inclusion of 5% Camelina meal in rations also significantly increases oil content, but only about twofold, indicating a dose-dependent increase. Again, as reported by others, we observed no changes in egg production, in egg storability, or in egg sensory parameters. In contrast, inclusion of 10% Camelina meal increases egg production in brown laying hens, while also increasing n-3 fatty acid content (Azizal et al., 2013). Hence, inclusion up to 10% Camelina meal in laying hen rations increases n-3 fatty acid content without reducing egg production, making Camelina meal an important high-quality additive meal to dietary regimens for hens.

In ruminates, replacing soy meal with Camelina meal (2.04 kg of dry matter (DM)) in beef steer diets results in a marked reduction in stress response hormones (Cappellozza et al., 2012). Further, no changes in thyroid function are noted, but this meal reduces acute-phase reactive protein reaction, which is known to increase when cattle are transported or when cattle are introduced to a feedlot setting (Cappellozza et al., 2012). These results are indicative of a positive impact of Camelina meal in reducing stress response in cattle. In dairy cows, Camelina meal (2 kg of DM) results in a marked decrease in milk fat without an impact on total milk production (Hurtaud and Peyraud, 2007). Interestingly, inclusion of Camelina meal results in a marked increase in *trans* fatty acids, *t*-10 18:1 and *t*-11 18:1, 11- and 2.6-fold, respectively, and a marked sevenfold increase in *c*-9, *t*-11 18:2 (conjugated linoleic acid). This may account for the enhanced spreadability of butter made from milk derived from Camelina meal–fed cows (Hurtaud and Peyraud, 2007). Hence, in a limited number of studies, Camelina meal is observed to have a number of positive effects on the health of cattle, as well as in the parameters associated with milk production and quality.

In monogastric mammals, a number of studies have been published using either Camelina seed or oil. In rabbits, Camelina seed (5%, 10%, or 15%) was mixed with alfalfa, pelleted, and then fed to rabbits for 50 days (Peiretti et al., 2007). Under this regimen, there is a dose-dependent increase in PUFA and a reduction in the n-6/n-3 ratio along with a corresponding decrease in saturated fatty acids in muscle and in perirenal fat. Hogs supplemented with 5% or 10% Camelina oil for 30 days have increased n-3 fatty acids and reduced n-6 fatty acids in plasma with a corresponding increase in n-3 fatty acids in muscle (Eidhin et al., 2003a), indicating that supplementing Camelina oil results in n-3 fatty acid accumulation. In this trial, the authors compared the results to an inclusion of 5% fish oil as well and noted that plasma triacylglycerol (TAG) levels are reduced in both Camelina oil groups but not in the fish oil group, although the fish oil group had higher levels of 20:5n-3 and 22:6n-3 in the plasma.

Similarly, in hypercholesterolemic humans, Camelina oil (30 g/day for 6 weeks), but not olive oil or rapeseed oil, reduces low-density lipoproteins (LDLs) by 13%, whereas less reduction is seen in the other groups (Uusitupa and Sarkkinen, 2002). Further, longer chain n-3 fatty acids, such as 20:5n-3 and 22:5n-3, are increased in the Camelina oil group, and 18:3n-3 is increased 2.5-fold greater than in the rapeseed oil group and fourfold greater than in the olive oil group. Thus, in monogastrics, Camelina oil has a positive impact in reducing LDL and TAG levels, reductions that are associated with improved cardiovascular health in humans.

ANTINUTRITIVE COMPOUNDS IN CAMELINA MEAL

Although Camelina meal is a high-quality meal that is similar in amino acid composition to soy meal, there are a number of antinutritive compounds associated with it (Matthäus and Angelini, 2005; Schuster and Friedt, 1998; Zubr, 1997). One family of these compounds is the glucosinolates, which are commonly found in oil seeds from the Brassicaceae family such as rapeseed, mustard, crambe, canola, and Camelina as well as vegetables from the Cruciferae family, such as broccoli, Brussels sprouts, cabbage, and cauliflower (Daxenbichler et al., 1991; Fahey et al., 2001). Glucosinolates are a rather diverse group of compounds; they often are degraded through the cooking of vegetables (Oerlemans et al., 2006; Song and Thornally, 2007) and may have numerous positive benefits on health (Fahey et al., 2001).

Camelina contains glucosinolates (14–23 µmol/g), at a level similar to that observed in rapeseed but at much lower levels than found in other crops such as crambe and mustard (>120 µmol/g) (Matthäus and Zubr, 2000; Zubr, 1997). Others have reported a level for glucosinolates in Camelina up to about 35 µmol/g (Schuster and Friedt, 1998). Camelina has a unique glucosinolate (see Fig. 8.1), 10-methylsulfinyldecyl (10-MSG), which comprises 60–65% of glucosinolates found in Camelina, while 9-methylsulfinylnonyl (9-MSG) comprises about 30% and 11-methylsulfinylundecyl (11-MSG) comprises about 10% of the glucosinolates found in Camelina seed (Matthäus and Angelini, 2005; Schuster and Friedt, 1998). Interestingly, winter varieties of Camelina have less glucosinolates present than in the typically planted spring varieties by about 10 µmol/g (Schuster and Friedt, 1998). Typical commercially available spring varieties planted in the United States and in Canada are Suneson, Blaine Creek, Celine, and Calena (USDA plant guide, http://plants.usda.gov/plantguide/pdf/pg_casa2.pdf). Further, the glucosinolate content of Camelina varies depending on the soil type and the sulfur content of the soil (Schuster and Friedt, 1998). The overall concern regarding glucosinolates is that their breakdown following activation of myrosinase, a thiogluocosidase, occurs in the seed during processing and results in the formation of isothiocynate, nitrile, and thiocynate (Daxenbichler et al., 1991). These compounds at sufficiently high concentrations are well known to inhibit iodine uptake into the thyroid, resulting in goiter and a

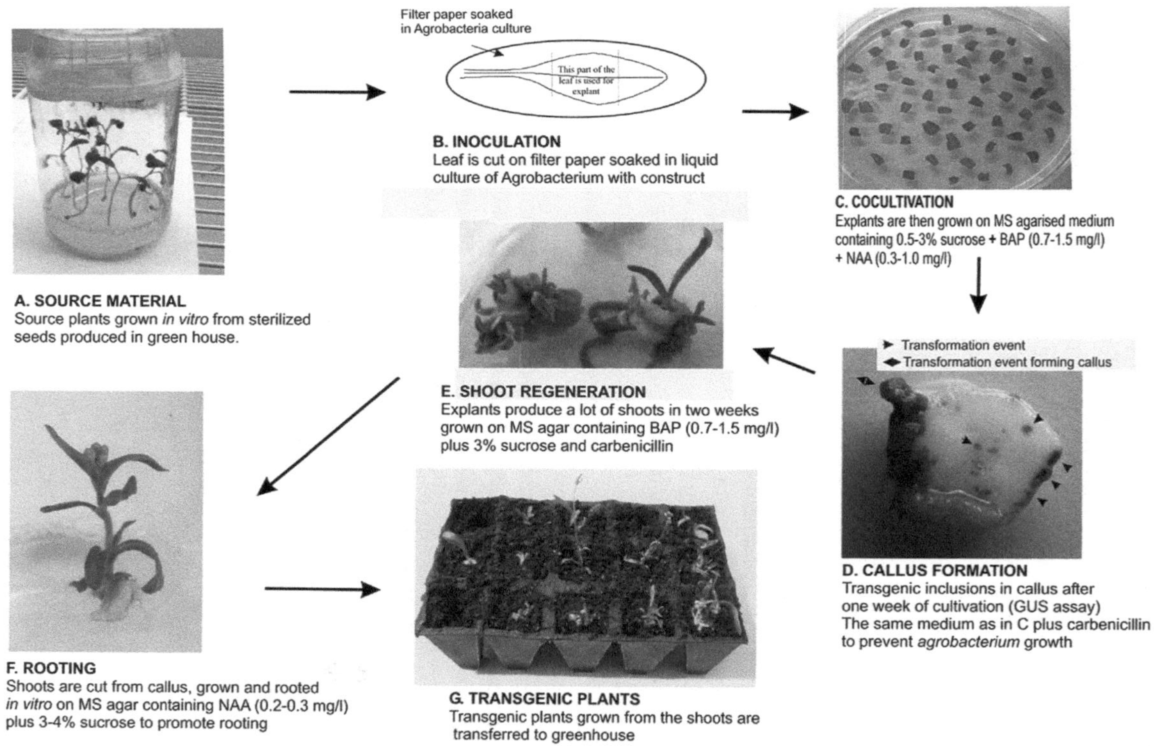

FIGURE 8.1 Patented Camelina transformation system using a nonselection technique.

reduction in growth. However, it is important to note that in hens-fed Camelina meal up to 10% of the ration, no measurable glucosinolates or their breakdown products are found in eggs (Kakani et al., 2012), indicating the likelihood of these compounds having a negative impact on poultry is low.

Other than glucosinolates, there are other antinutritive compounds found in Camelina. Sinapine is found in Camelina at concentrations in the range of 4 mg/g of seed (Matthäus and Angelini, 2005; Matthäus and Zubr, 2000), which is about 30% less than that found in rapeseed. This level of sinapine is sufficient to result in a classic fishy odor in products, such as eggs, if excessive amounts of Camelina meal are included in rations. This was a significant concern in our poultry trial, but a blinded, untrained taste panel did not note any negative organoleptic traits in eggs from hens fed up to 10% Camelina meal (Kakani et al., 2012).

Inositol hexaphosphate or phytic acid is another antinutritive found in Camelina and is a concern due to its strong chelating properties. The total amount of inositol phosphates, containing both inositol pentaphosphate and inositol hexaphosphate, range between 22 and 20 mg/g, with the highest levels found in winter cultivars and dependent on location of cultivation (Matthäus and Zubr, 2000). Others report phytic acid levels (>95% inositol hexaphosphate) between 22 and 30 mg/g, which is similar to the level found in sunflower, but greater than that found in mustard, soy bean, crambe, and rapeseed (Matthäus and Zubr, 2000). However, the amount of phytic acid found in Camelina is only about 1.5-fold that found in rapeseed (17.4 mg/g) (Matthäus and Angelini, 2005), a well-tolerated and well-used crop.

Condensed tannins are found in all seeds but represent a potential antinutritive compound that may reduce digestibility of rations in ruminants and nonruminants (Matthäus and Angelini, 2005). In Camelina, tannin levels are 1.1 mg/g, which is considerably less than rapeseed (4 mg/g) (Matthäus and Angelini, 2005). Others report a range from 1 to 2 mg/g, but still less than rapeseed although similar to mustard and crambe (Matthäus and Zubr, 2000).

Thus, a number of antinutritive compounds, such as sinapine, phytic acid, tannins, and glucosinolates, are found in Camelina seed and hence in its meal. However, in Camelina, most of these antinutritive compounds are found at levels on par or lower than in rapeseed, a well-tolerated oil seed crop that is often used as a supplement meal in rations for a variety livestock. Hence, the lack of an impact in trials using Camelina meal in a number of different animals, including poultry, ruminants, and nonruminants, indicate that at the appropriate level Camelina meal is a safe alternative for livestock producers.

CAMELINA GENOME

There has been rapid progress in mapping of the Camelina genome, imparting a greater understanding of its complexity and, more importantly, its similarity to *Arabidopsis thaliana*, as recently confirmed in the first large-scale sequencing of the Camelina genome (Kagale et al., 2014). It is widely accepted that the Camelina genome highly resembles that of *Abrabidopsis*, but until recently this was not substantiated in the literature (Hutcheon et al., 2010; Kagale et al., 2014; Liang et al., 2013). Further, while it was long suspected that Camelina has a polyploidy nature (Gehringer et al., 2006), evidence in the literature suggested it is a allohexaploid plant with three redundant genomes and differentiated copies of genes contributing to its genomic makeup (Hutcheon et al., 2010). However, the large-scale sequencing effort demonstrates that while Camelina is hexaploid, the genome is well preserved and similar to that observed in other Brassica crops (Kagale et al., 2014). This most recent study supports a whole genome triplication event for Camelina, similar to that seen in *Arabidopsis* (Kagale et al., 2014). The polyploidy nature of Camelina is supported by its clear diploid inheritance seen by us as well as others (Gehringer et al., 2006; Lu and Kang, 2008). Because studies demonstrate conservation of the genes encoding *fatty acid desaturase* (*FAD*) *2* and *fatty acid elongase* (*FAE*) *1* and evidence indicates all three copies are expressed in developing seeds, it suggests that these genes are all functional. Further, these genes have 90% sequence homology to *FAD2* and *FAE1* in *Arabidopsis*, demonstrating the close relationship between the two genomes. Importantly, it appears based on *FAD2* and *FAE1* sequences that *Camelina sativa* may contain one copy of the *Camelina microcarpa* genome, indicating that these two plants share a parental genome at some point (Hutcheon et al., 2010), although a recent thorough large-scale sequencing of the Camelina genome does not support this observation (Kagale et al., 2014). The existence of three separate copies of the genome in Camelina is further substantiated by the triplication of *LEAFY* (*LFY*), a gene known to be a single copy in many plants. Importantly, the polyploidy genome observed in Camelina may account for its adaptability to environmental conditions and the wide range where it is naturally found (Hutcheon et al., 2010; Kagale et al., 2014).

Why is it important to understand the genomic makeup of Camelina? Clearly, unlike a lot of crops, until 2005 there was extremely limited knowledge of the Camelina genome and the overall genetic diversity of Camelina. In 2005, rapid amplified polymorphic DNA analysis (RAPD) was used to examine the diversity of Camelina grown in three different microenvironments in Austria and accessions divided into four phenotypic groups (Vollman et al., 2005). In this study, only 63% of the 30 marker loci examined are polymorphic, which indicates significantly less diversity than other crops that are self-pollinating like Camelina (Vollman et al., 2005). The natural diversity of Camelina is important with regards to a productive breeding program for positive traits. A more recent study using an amplified fragment length polymorphism (AFLP) strategy to molecular fingerprint 53 accessions from the Ukraine and Russia found a high amount of genetic diversity in this group of accessions (Ghamkhar et al., 2010). Another important observation from this study is that the biggest influence on oil content, as previously mentioned, are various environmental factors. An important limitation for consideration in any study examining genetic and phenotypic diversity such as these studies is to ensure sufficient ecogeographic diversity of the collection examined. However, our own experience is that much of the Camelina demonstrates limited diversity in oil and protein content as well as fatty acid composition. That said, we have observed accessions that are outliers and these hold promise to offer the diversity that will be necessary to produce Camelina for use under diverse environmental conditions.

In 2006, the first quantitative trait loci (QTL) mapping experiments were performed on Camelina grown with and without added nitrogen (80 kg N/ha) and the QTL for oil content, for seed yield, for plant height, and for 1000 seed mass mapped (Gehringer et al., 2006). This is a first important step in beginning to use molecular markers in breeding programs focused at enhancing Camelina agronomic properties.

AGROBACTERIUM-MEDIATED TRANSFORMATION OF CAMELINA

Because of the limited breeding efforts and the difficulty in breeding Camelina due to its self-pollinating nature, one approach to alter phenotypic properties of the plant is through the use of biotechnology. While we have pioneered the use of an *Agrobacterium*-mediated transformation of Camelina using explants, others have used a floral dip procedure to introduce the bacterium to the plant material. We have had no success using a Gene Gun approach as the transformation efficiency with Camelina is low and not feasible even on a noncommercial scale.

The initial transformation of Camelina leaf explants demonstrated that inclusion of the auxin naphthaleneacetic acid (NAA, 5.4 μmol/L) is essential to increase the rooting response in nodal explants, increasing the rooting response from 26% to 47% (Tattersall and Millam, 1999). In addition, adding the cytokinin 6-benzylaminopurine (BAP, 4.44 μmol/L) in combination with NAA (0.54 μmol/L) to the explants induces shoot regeneration from the leaflet at rates of over 10 shoots per explant (Tattersall and Millam, 1999). We have further perfected this method (Fig. 8.1, Camelina Nonselection Transformation) and found that 2–6% sucrose in the cultivation medium is critical and that the frequency for callus formation bearing a transformation event is 70–100% (Kuvshinov et al., 2011). Of these events that become successfully rooted plants when using a selection marker, such as kanamycin, is in the range of 30–35%. In absence of a selection marker, this efficiency is still 11–14%. Hence, a key issue besides using BAP and NAA, is to use sucrose in excess of 2%, and for the best rooting efficiency, 4% sucrose is ideal at the rooting stage. Unlike for other crops such as *Brassica napus*, we have noted that 2,4-dichlorophenoxyacetic acid (2,4-D) is highly detrimental to Camelina and should not be used as an auxin. Further, like others, we found that hypocotyls and cotyledons are not useful for explant transformation and the leaves from surface sterilized seeds are best used 5–7 days after germination. Thus, using our explant protocol, we have successfully transformed Camelina with constructs containing the coding regions for up to 11 genes and up to 25 kb in size, which has resulted in stably transformed Camelina.

While we use an explant approach, many others have used a floral dip protocol for transformation of Camelina (Cappellozza et al., 2012; Horn et al., 2013; Iven et al., 2015; Kim et al., 2015; Liu et al., 2012; Lu and Kang, 2008; Mansour et al., 2014; Nguyen et al., 2015; Petrie et al., 2014; Ruiz-Lopez et al., 2014; Sayanova et al., 2012). The initial efforts with floral dip using Camelina at the early flowering stage required vacuum infiltration for successful transformation (Lu and Kang, 2008). Using this method, an efficiency of 1.3% is observed and that is calculated as the number of seeds bearing the *Discosoma* sp. red fluorescent proteins (DsRed) as a marker (Lu and Kang, 2008). This method produced Camelina that is stably transformed with a single gene, Δ-*12 Fatty Acid Hydroxylase* (*FAH12*), and results in an altered fatty acid profile (Lu and Kang, 2008). Refinement of this method was accomplished by using flowers in full bloom rather than at the beginning of the flowering stage as previously used (Lu and Kang, 2008). After dipping, the plants are covered with plastic for 24h and under these conditions the efficiency is 0.33%, but by dipping the plants a second time 1 week later increases the efficiency to 0.8%. Despite the low efficiency of floral dip in transforming Camelina, the relative ease of using this method has made it the method of choice for transforming Camelina.

It is important to note that in some jurisdictions, the use of floral dip is restricted due to claims for using *Agrobacterium*-mediated transformation of any organ system in the intact plant as well as use in explant systems (EP 1334199, designated in Germany, Finland, France, Great Britain, and The Netherlands). This is an important consideration for commercialization of genetically modified Camelina.

OUTCROSSING POTENTIAL FOR CAMELINA

An important consideration for GM Camelina is its limited ability to outcross to other plants and to non-GM Camelina. We, and others, have observed that under greenhouse conditions, no crossing of GM Camelina with non-GM Camelina occurs, even in plants that are in close proximity or even touching. However, this anecdotal evidence has been confirmed as DsRed-bearing plants could only be crossed to adjacent non–DsRed-bearing plants by emasculation of the flowers (Julié-Galau et al., 2014).

Under open field condition, the outcrossing rate of GM-DsRed–bearing Camelina to non-GM Camelina is very low, between 0.09% and 0.28% (Walsh et al., 2012). This rate is dependent on flowering synchrony and with a distance of 20cm having the highest level of outcrossing, which is significantly reduced at a distance of 60cm between plants. However, this distance also is dependent on prevailing winds, but under all of these conditions the outcrossing rate is less than that observed for other self-pollinating plants such as flax (1.85% at 1m (Jhala et al., 2011)) and less than soy bean (Ahrent and Caviness, 1994). Thus, under open field conditions, Camelina has an acceptable outcrossing rate to itself and pollen flow will be significantly reduced by asynchronous flowering and increased distance.

Despite an extremely low level of outcrossing to itself, does Camelina outcross to other Brassicaceae species that may enable the unintended establishment of transgenes in other close, naturally occurring relatives of Camelina? One study demonstrates that while Camelina did not outcross to *Arabidopsis*, it did outcross to *Capsella bursa-pastoris* (shepherd's purse) and to *Cardamine hirsuta* (hairy bittercress) when the flowers are emasculated (Julié-Galau et al., 2014). The crosses to *C. hirsute* result in only a small number of seeds for which the embryos are aborted at an early stage of development. However, the crosses to *C. bursa-pastoris* did make hybrids, but as these hybrids are sterile, its propagation is not possible. This study was recently extended by examining outcrossing of five Camelina accessions with 10 accessions of *C. bursa-pastoris* (Martin et al., 2015). Again, this much more robust study demonstrates a low rate of hybridization (1.5 hybrids per 10,000 pollination events) and, as previously demonstrated, that these hybrids have very low pollen fertility (Martin et al., 2015). This suggests that these hybrids will not establish a persistent hybrid that can present an issue with regards to outflow of genes from genetically modified Camelina to *C. bursa-pastoris*. Thus, the potential of GM-Camelina outcrossing via a pollen-mediated process to close relatives in the open field are very low, as in this instance the outcrossing occurred only after the flowers were emasculated.

GENETICALLY ENGINEERING CAMELINA FATTY ACID COMPOSITION

The major uses of Camelina in the past and future will be principally associated with its properties as a low-input oil seed crop. Camelina-derived oil will be used as feedstock for either biofuel or chemical process (Horn et al., 2013; Iven et al., 2015; Kim et al., 2015; Lu and Kang, 2008; Nguyen et al., 2013, 2015), while a rapidly emerging use is to produce bioactive lipids for use in humans (Mansour et al., 2014; Nguyen et al., 2015; Petrie et al., 2014; Ruiz-Lopez et al., 2014) and as a feed source in aquaculture (Betancor et al., 2015). The first demonstration that Camelina is amenable to these changes was the transformation with *FAH12* from caster bean driven by the seed-specific phaseolin promoter to produce a high level of 12-hydroxy fatty acids (12–14%) ranging from 18 to 20 carbons in length (Lu and Kang, 2008). Interestingly, this transformation also increases the 18:1n-9 content in the seed, suggesting a block of its sequential desaturation by *FAD2* and *FAD3* to form 18:2n-6 and 18:3n-3, respectively.

An intriguing approach to modify Camelina fatty acid composition is by the use of RNAi and antisense approaches. The first use of antisense in Camelina was to reduce the flux of fatty acids into 18:2n-6 by using the antisense message for *FAD2* (Kang et al., 2011). This eloquent approach results in a dramatic reduction of 18:2n-6 from 16.8% to 6.3% and 18:3n-3 from 33.2% to 11%, while increasing 18:1n-9 from 15.5% to 51.2% (Kang et al., 2011).

Interestingly, there is a significant increase in elongation of 18:1n-9 to 20:1n-9, suggesting enhanced elongation under these conditions. A similar approach using RNAi to reduce *FAD2* and *FAE1* results in a reduction of 18:3n-3 from 36% to 8% and 18:2n-6 from 17% to 4%, while enhancing the 18:1n-9 content up to nearly 70% (Nguyen et al., 2013). Unlike the antisense approach reducing *FAD2* expression, the use of RNAi to reduce both *FAD2* and *FAE1* reduces 20:1n-9 content from 12% to 3%, thereby reducing the flux of carbons to longer-chain fatty acids (Nguyen et al., 2013). In addition to this first demonstrated use of RNAi, this report also identifies the major lipid biosynthetic routes in Camelina, providing an important foundation for efforts to use a biotechnological approach to enhance oil formation in Camelina. Using a similar RNAi approach to reduce *FAD3* and *FAE1* levels, an 80% reduction in 18:3n-3 content is observed with an increase in 18:2n-6 content (Horn et al., 2013), demonstrating the RNAi is an important method that can be used to alter carbon flux thorough fatty acid biosynthetic pathways as a means to alter the fatty acid composition of Camelina oil.

Although this is one approach, an alternative strategy is to express enzymes that truncate fatty acid synthesis to produce shorter chain fatty acids. We have transformed Camelina with the acyl–acyl carrier protein thioesterase (FatB) from *Umbellularia californica* combined with the lysophosphatidic acid acyltransferase from *Cocus nucifera* to increase the amount of 12:0 and 14:0 fatty acids. In this instance, the highest producing plant lines had 23% 12:0 and 4% 14:0, with reduced 18:0, 18:1n-9, and 18:2n-6, while 18:3n-3 remains nearly unchanged (unpublished data). A similar approach was used to transform Camelina to express the thioesterase from *Cuphea pulcherrima*, which results in a dramatic increase in 16:0 from 6.6% to 41% (Horn et al., 2013). This effort was recently expanded by using the *FatB* gene encoding the acyl–acyl carrier protein thioesterase from *Cuphea pulcherrima* and *Cuphea viscosissima* to transform Camelina to increase shorter chains saturated fatty acids from 10:0 to 16:0 (Kim et al., 2015). These two approaches may be critical in reducing the amount of PUFA found in Camelina oil to make it more desirable for first generation biodiesel and biojet as well as a feedstock for consumer products and as heart-healthy oil–expressing medium-chain saturated fatty acids (12:0–14:0).

The potential for Camelina to be transformed to produce a sustainable, terrestrial source of longer-chain n-3 fatty acids and other bioactive lipids for use in human nutrition and as a feedstock for aquaculture has come closer to commercial reality. Transforming Camelina with the Δ6-desaturase from the green algae *Ostreococcus tauri* results in an increase in 18:3n-6 (GLA, ~18%) and in 18:4n-3 (SDA, ~12%), with an approximately 30% reduction in 18:1n-9 and in 18:3n-3 (Sayanova et al., 2012). This same group used two different constructs, one containing five genes to produce a high 20:5n-3 (EPA)–yielding Camelina (31% EPA for individual seed levels) and the other containing seven genes to make a high EPA and 22:6n-3 (DHA)–yielding Camelina (12% and 14% EPA and DHA, respectively, for individual seed levels) (Ruiz-Lopez et al., 2014).

These transformed plants were stable through three generations with no observed alteration in the biochemical and physical phenotype of the plant relative to the wild-type parent. The ability to produce a plant with a high level of EPA and one with an EPA and DHA content very similar to fish oil is truly exciting and has vast potential for nutritional and pharmaceutical use in humans as well as a long-chain n-3 fatty acid source in aquaculture and other livestock applications. The use in aquaculture has already been demonstrated in salmon with rations containing the oil derived from EPA Camelina demonstrating no adverse growth characteristics compared to fish fed rations containing fish oil (Betancor et al., 2015). In this study, the salmon muscle has the same levels of EPA as salmon fed standard fish oil–containing rations, although the DHA levels are lower. This is expected as the DHA content of the rations is less that the fish oil–containing rations.

Another group has taken a similar approach using a seven-gene construct with genes encoding for a Δ12-, Δ15-, Δ6-, Δ5-, and Δ4-desaturase and Δ5- and Δ6-elongases (Mansour et al., 2014; Petrie et al., 2014). Although this transformed Camelina has similar level of 22:6n-3 (DHA, 9.6–12.4%) to the other approach, it has only low to extremely modest levels of 20:5n-3 (EPA, 0.8–3.8%). Further, this plant contains significant amounts of 18:4n-3 (SDA, 7.4–8.9%), indicating a metabolic blockade in the flux of fatty acids through this pathway. This is further demonstrated by the high levels of 18:3n-3 (ALA, 29%), which are not significantly different from the parent, non-GM Camelina, indicative of a metabolic block in the desaturation and elongation of 18:3n-3 to longer-chain n-3 fatty acids. Nonetheless, production of a plant with high levels of SDA and DHA is certainly an interesting finding and again useful for supplying these bioactive lipids in a sustainable manner.

Two recently published studies further show the adaptability of Camelina to dramatic change in its oil composition. By suppressing the native Camelina FatB and 3-keto-acyl-ACP synthase II (KASII) protein levels via an antisense strategy combined with expression of a Δ9 16:0-ACP desaturase significantly increases the levels of vaccenic acid (18:1n-7) in Camelina to nearly 30% of the total fatty acid (Nguyen et al., 2015). This is important as 18:1n-7 is an important feedstock for polyethylene production and may be an important fatty acid for pharmaceutical use. Another study reports transforming Camelina with a combination of genes encoding for mouse fatty alcohol acyl-CoA reductase and wax ester synthase to produce TAG containing up to 21% wax esters (Iven et al., 2015). This is a commercially viable level of wax esters with a profile similar to that from sperm whales, which is a significant and important use of a GM approach to reduce an environmentally unfriendly natural source of these wax esters.

Overall, these studies demonstrate that Camelina is amenable to significant and diverse alterations in the fatty acid composition of its oil in a manner that does not impact the phenotype or germination of the seed. This is important as the low cost nature of Camelina production and its extremely low potential for outcrossing coupled with these examples of oil modification demonstrates that Camelina is an ideal platform for production of bioactive fatty acids for human use and for use in industrial applications in a sustainable manner.

INCREASING THE AGRONOMIC POTENTIAL OF CAMELINA

A significant issue with Camelina is the limited efforts made to date to enhance the agronomic potential via breeding. While some elite germplasm has been produced, and certainly we have ongoing conventional breeding efforts, the reality is very little has been done to enhance Camelina. We and others have noted that Camelina is highly susceptible to residual class 2 herbicide in fields (Plessers et al., 1962). To address this issue, we transformed Camelina with *Arabidopsis* acetolactate synthase (ALS) engineered with different specific mutations in its active site (Ala122Thr, Pro197Ser, and Trp574Leu) in various combinations (unpublished data). While the Pro197Ser mutation enhances tolerance to clorsulfueron (Glean) by about 1000-fold, the combination of the Pro197Ser and Trp574Leu mutation enhances tolerance by about 10,000-fold. More importantly, this double mutation enhances tolerance to imazamox beyond, by about 1000-fold. The triple mutation did not produce any added effect and the least effective single mutation is the Ala122Thr mutation. While others have also produced class 2 herbicide–tolerant Camelina using mutations in ALS, this effort was primarily focused on producing a herbicide selection marker and the enhanced tolerance to class 2 herbicides is not reported (Liu et al., 2012).

Two different approaches have been used to enhance Camelina seed and oil yield. Paclobutrazol (PBZ) treatment, an inhibitor of gibberellin synthesis, when applied during the flowering stage under field conditions with 100 mg/L application concentration, nearly doubles seed yield and increases seed oil content 15% (Kumar et al., 2012). With PBZ, there is a significant increase in chlorophyll content and photosynthesis, presumably causing an increase in carbon utilization as demonstrated by the enhanced biomass, seed yield, and seed oil content. Another approach is the overexpression of *Arabidopsis* purple acid phosphatase-2, which results in a twofold increase in seed yield (Zhang et al., 2012). Similar to PBZ treatment, the mechanism appears to be an enhanced rate of photosynthesis that causes the plant to flower 20% earlier and increases the seed yield. These and other approaches will be needed to enhance the field performance of Camelina with an emphasis on enhancing seed yield and oil content and composition.

CONCLUSION

Overall, the potential for Camelina to become a significant oil seed crop is vast, but this potential is somewhat limited by its limited breeding in the past. On the other hand, the ability to efficiently and stably transform Camelina in a manner that alters the fatty acid composition and content suggests that it is an ideal platform for production of specific fatty acids for use in human nutrition, in aquaculture and livestock feeding regimens, and in industrial applications. The evidence that these alterations do not have a negative impact on its phenotypic properties adds additional promise. In addition, the low outcrossing potential combined with the ability to manipulate its fatty acid composition makes Camelina an important candidate for sustainable custom oil biosynthesis in a manner that does not impact the amount of land used for crucial food crops. Thus, the production of Camelina for these various applications avoids the classic food for fuel debate that has plagued the use of other food oils in similar industrial applications.

ACKNOWLEDGMENTS

I thank Cindy Murphy for the typed preparation of this book chapter and her patience in supporting the publication process. I also thank our fine former Unicrop OY and current Agragen OY employees who contributed to this work over the years: in particular, Viktor Kuvshinov and Sveta Kuvshinov of Unicrop OY and Agragen OY and Anne Kanerva, Seppo Kaijalainen, and Kimmo Koivu former Unicrop OY employees. Conflict of Interest: note the author is an officer in a company with commercial interest in Camelina.

REFERENCES

Abramovic, H., Butinar, B., Nikolic, V., 2007. Changes occurring in phenolic content, tocopherol composition and oxidative stability of *Camelina sativa* oil during storage. Food Chem. 104, 903–909.

Acamovic, T., Gilbert, C., Lamb, K., Walker, K.C., 1999. Nutritive value of *Camelina sativa* meal for poultry. Br. Poult. Sci. 40, S27–S41.

Ahrent, D.K., Caviness, C.E., 1994. Natural cross-pollination of twelve soybean cultivars in Arkansas. Crop Sci. 34, 376–378.

Azizal, A.E., Panda, A.K., Quezada, N., Cherian, G., 2013. Nutrient digestibility, egg quality, and fatty acid composition of brown laying hens fed camelina or flaxseed meal. J. Appl. Poult. Res. 22, 832–841.

Azizal, A.E., Quezada, N., Cherian, G., 2010a. Antioxidative effect of dietary *Camelina* meal in fresh, stored, or cooked broiler chicken meat. Poult. Sci. 89, 2711–2718.

Azizal, A.E., Quezada, N., Cherian, G., 2010b. Feeding *Camelina sativa* meal to meat-type chickens: effect on production performance and tissue fatty acid composition. J. Appl. Poult. Res. 19, 157–168.

Betancor, M.B., Sprague, M., Usher, S., Sayanova, O., Campbell, P.J., Napier, J.A., Tocher, D.R., 2015. A nutritionally-enhanced oil from transgenic *Camelina sativa* effectively replaces fish oil as a source of eicosapentaeonic acid for fish. Sci. Rep. 5. http://dx.doi.org/10.1038/srep08104.

Bouby, L., 1998. Two early finds of gold-of-pleasure (*Camelina* sp.) in middle neolithic and chalcolithic sites in western France. Antiquity 72, 391–398.

Budin, J.T., Breene, W.M., Putnam, D.H., 1995. Some compositional properties of camlina (*Camelina sativa* L. Crantz) seed and oils. J. Am. Oil Chem. Soc. 72, 309–315.

Cai, X., Zhang, X., Wang, D., 2011. Land availability for biofuel production. Environ. Sci. Technol. 45, 334–339.

Cappellozza, B.I., Cooke, R.F., Bohnert, D.W., Cherian, G., Carroll, J.A., 2012. Effects of camelina meal supplementation on ruminal forage degradability, performance, and physiological responses of beef cattle. J. Anim. Sci. 90, 4042–4054.

Cherian, G., Campbell, A., Parker, T., 2009. Egg quality and lipid composition of eggs from hens fed *Camelina sativa*. J. Appl. Poult. Res. 18, 143–150.

Daxenbichler, M.E., Spencer, G.F., Carlson, D.G., Rose, G.B., Brinker, A.M., Powell, R.G., 1991. Glucosinolate composition of seeds from 297 species of wild plants. Phytochemistry 30, 2623–2638.

Eidhin, D.N., Burke, J., Lynch, B., O'Beirne, D., 2003a. Effects of dietary supplementation with camelina oil on porcine blood lipids. J. Food Sci. 68, 671–679.

Eidhin, D.N., Burke, J., O'Beirne, D., 2003b. Oxidative stability of ω3-rich camelina oil and camelina oil-based spread compared with plant and fish oils and sunflower spread. J. Food Sci. 68, 345–353.

Fahey, J.W., Zalcmann, A.T., Talalay, P., 2001. The chemical diversity and distribution of glucosinolates and isothiocyanates among plants. Phytochemistry 56, 5–51.

Frame, D.D., Palmer, M., Peterson, B., 2007. Use of *Camelina sativa* in the diets of young turkeys. J. Appl. Poult. Res. 16, 381–386.

Francis, A., Warwick, S.I., 2009. The biology of canadian weeds. 142. *Camelina alyssum* (Mill.) Thell.; *C. microcarpa* Andrz. ex DC.; *C. sativa* (L.) Crantz. Can. J. Plant Sci. 89, 791–810.

Gehringer, A., Friedt, W., Lühs, W., Snowdon, R.J., 2006. Genetic mapping of agronomic traits in false flax (*Camelina sativa* subsp. *sativa*). Genome 49, 1555–1563.

Ghamkhar, K., Croser, J., Aryamanesh, N., Campbell, M., Kon'kova, N., Francis, C., 2010. Camelina (*Camelina sativa* (L.) Crantz) as an alternative oilseed: molecular and ecogeographic analyses. Genome 53, 558–567.

Gugel, R.K., Falk, K.C., 2006. Agronomic and seed quality evaluation of *Camelina sativa* in western Canada. Can. J. Plant Sci. 86, 1047–1058.

Horn, P.J., Silva, J.E., Anderson, D., Fuchs, J., Borisjuk, L., Nazarenus, T.J., Shulaev, V., Cahoon, E.B., Chapman, K.D., 2013. Imaging heterogeneity of membrane and storage lipids in transgenic *Camelina sativa* seeds with altered fatty acid profiles. Plant J. 76, 138–150.

Hurtaud, C., Peyraud, J.L., 2007. Effects of feeding camelina (seeds or meal) on milk fatty acid composition and butter speadability. J. Dairy Sci. 90, 5134–5145.

Hutcheon, C., Ditt, R.F., Beilstein, M., Comai, L., Schroeder, J., Goldstein, E., Shewmaker, C.K., Nguyen, T., De Rocher, J., Kiser, J., 2010. Polyploid genome of *Camelina sativa* revealed by isolation of fatty acid synethsis genes. BMC Plant Biol. 10, 233.

Iven, T., Hornung, E., Heilmann, M., Feussner, I., 2015. Synthesis of oleyl oleate was esters in *Arabidopsis thaliana* and *Camelina sativa* seed oil. Plant Biotechnol. J. 1–8. http://dx.doi.org/10.1111/pbi. 12379.

Jhala, A.J., Bhatt, H., Topinka, K., Hall, L.M., 2011. Pollen-mediated gene flow in flax (*Linum usitatissimum* L.): can genetically engineered and organic flax coexist? Heredity 106, 557–566.

Julié-Galau, S., Bellec, Y., Faure, J.-D., Tepfer, M., 2014. Evaluation of the potential for interspecific hybridization between *Camelina sativa* and related wild Brassicaceae in anticipation of field trials of GM camelina. Transgenic Res. 23, 67–74.

Kagale, S., Koh, C., Nixon, J., Bollina, V., Clarke, W.E., Tuteja, R., Spillane, C., Robinson, S.J., Links, M.G., Clarke, C., Higgins, E.E., Huebert, T., Sharpe, A.G., Parkin, I.A.P., 2014. The emerging biofuel crop *Camelina sativa* retains a highly undifferentiated hexaploid genome structure. Nat. Commun. 5 (3706). http://dx.doi.org/10.1038/ncomms4706.

Kakani, R., Fowler, J., Haq, A.-U., Murphy, E.J., Rosenberger, T.A., Berhow, M., Bailey, C.A., 2012. Camelina meal increases egg n-3 fatty acid content without altering quality or production in laying hens. Lipids 47, 519–526.

Kang, J., Snapp, A.R., Lu, C., 2011. Identification of three genes encoding microsomal oleate desaturases (FAD2) from the oilseed crop *Camelina sativa*. Plant Physiol. Biochem. 49, 223–229.

Kim, H.J., Silva, J.E., Vu, H.S., Mockaitis, K., Nam, J.-W., Cahoon, E.B., 2015. Toward production of jet fuel functionality in oilseeds: Identification of FatB acyl-acyl carrier protein thioesterases and evaluation of combinatorial expression strategies in *Camelina* seeds. J. Exp. Bot. Adv. Access. http://dx.doi.org/10.1093/jxb/erv225.

Kumar, S., Ghatty, S., Satyanarayana, J., Guha, A., Chaitanya, B., Reddy, A.R., 2012. Paclobutrazol treatment as a potential strategy for higher seed and oil yield in field-grown *camelina sativa* L. Crantz. BMC Res. Notes 5, 137.

Kuvshinov, V., Kaneava, A., Koivu, K., Kuvshinova, S., Pehu, E., 2011. Transformation in *Camelina sativa*. 12/288,791 Finland.

Liang, C., Liu, X., Yiu, S.-M., Lim, B.L., 2013. De novo assembly and characterization of *Camelina sativa* transcriptome by paired-end sequencing. BMC Genomics 14, 146.

Liu, X., Brost, J., Hutcheon, C., Guilfoil, R., Wilson, A.K., Leung, S., Shewmaker, C.K., Rooke, S., Nguyen, T., Kiser, J., De Rocher, J., 2012. Transformation of the oilseed crop *Camelina sativa* by *Agrobacterium*-mediated floral dip and simple large-scale screening of transformants. In Vitro Cell. Dev. Biol. 48, 462–468.

Lu, C., Kang, J., 2008. Generation of transgenic plants of a potential oilseed crop *Camelina sativa* by *Agrobacterium*-mediated transformation. Plant Cell Rep. 27, 273–278.

Mansour, M.P., Shrestha, P., Belide, S., Petrie, J.R., Nichols, P.D., Singh, S.P., 2014. Characterization of oilseed lipids from "DHA-producing *Camelina sativa*": a new transformed lipid plant containing long-chain omega-3 oils. Nutrients 6, 776–789.

Martin, S.L., Sauder, C.A., James, T., Cheung, K.W., Razeq, F.M., Kron, P., Hall, L., 2015. Sexual hubridization between *Capsella bursa-pastoris* (L.) Medik (female) and *Camelina sative* (L.) Crantz (male) (Brassicaceae). Plant Breed. 134, 212–220.

Matthäus, B., 2002. Antioxidant activity of extracts obtained from residues of different oilseeds. J. Agric. Food Chem. 50, 3444–3452.

Matthäus, B., Angelini, L.G., 2005. Anti-nutritive constituents in oilseed crops from Italy. Ind. Crops Prod. 21, 89–99.

Matthäus, B., Zubr, J., 2000. Variability of specific components in *Camelina sativa* oilseed cakes. Ind. Crops Prod. 12, 9–18.

Murphy, E.J., 2011. Versatile *Camelina*: the future of biofuel and much more. Inform 22, 604–608.

Nguyen, H.T., Park, H., Koster, K.L., Cahoon, R.E., Nguyen, H.T.M., Shanklin, J., Clemente, T.E., Cahoon, E.B., 2015. Redirection of metabolic flux for high levels of omega-7 monounsaturated fatty acid accumulation in camelina seeds. Plant Biotechnol. J. 13, 38–50.

Nguyen, H.T., Silva, J.E., Podicheti, R., Macrander, J., Yang, W., Nazarenus, T.J., Nam, J.-W., Jaworski, J.G., Lu, C., Scheffler, B.E., Mockaitis, K., Cahoon, E.B., 2013. Camelina seed transcriptome: a tool for meal and oil improvement and translational research. Plant Biotechnol. J. 11, 759–769.

Oerlemans, K., Barrett, D.M., Suades, C.B., Verkerk, R., Dekker, M., 2006. Thermal degradation of glucosinolates in red cabbage. Food Chem. 95, 19–29.

Peiretti, P.G., Mussa, P.P., Prola, L., Meineri, G., 2007. Use of different levels of false flax (*Camelina sativa* L.) seed in diets for fattening rabbits. Livestock Sci. 107, 192–198.

Pekel, A.Y., Patterson, P.H., Hulet, R.M., Acar, N., Cravener, T.L., Dowler, D.B., Hunter, J.M., 2009. Dietary camelina meal versus flaxseed with and without supplemental copper for broiler chickens: live performance and processing yield. Poult. Sci. 88, 2392–2398.

Petrie, J.R., Shrestha, P., Belide, S., Kennedy, Y., Lester, G., Liu, Q., Divi, U.K., Mulder, R.J., Mansour, M.P., Nichols, P.D., Singh, S.P., 2014. Metabolic engineering *Camelina sativa* with fish oil-like levels of DHA. PLoS One 9.

Plessers, A.G., McGregor, W.G., Carson, R.B., Nakoneshny, W., 1962. Species trials with oilseed plants II. Camelina. Can. J. Plant Sci. 42, 453–459.

Quezada, N., Cherian, G., 2012. Lipids characterization and antioxidant status of the seeds and meals of *Camelina sativa* and flax. Eur. J. Lipid Sci.Technol. 114, 974–982.

Rodriquez-Rodriquez, M.F., Sánchez-García, A., Salas, J.J., Garcés, R., Martínez-Force, E., 2013. Characterization of the morphological changes and fatty acid profile of developing *Camelina sativa* seeds. Ind. Crops Prod. 50, 673–679.

Rokka, T., Alén, K., Valaja, J., Ryhänen, E.-L., 2002. The effect of a *Camelina sativa* enriched diet on the composition and sensory quality of hen eggs. Food Res. Int. 35, 253–256.

Ruiz-Lopez, N., Haslam, R.P., Napier, J.A., Sayanova, O., 2014. Successful high-level accumulation of fish oil omega-3 long-chain polyunsaturated fatty acids in a transgenic oilseed crop. Plant J. 77, 198–208.

Ryhänen, E.-L., Perttilä, S., Tupasela, T., Valaja, J., Eriksson, C., Larkka, K., 2007. Effect of *Camelina sativa* expeller cake on performance and meal quality of broilers. J. Sci. Food Agric. 87, 1489–1494.

Saucke, H., Ackermann, K., 2006. Weed suppression in mixed cropped grain peas and false flax (*Camelina sativa*). Weed Res. 46, 453–461.

Sayanova, O., Ruiz-Lopez, N., Haslam, R.P., Napier, J.A., 2012. The role of Δ6-desaturase acyl-carrier specificity in the efficient synthesis of long-chain polyunsaturated fatty acids in transgenic plants. Plant Biotechnol. J. 10, 195–206.

Schuster, A., Friedt, W., 1998. Glucosinolate content and composition as parameters of quality of *Camelina* seed. Ind. Crops Prod. 7, 297–302.

Scoggan, H.J., 1957. Flora of manitoba. Natl. Mus. Can. Bull. 140, 619.

Séguin-Swartz, G., Eynck, C., Strelkov, S.E., Olivier, C.Y., Li, J.L., Klein-Gebbinck, H., Borhan, H., Caldwell, C.D., Falk, K.C., 2009. Diseases of *Camelina sativa* (false flax). Can. J. Plant Pathol. 31, 375–386.

Shukla, V.K.S., Dutta, P.C., Artz, W.E., 2002. Camelina oil and its unusual cholesterol content. J. Am. Oil Chem. Soc. 79, 965–969.

Song, L., Thornally, P.J., 2007. Effect of storage, processing and cooking on glucosinolate content of *Brassica* vegetables. Food Chem. Toxicol. 45, 216–224.

Szterk, A., Roszko, M., Sosinska, E., Derewiaka, D., Lewicki, P.P., 2010. Chemical composition and oxidative stability of selected plant oils. J. Am. Oil. Chem. Soc. 87, 637–645.

Tattersall, A., Millam, S., 1999. Establishment and in vitro regeneration studies of the potential oil crop species *Camelina sativa*. Plant Cell Tissue Organ Cult. 55, 147–149.

Torrey, J., Gray, A., 1838. In: A Flora of North America, vol. 1. Wiley & Putnam, New York, p. 711.

Urbaniak, S.D., Caldwell, C.D., Zheljazkov, V.D., Lada, R., Luan, L., 2008. The effect of cultivar and applied nitrogen on the performance of *Camelina sativa* L. in the maritime provinces of Canada. Can. J. Plant Sci. 88, 111–119.

Uusitupa, M.I.J., Sarkkinen, E.S., 2002. Effect of α-linolenic acid-rich *Camelina sativa* oil on serum fatty acid composition and serum lipids in hypercholesterolemic subjects. Metab. Clin. Exp. 51, 1253–1260.

Vollman, J., Damboeck, A., Eckl, A., Schrems, H., Ruckenbauer, P., 1996. Improvement of *Camelina sativa*, an underexploited oilseed. In: Janick, J. (Ed.), Progress in New Crops. ASHS Press, Alexandria, VA, pp. 357–362.

Vollman, J., Grausgruber, H., Stift, G., Dryzhyruk, V., Lelly, T., 2005. Genetic diversity in camelina germplasm as revealed by seed quality characteristics and RAPD polymorphism. Plant Breed. 214, 446–453.

Walsh, K.D., Puttick, D.M., Hills, M.J., Yang, R.-C., Topinka, K.C., Hall, L.M., 2012. Short Communication: first report of outcrossing rates in camelina [*Camelina sativa* (L.) Crantz], a potential platform for bioindustrial oils. Can. J. Plant Sci. 92, 681–685.

Zhang, Y., Yu, L., Yung, K.-F., Leung, D.Y.C., Sun, F., Lim, B.L., 2012. Over-expression of AtPAP2 in *Camelina sativa* leads to faster plant growth and higher seed yield. Biotechnol. Biofuels 5, 19.

Zubr, J., 1997. Oil-seed crop: *Camelina sativa*. Ind. Crops Prod. 6, 113–119.

Zubr, J., 2003. Qualitative variation of *Camelina sativa* seed from different locations. Ind. Crops Prod. 17, 161–169.

Zubr, J., Matthäus, B., 2002. Effects of growth conditions on fatty acids and tocopherols in *Camelina sativa* oil. Ind. Crops Prod. 15, 155–162.

Chapter 9

Coconut (*Cocos nucifera*)

Laura J. Pham
Oils and Fats Laboratory, National Institute of Molecular Biology and Biotechnology (BIOTECH), University of the Philippines at Los Baños, Laguna, Philippines

INTRODUCTION

One of the most lavish gifts of nature to man is the coconut tree. It is one of the most important tropical tree crops in the world and is widely known as the "tree of life" for many communities in different nations because it provides a large number of products that can be used to support the local economy. It is grown in more than 90 countries in the world, with production concentrated in Asia and the Pacific Region. It is also an important export for some countries in this region, eg, the Philippines and countries of Southeast Asia. Thus it plays an important role in the local economies and cultures of these countries.

The coconut is a very versatile crop which requires little care. Every part of the coconut palm is used, and it produces a range of valuable products, many of which contribute to health maintenance and improvement. The coconut industry is founded on producing coconut oil and was the foundation of vegetable oil production for importation by developing industrial economies until other competitive crops entered the market for edible oils. Industrial uses include separation of fatty acids and glycerol, soap production, cosmetics, detergent and biofuels. A recent focus has been on the key nutritional benefits of coconut oil for human health (Fife, 2003).

In countries where the coconut plays an important role in the economy, there is a need to revitalize the coconut industry. In the 21st century, the challenge for the coconut industry is that of survival in the midst of global competition. The coconut industry has to meet the standards of safety and quality that now guide international trade in all aspects from production to processing.

BOTANY, CULTIVATION AND PRODUCTION

Botanical Features

The coconut palm *Cocos nucifera* belongs to the palm family, one of the largest in the monocotyledon group. It is an economically important plant in the family

Industrial Oil Crops. http://dx.doi.org/10.1016/B978-1-893997-98-1.00009-9

and is used as both an ornamental and a food crop. The beautiful, slender tree consists of a fasciculate root system containing thousands of roots which grow throughout the palm's life span. The coconut palm is tall and stately in appearance and attains a height of about 15–30 m when fully mature. A crown of large pinnate leaves tops the trunk, which is characteristically unbranched. An adult palm usually carries 30–40 paripinnate leaves in its crown. Each leaf measures 5–6 m in length and weighs 10–15 kg. Leaf emergence to shedding takes about 2.5–3 years. A new leaf appears in the crown every 3 or 4 weeks in a healthy palm. The palm produces an average of about 12 leaves annually. The inflorescence develops within a strong, tough pointed double sheath called spathe. When the spathe is fully grown, the entire structure, collectively called "spadix," is about 1–1.2 m in length and 14–16 cm in diameter. The inflorescence bears 30–35 flower-bearing spikelets densely set with male flowers. The male flowers number about 250–300 per spikelet. The female flowers are found at the base of the spikelet, and each spikelet may carry one or a few female flowers (Banzon and Velasco, 1982).

Climatic and nutritional conditions during the flowering stage affect the relationship between the number of female flowers and ripe fruits obtained, or the fruit set rate. Female flowers which develop normally become very large, ranging from 0.5 to more than 2.5 kg. Since the fruit is monospermous, it encloses a single seed. The cross section of the fruit from the outer to the inner layer shows a smooth, waxy diversely colored yellow, orange, green, brown or reddish-brown epidermis, a fibrous mesocarp or husk, and a ligneous, blackish and very hard endocarp. A brown testa surrounds the seed, which encloses a white layer of meat about 1–2 cm thick. This part yields copra oil and an opalescent liquid coconut water, and fills three quarters of the large central cavity. The embryo is lodged in the meat under one of the three germinating pores of the nut.

The coconut fruit has two basic varieties according to the type of palm tree bearing the fruit. These are the "tall" and "dwarf" coconut trees. The tall variety of coconut is slow growing. The tree produces fruits (Fig. 9.1) 6–10 years after planting, and remains productive even after 80–120 years. The fruits of the palm mature in 12 months. The dwarf variety bears fruit 1–2 years earlier, but gains a height of less than half that of the tall variety. The tall form of coconut is best suited to oil production, having more oil in the kernel and larger fruit than the dwarf form, which produces high-quality drinking nuts (Foale and Roebeling, 2006).

Cultivation

Coconuts are seed propagated. Research on vegetative multiplication has not produced positive results. Production starts 4–5 years after planting for hybrids and 6–10 years after planting for local tall varieties. One tree can yield an average of 70–150 nuts per year. Fruit production continues throughout the year with more or less marked seasonal variations. The fruit, which is well protected by its layers, can be stored for 2–3 months. However, the seed is not

FIGURE 9.1 Laguna Tall coconut fruits from the Philippines.

dormant, and beyond this period the germination process uses the oil reserves. The coconut palm requires very little by way of fertilizers. Commonly used fertilizers supply the four nutrients widely deficient in many coconut soils and these are nitrogen (N), potassium (K), chlorine (Cl) and sulfur (S). N- and Cl-containing fertilizers are needed to ensure good growth and development. Applying ammonium sulfate as a source of N and S nutrients, plus either muriate of potash or common salt (NaCl) as a source of K and Cl nutrients, was found to be effective (Foale, 2003).

The coconut palm's needs for water, light and chloride ions are easily met because it can be found throughout the intertropical zone, mainly along seacoasts. The most favorable conditions for its cultivation are: temperature of 25°C–30°C, sunshine period of 1800–2000 h/year, rainfall of 1800–2000 mm/year well distributed (around 150 mm/month), relative humidity of 75% and an altitude <400 m above sea level. The coconut can adapt to a wide range of soil types. Coarse sand is its natural habitat, but best growth is obtained in deep soils with good physical structure and chemical properties. Soils commonly planted to coconut have adequate amounts of Na, Ca, Mg, Fe, Mn and Zn, but most inland and coastal soils are severely deficient (very low in N and Cl and inadequate K, P, S and B). Such soils require fertilizer application to sustain high yields. The coconut palm can grow on light, medium and heavy soils, loams and well-drained clay. Coconuts tolerate alkaline soils up to pH 8 and acid soils with pH 4.5, but the ideal pH range is 5–5.7. The coconut palm is a very hardy plant that can withstand low temperatures (Hainan island in China: minimum <15°C and <10°C in winter) and high altitudes (up to 1000 m in India). When coconuts are fully mature, they fall from the tree and are easily collected from the ground. Coconuts intended for copra or oil production are split open, discarding the water and exposing the endosperm to the sun for drying.

An estimated 11.8 million hectares of coconut palms are planted worldwide. They are extensively grown in the tropics, with commercial production concentrated mostly in Asia and the Pacific region. The Philippines, Indonesia, India, Sri Lanka, Thailand, Malaysia and Papua New Guinea together account for about 80% of the total area planted to coconuts worldwide. In spite of relatively significant planting programs, production potential is stagnating because of the aging and reduced number of the plantations.

YIELDS AND PRODUCTION

Coconut has high productivity and production consistency compared with other oilseed crops and is also less susceptible to abnormal climatic conditions. World production of coconut has been estimated at 61.7 million tons (MT) with an average yield of 5.2 ton/ha. The top three producers are Indonesia, the Philippines and India, which account for 75% of global production. Of the 77 African, Caribbean and Pacific member countries, 47 produced 4.59 MT of coconuts on 1.7 million hectares, representing 7.4% of world production.

The bulk of the world's coconut is dried to produce copra which is processed into coconut oil and cake. Indonesia is the largest coconut-producing country, with 18 MT of production in 2012, or 30% of total world production, followed by the Philippines with 15.9 MT of production and India with 10.6 MT of production, which is 26.4% and 17.0% of world production, respectively (FAOSTAT, 2015).

Copra is the dried meat of the coconut, the moisture content of which has been brought down from 45% to 6%. It is the article of commerce from which coconut oil is obtained. Between 4000 and 6000 nuts are required to obtain 1 ton of copra. The oil content of the fresh meat is about 40%, that of the dry meat 68% and that of copra 64%–65%. These averages can vary substantially from one variety to another. With this oil yield, copra has a higher oil content than that of any other commercially available oilseed.

With progress in breeding and agronomic research, yields may exceed 5 t copra/ha at research stations and in large well-managed plantations. But copra and coconut oil have been facing financial difficulties for decades. Nearly two-thirds of the palms being harvested for copra are now becoming too old, and as a consequence are becoming unproductive. In addition, a number of serious new pests, diseases and nutritional deficiencies are also reducing yield. A further issue is that any new seedlings being planted are coming from unimproved germplasm with inherently low productivity. In many traditional coconut production areas, fruit harvesting is not being undertaken, and only the fallen fruit is collected and sold for a very low price. These farmers have limited access to new fruit-processing technologies and are unable to add value to traditional copra production. However, most of the problems listed above are being addressed by research-driven activities undertaken at the international, national or regional level (Batugal, 2006).

COCONUT OIL

Processing

Coconut oil produced by crushing copra is edible and has been consumed in tropical countries for thousands of years. World coconut oil production has been increasing over the past decade mainly because of greater global demand for the important characteristics of coconut oil. Consumption of coconut oil is now estimated at 3.5 MT/annum. This accounts for 2.5% of world vegetable oil production. Over 70% of global coconut oil production comes from the Philippines and Indonesia. The Philippines is the major exporter of coconut oil, accounting for 42% of world exports. Some formerly high-oil-producing countries have significantly reduced their production of oil because of the higher price that can be obtained for fresh nuts. Others are actually importing fresh nuts to satisfy their demand while it has become unprofitable for other countries to produce coconut oil.

Copra, either shipped to overseas countries or processed locally, is the almost-exclusive raw material for producing coconut oil. Oil processing involves three stages: pretreatment, extraction and postextraction treatments. Conventional extraction methods are pressing or solvent extraction. A combination of both—the expeller with a solvent—is also carried out in some plants, as it is difficult to quantitatively extract the oil from copra in a single process. Another frequently used method of extraction to obtain high oil yield is double pressing with either a hydraulic press or expellers (Gopala et al., 2010). The wet process of coconut oil extraction has been gaining in popularity, and studies have been carried out to compare oils extracted by dry and wet processes (Rajamohan and Nevin, 2009).

Composition and Characteristics

The quality of coconut oil obtained from copra depends considerably on drying, storage and transport conditions. When coconut oil is obtained from copra through continuous pressing, its color may be brownish yellow if high temperature is used in processing. Coconut oil obtained from the fresh meat after going through the coconut "milk" stage is practically white and has very low acidity (0.1%–0.2%). Acidity is related to copra quality and storage conditions (moisture). Oil from dry, clean copra seldom has more than 2% free fatty acids in terms of lauric acid. Rancidity due to oxidation of unsaturated fatty acids is very limited given the low levels of these acids in coconut oil. However, coconut oil can have "fragrance" or "ketonic" rancidity. This is due to the presence of methyl ketones, which have a penetrating and sharp odor. These compounds develop in the presence of water and are the result of β-oxidation of medium chain fatty acids released from the triacylglycerol. During deodorization, the distillate of the condensed vapor contains significant quantities of methyl ketones, particularly methyl-heptyl and methyl-nonyl ketones obtained from capric and lauric acids respectively.

Coconut oil belongs to the group of "lauric" oils (palm kernel oil, babassu kernel oil, and cuphea oil) because of its high level of medium chain fatty acids and low content of unsaturated fatty acids. These oils are characterized by a high lauric acid content of about 40%–50%. While the main properties of coconut oil, such as melting behavior, depend on its fatty acid composition, it is the distribution of the fatty acids in the glycerol backbone that gives coconut oil its properties. With advances in the chemistry and biotechnology of fats and oils, it is now possible to analyze the molecular composition of the oil and the distribution of its fatty acids.

Table 9.1 shows the fatty acid profile of coconut oil obtained by gas chromatography. This profile indicates that coconut oil is a saturated oil with about 90% saturated fatty acid content. With its very low (2.2%) content of polyunsaturated fatty acids, coconut oil is very stable and resistant to oxidative rancidity. Coconut oil contains medium chain fatty acids (caprylate, laurate and myristate), which most oils do not have. The soaps of these fatty acids are more soluble, retaining their cleansing power in hard water and even

TABLE 9.1 Fatty Acid Composition of Coconut Oil as Determined by Gas Chromatography

Fatty Acid	% Composition
Caproic (C_6)	Trace
Caprylic (C_8)	2.4
Capric (C_{10})	4.4
Lauric (C_{12})	44.5
Myristic (C_{14})	18.6
Palmitic (C_{16})	12.0
Stearic (C_{18})	4.8
Oleic ($C_{18:1}$)	11.0
Linoleic ($C_{18:2}$)	2.2

seawater. The glyceride oils of medium chain fatty acids are more digestible. Low molecular weight glycerides yield a greater percentage of glycerol than those of high molecular weight. Coconut oil produces approximately 14 g of glycerol from 100 g oil, which is the highest yield among commercially available oils.

Results of high-performance liquid chromatography determination of the triacylglycerol composition of coconut oil are shown in Table 9.2. Several studies have investigated its triacylglycerol composition for transformation into high-value products. Coconut oil contains about 84% trisaturates, meaning that the three fatty acids bound to the glycerol moiety are saturated.

$$\text{ECN} = \text{Carbon number-}2 \times \text{total number of double bonds.}$$

Of the other triacylglycerols, 12% contain two saturated and one unsaturated fatty acid, while only 4% contain two unsaturated and one saturated fatty acid. Based on equivalent carbon number (ECN), the major triacylglycerol of coconut oil is trilaurin, which is understandable because the dominant fatty acid is lauric acid (Laureles et al., 2002). Studies on stereospecific analysis, which gives information on the fatty acids attached to the three positions of the glycerol backbone, have been carried out for coconut oil. This information is of value for the production of structured lipids (Pham and Gregorio, 2008).

TABLE 9.2 Triacylglycerol Profile of Coconut Oil in Terms of Equivalent Carbon Number as Determined by High-Performance Liquid Chromatography

ECN	% Composition (area)
28	0.46
30	2.69
32	11.50
34	16.65
36	21.39
38	18.82
40	14.17
42	8.28
44	4.36
46	0.58
48	1.11

Utilization

Coconut oil remains the most important form of consumption of coconuts. The European Union group of 27 countries is the largest consumer of coconut oil in the world, currently utilizing about 443,000 metric ton/annum. Soap is made from a mixture of fatty acids of various chain lengths to give it the required balance for solubility, detergent and foaming capacity and a suitable degree of saturation. For this, 80%–85% C16 and C18 fatty acids are used, the rest coming from lauric oils including coconut oil. The coconut oil can be subjected to splitting, a process that uses either lipases or water at 250°C–255°C under 50–55 bar pressure to release free fatty acids from the triglycerides. The free fatty acids are then purified to 90%–95% purity by fractional distillation (Ooi and Pee, 1985).

Coconut oil contains the most concentrated commodity source of medium chain fatty acids available. These are widely used in the flavor industries, because they are more polar than long chain fatty acids and thus more hydrophilic and able to dissolve a variety of polar substances that are insoluble in conventional fats and oils. The medium chain triglycerides (MCTs) are miscible with hydrocarbons, esters, alcohols, acids, ketones and natural oils, and this property makes coconut oil a superior carrier for vitamins, flavor and color when compared with conventional oils. The metabolism of medium chain fatty acids differs from that of long chain fatty acids. When medium chain fatty acids are released from triglycerides, the fatty acids enter directly into the portal vein and are transported directly to the liver to provide energy immediately. As a result of this type of fat metabolism, MCTs now find use when added to sports drinks and energy bars to provide a quick source of energy. They are also of pharmaceutical importance for providing caloric intake in treating patients affected by cachexia and other conditions that cause loss of appetite.

Coconut oil is a major source of medium chain saturated fatty acids used in the lipo chemical industry. Their fatty acid derivatives also provide good solubility of detergents in water and good foaming action, and these are mainly used in the fields of plastics, cosmetics and creams.

A large proportion of copra oil is used directly for human consumption. In particular, the medium chain fatty acids of which lauric and myristic acids are dominant lead to a higher melting point fat that is solid at room temperature and relatively hard and brittle at low temperatures. It is a useful ingredient for margarine. Lauric oils are much in demand for confectionery, for which the melting point is approximately 32°C–37°C. It is thus suitable as a cocoa butter substitute for chocolate coatings, or as synthetic creams or cream substitutes.

EMERGING APPLICATIONS

Virgin Coconut Oil

The standard commercial type of coconut oil passes through the "copra" stage and undergoes the refining, bleaching and deodorization processes that include high temperature treatment. There is also virgin coconut oil (VCO), which is derived by cold pressing freshly ground copra meat with little or no heating, and separating the coconut protein from water in various ways. VCO is suitable for consumption without the need for further processing. It is colorless and clear, and cold process oil has the aroma of fresh coconut. The traditional process heats the coconut milk to boiling to remove water from the oil mixture, with concomitant high recovery of oil from the coconut milk. This heat process preserves the essential fatty acid linoleic and tocopherol content. However, VCO may contain a higher content of antipathogenic monoacylglycerols, as suggested by reports that credit VCO with faster cures of skin ailments when directly applied (Fife, 2006).

VCO is a relatively new value-added product which has become available in international commercial markets only during the past decade. It is gaining popularity as a healthy nutritive oil, and has broad use in the nutraceutical industry. Natural VCO may be exploited in niche health markets for premium prices. In recent years, the export quantity of VCO from the Philippines has increased significantly. Countries in the Pacific region such as Fiji and Samoa are exporting VCO to Australia. Indonesia is also producing VCO, and the domestic demand for this commodity is booming, with about 200 small and medium manufacturers in the country. Most of this production is going to the domestic market, but some producers are exporting to world markets (Bawalan and Chapman, 2006).

Biofuels/Biolubricants

Increasing attention is being given to the use of coconut oil for energy generation, either mixed with or as a substitute for diesel (Hossain et al., 2012). Various incentives and subsidies have led to biofuels becoming increasingly popular in the United States and Europe, and many countries now are in the process of making biofuels. When price differences between petroleum and edible oils widen, it becomes more attractive to use edible oils for fuel. Both environmental and energy security perspectives have become driving forces for the use of vegetable oil–based biodiesel fuels. However, there are three challenges that the biofuel sector must overcome: price considerations, limited awareness of the fuel feedstock and negative impact on the glycerin supply to existing markets such as Malaysia and Singapore.

Biolubricants are functional fluids made from vegetable oils and downstream esters. For example, coconut oil as a biolubricant has been used in India for three-wheeled vehicles. The overall global usage of renewable raw materials in lubricants and related functional fluid applications is about 250,000 t, comprising about 0.7% of the total lubricant market and 0.25% of total oils and fats produced annually. The Philippines is moving forward, followed by Thailand, Vanuatu and the Marshall Islands. Double-filtered coconut oil is used directly in the Marshall Islands to run cars, fishing boats, trucks, etc., while in the Philippines a mixture of diesel and methyl ester from coconut oil at a 99:1 ratio is used, with plans to increase the level of coconut oil by using a ratio of 95:5 (Corpuz, 2009).

In the Pacific, there are opportunities to use coconut oil and other vegetable oils as fuel for transport and in the generation of electrical power. Technologies already exist which can combust crude coconut oil in adapted compression engines, or by means of methyl esterification for biodiesel using a standard compression engine (see chapter: Biodiesel and Its Properties). The initiatives undertaken in the Pacific to use coconut oil in cars and generators have had different degrees of success. The technological challenges can be solved and economic applications can be found if the supply of oil is secured. Economically attractive niches are available for the use of raw coconut oil in adapted engines in remote communities with an abundant supply of coconuts and milling capacity. The production of biodiesel from coconut oil in combination with other vegetable oils can be set up commercially in larger communities by the provision of cleaner fuels for cars and electrical generators.

Cosmeceuticals

Production of cosmetics and personal products in Asian and Pacific countries is rapidly developing, with emerging popularity of coconut oil–based products. With annual demand growth of 10% for cosmetics and personal care products and 5%–19% for soap in Asia and the Pacific, the demand for coconut oil is enormous. Many Asian and Pacific countries are importing these products. Coconut oil, which is rich in C12 and C14 fatty acids (lauric and myristic), is good for skin care when applied as a pure product, and also for manufactured cleaning products having wide acceptance (Gopala et al., 2010).

PREMIUM GRADE MONOLAURIN AND HIV/AIDS

Coconut oil is emerging as both a nutritious health food and a promising new natural treatment for many health problems (Fife, 2008). Past research has provided the groundwork for healthful benefits, and the few clinical studies that have been done on human subjects have shown promise in a variety of areas (Gursche, 2008).

Over a period of 22 years, 42 million people in the world have been affected by HIV/AIDS. This viral disease affects 7.2 million people in Asian and Pacific countries, while India is reported to have 3.5–5.0 million sufferers. Coconut oil has been found to have antiviral properties due to its content of 48%–50% lauric acid. Monolaurin derived from coconut oil has virucidal effects on RNA and DNA viruses that are surrounded by a lipid membrane, including HIV. Capric acid, another constituent of coconut oil, has also been shown to have antimicrobial activity. When it was metabolized into monocaprin by the human digestive system, it was shown to have antiviral activity (Sy et al., 2006). It was established that lauric acid (C_{12}) from coconut oil has potent antimicrobial activity (Hess et al., 2014) and was experimentally found to reduce HIV levels (Kaunitz and Dayrit, 1992). In experiments using the macaque model for HIV-1 transmission in women, it was demonstrated that the use of monolaurin was highly effective in preventing vaginal transmission of the HIV analog, the simian immunodeficiency virus. Its modes of action were thought to include membrane stabilization including its microbicide activity—ie, blocking bacterial pore-forming toxins and T-cell activation, as well as blocking expression of the signaling pathway that leads to establishment of infected T-cells (Haase et al., 2015). Dr. Jon Kabara has undertaken some preliminary work, but pilot-scale testing with a large number of AIDS patients needs to be undertaken (Rethinam, 2006). The results of this research suggest that it may be possible to develop a cost-effective control measure to control the spread of HIV. If this happens, current world coconut production may not be adequate to meet the needs of all those afflicted.

An initial study conducted in the Philippines using coconut oil to treat HIV/AIDS patients demonstrated a 60% success rate. A much larger study is now being organized in Africa, where more HIV-infected subjects are available. A lack of funding has limited the number of clinical studies to date. More clinical studies are needed in order to verify laboratory, theoretical and anecdotal evidence (Dayrit, 2000).

ACKNOWLEDGMENT

The author is grateful for the essential contribution of Dr. Chay Binh Pham of Biozym Systems Enterprises and the BIOTECH Oils and Fats Laboratory, University of the Philippines at Los Baños.

REFERENCES

Banzon, J.A., Velasco, J.R., 1982. Coconut Production and Utilization. Philippine Coconut Research and Development Foundation, Manila, pp. 14–38.

Batugal, P., 2006. In: Adkins, S.W., Foale, M., Samosir, Y.M.S. (Eds.), Coconut Revival—New Possibilities for the 'Tree of Life.' Proceedings of the International Coconut Forum, Cairns, Australia, November 22–24, 2005. Australian Centre for International Agricultural Research, Canberra, pp. 11–47 (The International Coconut Genetic Resources Network (COGENT): its history, achievements and future plans).

Bawalan, D., Chapman, K.R., 2006. Virgin Coconut Oil – Production Manual for Micro- and Village-Scale Processing. FAO Regional Office for Asia and the Pacific, Bangkok, Thailand.

Corpuz, P., 2009. Philippine Bio-fuel Industry Outlook Gain Report. Global Information Network (Biofuels Annual).

Dayrit, C.S., 2000. Coconut oil in health and diseases and its potential cure for HIV/AIDS. In: Paper Presented at the 37th Annual Cocotech Meeting, India.

FAOSTAT, 2015. The Top 5 Coconut Producing Countries. (accessed January 2015) http://www.faostat.fao.org.

Fife, B., 2003. The Healing Miracles of Coconut Oil. PiccadillyPress, Colorado Springs, USA.

Fife, B., 2006. In: Adkins, S.W., Foale, M., Samosir, Y.M.S. (Eds.), In Coconut Revival—New Possibilities for the 'Tree of Life.' Proceedings of the International Coconut Forum, Cairns, Australia, November 22–24, 2005. Australian Centre for International Agricultural Research, Canberra, pp. 49–56 (Coconut Oil and Health).

Fife, B., 2008. Coconut and its healing properties. Acta Horticul. 787, 27–32.

Foale, M., 2003. The Coconut Odyssey: The Bounteous Possibilities of the Tree of Life. ACIAR, Canberra, Australia.

Foale, M., Roebeling, P., 2006. In: Adkins, S.W., Foale, M., Samosir, Y.M.S. (Eds.), In Coconut Revival—New Possibilities for the 'Tree of Life.' Proceedings of the International Coconut Forum, Cairns, Australia, November 22–24, 2005. Australian Centre for International Agricultural Research, Canberra, pp. 11–47 (The International Coconut Genetic Resources Network (COGENT): its history, achievements and future plans).

Gopala, R.A.G., Raj, G., Bhatnagar, A.S., Prasanth, P.K., Chandrashekar, P., 2010. Coconut oil: chemistry, production and its applications – a review. Indian Coconut J. 73 (3), 15–27.

Gursche, S., 2008. Coconut Oil: Discover the Key to Vibrant Health, Alive Natural Health Guides. Book Publishing Company.

Haase, et al., 2015. Glycerol monolaurate microbicide protection against repeat high-dose SIV vaginal challenge. PLoS One. http://dx.doi.org/10.1371/journal.pone.0129465.

Hess, D.J., Henry-Stanley, M.J., Wells, C.L., 2014. Antibacterial synergy of glycerol monolaurate and aminoglycosides in *Staphylococcus aureus* biofilms. Antimicrob. Agents Chemother. 58, 6970–6973.

Hossain, M.D.A., Choudhury, S.M., Rekhu, Y., Faraz, S.K., Islam, M.U.I., 2012. Biodiesel from coconut oil: a renewable alternative fuel for diesel engine. World Academy Sci. Eng. Technol. 68.

Kaunitz, H., Dayrit, C.S., 1992. Coconut oil consumption and coronary heart disease. Philipp. J. Intern. Med. 30, 165–171.

Laureles, L.R., Rodriguez, F.M., Reaño, C.E., Santos, G.A., Laurena, A.C., Mendoza, E.M., 2002. Variability in fatty acid and triacylglycerol composition of the oil of coconut (*Cocos nucifera* L.) hybrids and their parentals. J. Agric. Food Chem. 50 (6), 1581–1586.

Ooi, S.L., Pee, H.K., 1985. Processing for industrial fatty acids–II. J. Am. Oil Chem. Soc. 62, 348–351.

Pham, L.J., Gregorio, M.A., 2008. The triacylglycerol structure of coconut oil determined by chromatography combined with stereospecific analysis. Philipp. Agric. Sci. 91 (3), 343–347.

Rajamohan, T., Nevin, K.G., 2009. Wet and dry extraction of coconut oil: impact on lipid metabolic and antioxidant status in cholesterol coadministered rats. Can. J. Physiol. Pharmacol. 87 (8), 610–616.

Rethinam, P., 2006. Asian Pacific coconut community activities, achievements and future outlook. In: Adkins, S.W., Foale, M., Samosir, Y.M.S. (Eds.), In Coconut Revival—New Possibilities for the 'Tree of Life.' Proceedings of the International Coconut Forum, Cairns, Australia, November 22–24, 2005. Australian Centre for International Agricultural Research, Canberra, pp. 11–47 (The International Coconut Genetic Resources Network (COGENT): its history, achievements and future plans).

Sy, R.G., Gonzales-Santos, L., Salunat Flores, J., 2006. Fats and oils in human disease in coconut oil: issues and prospects. In: Florentino, R.F., Santos-Ocampo, P. (Eds.), Proceedings of the NAST Table Discussion on the Controversy of Coconut Oil. Manila, Philippines. June 8, 2006 (NAST, DOST), pp. 11–24.

Chapter 10

Tung (*Vernicia fordii* and *Vernicia montana*)

Jay Shockey
United States Department of Agriculture-Agricultural Research Service, Southern Regional Research Center, New Orleans, LA, United States

Timothy Rinehart
United States Department of Agriculture-Agricultural Research Service, Thad Cochran Southern Horticultural Laboratory, Poplarville, MS, United States

Yicun Chen, Wang Yangdong, Zhan Zhiyong, Hu Lisong
Institute of Subtropical Forestry, Chinese Academy of Forestry, Fuyang, Zhejiang, China

INTRODUCTION

The oils produced from the seeds of the tung tree (also known as the China wood oil tree) [*Vernicia fordii* (Hemsl.) and *Vernicia montana* (Lour.)] (previously *Aleurites fordii* and *Aleurites montana*) (Govaerts, 1966) occupy an important niche in the industrial vegetable oils market. The trees likely originated in China and have been grown and cultivated for oil production there for thousands of years; descriptions of tung oil use appear in the writings of Confucius from about 400 BC. For a more comprehensive description of the industrial development and history of tung production in China, please see Zhan et al. (2012). China still accounts for at least 70% of global tung oil production annually, but over time production has spread to various other countries (see Table 10.1). This crop was established in the United States nearly a century ago, and the supply of tung oil quickly rose to fill the demand for high-quality fast-drying oils for use in the manufacturing of paints, varnishes, inks, linoleum, and numerous other products. In recent years, tung oil has been used in synthesis schemes for the production of modern high-value/high-performance materials such as specialized resins and biodiesel. Recent advances in biochemistry and molecular biology have provided the means to dissect the tung oil biosynthetic pathway, an important step toward the long-term goal of engineering tung-like oils in the seeds of other oilseed species. The progression of tung tree propagation and oil utilization along this historical timeline, from the primitive physical processing of tung oil from a few trees in 1900s Florida to commodity level production and

Industrial Oil Crops. http://dx.doi.org/10.1016/B978-1-893997-98-1.00010-5

TABLE 10.1 Leading Global Producers of Tung Nuts and Estimated Annual Oil Yields for 2013

Country	Tonnes	Estimated Oil Production (1000 gallons)
China	418,294	20,915
Paraguay	35,980	1799
Argentina	6800	340
Malawi	4100	205
Madagascar	2600	130
Brazil	264	13.2

From: http://faostat3.fao.org/Q/QC/E (Production; Crops; Tung Nut). Tung oil production estimated based on 20% yield from dried pressed nuts.

utilization of the oil in sophisticated organic synthesis designs and molecular breeding programs is an interesting vignette in the history of American agricultural science. In this review, we will present a brief history of the tung industry in the United States, describe some of the modern uses of tung oil in industrial processes, and summarize the current status of genetic and biochemical analysis of the genes and enzymes in the tung oil biosynthetic pathway.

THE HISTORY OF TUNG IN THE UNITED STATES: THE TUNG BELT

Tung trees may have been introduced to the United States as early as 1859, but the first attempt at commercial introduction was by Consul General L.S. Wilcox, who sent seed from his post in Hankow, China to the State Department in 1904 (Fry, 1973; Lagasse and Sell, 1941; Potter and Crane, 1957; Potter, 1959, 1968). The US Department of Agriculture (USDA) assumed responsibility for the seed, and it was distributed by the Section of Foreign Seed and Plant Introduction. Dr. David Fairchild, senior agricultural engineer for the USDA, distributed seedlings to cooperators and researchers in the warmer regions of the United States in 1906 and 1907 (Fairchild, 1913). Despite poor results in California, tung trees planted at experimental sites in Southern states, particularly those sharing a shoreline with the Gulf of Mexico, performed well. The first tung oil was extracted in 1913 from a single tree growing in Florida that was planted in 1907 by a farmer named W.H. Raynes (Potter, 1959, 1968). Total yield was 2.2 gallons of domestic tung oil, which was tested by officials from the National Paint, Varnish, and Lacquer Association. In 1912 the USDA Bureau of Plant Industry, with approval of Secretary of Agriculture James Wilson, issued a notice that recommended farmers in the Gulf Coast region plant tung trees as a new crop. A limited number of free 1-year-old

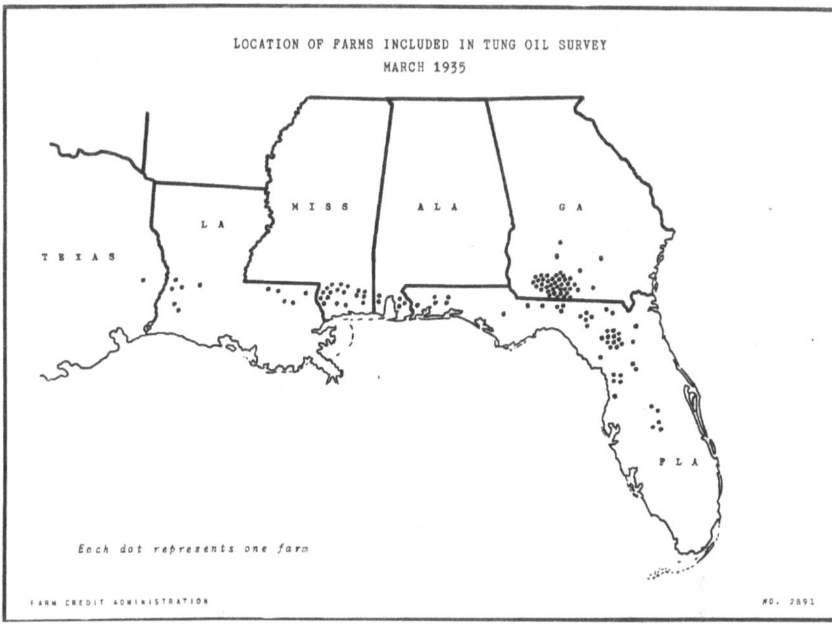

FIGURE 10.1 Distribution of tung orchards in the southern United States in 1935 shows the area along the Gulf Coast that was most adapted for tung oil production (Farm Credit Administration, 1935). The number of tung orchards increased in subsequent years, but the geographic distribution remained the same.

trees were distributed to interested parties, and the first tung tree orchards were planted in the United States in 1913.

Florida was the epicenter of enthusiasm for domestic tung oil production, particularly the areas around Gainesville and Tallahassee. The National Paint, Varnish, and Lacquer Association planted trial orchards near Paradise, Florida, in 1924 (Farm Credit Administration, 1935). Successful commercial orchards were planted in Florida shortly thereafter, including 216 acres planted by the American Tung Oil Corporation, a company financed by the National Paint, Varnish, and Lacquer Association, and 700 acres planted in 1925 by the Alachua Tung Oil Company, a subsidiary of Benjamin Moore Paint (Farm Credit Administration, 1935). B.F. Williamson, a large landholder near Gainesville, Florida, planted 23 acres in 1924 and 1925, and went on to manufacture tung oil-based paints (Farm Credit Administration, 1935). L.P. Moore, the nephew of Benjamin Moore, founder of Benjamin Moore Paints, built the first mechanized tung oil compressing mill in 1928 near Gainesville, Florida (Brown and Keeler, 2005). Tung nuts from the first orchards were pressed in this mill in 1928 and signified the first commercial-scale production of domestic tung oil (Potter and Crane, 1957). Plantings spread from northern Florida across the Gulf Coast region through Alabama, Mississippi, and Louisiana in the late 1920s (Fig. 10.1).

The tung belt of the United States. (Dots indicate 1950 acreages.)

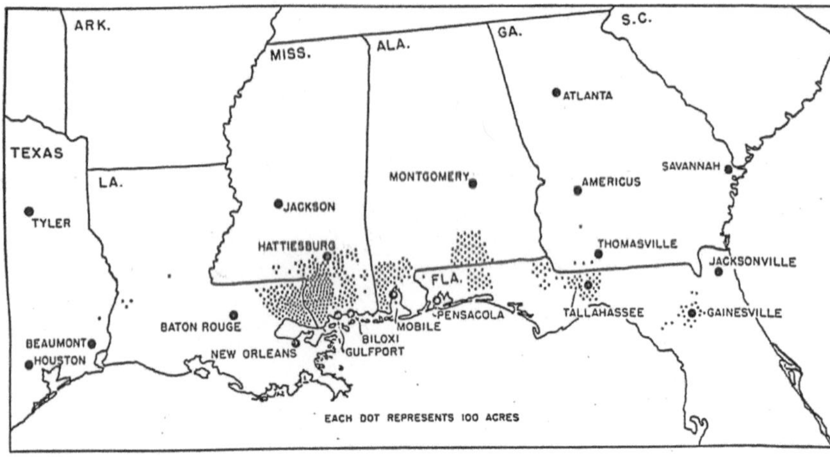

N–20323

FIGURE 10.2 Tung production was closely associated with tung processing mills, which accounts for the increased density of tung orchards in Louisiana and Mississippi (United States Tariff Commission, 1960). *U.S. Department of Agriculture, Tung Production, Farmers' Bulletin No. 2031, revised July 1957.*

The Tung Belt is a swath of land 100 miles wide that runs east to west across the top of the Gulf Coast from Texas to the Atlantic Ocean. It passes through Louisiana, Mississippi, Alabama, a small part of Georgia, and across northern Florida (Fig. 10.2). This land is highly favorable to tung production because of its warm weather and sandy soils. In the early 1900s, large tracts of land in this region were cleared of old growth pine. Farmers and landowners who had made substantial money from timber sales were looking for a crop to plant on the recently cleared land, which was marginal and largely unsuitable for growing food crops. In the late 1800s when US manufacturers began using tung oil in large amounts, it was primarily imported from China, although quality and availability was unpredictable. Tung oil production in China at the time was mainly from tung nuts collected in groves of trees that were planted in marginal soils and not extensively cultivated (Blackmon, 1943). Domestic production in the United States was enthusiastically promoted as a cultivated crop with the potential to diversify small farm income or be the primary income from large operations. As such, investors in tung orchards included small farmers, hobbyists, wealthy investors, enthusiasts, speculators, grower collectives, and industrial or commercial enterprises.

Most of the infrastructure to establish the industry was centered on mills that pressed, stored, and shipped the oil in barrels or railcar containers. Some operations had their own mills, but most brought their crops to a regional mill owned by a large farm or private company. Farmers either sold the nuts to the mill, which then assumed responsibility for selling the oil, or contracted with the mill for oil extraction and retained the rights to storing and selling the product. In addition to investing in orchards, growers sometimes built storage barns for the nuts after harvest. With increased storage of tung nuts and tung oil, the seasonal spike in sales after harvest time decreased, and oil was sold year-round with market prices and available supplies monitored like many other agricultural crops.

The boom in tung oil production from 1925 to 1940 was driven by domestic demand for tung oil and the idea that domestic production could satisfy that demand at world market prices. Annual imports into the United States ranged from 83,004,000 to 174,885,000 pounds of oil during 1925–40, according to the US Bureau of Foreign and Domestic Commerce, US Department of Commerce (Blackmon, 1943). Domestic production accounted for 1/40 of domestic consumption, so the outlook for US tung oil producers was positive. During this time there was considerable variation in international supply, mostly due to war conditions.

Tung tree orchards spread across the Tung Belt. Mississippi replaced Florida as the number one state for tung oil production. In 1935, there were 44,162 acres of tung production, and 21,000 were in Mississippi, slightly more than the 17,000 in Florida, according to a survey in the 1935 Farm Credit Administration Report (Farm Credit Administration, 1935). Lumber companies made up 21.5% of the industry and were mostly located in Mississippi. Promoters made up 7.2% of the industry. Tung farmers and tung oil users made up only 4.3% and 5.2% of the industry, respectively. Thus, most industry expansion was due to persons who did not rely on tung oil as their major source of income. As expected for this group of investors, farm size and percentage of area devoted to tung oil production varied widely.

The age of the tung orchards varied as well, but most were young and not yet in peak production. On 122 farms surveyed in six states in 1935, more than 90% of the tung trees planted for commercial production were 5 years old or less (Farm Credit Administration, 1935). It takes 4–5 years for a tung tree to produce a crop commercially. Therefore, most of the area planted was just reaching production stage and would not see peak production until after 1940. With the bulk of the trees not yet at peak production, the yields per acre were low—270 to 400 pounds of whole fruit per acre for 6-year-old trees—but expected to climb as the trees matured (Farm Credit Administration, 1935). Most orchards planted before 1940 were derived from random seed collections, which also contributed to variation in yields. Despite differences between trees, between years, and between locations, the average yield for each state could be compared with its costs such as clearing, planting, fertilizing, labor, and harvesting.

In 1935, Florida had the highest average cultivation costs ($72.57 per acre by the fifth year), followed by Georgia and Alabama ($32.58) and finally Mississippi, Louisiana, and Texas at $21.20 (Farm Credit Administration, 1935). At a market price of $24 per ton of whole fruit, a yield of nearly 2000 pounds per acre would be needed to cover costs in Mississippi, and three times that in Florida (Farm Credit Administration, 1935). Fruit sold as seed to further expand the industry commanded higher prices, so it not surprising that the promotion of tung was a profitable business. The most rapid expansion took place in Mississippi and Louisiana, which was expected given their lower production costs.

Census reports document the significant growth of tung from 1930 to 1945, during which time the industry expanded from 400,000 to 9.6 million tung trees in the United States (United States Tariff Commission, 1960; includes all trees of any age). The tree total was 13 million in 1950 and 13.7 million in 1954. Many small farms that fueled the initial enthusiasm for tung oil were consolidated into larger ones in the early 1950s. This was observed as a decrease in the number of farms growing tung. By 1954, only 59% of the farms reported as growing tung in 1950 were still doing so (United States Tariff Commission, 1960). Approximately 50 large-scale producers accounted for half the domestic production in the United States, and most produced tung oil as their principal cash crop (United States Tariff Commission, 1960). This was a considerable shift in the makeup of the industry from its beginnings in the 1920s and 1930s. US growers also formed commodity organizations such as the Tung Growers Council and the American Tung Oil Association, which supplied research and marketing services to their members through groups such as the Tung Research and Development League and the National Tung Oil Marketing Cooperative.

RESEARCH ON TUNG

Tung research facilities were located in the heart of the major production zones of the Tung Belt. The USDA began tung production research in 1938 in Louisiana, Georgia, Florida, and Alabama. Federal researchers had expertise in horticulture, plant physiology, biochemistry, soil technology, genetics, histology, plant breeding, cytology, and orchard management. Initial research focused on tung tree responses to cultivation practices and the best orchard management solutions for lowering production costs. Published work includes tung tree responses to fertilizer amounts, types, and placement, the use of cover crops during orchard establishment, and optimized methods of cultivation (Fig. 10.3). By 1967, after almost 30 years of research, 120 technical reports had been generated and more than 160 reports had been made to growers' meetings. A thorough understanding of the tung tree was the highest priority, because it was an introduced crop and unfamiliar to most farmers and scientists in the United States. Of particular importance to the tung oil industry was oil content and flower production including bud production, dormancy, pollination, and fruit set. For example, some initial findings include the inverse relationship between fruiting and new shoot growth. Inflorescences contain both male and female flowers, with the latter giving rise to fruits after pollination. Tung researchers observed that the presence of five or more fruit per terminal could completely inhibit new shoot growth from that terminal, especially if nitrogen was limited.

FIGURE 10.3 *The Farmers Bulletin on Tung Production* was issued in December 1951 to educate growers and prospective farmers on the benefits and profitability of growing tung in the Gulf Coast region of the United States. It was compiled by George F. Potter and Harley L. Crane, scientists with the Agricultural Research Service.

Lack of shoot growth limited carbohydrate production to the few leaves that emerged from the terminal, which reduced oil production in the tung nut kernels at that branch. Conversely, fewer fruit per terminal meant an increase in shoots, and the number and size of the leaves on a shoot were closely correlated with shoot length. Longer shoots meant more leaf area, which was proportional to photosynthesis and directly affected oil production and storage in the tung nut kernels, which mature in early October with up to 65% oil by weight in a good season. Coincidentally, the timing of oil formation coincides with flower bud initiation. If carbohydrates were not plentiful enough for both processes, the oil yield was low and the initiation of flower buds was also limited, which reduced the crop in the next season.

Science-based recommendations on soil requirements, cultivation, fertilization, and prevention of cold injury helped to increase yield and maximize profitability for the entire Tung Belt. Average yields climbed to 1.5 t of whole fruit per acre when measured over a 5-year window (Spiers and Kilby, 1973). Test plots using the latest methods and cultivars could produce 4–5 tons per acre in a given year (Spiers and Kilby, 1973). If environmental and weather-related impacts could be lessened by genetics or cultivation practices, goals to double the average yields to as much as 4 tons per acre seemed reasonable given the quality of the science and number of researchers who were focused on tung oil.

After the rapid expansion of commercial tung orchards, research on genetics, plant breeding, and cultivar improvement began in earnest. Most orchards were planted with seed obtained from other orchards or unknown sources. Naturally, trees varied widely in size, production, and fruit qualities. Because trees were propagated by seed, early work on tung genetic improvement focused on selecting mother trees that produced uniform seedlings in terms of growth habit, fruit production, and oil content of fruit (Fig. 10.4). Techniques for asexual propagation such as bud grafting were refined, but usually only the mother trees were

FIGURE 10.4 *The Farmers Bulletin on Tung Production* was reissued several times with updated information on cultural practices, propagation, cover crops, and harvesting methods perfected by scientists with the Agricultural Research Service.

clonally propagated. Evaluation and distribution was done using seed produced lines, some of which were named and released by the USDA. Cultivars introduced include "Gahl" (L51), "LaCrosse" (F99), "Lamont" (F452), "Cooter" (F4), "Isabel" (L2), "Folsom" (L47), and "Lampton" (F578). In the latter stages of the USDA tung breeding program, controlled hybridization was used to breed new combinations of useful traits. Parents with desired traits, generally a combination of late blooming, high oil content, cold hardiness, and productivity, were intercrossed and seedlings were evaluated for superior combinations of traits. Selections were then self-fertilized to fix the desired combination of traits. Seed lines derived from F_2 seedlings were evaluated for stability and uniformity and compared with existing commercial cultivars, typically "Isabel" and "Folsom," before making a final determination about their value to the industry. Considering that it takes up to 4 years to generate fruit and that accurate measures of pounds of oil per tree required 6 years of data, a tung breeding cycle to release a new cultivar took at least 10 years, and oftentimes much longer. The USDA breeding program for improved tung also experimented with interspecific hybridizations between *V. fordii* and *V. montana*.

The fruits of the USDA's breeding efforts were not fully realized because research into tung declined rapidly after 1969 with most research stations closed or redirected to other crops. The USDA Tung Research facility in Poplarville, Mississippi, switched to small fruits, mostly blueberries, and was renamed the Small Fruit Research Station in 1971. Later this became the Southern Horticultural Laboratory (TCSHL) with an emphasis on small fruits, ornamentals, and nursery crops. To this day it maintains the largest, most complete *V. fordii* germplasm collection in North America. Elite selections and higher-yielding frost-tolerant cultivars that were destined to be released in the 1970s have been conserved and maintained at TCSHL for the last 50 years. The potential for these new breeding lines has never been explored because the tung oil industry in the United States never recovered from its slow collapse in the late 1960s.

TUNG'S DEMISE IN THE UNITED STATES

Early investors in tung orchards expected to sell their oil on the world market, which included US manufacturers. Prior to World War II, domestic tung oil production was still a developing industry and profitable years were unpredictable for young orchards. Prices declined from 1928 to 1940. Fortunately for growers, the entire crop of domestic tung oil was devoted to war efforts and purchased by the federal government during World War II. The price set by the government was profitable and demand was high for military purposes. Private manufacturers that had relied on tung oil prior to the war were forced to seek alternatives during this same time period because military uses dominated available supplies. Alternative oils, while not of the same quality as tung, were developed by the chemical industry and ultimately reduced demand for tung oil after the military discontinued its use. Also, the United States encouraged growth of the tung oil industry in South America to help domestic production, which could not keep pace with the military's demand for tung oil at the time. This investment eventually became a competing industry as shipping costs were reduced and the global market for tung oil included China and beyond. The net effect of cheap imported tung oil, competing alternative oils and chemistries, and decreased demand after World War II was a fragile and unstable domestic tung oil industry that eventually collapsed.

In 1942, the USDA began a price support program for domestic tung oil to stabilize profits for the young industry and help ensure long-term viability. A system for loans and seed/oil purchases was administered by the Commodity Credit Corporation (CCC) of the USDA. The CCC often acquired large stocks of tung oil that it could then sell on the world market. Selling CCC tung oil on the US market would only drive down domestic prices and shift more of the growers to price support. Since the world market price for tung oil was considerably less than the US market price, selling on the world market meant the CCC sold at a loss.

After World War II, increased levels of foreign imports and expanded development and use of synthetic drying oils reduced demand considerably. Consumption averaged about 100 million pounds of tung oil per year until 1951 (United States Tariff Commission, 1960). From 1951 to 1957, consumption declined to only 51 million pounds. To combat decreasing demand, research was initiated to find new innovative uses of tung oil in manufacturing. The Pan American Tung Research and Development League explored tung oil as anti-spalling for concrete and fast-drying automotive paints that could be electro-deposited on metal. Tung meal, the byproduct of processing, was researched for new uses in agriculture as feed, insecticides, and even explosives. None of the new uses proved enough to expand the domestic market for tung oil. At the same time, research on mechanization was also initiated to offset the rising cost of production and labor. Lowering production costs included amendments to the Fair Labor Act of 1933, but these had little effect on the bottom line for tung oil producers, who saw little increase in domestic demand. From 1948 to 1960 the CCC sustained losses of $3.5 million due to its support of the price of tung oil in the United States (United States Tariff Commission, 1960). While price supports did stabilize the tung oil industry in the United States, they did not solve long-term industry solvency. Because US market prices were high compared with those of the world market, foreign producers, particularly those in South America, shipped tung oil to the United States. The importation of tung oil helped satiate US demand, but it also kept domestic prices low and decreased the profitability of US growers. The CCC continued to acquire oil and sell it at a loss. In 1967 the CCC still had a supply of 65 million pounds of tung oil that it had purchased from domestic growers; this supply was sold at a loss on the world market (Fig. 10.5).

Weather, particularly frosts, contributed to volatility in the tung oil industry. For example, late frosts and freezes significantly disrupted supply in the

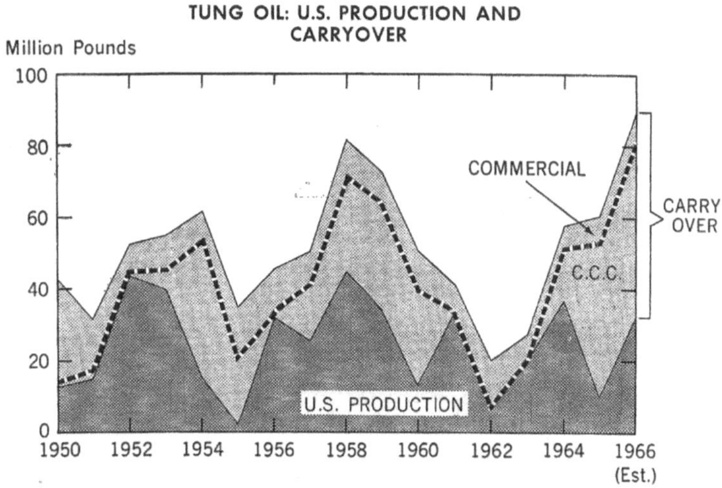

FIGURE 10.5 Tung oil supplies from 1950 to 1966 demonstrate the increase in surplus tung oil that the CCC stockpiled from year to year. These supplies helped stabilize domestic production, but were eventually sold for a loss on the world market. *The graph above was reprinted in the Proceedings of the 34th Annual Tung Industry Convention held in Edgewater Park, Mississippi in 1967.*

FIGURE 10.6 Damage from hurricane Camille in 1969 was extensive along the Gulf Coast, as these before and after images from Mississippi demonstrate.

years between 1950 and 1959, with almost complete crop loss in 1955 and aggregate loss of 20% of production due to late frost events over the 21 years recorded (Potter, 1959). Hurricanes, a weather phenomenon associated with the Gulf Coast location of the Tung Belt and less predictable than frost, were a significant factor in the industry's demise. Hurricanes Betsy in 1965 and Camille in 1969 caused widespread destruction of orchards and crops (Fig. 10.6). Tung trees are soft wood and prone to fracturing in high winds. Damage estimates from 1969 indicate that all orchards, including the more established groves in Mississippi and Florida, experienced significant damage that required replanting. The US federal government offered disaster relief funds to farmers after Hurricane Camille, but many tung growers decided not to use the funds to replant tung, instead converting tung orchards into pasture for cattle or other agronomic crops. Given the deteriorating state of the tung oil industry at the time, many growers were already looking at alternative land uses, including sales from timber and cattle operations, that had higher profits per acre. Damaged orchards were not replaced, and with no major increase in the demand for domestic tung oil, domestic production began to dry up. The USDA discontinued tung breeding research in 1969. Without price support from the federal government, domestic production had completely dried up by 1972. The Tung Research Laboratory in Poplarville, Mississippi, was renamed the Small Fruit Laboratory in 1972 to indicate the switch from tung oil research to other crops.

INDUSTRIAL UTILIZATION OF TUNG OIL

The Composition of Tung Oil: "The Secret Ingredient"

Tung oil contains minor proportions of the familiar fatty acids palmitate, stearate, oleate, linoleate, and linolenate. What makes tung oil special is α-eleostearate ($18{:}3\Delta^{9cis,11trans,13trans}$). A few other plant species, such as Chinese melon (*Momordica charantia)*, contain significant amounts of eleostearate; however, processed oil from *V. montana* and *V. fordii* (up to 80–82 mol%) is the purest available natural source of this novel conjugated trienoic fatty acid.

The typical appearances of a tree from *V. fordii* cultivar L2 and maturing fruit are shown in Figs. 10.4 and 10.3, respectively. Fig. 10.7A displays a cross section of a mature tung seed containing up to 60% oil by weight, which is present in numerous oil bodies, interspersed among less-abundant protein bodies. The structure of α-eleostearic acid is shown in Fig. 10.7B.

The unique reactivity of eleostearic acid enhances its participation in a variety of chemical reactions, such as carbon–carbon double bond addition, acid–base reactions, esterifications, reduction reactions, Diels–Alder reactions, and thermal polymerization. Recent descriptive examples of a few of these reactions, and how they can be applied to the production of various useful products, are listed in the following section.

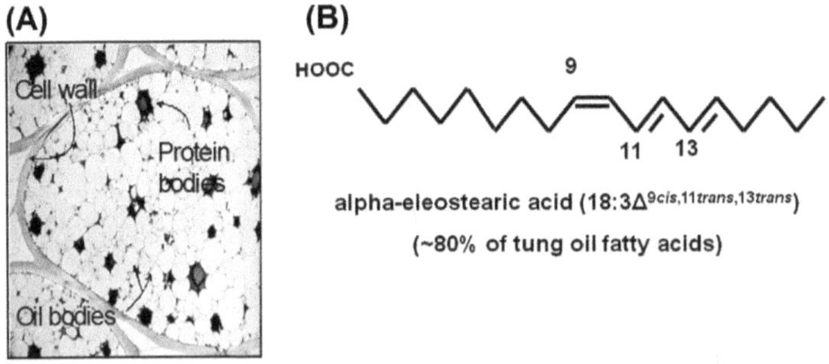

FIGURE 10.7 (A) Light microscope cross section of a mature tung seed. Note the numerous light-colored oil bodies interspersed among dark-colored protein bodies. (B) The structure of alpha-eleostearic acid. The positions of the double bonds are numbered starting from the carboxyl end. *Microscope image courtesy of Tom Jacks (USDA-ARS, SRRC).*

Resins from Unmodified Tung Oil: Thermal Polymerization of Eleostearic Acid

A specific example of the utility of vegetable oils in general and tung oil specifically in the production of modified polymer products was recently described by Das et al. (2011), who tested the effects of unmodified tung oil on unsaturated polyester resins. These petroleum-based resins are inexpensive and contain a proper balance of mechanical, electrical, and chemical properties, but often form excessively brittle end products after curing. The addition of vegetable oils into this type of polymer allows for modification of the properties of the matrix, as the oils act as plasticizers and reduce the use of petroleum-based inputs in the production process. Incorporation of 1% tung oil (by weight) into the unsaturated polyester matrix significantly increased various mechanical properties of the resin, including impact strength, modulus, hardness, and flexural properties. Tung oil was used in its unmodified form, unlike most similar studies (cited in Das et al., 2011) that required pretreatment of the vegetable oil components, which increases the expected input costs of such synthesis schemes. Fig. 10.8 displays the probable crosslinking reaction of all three components as suggested by Das et al. (2011).

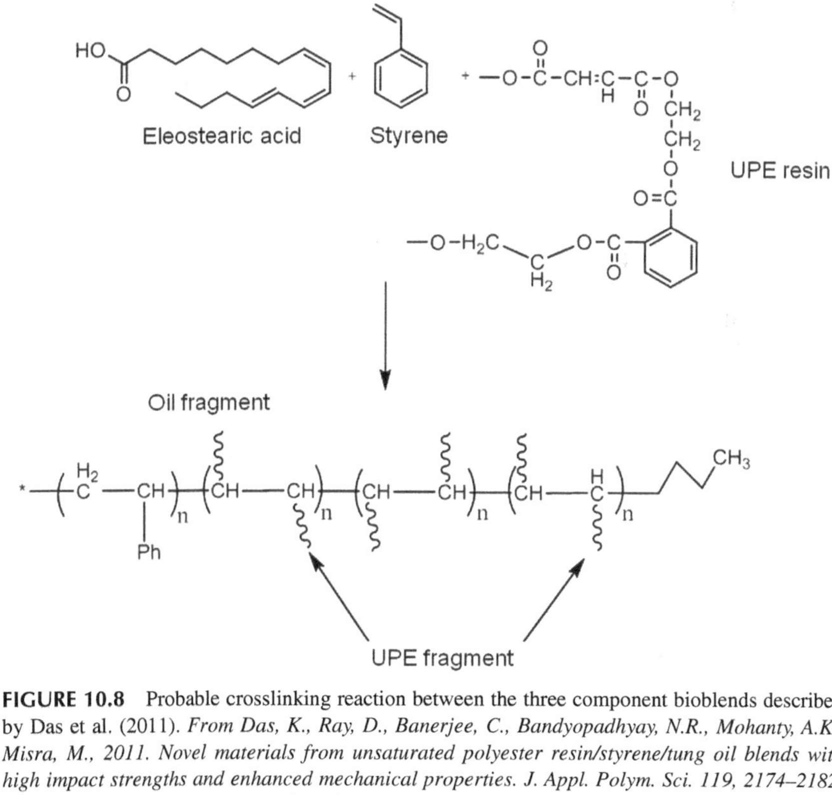

FIGURE 10.8 Probable crosslinking reaction between the three component bioblends described by Das et al. (2011). *From Das, K., Ray, D., Banerjee, C., Bandyopadhyay, N.R., Mohanty, A.K., Misra, M., 2011. Novel materials from unsaturated polyester resin/styrene/tung oil blends with high impact strengths and enhanced mechanical properties. J. Appl. Polym. Sci. 119, 2174–2182, reproduced with permission of John Wiley and Sons, Inc.*

Epoxy Resins Containing Tung Oil: Epoxidation and Diels–Alder Reactions

Epoxies are a multi-billion-dollar industry covering a wide range of applications including structural adhesives, metal coatings, electronics, high-tension electrical insulators, reinforced plastics materials, and dental supplies. Curing of the component epoxide resins, either alone via homopolymerization or in the presence of a wide range of curatives such as phenols, thiols, alcohols, and acids, forms the class of compounds generally called thermosets. Epoxy resins are important thermosetting resins because of their rigidity and high chemical and temperature tolerances. One of the most common types of epoxy resins is formed from the reaction of epichlorohydrin and bisphenol A to form diglycidyl ethers of bisphenol A. The rigid aromatic structure of bisphenol A confers beneficial properties to the final product. However, bisphenol A has recently come under intense scrutiny due to its role as a possible estrogen mimic and the attendant disruption it causes to endocrine signaling and metabolism, such as causing precocious puberty in young girls (Chakraborty et al., 2012). Many countries have strictly limited its production. Therefore, it is of great interest for resin manufacturers and end users to explore low toxicity substitutes, particularly from renewable alternatives that have comparable properties to bisphenol A-type epoxies. Therefore, both manufacturers and consumers would greatly benefit from the development of new epoxy resin chemistries developed with bio-based low-toxicity alternatives without sacrificing the physical and chemical properties found in products derived from bisphenol A. Epoxidized vegetable oils are often used for this purpose, but suffer from some disadvantages. The majority of the epoxide groups in epoxidized vegetable oil are internal oxiranes, which are substantially less reactive than the terminal epoxide groups found in other types of monomers. The low reactivity of the internal oxirane may inhibit complete curing, thus leading to incomplete crosslinking and unsatisfactory thermal and/or mechanical properties of the final resins (Huang et al., 2014).

Abundant inexpensive plant-based feedstocks such as lignins and rosins have been tested for use in epoxy resin synthesis, but display negative characteristics including slow curing rates and severe brittleness (Liu et al., 2012). Huang et al. (2014) synthesized the glycidyl ester of eleostearic acid (GEEA) using tung oil. Unlike other types of epoxy monomers, this compound shows great promise because it can be cured independently with both anhydrides, which react with the carboxy-terminal epoxide group formed after activation with epichlorohydrin, and dienophiles, which react with the conjugated double bonds in the center of the fatty acid chain via a Diels–Alder reaction. Because these two curing methods are independent of each other, anhydrides and dienophiles with different structures or special properties could be used together as cocuring agents in order to adjust the structure and the crosslink densities of the resulting thermosets. Depending on the structures of the anhydrides and dienophiles and the ratios of the two curing agents, a wide variety of thermosets with various properties may be produced (Huang et al., 2014).

Coatings and Paints

Tung oil, in pure form and in formulations with other oils and catalysts, has been commercialized by various companies, including its use as a below-water marine vessel finish that resists salt corrosion [Biopin boat oil (http://www.biopin.com/bootsoel.html)] and as a quick-drying nondarkening wood treatment [Waterlox (www.waterlox.com), and Real Milk Paint Company tung oil wood finishes (http://www.realmilkpaint.com/oil)]. The seed oil of *Calendula officinalis* [rich in calendic acid (8,10-*trans*,12-*cis*-octadecatrienoic acid), an isomer of α-eleostearic acid], compares favorably with tung oil as a reactive diluent in some types of alkyd resin formulations (Biermann et al., 2010). Cultivation of *C. officinalis* as a nonfood crop may offer the advantage of expanding the industrial oil growing area to certain regions of Europe and other places where tung trees will not grow.

Another compelling field of materials science that utilizes tung oil is the production of UV-curable coatings. The use of ultraviolet light to initiate curing offers the advantages of reducing both energy usage and some of the hazardous emissions typical of conventional coating/curing combinations. Resins made from fatty acid–modified polyesters, otherwise known as alkyds, are a very important class of oil-based coating resins. Alkyds are often compatible with UV curing and can be produced from a variety of plant oils including palm, soy, and linseed (Cheong et al., 2009; Chittavanich et al., 2012, and references therein), thus lessening the environmental impact even further.

Generally speaking, many plant-based alkyds work best in unpigmented formulations. Most do not function as well in the presence of pigments, which hinder penetration of UV energy and reduce the efficiency of the photo-activated crosslinkers. This effect leads to what is called a "cure gradient," wherein the top surface of the film cures more rapidly than the lower levels, leading to wrinkles in the film surface. Recent studies have taken advantage of the increased reactivity of tung oil to address this problem. First described by Thanamongkollit and Soucek (2012) and Chittavanich et al. (2012) described further improvements in the synthesis and use of a UV-curable tung oil alkyd system, which was prepared using a significantly milder catalyst (lithium hydroxide) instead of lead-based catalysts. When blended with optimized mixes of additives, pigments, and reactive diluents, coatings containing up to 14% pigment were produced that still retained excellent film properties.

Biodiesel

Petroleum has helped mankind conquer the world, in its various forms fueling and lubricating the vehicles that have transported us and the products we consume across the skies, oceans, and roads, in addition to being used to produce various pieces, parts, and accessories of the vehicles themselves. However, combustion and other uses of petroleum products have generated immeasurable waste products that have often poisoned the land, sky, oceans, and lakes as we travel across and through them. Petroleum is also a finite resource, the remaining portions of which will only be extracted from the earth with increasing cost, environmental impact, and political instability. For these reasons, there is ample motivation to explore a more diversified profile of possible petroleum alternatives, including renewable bio-based sources. Biodiesel is one such promising alternative (see chapter: Biodiesel and Its Properties). Biodiesel offers numerous advantages including its reduced toxicity and biodegradability. Biodiesel is composed of fatty acid methyl esters produced via the transesterification of fatty acids with methanol or other low-molecular-weight alcohols. Many types of vegetable oils are compatible for use as biodiesel, but the burgeoning global population and accompanying caloric needs create an untenable conflict between food and non-food demands. Nonedible vegetable oils that could be grown at large scale and/or grown on marginal lands would be an ideal alternative.

Tung oil displays low oxidative stability, and is thus generally not considered a good biodiesel feedstock. However, biodiesel derived from tung oil does display a very low cold flow plugging point (CFPP) and some other beneficial properties that complement some of the weaknesses of fuels with higher saturated fatty acid content that are derived from more common and abundant oils such as palm and canola. Chen et al. (2010b) tested the properties of different blends of tung, palm, and canola oils. Methanolysis of the different oils to produce fatty acid methyl esters, followed by blending at an optimal ratio of 60:20:20 (canola:palm:tung), created a fuel that maintained the low CFPP while improving other parameters such as density and kinematic viscosity. With the addition of normal levels of antioxidants, this blend should satisfy European biodiesel specification EN 14214 (Chen et al., 2010b).

TUNG TREES MEET THE MOLECULAR AGE

Despite the discontinuation of traditional research into tung tree germplasm improvement in the late 1960s, interest remained in approaching the "tung problem" from a different perspective: biochemical and molecular genetic analysis of tung seed oil biosynthesis. Interest in maintaining or expanding the cultivated acreage dedicated to tung tree cultivation may have waned, but the scientific and economic intrigue associated with commercial production of tung oil had not. The development of polymerase chain reaction (PCR) and molecular cloning protocols in the 1970s and early 1980s provided the means to identify and manipulate the genes and enzymes responsible for production of eleostearic acid. Such programs were initiated at the USDA-ARS location in New Orleans, Louisiana and at the Chinese Academy of Forestry's Institute of Subtropical Forestry. The early focus of this work was the identification of the novel enzyme activity that converted an otherwise "normal" fatty acid (oleate, linoleate, or linolenate) into the novel conjugated trienoic fatty acid α-eleostearate. Initially, attempts were made to extract this and other enzyme activities associated with oil production from developing tung tissues in proteinaceous form

(Dyer et al., 1998, 1999) for use in in vitro immobilized enzymatic systems, as had been successfully demonstrated for lipase-mediated enzymatic conversion of evening primrose oil by Akoh et al. (1996). The recalcitrance of this tung seed activity to traditional purification methods quickly led to a revised approach wherein the necessary genes from tung could be identified, studied in model systems in the laboratory, and then introduced into other commodity oilseed crops with amenable agronomic characteristics.

Fatty Acid Desaturases and Conjugases

The first major molecular insight into the biochemical analysis of tung oil synthesis was the cloning of genes for *fatty acid desaturase 2* (*FAD2*) which is common to all higher plants, and *FADX*, a diverged relative of *FAD2* (Dyer et al., 2002). Like other plant FAD2s characterized at the time, tung FAD2 converted oleic acid to linoleic acid, by addition of a cis double bond to the Δ12 position of the acyl chain. Consistent with a role in the production of a ubiquitous fatty acid, tung FAD2 was expressed in both seed and vegetative tissues (Dyer et al., 2002). A recent analysis of tung FAD2 from multiple tung tree accessions revealed a single nucleotide polymorphism that may be linked to higher seed lipid polyunsaturated fatty levels, including eleostearic acid (Ren et al., 2013). Expression of tung FAD2, especially when combined with coexpressed tung DGAT2 (discussed below) leads to large increases in cellular linoleic acid content in the fungus *Aspergillus fumigatus* (Chen et al., 2012).

FADX was expressed specifically in seed tissues, suggestive of a role exclusive to seed lipid metabolism. Biochemical analysis confirmed that yeast-expressed FADX produced α-eleostearic acid from linoleic acid (Dyer et al., 2002), combined with studies of the analogous conjugase from bitter gourd (*M. charantia*) (Cahoon et al., 2006), demonstrated that these fatty acid conjugases, like some of their related desaturases, act on linoleoyl groups esterified primarily to phosphatidylcholine. Besides the novelty of being the first known plant desaturase-like enzyme that introduced a carbon–carbon double bond in the *trans* configuration, tung FADX showed remarkable enzymatic plasticity. Feeding of various types of fatty acids to yeast expressing *FADX* led to the production of a wide variety of unusual conjugated and desaturated fatty acid products (Dyer et al., 2002). The cloning of the tung *FADX* gene was an essential first step in the development of metabolic engineering efforts aimed at production of eleostearic acid–containing lipids in transgenic organisms.

Tung *FADX* recently has been expressed in transgenic *Arabidopsis thaliana* lines. Expression of the gene was optimized using a variety of seed-specific promoters and strong transcriptional terminators, among other variables, as a means of exploring the parameters of using *FADX* as a technological tool to drive eleostearate production in transgenic crops (Shockey et al., 2015). Promoter strength and timing proved to be very important factors in controlling the level of eleostearic acid produced in transgenic seeds (Fig. 10.9).

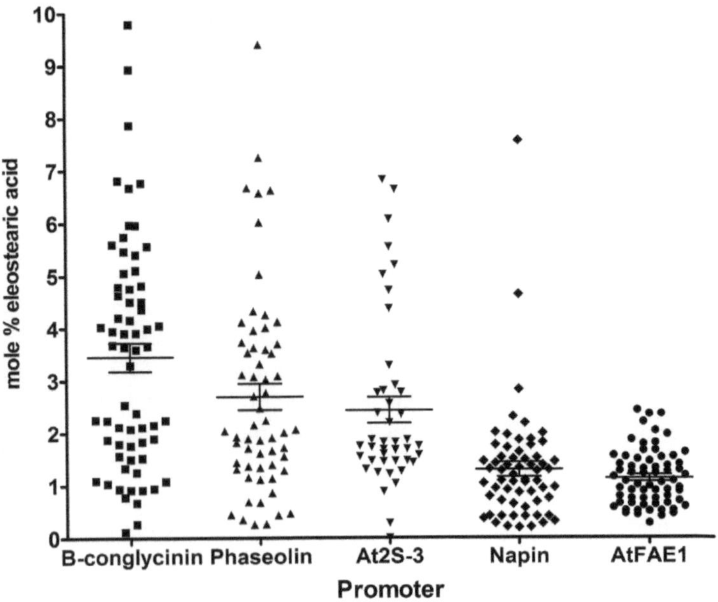

FIGURE 10.9 Comparison of α-eleostearic acid production in populations of transgenic *Arabidopsis* plants expressing tung *FADX* driven by different seed-specific promoters. The eleostearic acid levels measured in each of several independent segregating T_2 transgenic seed samples were plotted. Each group of points represents the level of product formed by tung FADX fused to a different seed-specific promoter. The mean levels of eleostearic acid and SEM are shown for each population. *From Shockey, J., Mason, C., Gilbert, M., Cao, H., Li, X., Cahoon, E., Dyer, J., 2015. Development and analysis of a highly flexible multi-gene expression system for metabolic engineering in Arabidopsis seeds and other plant tissues. Plant Mol. Biol. 89. doi:10.1007/s11103-015-0355-5, reproduced with permission from Elsevier.*

Dyer et al. (2004) also cloned tung FAD3, the enzyme that converts linoleic acid to linolenic acid, the final step of the traditional trienoic polyunsaturated fatty acid biosynthetic pathway in plants. Linolenate is a major constituent of lipids in tung vegetative tissues, but makes up only 1% of tung oil fatty acids, despite being expressed at substantial levels in developing tung seeds (Dyer et al., 2004). The surprisingly low levels of linolenate in tung seeds raised a number of questions regarding the control of relative gene expression and enzyme activity levels of the various desaturases and conjugases.

Both tung FAD2 and tung FAD3 were active in yeast cells, thus providing a facile transgenic expression system in which to explore the regulatory mechanisms at play. Comparisons of tung FAD2 and *Arabidopsis* FAD2 enzymes, in their native and epitope-appended forms, revealed that addition of an epitope tag to the 5′ end of the open reading frames for both *FAD2*s increased protein steady-state amounts via an increase in the rate of mRNA translation (O'Quin et al., 2009) (Fig. 10.10). Comparative analysis of FAD3 enzymes from tung and rape seed (*Brassica napus*) uncovered additional modes of control.

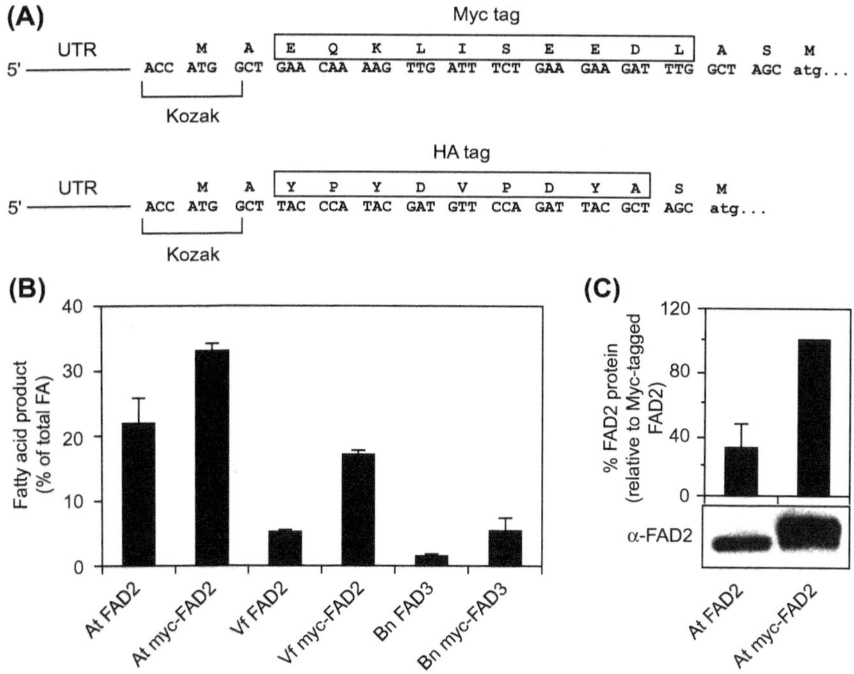

FIGURE 10.10 Expression of native and epitope-tagged plant FAD enzymes in yeast cells. (A) Diagrams illustrating the myc and HA epitope tag sequences used in this study. The translation start codon corresponding to the native FAD sequences is shown in lowercase (atg). (B) Gas chromatography analysis of yeast fatty acid composition showing the accumulation of fatty acid products (ie, linoleic acid for FAD2 and linolenic for FAD3) synthesized by each yeast strain. Bars represent the average and standard deviation of three independent experiments. (C) Protein immunoblot analysis of yeast cells expressing native or myc-epitope-tagged *Arabidopsis* (At) FAD2, probed with a native anti-FAD2 antibody. Shown are the representative immunoblot and the corresponding bar graph illustrating the average and standard deviation of relative protein amounts determined from quantitative densitometry of three independent experiments. *From O'Quin, J.B., Mullen, R.T., Dyer, J.M., 2009. Addition of an* N*-terminal epitope tag significantly increases the activity of plant fatty acid desaturases expressed in yeast cells. Appl. Microbiol. Biotechnol. 83, 117–125, reproduced with permission of Springer-Verlag.*

The tung protein produced approximately threefold more steady state protein and linolenic acid than did *B. napus* FAD3 when these enzymes were expressed in yeast fed linoleate (Fig. 10.11A and B) (O'Quin et al., 2010). The two enzymes displayed high amino acid identity in all regions of the proteins, except near the extreme *N*-termini. Domain swapping and mutagenesis experiments demonstrated that the *N*-terminal regions of these proteins contained degradation signals that interacted, at different levels of strength, with components of the ubiquitin-proteasomal protein degradation apparatus. The different rates of probable ubiquitination and proteasome recognition led to substantially different rates of protein turnover, which explains the differences in steady-state

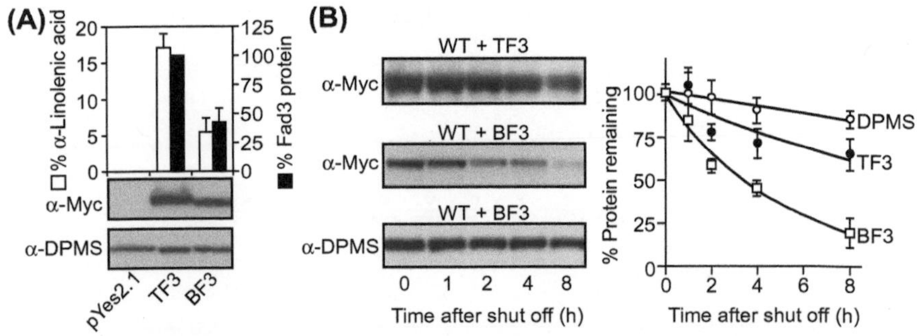

FIGURE 10.11 Relative stability of different plant FAD3 enzymes in yeast cells. (A) Measurement of protein amounts and polyunsaturated fatty acid enzyme products from tung FAD3 (TF3) or *Brassica napus* FAD3 (BF3) in yeast fed linoleic acid. Then the cells were processed for either immunoblotting or GC. *White bars*, % linolenic acid (the product of FAD3) present in total yeast lipids. *Black bars* represent different FAD3 protein amounts normalized to the amount of TF3. Below, representative protein immunoblots of protein extracts probed with either anti-myc antibody, which recognizes both TF3 and BF3 proteins, or anti-DPMS antibody, which recognizes the endogenous yeast ER membrane protein DPMS (load control). (B) Measurement of protein half-life in yeast cells through expression shutoff experiments. Transcription and translation were halted at time zero by the addition of dextrose and cycloheximide. Aliquots were removed over time and processed for immunoblotting. Left: representative blots; right: graph showing average and standard deviation of protein amounts from three independent experiments. *This research was originally published in* The Journal of Biological Chemistry. *O'Quin, J.B., Bourassa, L., Zhang, D., Shockey, J.M., Gidda, S.K., Fosnot, S., Chapman, K.D., Mullen, R.T., Dyer, J.M., 2010. Temperature-sensitive, posttranslational regulation of plant omega-3 fatty acid desaturases is mediated by the ER-associated degradation pathway. J. Biol. Chem. 285, 21781–21796. © The American Society for Biochemistry and Molecular Biology.*

protein levels (Fig. 10.11B) (O'Quin et al., 2010). This process was also controlled in part by temperature; both enzymes accumulated to higher steady-state levels and produced more linolenic acid at cooler growth temperatures (O'Quin et al., 2010). Collectively, these findings demonstrated some of the first significant insights into how lipid production in plants is regulated at the molecular and cellular levels.

Tung FAD2, FAD3, and FADX also made up part of the data set of enzymes that contributed to the discovery of a new protein sequence motif that is responsible for retrieval of ER-resident membrane-bound plant proteins. McCartney et al. (2004) characterized a short, aromatic amino acid-rich motif located very near the extreme *C*-terminus of various known or suspected intrinsic ER membrane proteins (including FAD2, FAD3, and FADX) that is both necessary and sufficient for ER retrieval and retention. This motif was distinct from the well-known "KDEL" signal present in resident ER lumen proteins, and the dilysine motif present in many other ER membrane proteins (McCartney et al., 2004). The aromatic retrieval motif has since been discovered in an increasingly broad array of other ER-localized enzymes, and has become an important tool in the list of quick, reliable diagnostic sequences that can be used for rapid, accurate predictions of subcellular targeting.

Electron Transport Chain: Cytochrome b5 and NADH:cytochrome b5 Reductase

Both fatty acid desaturases and conjugases extract two hydrogen atoms from acyl chain carbons to convert a single bond to a double bond. The hydrogens are transferred to an oxygen atom to produce water. Reduction of oxygen requires a pair of electrons, which are provided by the microsomal electron transport chain composed of NADH:cytochrome b5 reductase (CBR) and cytochrome b5 (cyt b5). Though expression of transgenic desaturases and conjugases from various plant species most often leads to production of the desired product, these levels are usually modest (van de Loo et al., 1995; Lee et al., 1998; Cahoon et al., 2006), raising the question of possible inefficient interactions between desaturases and cyt b5. The ability of cyt b5 and CBR to accentuate novel fatty acid production in transgenic systems has been addressed in some recent studies, with strikingly mixed results (Hatanaka et al., 2004; Kumar et al., 2006; Dahmen et al., 2013; Wayne and Browse, 2013). The collective uncertainty generated from these studies suggests that more research is required to determine the specific conditions in which electron transport chain components exert a noticeable influence over the types and rates of fatty acid biosynthesis in transgenic systems.

Cytochrome b5 proteins are members of the class of tail-anchored (TA) membrane proteins. TA proteins are widely distributed across nature, and are thought to be present in nearly all types of cellular membranes. This class of proteins shares structural features, the most prominent of which are a short *C*-terminal "tail" sequence that protrudes into the organellar lumen and a hydrophobic membrane-spanning domain immediately *N*-terminal to the tail. Hwang et al. (2004) cloned and characterized four isoforms of cyt b5 from tung tree. Three were found to localize to ER membranes, while the fourth (cyt b5D) represented the first characterized example of exclusive mitochondrial targeting of cyt b5. Extensive analysis revealed that while mammalian and plant cyt b5s share a number of biological, structural, and physicochemical properties, plant cyt b5s evolved additional features that ensure association only with the correct organelle, even in transgenic or in vitro systems, thus avoiding the spurious incorrect membrane interactions often seen with cyt b5s from other types of organisms. Henderson et al. (2007) later showed that the ER membrane specificity of tung cyt b5A and cyt b5B was conferred by the residues present in the charged terminal sequence, the portion of the protein downstream of the transmembrane spanning region.

Shockey et al. (2005) cloned and characterized two genes for tung *CBR*. Both enzymes were active in in vitro assays after expression in baker's yeast, and both genes were expressed in all tung tissues tested and were targeted to the ER membrane. These data clearly indicate that one or both tung CBRs could contribute to the process of cyt b5 reduction to help supply electrons for fatty acid desaturation throughout the tung tree, and for eleostearic acid production in developing tung seeds.

Acquisition of the gene set described in the previous paragraphs represented a major achievement in the dissection of the "front-end" reactions of polyunsaturated/conjugated fatty acid synthesis in tung tree tissues. However, the gene discovery efforts had only begun; equally important to the long-term goal of tung-like drying oil engineering was the need to understand the "back-end" reactions of TAG synthesis and packaging—that is, those that occur after fatty acid synthesis and desaturation/conjugation is complete. While the numbers and types of genes necessary for production of reduced cyt b5 and the fatty acid desaturation steps were substantial, the biochemistry underlying the back-end reactions was even more complex. Fig. 10.12 diagrams each of the possible reactions that could affect the flux of a molecule of eleostearic acid away from its site of synthesis in PC into TAG (Cahoon et al., 2007). Many genes representing the various steps shown in Fig. 10.12 have been isolated from tung. The descriptions of some of these genes and the insights gained from them are summarized below.

Diacylglycerol Acyltransferases

Diacylglycerol acyltransferase (DGAT) catalyzes the final acylation step in the TAG biosynthetic pathway by transferring a fatty acyl group from acyl-CoA to the *sn*-3 position of diacylglycerol to form TAG. Long before any gene for DGAT was known, biochemical studies hinted that DGAT may be a rate-limiting step in this pathway (Ichihara et al., 1988). The evolutionary importance of this reaction is reflected in the fact that no less than five distinct classes of plant enzymes can catalyze this reaction in vitro (Routaboul et al., 1999; Lardizabal et al., 2001; Saha et al., 2006; Rani et al., 2010; Durrett et al., 2010). The specific roles and quantitative significance of many of these enzyme types is not yet known, but the focus of most seed oil biotechnologists has been directed toward DGAT1 and DGAT2.

Both of these unrelated families of DGAT are ancestral to the evolutionary divergence of plants from other kingdoms of organisms. Tung was one of the first plant systems in which both *DGAT1* and *DGAT2* were cloned and analytically compared. Tung *DGAT2* is expressed at very low levels in the vegetative organs of the tree, but is strongly upregulated during seed development when eleostearic acid and total seed oil synthesis occur. *DGAT1*, on the other hand, is expressed at low levels in all tissues tested, and is not appreciably upregulated during seed development (Shockey et al., 2006). These general relative patterns of expression were later confirmed and quantified using sensitive real-time quantitative PCR (qPCR) (Cui et al., 2013; Cao et al., 2013). An in vivo functional assay in yeast showed clear substrate selectivity differences; the profile of TAGs produced by DGAT1 did not differ significantly from those produced by yeast DGAT, an enzyme that never encounters eleostearate under normal growth conditions. DGAT2, on the other hand, displayed a strong preference for eleostearate-containing substrates and produced more tri-eleostearin than tung DGAT1 or yeast DGAT. Perhaps most importantly, subcellular

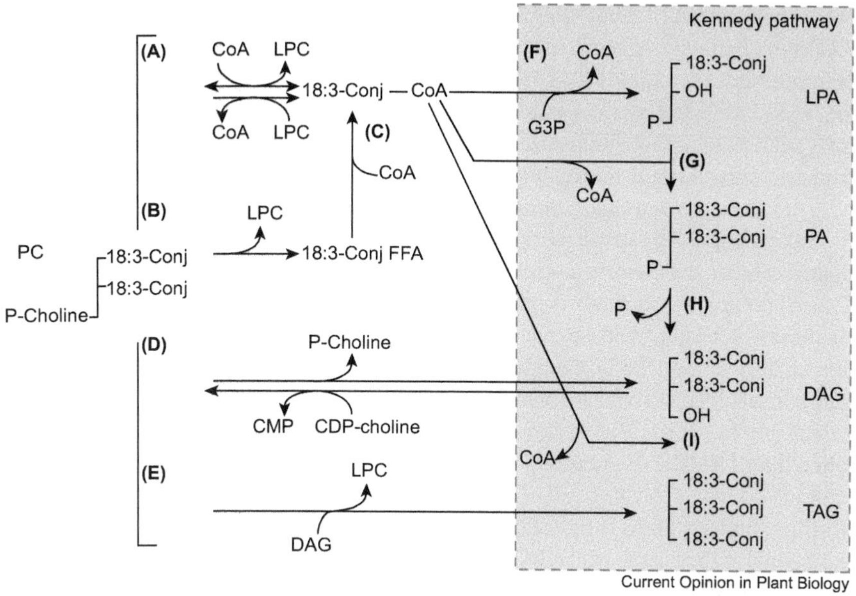

FIGURE 10.12 Multiple routes exist for the transfer of conjugated fatty acids from phospholipids to triacylglycerol (TAG). After the synthesis of conjugated (18:3-Conj) fatty acids on phosphatidylcholine (PC) (structure shown on the left), several different enzymes contribute to the channeling of these fatty acids into TAG. (A) Reversible lysophosphatidylcholine acyltransferase (LPCAT) transfers a fatty acid from PC to CoA, yielding a fatty acyl-CoA (18:3-Conj-CoA) and lysophatidylcholine (LPC). (B) Phospholipase A2 (PLA2) cleaves a fatty acid from PC, yielding a "free" (nonesterified) fatty acid (18:3-Conj FFA) and LPC. (C) Long-chain acyl-CoA synthetase (LACS) ligates an FFA to CoA, yielding a fatty acyl-CoA (18:3-Conj-CoA). (D) Reversible choline phosphotransferase (CPT) catalyzes the release of the PC head group (PCholine) to produce to diacylglycerol (DAG). (E) Phospholipid:diacylglycerol acyltransferase (PDAT) transfers a fatty acid from PC to DAG to produce TAG and LPC. Fatty acids in the form of acyl-CoA (18:3-Conj-CoA) can be incorporated into TAG by the enzymes of the Kennedy pathway (reactions highlighted on the right in the red box). (F) Glycerol-3-phosphate acyltransferase (GPAT) transfers a fatty acid from fatty acyl-CoA (18:3-Conj-CoA) to glycerol-3-phosphate (G3P), yielding CoA and lysophosphatidic acid (LPA). (G) Lysophosphatidic acid acyltransferase (LPAT) transfers a fatty acid from the fatty acyl-CoA (18:3-Conj-CoA) pool to LPA, forming CoA and phosphatidic acid (PA). (H) Phosphatidic acid phosphatase (PAP) cleaves the phosphate head group from PA, yielding phosphate and DAG. (I) Diacylglycerol acyltransferase (DGAT) transfers a fatty acid from the fatty acyl-CoA (18:3-Conj-CoA) pool to DAG, yielding CoA and TAG. *From Cahoon, E.B., Shockey, J.M., Dietrich, C.R., Gidda, S.K., Mullen, R.T., Dyer, J.M., 2007. Engineering oilseeds for sustainable production of industrial and nutritional feedstocks: solving bottlenecks in fatty acid flux. Curr. Opin. Plant Biol. 10, 236–244, reproduced with permission from Elsevier.*

localization of tung DGAT1 and DGAT2 transiently expressed in tobacco BY-2 cells revealed that these proteins are enriched in distinct subdomains of the ER (Shockey et al., 2006). The ER is a complex organelle, with at least 16 morphologically different membrane domains as determined by microscopic analysis (Staehelin, 1997), and discovery of tung DGATs in such domains was one of the first examples of determination of specific enzyme content in specialized domains. The enrichment of tung DGAT proteins in distinct subdomains of the ER suggests that TAG synthesis occurs in at least two of these domains. The functional specialization between the two remains a focus of current studies, but it is tempting to speculate that membrane and storage lipid synthesis occur independently, at least in organisms such as tung that synthesize unusual fatty acids in TAG that must be excluded from membrane lipids (Vogel and Browse, 1996) in order to minimize interference with the structure and function of the membrane lipid bilayer.

Taken together, these earlier data suggested that DGAT2 might play a major role in the determination of eleostearic acid content in tung seed oils. This hypothesis was confirmed when DGAT2 was overexpressed in transgenic *Arabidopsis* engineered to overexpress tung FADX. Lines that previously produced only about 7.5% eleostearic acid (utilizing only the endogenous *Arabidopsis* DGAT activities) produced ~11% when the tung DGAT2 gene was expressed in seeds. Eleostearic acid levels were increased further (to ~14%) when tung DGAT2 overexpression was combined with suppression of the dominant endogenous *Arabidopsis* DGAT activity (Fig. 10.13; van Erp et al., 2015). Similar findings from tung DGAT2 expression in fungi (Chen et al., 2012) and transgenic expression of DGATs from other plant species that produce TAG containing unusual fatty acids (Burgal et al., 2008; Li et al., 2010) confirm that DGAT2s are often a major determinant of seed oil composition in certain classes of plants. However, future studies in other crops should rightfully always address the relative roles of DGAT1, DGAT2, and other enzymes that actively participate in TAG biosynthesis.

Other Genes of Interest

Acyl-CoA binding proteins (ACBP) have long been implicated in various aspects of lipid metabolism. A long series of elegant studies in yeast has confirmed the necessity of ACBP for normal lipid biosynthesis in *Saccharomyces cerevisiae* (Færgeman et al., 2007). The study of ACBP in plants is not as thorough (Yurchenko et al., 2009). Genetic analysis has been complicated by the fact that unlike yeast, most plants surveyed contain a large complex family of *ACBP* genes containing many duplicated genes and functional subsets (Meng et al., 2011). Tung adds to the complexity of plant ACBP by virtue of containing two subclass III *ACBP* genes, called *VfACBP3A* and *3B* (Pastor et al., 2013). Gene expression profiling suggests genetic redundancy between these two genes. However, differences in predicted folding structures and acyl-CoA binding profiles cast some doubt on this assertion. Both proteins are also targeted to the ER, in contrast to prior data on other plant ACBP3 proteins (Pastor et al., 2013). The ER localization, along with prior determination of phosphatidylcholine (PC) binding by *Arabidopsis* ACBP3 (Xiao et al., 2010) may help to explain the fact that coexpression of *VfACBP3A*, along with tung *FADX* in transgenic *Arabidopsis*, appears to result in slight increases in the average amount of eleostearic acid produced in seed lipids (Pastor et al.,

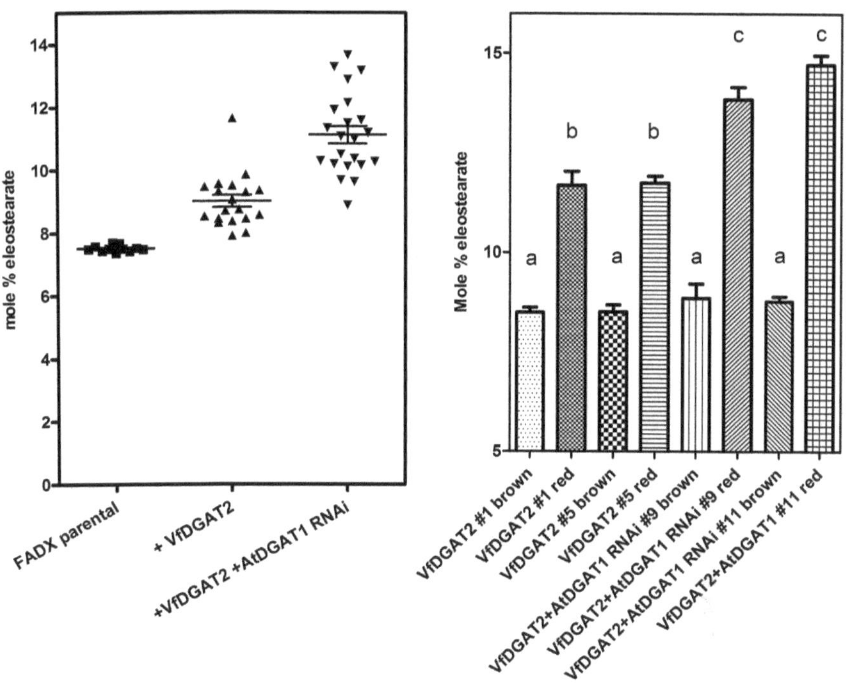

FIGURE 10.13 Left: *Arabidopsis* seed eleostearic acid levels in transgenic plants expressing tung FADX alone or with either tung DGAT2 alone or tung DGAT2 plus AtDGAT1 RNAi. A parental FADX line was retransformed with VfDGAT2 or VfDGAT2+AtDGAT1 RNAi, which suppresses expression of endogenous AtDGAT1 activity. Segregating T_2 seed pools from independent lines were analyzed by GC; the data are plotted as a scatter plot, with the mean and standard error shown. The mean value for each data set is significantly different than each of the other two (two-tailed t test, $P<0.0001$). Right: Determination of eleostearic acid levels in homozygous transgenic plants from left panel. Seeds from two high-performing lines from each of the transgenic genotypes were sown on soil and grown to maturity. Seed samples from progeny of two high-performing lines from each of the transgenic genotypes (representing at least four homozygous double transgenics or *FADX* parental revertants) were collected and analyzed by GC analysis, and plotted as mean eleostearate levels and standard error. Please note that the Y-axis does not begin at zero. Lowercase letters indicate different statistically significant groups (two-tailed t-test, $P<0.01$). *Reproduced with permission from van Erp, H., Shockey, J., Zhang, M., Adhikari, N.D., Browse, J., 2015. Reducing isozyme competition increases target fatty acid accumulation in seed triacylglycerols of transgenic* Arabidopsis. *Plant Physiol. 168, 36–46. Copyright American Society of Plant Biologists (*www.plantphysiol.org*).*

2013). Given the ability of ACBP3s to bind PC (the phospholipid substrate of FADX), does ACBP3A help to channel eleostearate out of PC and into TAG via a traditional acyl-CoA binding role, or a PC-binding role? These questions await future research.

Tung also contains several genes that encode oleosins. These small structural proteins adopt an unusual "thumbtack" structure, which allows for the hydrophobic central region to protrude into the lipid core of oil bodies in seeds, pollen, and other plant tissues. The termini of oleosin proteins are incorporated into the phospholipid monolayer that surrounds the TAG core of the oil body. This overall structure is thought to help maintain proper oil body number and size, thus providing optimal available surface area for mobilization of stored lipid reserves during germination (Huang, 1996). The precise roles of specific isoforms of oleosin are not well known at this time, but given their intimate association with stored TAG in vivo, they may be good targets for genetic manipulation. Tung contains several oleosin genes (Chen et al., 2010a; Cao et al., 2014) that have been cloned and analyzed, and will be the focus of further investigation.

Glycerol-3-phosphate acyltransferase (GPAT) catalyzes the acylation of the *sn*-1 position of glycerol-3-phosphate, the first step in the biosynthesis of membrane phospholipids, triacylglycerols, and epicuticular waxes. Two tung GPATs were cloned: *GPAT8* and *GPAT9* (Gidda et al., 2009). *Arabidopsis* GPAT8 is involved in the biosynthesis of cutin, and is a member of a large family of related genes that play various overlapping roles in the synthesis of cutin, suberin, waxes, and other epidermal lipid compounds (Zheng et al., 2003; Beisson et al., 2007; Li et al., 2007; Yang et al., 2010). Intriguingly, to date none of the *GPAT1–GPAT8* family have been implicated in oil synthesis. *GPAT9* is a novel putative GPAT; it is essentially nonhomologous to the *GPAT1–GPAT8* gene family, but shares significant identity with a mammalian GPAT that is critical for TAG biosynthesis (Cao et al., 2006), suggesting *GPAT9* as a candidate for this role in plants. Both GPAT8 and GPAT9 were targeted to DGAT2-type ER subdomains in tobacco BY-2 cells (as described for DGATs above), and interacted more strongly with DGAT2 than DGAT1 in yeast two-hybrid assays that measure protein–protein interactions. Despite the described role in cutin synthesis for *Arabidopsis* GPAT8 and the undetermined role for tung GPAT9, the current results suggest that either or both of these enzymes may functionally interact with DGAT2 in vivo and could supply some of the precursors for the TAG biosynthetic pathway in these membrane domains (Gidda et al., 2011).

CONCLUSIONS

Tung trees have held a unique place in the industrial oils industry for hundreds of years, from simple beginnings as a finishing product in ancient China to modern utility in complex organic chemistry schemes today. The high reactivity and ease of crosslinking via the conjugated double bond system present in the α-eleostearic acid-rich seed oils continue to present new and exciting opportunities to use tung oil as a valuable industrial commodity. The future for cultivation of tung trees as a crop species in the United States is in doubt, but efforts are being made to revive it (Greg Frost, Gulf Coast Tung Oil, Tallahassee, Florida). Modern molecular techniques will continue to reveal many of the puzzle pieces necessary for reconstitution of the tung oil biosynthetic machinery in other organisms. Future efforts will be directed toward continued progress on both fronts. Between the two approaches, domestic production of high-quality drying oils in the United States can hopefully still become a reality.

REFERENCES

Akoh, C.C., Jennings, B.H., Lillard, D.A., 1996. Enzymatic modification of evening primrose oil: incorporation of n-3 polyunsaturated fatty acids. J. Am. Oil Chem. Soc. 73, 1050–1062.

Beisson, F., Li, Y., Bonaventure, G., Pollard, M., Ohlrogge, J.B., 2007. The acyltransferase GPAT5 is required for the synthesis of suberin in seed coat and root of *Arabidopsis*. Plant Cell 19, 351–368.

Biermann, U., Butte, W., Holtgrefe, R., Feder, W., Metzger, J.O., 2010. Esters of calendula oil and tung oil as reactive diluents for alkyd resins. Eur. J. Lipid Sci. Technol. 112, 103–109.

Blackmon, G.H., 1943. The tung-oil industry. Bot. Rev. 9, 1–40.

Brown, K., Keeler, K., 2005. The history of tung oil. Wildland Weeds 9, 4–6.

Burgal, J., Shockey, J., Lu, C., Dyer, J., Larson, T., Graham, I., Browse, J., 2008. Metabolic engineering of hydroxy fatty acid production in plants: RcDGAT2 drives dramatic increases in ricinoleate levels in seed oil. Plant Biotechnol. J. 6, 819–831.

Cahoon, E.B., Dietrich, C.R., Meyer, K., Damude, H.G., Dyer, J.M., Kinney, A.J., 2006. Conjugated fatty acids accumulate to high levels in phospholipids of metabolically engineered soybean and *Arabidopsis* seeds. Phytochemistry 67, 1166–1176.

Cahoon, E.B., Shockey, J.M., Dietrich, C.R., Gidda, S.K., Mullen, R.T., Dyer, J.M., 2007. Engineering oilseeds for sustainable production of industrial and nutritional feedstocks: solving bottlenecks in fatty acid flux. Curr. Opin. Plant Biol. 10, 236–244.

Cao, H., Shockey, J.M., Klasson, K.T., Chapital, D.C., Mason, C.B., Scheffler, B.E., 2013. Developmental regulation of diacylglycerol acyltransferase family gene expression in tung tree tissues. PLoS One 8, e76946.

Cao, H., Zhang, L., Tan, X., Long, H., Shockey, J.M., 2014. Identification, classification and differential expression of oleosin genes in tung tree (*Vernicia fordii*). PLoS One 9, e88409.

Cao, J., Li, J.-L., Li, D., Tobin, J.F., Gimeno, R.E., 2006. Molecular identification of microsomal acyl-CoA:glycerol-3-phosphate acyltransferase, a key enzyme in de novo triacylglycerol synthesis. Proc. Natl. Acad. Sci. U. S. A. 103, 19695–19700.

Chakraborty, T.R., Alicea, E., Chakraborty, S., 2012. Relationships between urinary biomarkers of phytoestrogens, phthalates, phenols, and pubertal stages in girls. Adolesc. Health Med. Therap. 3, 17–26.

Chen, Y., Zhou, G., Wang, Y., Xu, L., 2010a. F-BOX and oleosin: additional target genes for future metabolic engineering in tung trees? Ind. Crops. Prod. 32, 684–686.

Chen, Y.-H., Chen, J.-H., Chang, C.-Y., Chang, C.-C., 2010b. Biodiesel production from tung (*Vernicia montana*) oil and its blending properties in different fatty acid compositions. Bioresour. Technol. 101, 9521–9526.

Chen, Y., Yang-Dong, W., Cui, Q.Q., Zhan, Z.-Y., 2012. FAD2-DGAT2 genes coexpressed in endophytic *Aspergillus fumigatus* derived from tung oilseeds. Sci. World J. 2012, 4:390672.

Cheong, M.Y., Ooi, T.L., Ahmad, S., Yunus, W.M.Z.W., Kuang, D., 2009. Synthesis and characterization of palm-based resin for UV coating. J. Appl. Polym. Sci. 111, 2353–2361.

Chittavanich, P., Miller, K., Soucek, M.D., 2012. A photo-curing study of a pigmented UV-curable alkyd. Prog. Org. Coat. 73, 392–400.

Cui, Q.Q., Chen, Y.C., Han, X.J., Zhan, Z.Y., Lin, L.Y., Si, L.L., Wang, Y.D., 2013. Expression analysis of VfDGAT2 in various tissues of the tung tree and in transgenic yeast. Genet. Mol. Res. 12, 6554–6564.

Dahmen, J.L., Olsen, R., Fahy, D., Wallis, J.G., Browse, J., 2013. Cytochrome b5 coexpression increases *Tetrahymena thermophila* Δ6 fatty acid desaturase activity in *Saccharomyces cerevisiae*. Eukaryot. Cell 12, 923–931.

Das, K., Ray, D., Banerjee, C., Bandyopadhyay, N.R., Mohanty, A.K., Misra, M., 2011. Novel materials from unsaturated polyester resin/styrene/tung oil blends with high impact strengths and enhanced mechanical properties. J. Appl. Polym. Sci. 119, 2174–2182.

Durrett, T.P., McClosky, D.D., Tumaney, A.W., Elzinga, D.A., Ohlrogge, J., Pollard, M., 2010. A distinct DGAT with sn-3 acetyltransferase activity that synthesizes unusual, reduced-viscosity oils in *Euonymus* and transgenic seeds. Proc. Natl. Acad. Sci. U. S. A. 107, 9464–9469.

Dyer, J.M., Chapital, D.C., Lax, A.R., Pepperman, A.B., 1999. Identification of a subtilisin- like protease in seeds of developing tung fruits. J. Plant Physiol. 155, 802–805.

Dyer, J.M., Chapital, D.C., Kuan, J.-C., Mullen, R.T., Turner, C., McKeon, T.A., Pepperman, A.B., 2002. Molecular analysis of a bifunctional fatty acid conjugase/desaturase from tung. Implications for the evolution of plant fatty acid diversity. Plant Physiol. 130, 2027–2038.

Dyer, J.M., Chapital, D.C., Kuan, J.-C.W., Shepherd, H.S., Tang, F., Pepperman, A.B., 2004. Production of linolenic acid in yeast cells expressing an omega-3 desaturase from tung (*Aleurites fordii*). J. Am. Oil Chem. Soc. 81, 647–651.

Dyer, J.M., Fuqiang, T., Chapital, D.C., Lax, A.R., Shepherd, H.S., Shih, D.S., Pepperman, A.B., 1998. Differential extraction of eleostearic acid-rich lipid–protein complexes in tung seeds. J. Am. Oil Chem. Soc. 75, 1687–1690.

Færgeman, N.J., Wadum, M., Feddersen, S., Burton, M., Kragelund, B.B., Knudsen, J., 2007. Acyl-CoA binding proteins; structural and functional conservation over 2000 MYA. Mol. Cell. Biochem. 299, 55–65.

Fairchild, D., 1913. The Chinese Wood Oil Tree. U.S. Department of Agriculture. Bureau of Plant Industry. Circular no. 108. 7 p.

Farm Credit Administration, May 8, 1935. Some Results of a Study of the Tung-oil Industry in the United States. Division of Finance and Research, Washington, DC. 53 p.

Fry, V.K., 1973. Factors contributing to the demise of tung production in the United States. Econ. Bot. 27, 131–136.

Gidda, S.K., Shockey, J.M., Falcone, M., Kim, P.K., Rothstein, S.J., Andrews, D.W., Dyer, J.M., Mullen, R.T., 2011. Hydrophobic-domain-dependent protein-protein interactions mediate the localization of GPAT enzymes to ER subdomains. Traffic 12, 452–472.

Gidda, S.K., Shockey, J.M., Rothstein, S.J., Dyer, J.M., Mullen, R.T., 2009. *Arabidopsis thaliana* GPAT8 and GPAT9 are localized to the ER and possess distinct ER retrieval signals: functional divergence of the dilysine ER retrieval motif in plant cells. Plant Physiol. Biochem. 47, 867–879.

Govaerts, R., 1966. *Vernicia fordii* (Hemsl.) Airy Shaw. Kew Bull. 20, 394. World Checklist of Selected Plant Families. Facilitated by the Royal Botanic Gardens, Kew. Published on the Internet; http://apps.kew.org/wcsp/ (retrieved 18.08.15.).

Hatanaka, T., Shimizu, R., Hildebrand, D., 2004. Expression of a *Stokesia laevis* epoxygenase gene. Phytochemistry 65, 2189–2196.

Henderson, M., Hwang, Y.T., Dyer, J.M., Mullen, R.T., Andrews, D.W., 2007. The C-terminus of cytochrome b5 confers endoplasmic reticulum specificity by preventing spontaneous insertion into membranes. Biochem. J. 401, 701–709.

Huang, A.H., 1996. Oleosins and oil bodies in seeds and other organs. Plant Physiol. 110, 1055–1061.

Huang, K., Liu, Z., Zhang, J., Li, S., Li, M., Xia, J., Zhou, Y., 2014. Epoxy monomers derived from tung oil fatty acids and its regulable thermosets cured in two synergistic ways. Biomacromolecules 15, 837–843.

Hwang, Y.T., Pelitire, S.M., Henderson, M.P., Andrews, D.W., Dyer, J.M., Mullen, R.T., 2004. Novel targeting signals mediate the sorting of different isoforms of the tail- anchored membrane protein cytochrome b5 to either endoplasmic reticulum or mitochondria. Plant Cell 16, 3002–3019.

Ichihara, K., Takahashi, T., Fujii, S., 1988. Diacylglycerol acyltransferase in maturing safflower seeds: its influences on the fatty acid composition of triacylglycerol and on the rate of triacylglycerol synthesis. Biochim. Biophys. Acta 958, 125–129.

Kumar, R., Wallis, J.G., Skidmore, C., Browse, J., 2006. A mutation in *Arabidopsis* cytochrome b5 reductase identified by high-throughput screening differentially affects hydroxylation and desaturation. Plant J. 48, 920–932.

Lagasse, F.S., Sell, H.M., 1941. Tung investigations by the U.S. Department of Agriculture Laboratory. Annu. Rep. Fla. Agric. Exp. Stn. 91, 90–91.

Lardizabal, K.D., Mai, J.T., Wagner, N.W., Wyrick, A., Voelker, T., Hawkins, D.J., 2001. DGAT2 is a new diacylglycerol acyltransferase gene family: purification, cloning, and expression in insect cells of two polypeptides from *Mortierella ramanniana* with diacylglycerol acyltransferase activity. J. Biol. Chem. 276, 38862–38869.

Lee, M., Lenman, M., Banas, A., Bafor, M., Singh, S., Schweizer, M., Nilsson, R., Liljenberg, C., Dahlqvist, A., Gummeson, P.O., et al., 1998. Identification of non-heme diiron proteins that catalyze triple bond and epoxy group formation. Science 280, 915–918.

Li, R., Yu, K., Hatanaka, T., Hildebrand, D.F., 2010. *Vernonia* DGATs increase accumulation of epoxy fatty acids in oil. Plant Biotechnol. J. 8, 184–195.

Li, Y., Beisson, F., Koo, A.J.K., Molina, I., Pollard, M., Ohlrogge, J., 2007. Identification of acyltransferases required for cutin biosynthesis and production of cutin with suberin-like monomers. Proc. Natl. Acad. Sci. U. S. A. 104, 18339–18344.

Liu, X.Q., Huang, W., Jiang, Y.H., Zhu, J., Zhang, C.Z., 2012. Preparation of a bio-based epoxy with comparable properties to those of petroleum-based counterparts. Express Polym. Lett. 6, 293–298.

McCartney, A.W., Dyer, J.M., Dhanoa, P.K., Kim, P.K., Andrews, D.W., McNew, J.A., Mullen, R.T., 2004. Membrane-bound fatty acid desaturases are inserted co-translationally into the ER and contain different ER retrieval motifs at their carboxy termini. Plant J. 37, 156–173.

Meng, M., Su, Y.C., Saunders, R.M., Chye, M.-L., 2011. The rice acyl-CoA-binding protein gene family: phylogeny, expression and functional analysis. New Phytol. 189, 1170–1184.

O'Quin, J.B., Mullen, R.T., Dyer, J.M., 2009. Addition of an N-terminal epitope tag significantly increases the activity of plant fatty acid desaturases expressed in yeast cells. Appl. Microbiol. Biotechnol. 83, 117–125.

O'Quin, J.B., Bourassa, L., Zhang, D., Shockey, J.M., Gidda, S.K., Fosnot, S., Chapman, K.D., Mullen, R.T., Dyer, J.M., 2010. Temperature-sensitive, post-translational regulation of plant omega-3 fatty acid desaturases is mediated by the ER-associated degradation pathway. J. Biol. Chem. 285, 21781–21796.

Pastor, S., Sethumadhavan, K., Ullah, A.H., Gidda, S., Cao, H., Mason, C., Chapital, D., Scheffler, B., Mullen, R., Dyer, J., Shockey, J., 2013. Molecular properties of the class III subfamily of acyl-coenyzme A binding proteins from tung tree (*Vernicia fordii*). Plant Sci. 203–204, 79–88.

Potter, G.F., 1959. The domestic tung industry. I. Production and improvement of the tung tree. Econ. Bot. 13, 328–342.

Potter, G.F., 1968. The domestic tung industry today. J. Am. Oil Chem. Soc. 45, 281–284.

Potter, G.F., Crane, H.L., 1957. Tung Production. U.S. Department of Agriculture. Farmers' Bulletin 2031 35 pp.

Rani, S.H., Krishna, T.H., Saha, S., Negi, A.S., Rajasekharan, R., 2010. Defective in cuticular ridges (DCR) of *Arabidopsis thaliana*, a gene associated with surface cutin formation, encodes a soluble diacylglycerol acyltransferase. J. Biol. Chem. 285, 38337–38347.

Ren, J., Guo, R., Chen, L., Nevo, E., Zhang, Z., Sun, D., Peng, J., 2013. Single nucleotide polymorphisms of fad2 gene from tung tree, *Vernicia fordii*, a potential biodiesel plant. Euphytica 194, 93–107.

Routaboul, J.M., Benning, C., Bechtold, N., Caboche, M., Lepiniec, L., 1999. The TAG1 locus of *Arabidopsis* encodes for a diacylglycerol acyltransferase. Plant Physiol. Biochem. 37, 831–840.

Saha, S., Enugutti, B., Rajakumari, S., Rajasekharan, R., 2006. Cytosolic triacylglycerol biosynthetic pathway in oilseeds. Molecular cloning and expression of peanut cytosolic diacylglycerol acyltransferase. Plant Physiol. 141, 1533–1543.

Shockey, J.M., Dhanoa, P.K., Dupuy, T., Chapital, D.C., Mullen, R.T., Dyer, J.M., 2005. Cloning, functional analysis, and subcellular localization of two isoforms of NADH:cytochrome b5 reductase from developing seeds of tung (*Vernicia fordii*). Plant Sci. 169, 375–385.

Shockey, J.M., Gidda, S.K., Chapital, D.C., Kuan, J.C., Dhanoa, P.K., Bland, J.M., Rothstein, S.J., Mullen, R.T., Dyer, J.M., 2006. Tung tree DGAT1 and DGAT2 have nonredundant functions in triacylglycerol biosynthesis and are localized to different subdomains of the endoplasmic reticulum. Plant Cell 18, 2294–2313.

Shockey, J., Mason, C., Gilbert, M., Cao, H., Li, X., Cahoon, E., Dyer, J., 2015. Development and analysis of a highly flexible multi-gene expression system for metabolic engineering in Arabidopsis seeds and other plant tissues. Plant Mol. Biol. 89, 113–126.

Spiers, J.M., Kilby, W.W., 1973. Evaluation of Late Blooming Tung Selections, vol. 1219. Mississippi Agricultural and Forestry Experiment Station. 2 p.

Staehelin, L.A., 1997. The plant ER: a dynamic organelle composed of a large number of discrete functional domains. Plant J. 11, 1151–1165.

Thanamongkollit, N., Soucek, M.D., 2012. Synthesis of UV-curable tung oil and UV curable tung oil based alkyds. Prog. Org. Coat. 73, 425–434.

United States Tariff Commission, October 1960. Tung Oil and Tung Nuts Report to the President of Investigation No. 22–23, Under section 22 of the Agricultural Adjustment Act, as amended. Washington, DC. 31 p.

van de Loo, F.J., Broun, P., Turner, S., Somerville, C., 1995. An oleate 12-hydroxylase from *Ricinus communis* L. is a fatty acyl desaturase homolog. Proc. Natl. Acad. Sci. U. S. A. 92, 6743–6747.

van Erp, H., Shockey, J., Zhang, M., Adhikari, N.D., Browse, J., 2015. Reducing isozyme competition increases target fatty acid accumulation in seed triacylglycerols of transgenic *Arabidopsis*. Plant Physiol. 168, 36–46.

Vogel, G., Browse, J., 1996. Cholinephosphotransferase and diacylglycerol acyltransferase (substrate specificities at a key branch point in seed lipid metabolism). Plant Physiol. 110, 923–931.

Wayne, L.L., Browse, J., 2013. Homologous electron transport components fail to increase fatty acid hydroxylation in transgenic *Arabidopsis thaliana*. F1000 Res. 2, 203.

Xiao, S., Gao, W., Chen, Q.F., Chan, S.W., Zheng, S.X., Ma, J., Wang, M., Welti, R., Chye, M.-L., 2010. Overexpression of *Arabidopsis* acyl-CoA binding protein ACBP3 promotes starvation-induced and age-dependent leaf senescence. Plant Cell 22, 1463–1482.

Yang, W., Pollard, M., Li-Beisson, Y., Beisson, F., Feig, M., Ohlrogge, J., 2010. A distinct type of glycerol-3-phosphate acyltransferase with sn-2 preference and phosphatase activity producing 2-monoacylglycerol. Proc. Natl. Acad. Sci. U. S. A. 107, 12040–12045.

Yurchenko, O.P., Nykiforuk, C.L., Moloney, M.M., Ståhl, U., Banas, A., Stymne, S., Weselake, R.J., 2009. A 10-kDa acyl-CoA-binding protein (ACBP) from *Brassica napus* enhances acyl exchange between acyl-CoA and phosphatidylcholine. Plant Biotechnol. J. 7, 602–610.

Zhan, Z., Yangdong, W., Shockey, J., Chen, Y., Zhou, Z., Yao, X., Ren, H., 2012. Breeding status of tung tree (*Vernicia* sp.) in China, a multipurpose oilseed crop with industrial uses. Silvae Genet. 61, 265–271.

Zheng, Z., Xia, Q., Daux, M., Shen, W., Selvaraj, G., Zou, J., 2003. *Arabidopsis AtGPAT1*, a member of the membrane-bound glycerol-3-phosphate acyltransferase gene family, is essential to tapetum differentiation and male fertility. Plant Cell 15, 1872–1887.

Chapter 11

Emerging Industrial Oil Crops

Thomas A. McKeon
United States Department of Agriculture, Agricultural Research Service, Western Regional Research Center, Albany, CA, United States

INTRODUCTION

It is now the age of the biobased economy, with great efforts expended in both basic and applied research to develop useful products from agriculture that replace those derived from petroleum. Although the name, methods, and approaches have changed, this movement is far from new. Agricultural materials, including seed oils such as castor (see chapter: Castor (*Ricinus communis* L.)) and jojoba (see chapter: Jojoba (*Simmondsia chinensis*)), have been used for nonfood purposes since time immemorial. The motto of the US Department of Agriculture (USDA), dated from 1862, summarizes the general concept of biobased products: "Agriculture is the foundation of manufacture and commerce." For centuries, botanists and explorers have collected and classified plants from around the world. The 19th and 20th centuries witnessed organized seed collection programs to identify new crops with economic potential and to expand germplasm collections for existing crops. The first US government–sponsored plant collection in 1856 predated the establishment of the USDA and led to the founding of the Foreign Seed and Plant Division, a precursor of the National Genetic Resources Program (USDA-NGRP) (Stoner and Hummer, 2007). In the early 20th century, Nikolai Vavilov initiated a similar program for Imperial Russia, collecting plants and seeds from five continents and establishing a repository in St. Petersburg, continuing to expand the collection until Stalin incarcerated him as a scapegoat for crop failures and famine in 1940 (Stoner and Hummer, 2007).

Through the early part of the 20th century, agriculture was still a major feedstock for manufacturing. It has even been pointed out that oats were once considered to be biofuel (du Preez, 2007), fueling the horses and mules that carried people and goods, even after the advent of automobiles and trucks. In the 1920s, there was a conscious effort to expand the use of agricultural materials in nonfood applications. The research and development efforts of George Washington Carver (Wright, 1946), Thomas Edison, and Henry Ford (Finlay, 2004) formed a basis for the birth of the "chemurgy" movement, with the name derived from chemistry

Industrial Oil Crops. http://dx.doi.org/10.1016/B978-1-893997-98-1.00011-7

patterned after the term "metallurgy." The iconic representation of this movement was the photograph of Henry Ford using an axe to strike an automobile trunk made from soy and hemp to demonstrate its durability. It was during this movement that chemists and botanists noted the value of producing industrially useful fatty acids found in seed oils in hopes of replacing imported products. This era saw the introduction into the United States of tung trees and meadowfoam plants as well as interest in jojoba (see chapters: Tung (*Vernicia fordii* and *Vernicia montana*), Jojoba (*Simmondsia chinensis*), and Meadowfoam (*Limnanthes alba*)).

The underpinnings of the chemurgy movement included advances in chemical and mechanical science driven by economic nationalism based on a desire to be free from dependence on imported feedstocks (Finlay, 2004). In the United States, the chemurgical movement resulted in establishment of the USDA Regional Utilization Laboratories, intended to develop new industrial products from agricultural surpluses and byproducts (Wallace, 1939). With the onset and continuation of World War II, use and modification of agricultural materials took on great urgency for the Allies and Axis powers as both were forced to replace imported strategic materials with domestically developed sources (Finlay, 2004; Danquah, 2005). After the war, advances in chemistry and petroleum chemistry supplanted many applications previously derived from agriculture, such as lubricants, paints, and plastics (Jones and Wolff, 1960).

Although the chemurgy concept faded, it never quite disappeared. In the 1950s, the idea of reducing imports by replacing the imported feedstocks with domestically produced agricultural materials returned, if not under the name chemurgy. It was at this time that "new crops" research started at the USDA utilization laboratory in Peoria, IL, with the goal of introducing oilseed crops that could be grown in place of crops then in surplus or on set-aside land in order to provide new feedstocks for industry (Jones and Wolff, 1960; Wolff, 1966). The early stage of this program identified a number of oilseeds with interesting, unusual fatty acid composition, some of which are still of interest (Cruciferae), while others have fallen by the wayside (*Rudbeckia bicolor*, *Chrysanthemum coronaria*). As this new crops program evolved, similar programs arose worldwide. The European Union framework programmes from 1984 were geared to enhancing industrial seed oil production for sustainability (Zanetti et al., 2013; Deighton and O'Donnell, 2014). In India, research programs were initiated for cultivation of underused industrial crops for sustainability and cultivation of wastelands (Paroda and Mal, 1993). Success with breeding of canola (low-erucic acid rape) in Canada led to increased interest in other *Brassica* for increased oil and erucic acid content (Downey, 1971).

As noted in Chapter Introduction to Industrial Oil Crops, there are hundreds of plants that produce seed oils containing fatty acids with unusual functionalities. While there are a number of domesticated crops that are used primarily for industrial applications, many of the crops identified were not well-suited to cultivation and required breeding and selection in order to be cultivated as crops (White et al., 1971). Among those crops identified as having value as industrial oil crops, some have been brought into production, although none have approached the success of soybean introduction into the Americas. This chapter includes brief descriptions in the form of "mini-chapters" on a number of the crops that have been identified as industrially useful. Other industrial oil crops are described more extensively in the major crop chapters.

REFERENCES

Danquah, F.K., 2005. Reports on Philippine industrial crops in World War II from Japan's English language press. Agric. Hist. 79, 74–96.

Deighton, B., O'Donnell, P., 2014. Europe's Framework Programmes – A Key Element of Research Policy in Europe. Horizon. 12/16/2014 http://horizon-magazine.eu/article/europe-s-framework-programmes-key-element-research-policy-europe_en.html (9.23.2015).

Downey, R.K., 1971. Agricultural and genetic potentials of cruciferous oilseed crops. J. Am. Oil Chem. Soc. 48, 718–722.

Finlay, M.R., 2004. Old efforts at new uses: a brief history of chemurgy and the American search for biobased materials. J. Ind. Ecol. 7, 33–46.

Jones, Q., Wolff, I.A., 1960. The search for new industrial crops. Econ. Bot. 14, 56–68.

Paroda, R.S., Mal, B., 1993. Developing a national program for research on underutilized crops in India. Chapter 59. In: Buxton, D.R. et al. (Eds.), International Crop Science I. Crop Science Society of America. Madison, WI, USA, pp. 459–464.

du Preez, M., 2007. Oats for the Clydesdale: Biofuel on Farm. ABC Rural. http://www.abc.net.au/site-archive/rural/content/2007/s1934917.htm (23.09.15.).

Stoner, A., Hummer, K., 2007. 19th and 20th century plant hunters. Hort Sci. 42, 197–199.

Wallace, H.A., 1939. A report of a survey made by the department of agriculture relative to four regional research laboratories, one in each major farm production area. In: Government Printing Office Washington, DC, USA 76th Congress, Document 65 429 pp.

White, G.A., et al., 1971. Agronomic evaluation of prospective new crop species. Econ. Bot. 25, 22–43.

Wolff, I.A., 1966. New crops – visionary dream or practical reality. Econ. Bot. 20, 2–5.

Wright, C.W., 1946. George Washington Carver: an American scientist. J. Chem. Educ. 23, 268–270.

Zanetti, F., Monti, A., Berti, M.T., 2013. Challenges and opportunities for new industrial oilseed crops in EU-27: a review. Ind. Crops Prod. 50, 580–595.

Chapter 11.1

Chia (*Salvia hispanica* L.)

William Serson, Maythem AL–Amery, Shreya Patel, Tim Phillips, David F. Hildebrand
Department of Plant and Soil Sciences, University of Kentucky, Lexington, KY, United States

INTRODUCTION

Chia has experienced an agricultural resurgence in recent years. Although introduced to the American public in the 1980s as "Chia Pets®" with sprouts forming green "hair," this crop served as a vital food staple and experienced cultural and economic significance in pre-Columbian Mesoamerica among tribal peoples such as the Aztecs (Cahill, 2003). Chia is also known as Spanish sage; the roasted and ground seeds, known as chiapinolli, were eaten as a gruel, and the oil was used as a body emollient in those ancient cultures (Reyes-Caudillo et al., 2008). During the Spanish conquest, cultural differences caused a decline in the cultivation of chia as it was replaced with European alternatives (Acuña, 1986). Chia production has increased significantly in the past two decades due to its potential as a renewable, low impact source of omega-3 fatty acids (ω-3 FA), having one of the highest levels of ω-3 FA s of any known crop plant (Cahill, 2004) (Table 11.1.1).

Chia's most valued biomass product is the nutlet, or dried fruit, it produces. The seeds are egg shaped and flow easily in bulk handling making chia seeds similar to canola among major oilseeds, with mean seed mass ranging from one to 1.3 mg/seed (Ixtaina et al., 2008). So-called chia "seeds" are actually dried fruits, or nutlets, but are handled and treated like seeds and will be referred to as seeds for the remainder of this chapter (Bueno et al., 2010). The seed (fruit) pericarp has three distinct cell layers between the transparent cuticle and testa. There is a single layer of epicarp cells, the mesocarp composed of several amorphous cell layers and the sclerenchyma, which is also composed of three cell types including the inner layer, the endocarp (Hedge, 1970). Chia seed coat colors are white or blue-grey with black spots. Brown or tan seed are also common and are apparently due to incompletely matured seeds in which full seed coat pigmentation is not developed. Seed coat color (except for brown seeds) is genetically determined with white coat color being recessive although traditional production in Mexico and Central America included both seed coat genotypes in the same field.

Chia (*Salvia hispanica* L.) is a member of the Laminaceae family. The center of genetic diversity of chia is in the highlands of western Mexico (Cahill, 2004), with a diploid chromosome number of 12 (Estilai et al., 1990). The genus *Salvia*

TABLE 11.1.1 Fatty Acid Composition of Chia Lines vs Flax

	Pinta	SE	G8	SE	Salba	SE	Flax 1	SE	Flax 2	SE
16:0	7.7	0.4	10.2	3.30	10.2	0.05	5.9	0.01	5.65	0.11
18:0	4.0	0.1	5.9	1.86	3.8	0.05	4.1	0.02	5.64	0.01
18:1Δ9	7.7	0.2	8.1	0.01	5.9	0.05	19.2	0.05	33.4	0.03
18:1Δ11	0.8	0.5	0.7	0.03	1.0	0.07	0.7	0.04	0.3	0.3
18:2	19.4	0.9	19.0	0.10	12.1	0.29	15.0	0	17	0.33
18:3	60.3	0.9	56.2	0.08	67.0	0.15	55.2	0.01	38	0.11

"Pinta" is a standard chia line produced in Mexico. "G8" is a long-day flowering chia line grown in Lexington, KY. "Salba" is a trademarked *Salvia hispanica* line from a GNC store in Lexington, KY. Flax 1 is "Arrowhead" flax seed from a Kroger grocery in Lexington, KY. Flax 2 is "Rehab 94" flax grown in Lexington, KY.

contains about 900 known species with some species being cultivated and used globally in folk medicine (Lu and Foo, 2002). Stems are about 1–2 m and obtusely quadrangular. Leaves are opposite, ovate, tapered and sharply serrated. The flowers are produced in terminal and axillary four-cornered spikes protected by small bracts with long sharp points. The blue or white corolla is tubular with four stamens, two of them larger and sterile. Seeds are present in groups of four.

Until recently, chia production was limited to subtropical regions such as Florida, California and Arizona in the United States, due to its photoperiod requirements (day length of 12.5 h or less for at least two months before frost), and only with the help of irrigation. Outside of the United States, chia can be grown in most of Mexico and parts of Central and South America, including Nicaragua, Chile, Argentina, and Ecuador (Cahill, 2004). Recently, long day flowering mutant chia lines have been developed. These mutants flower under long-day lengths, much earlier in a temperate growing season than chia populations from tropical latitudes, meaning these lines can reach maturity before frost damage can occur on maturing plants and seeds (Jamboonsri et al., 2012). While management practices continue to be optimized for this crop in Kentucky, great potential exists for chia as a major ω-3 fatty acid (ω-3 FA) source grown in the heartland of the United States.

SEED COMPOSITION

Many members of this family exhibit high levels of ω-3 FA s. Chia has oil composition levels similar to its Laminaceae relative *Perilla frutescens* (Ciftci et al., 2012) and *Dracocephalum moldavica,* averaging 68% ω-3 FA. However, only chia and perilla have high seed and oil yield potential, making them two of the most economically feasible members of the family to develop for large-scale crop production (Rao et al., 2008). Chia seeds contain about 20% protein, and oil content ranges from 28.5 to 32.7% (Ayerza and Coates, 2004, 2007). Chia is high in the ω-3 FA α-linolenic acid, 18:3. There are claims that white chia seeds are higher in oil and ω-3 FA levels than dark seeds, though genes regulating seed pigmentation are independent from those regulating oil composition (Ayerza, 2010).

Chia oil has high oxidation potential due to its very polyunsaturated FA nature, but the crude oil obtained by cold pressing initially has a low peroxide index value (2.6 mEq peroxide/kg) as well as high antioxidant activity due to the presence of phenolic compounds, mainly, myricetin, quercetin, kaempferol, chlorogenic acid, 3,4-dihydroxyphenylethanol-elenolic acid dialdehyde (3,4-DHPEA-EDA), tocopherols, (Sargi et al., 2013; Capitani et al., 2012; Reyes-Caudillo et al., 2008; Marineli et al., 2014) as well as other antioxidants such as carotenoids (Ixtaina et al., 2011; Alvarez-Chavez et al., 2008). The carotenoid content in chia varies from 0.5 mg/kg to 1.2 mg/kg (Ixtaina et al., 2011). Chia seed oil contains about 238–427 mg/kg tocopherols, mainly γ-tocopherol, which is similar to levels found in peanut, 398.6 mg/kg (Ixtaina et al., 2011), though less than crude soybean oil which typically ranges from 1205 to 2195 mg/kg on a dry basis (Medic et al., 2014) and flax oil, approximately 800 mg/kg oil (Bozan and Temelli, 2008). Tocopherols consist of four forms, α, β, γ and δ tocopherol (Kamal-Eldin and Appelqvist, 1996). α-Tocopherol has been studied extensively for its antioxidant activity mainly due to its high uptake (Brigelius-Flohe and Traber, 1999; Zingg and Azzi, 2004; Brigelius-Flohé, 2006). γ-Tocopherol has been reported to play an important role in reducing the risk of cardiovascular disease and cancer (Stones et al., 2012; Wagner et al., 2004; Nijveldt et al., 2001). The high phenolic content including tocopherol levels in chia oil contribute to the health value of chia.

The total dietary fiber from ground chia seeds is approximately 40%, and of that, 17% is reported to be soluble dietary fiber (SDF) while the remaining 83% is insoluble dietary fiber (IDF) (Reyes-Caudillo et al., 2008). The most abundant component of insoluble fiber is "Klason lignin" at 40% which is suggested to protect the unsaturated fatty acids from oxidation, and may contribute to a strong, impervious structure. The lignin may be a major contributor to the hypocholesterolemic effect of chia dietary fiber due to its ability to absorb bile acids (Reyes-Caudillo et al., 2008) Defatted chia flour has ~56.5 g/100 g total dietary fiber content composed of ~3% SDF and 53.5% IDF (Vázquez-Ovando et al., 2009). The fiber-rich fraction water-holding capacity is 15.4 g/g.

Copious production of mucilage is one of the properties that contributes to the value of the seeds. Chia mucilage can be considered to consist of non-cellulose slime and cellulose mucilage. When seeds are moistened, water is rapidly absorbed, causing the cuticle to rupture and the exocarp cell contents to be expelled as mucilage in twisted or coiled helical threads, often, if not universally, in tubes or sheaths. It is thought that the mucilage together with its very high water content provides an oxygen barrier, preventing germination and causes the seeds to adhere to the soil substrate tightly.

The seed coat is high in a fiber which becomes mucilaginous and expands considerably (>10-fold) when soaked in water. Mucilage yield is on the magnitude of 38 ± 1.0 g/kg of seed. Composition analysis of freeze-dried chia mucilage reveals 11.5% moisture, 11.3% protein, 3.1% fat, 8.4% ash, 13.5% crude fiber, and 52.3% carbohydrates (Capitani et al., 2013). The carbohydrates have been extensively analyzed, and found to be a polysaccharide mainly composed of xylose, glucose, and methyl-glucuronic acid residues. The molecular weight of the main polysaccharide ranges from $1–2 \times 10^6$ Da from linear repeats of a tetrasaccharide with 4-O-methyl-α-D-glucoronopyranosyl residues occurring as branches at O-2 of some β-D-xylopyranosyl residues in the main chain consisting of (1-->4)-β-D-xylopyranosyl-(1-->4)-α-D-glucopyranosyl(1-->4)-β-D-xylopyranosyl units (Lin et al., 1994).

HEALTH AND NUTRITION

Clinical trials, epidemiological investigations, and experimental studies suggest consumption of α-linolenic acid (ALA-18:3n3, ω-3) has a positive impact on cardiovascular disease (CVD) and may also have a positive impact on diabetes. Flaxseed with its high ALA content has been found, in a limited number of studies, to decrease the risk of CVD and diabetes (Prasad, 2009; Taylor et al., 2010). Most *Salvia hispanica* genotypes have a greater ALA content than flaxseed (~10–15% more ALA). One line of *S. hispanica*, Salba®, has shown some risk reduction benefits regarding cardiovascular health and blood glucose regulation (Vuksan et al., 2007, 2009) which may be due to its higher ALA, antioxidant and/or fiber content or composition. Another study demonstrates that use of chia oil compared to maize oil in rats which were fed a high sucrose diet and developed insulin resistance, is able to reduce the activity of liver enzymes and molecules responsible for adipose tissue accumulation, including liver triglycerides (TAG), fatty acid synthase (FAS), acetyl CoA carboxylase (ACC), and glucose six-phosphate dehydrogenase (G-6-PDH) activities, as well as increase fatty acid oxidation and carnitine palmitoyl transferase (CPT-1) activities, which oxidize existing fats (Rossi et al., 2013). A follow-up study found that the replacement of corn oil by chia seed in a sucrose-rich diet reduced adipocyte hypertrophy, cell volume and size distribution, improved lipogenic enzyme activities, lipolysis and the anti-lipolytic action of insulin. In the skeletal muscle lipid storage, glucose phosphorylation and oxidation were normalized. Chia seed reversed the impaired insulin stimulated glycogen synthase activity, glycogen, glucose-6-phosphate and glucose transporter type 4 (GLUT-4) protein levels as well as insulin resistance and dyslipidemia (abnormal amounts of lipids in blood) (Oliva et al., 2013).

Chia diets dramatically decreased TAG levels and increased high density lipoprotein cholesterol and ω-3 FA content in rat serum (Ayerza and Coates, 2005, 2007). Chia seed decreased TAG and cholesterol more than chia oil but both were significantly superior to the control diets. Dietary chia seed also improves adiposity and insulin resistance in dyslipidemic rats (Chicco et al., 2009). Diets supplemented with chia have been found to decrease risks from some types of cardiovascular disease (CVD), cancer, and diabetes. It has been reported that chia diet decreased the tumor weight and metastasis number and inhibited growth and metastasis in a murine mammary gland adenocarcinoma (Espada et al., 2007). Long-term supplementation with chia attenuated a major cardiovascular risk factor and emerging factors safely beyond conventional therapy, while maintaining good glycemic and lipid control in people with well-controlled type 2 diabetes (Vuksan et al., 2007, 2009). A report in Inpharma Weekly indicates benefits from *Salvia hispanica* seed in treating type 2 diabetes mellitus (Anon, 2008). However, this study conflicts with another study that reported no changes in a number of health parameters including C-reactive protein, interleukin 6, monocyte chemoattractant protein 1, tumor necrosis factor alpha, oxidative stress markers, and blood pressure with consumption of 50 g/day of chia seed compared to 50 g/day of soy for 12 weeks by overweight middle-aged men and women (Nieman et al., 2009). Plasma ALA increased from 2.8% to 24.4% but EPA and DHA levels did not show a significant change. This could be due to differences in regular chia vs. Salba®, the fact that the whole seeds that were consumed in a drink without mastication were poorly digested; a masking effect to the health benefits of soy itself in the placebo or that chia has few health benefits compared to more regular American diets. Chia is reported to exhibit some reduction on CVD risk factors (Ulbricht et al., 2009).

Chia may also be useful for treating other ailments. Chia seed oil was found to be beneficial as an adjuvant moisturizing agent for pruritic skin, including patients with end-stage renal disease (Jeong et al., 2010). Omega-3 FA are reported to have benefit in psychiatric disorders, specifically that ω-3 FAs have significant benefit in prevention and/or treatment of unipolar and bipolar depression (Freeman et al., 2006).

Though many studies find chia seeds to ease symptoms of diseases including diabetes, CVD, and cancer, evidence for in vivo safety and efficacy is limited (Ali et al., 2012). A study on the effect of chia seeds on disease risk factors in 62 overweight post-menopausal females (Nieman et al., 2012) found no significant difference in risk factors like body composition, inflammation, and blood pressure between the chia groups and placebo (poppy seeds) groups. However, the group that received milled chia showed a 58% increase in ALA and 39% increase in EPA in plasma than the group that received whole chia seed supplementation. Studies (Nieman et al., 2009) found no benefit of chia on body composition in overweight individuals when supplemented with 25 g chia seed in 250 mL of water twice a day. A randomized control trial using a beverage containing soy protein, nopal, chia seed, and oat was also carried out (Guevara-Cruz et al., 2012). In comparison to a control diet group, the beverage group experienced body weight loss and reduction of triglyceride and blood glucose levels. In another study (Vuksan et al., 2010) where chia was an ingredient in white bread, participants showed reduced postprandial glycemia at 120 min after consumption. In another report (Da Silva Marineli et al., 2015), chia seed and extracted chia oil can reduce oxidative stress in vivo, by improved antioxidant status and reduced lipid peroxidation in diet-induced obese rats.

Higher in ALA, fiber, and minerals than flaxseeds, chia is considered a "superfood." Recently, recognized as a novel food by the European Union in 2009 (http://eur-lex.europa.eu/LexUriServ/LexUriServ.do?uri=OJ:L:2009:294:0014:0015:EN:PDF), chia is being explored and used in food products, thereby adding nutritional value to bread products, cakes, smoothies, yogurt, etc. Chia gel has been found to be able to replace one-fourth of the oil content in cake with no significant difference in color, taste, or texture (Borneo et al., 2010). In addition to improving health and reducing diseases, chia is being studied for applications in sports nutrition. Though it still remains unclear if chia can provide a performance-enhancing food source, a crossover study (Illian et al., 2011) found there was no significant difference in level of performance in endurance events lasting >90 min in individuals who were subjected to CHO carbohydrate loading compared to participants who were loaded with a chia drink (50% chia seeds and 50% Gatorade). Thus, chia drinks can be a good substitute for high sugar drinks, thereby reducing dietary intake of sugar and providing athletes with ω-3 FA, which also have beneficial effect on psychiatric health. A feedback questionnaire submitted by a participant indicated that they performed their activity at a reduced heart rate with the chia drink compared to only Gatorade. Another reported that it was easier to consume the chia drink than straight Gatorade especially when a usual meal plan is followed. A study to determine the effect of chia seed oil supplementation on prolonged intense running (Nieman et al., 2015) concluded similar run and exhaustion times in both chia seed oil supplement group and flavored water group.

Feed Supplementation

In addition to direct human consumption, chia has utility as a feed ingredient. Feeding chia to egg-laying hens results in a significant accumulation of ω-3 FA in the eggs but without the fishy or off taste associated with other supplements (Ayerza and Coates, 1999). Chia also has the potential to be a much less expensive source of ω-3 FAs and is much more environmentally sound than harvesting krill for oil, one of the common sources of ω-3 FA feed today. Omega-3 FA–enriched eggs are now widely available in a wide variety of supermarkets and have been marketed successfully, and many farmers are switching to chia as a source of ω-3 FAs as a dietary supplement in egg-laying hens (Simopoulos and De Meester, 2009). Another study finds that feeding chia seed to chickens and quail to supplement their diet increases levels of ω-3 FAs in breast meat, thigh meat, and eggs (Komprda et al., 2013). Fish and krill do not biosynthesize ω-3 FAs but only accumulate them from dietary sources, making chia a potentially useful ω-3 FA source in fish diets (Silva et al., 2014).

SUMMARY

Since its introduction to American culture as the Chia Pet® over 30 years ago, *Salvia hispanica* has captured our culture's imagination. However, people are starting to take a more serious look at this small seed, which is starting to have a big impact on global and American agriculture. A member of the mint family, its production of large quantities of ω-3 FAs is causing agronomists and farmers to start taking it seriously as a viable crop. It is among the most sustainable source of ω-3 FAs available on the market. Combined with high fiber levels and its promise for improving cardiovascular and diabetic health, chia is sure to be a superfood people will talk about and consume for years to come.

ACKNOWLEDGMENTS

Our chia research is supported by the KY Small Grains Growers Association and the KY Agricultural Experiment Station.

REFERENCES

Acuña, R., 1986. Relaciones geográficas del siglo XVI. Serie antropológica. Universidad Nacional Autónoma de México. Instituto de Investigaciones Antropológicas, México City.

Ali, N.M., Yeap, S.K., Ho, W.Y., Beh, B.K., Tan, S.W., Tan, S.G., 2012. The promising future of chia, *Salvia hispanica* L. J. Biomed. Biotech. 2012.

Alvarez-Chavez, L.M., Valdivia-Lopez, M.D., Aburto-Juarez, M.D., Tecante, A., 2008. Chemical characterization of the lipid fraction of Mexican chia seed (*Salvia hispanica* L.). Int. J. Food Prop. 11, 687–697.

Anon, 2008. Food for thought for treating type 2 diabetes mellitus. Inpharma Weekly 18.

Ayerza, R., 2010. Effects of seed color and growing locations on fatty acid content and composition of two chia (*Salvia hispanica* L.) genotypes. J. Am. Oil Chem. Soc 87, 1161–1165.

Ayerza, R., Coates, W., 1999. An omega-3 fatty acid enriched chia diet: Influence on egg fatty acid composition, cholesterol and oil content. Can. J. Anim. Sci. 79, 53–58.

Ayerza, R., Coates, W., 2004. Composition of chia (*Salvia hispanica*) grown in six tropical and subtropical ecosystems of South America. Trop. Sci. 44, 131–135.

Ayerza, R., Coates, W., 2005. Ground chia seed and chia oil effects on plasma lipids and fatty acids in the rat. Nutr. Res. 25, 995–1003.

Ayerza, R., Coates, W., 2007. Effect of dietary alpha-linolenic fatty acid derived from chia when fed as ground seed, whole seed and oil on lipid content and fatty acid composition of rat plasma. Ann. Nutr. Metabol. 51, 27–34.

Borneo, R., Aguirre, A., León, A.E., 2010. Chia (*Salvia hispanica* L.) gel can be used as egg or oil replacer in cake formulations. J. Am. Diet. Assoc. 110, 946–949.

Bozan, B., Temelli, F., 2008. Chemical composition and oxidative stability of flax, safflower and poppy seed and seed oils. Bioresour. Technol. 99, 6354–6359.

Brigelius-Flohé, R., 2006. Bioactivity of vitamin E. Nutr. Res. Rev. 19, 174–186.

Brigelius-Flohe, R., Traber, M.G., 1999. Vitamin E: function and metabolism. FASEB J. 13, 1145–1155.

Bueno, M., Di Sapio, O., Barolo, M., Busilacchi, H., Quiroga, M., Severin, C., 2010. Quality tests of *Salvia hispanica* L. (Lamiaceae) fruits marketed in the city of Rosario (Santa Fe province, Argentina). Bol. Latinoam. Caribe Plant. Med. 9, 221–227.

Cahill, J.P., 2003. Ethnobotany of chia, *Salvia hispanica* L. (Lamiaceae). Econ. Bot. 57, 604–618.

Cahill, J.P., 2004. Genetic diversity among varieties of Chia (*Salvia hispanica* L.). Gen. Resour. Crop Evol. 51, 773–781.

Capitani, M., Spotorno, V., Nolasco, S., Tomás, M., 2012. Physicochemical and functional characterization of by-products from chia (*Salvia hispanica* L.) seeds of Argentina. LWT-Food Sci. Technol. 45, 94–102.

Capitani, M.I., Ixtaina, V.Y., Nolasco, S.M., Tomas, M.C., 2013. Microstructure, chemical composition and mucilage exudation of chia (*Salvia hispanica* L.) nutlets from Argentina. J. Sci. Food Agri. 93, 3856–3862.

Chicco, A.G., D'alessandro, M.E., Hein, G.J., Oliva, M.E., Lombardo, Y.B., 2009. Dietary chia seed (*Salvia hispanica* L.) rich in alpha-linolenic acid improves adiposity and normalises hypertriacylglycerolaemia and insulin resistance in dyslipaemic rats. Brit. J. Nutr. 101, 41–50.

Ciftci, O.N., Przybylski, R., Rudzińska, M., 2012. Lipid components of flax, perilla, and chia seeds. Eur. J. Lipid Sci. Technol. 114, 794–800.

Da Silva Marineli, R., Lenquiste, S.A., Moraes, É.A., Maróstica Jr., M.R., 2015. Antioxidant potential of dietary chia seed and oil (*Salvia hispanica* L.) in diet-induced obese rats. Food Res. Int. 76, 666–674.

Espada, C.E., Berra, M.A., Martinez, M.J., Eynard, A.R., Pasqualini, M.E., 2007. Effect of Chia oil (*Salvia hispanica*) rich in omega-3 fatty acids on the eicosanoid release, apoptosis and T-lymphocyte tumor infiltration in a murine mammary gland adenocarcinoma. Prostaglandins Leukot. Essent. Fatty Acids 77, 21–28.

Estilai, A., Hashemi, A., Truman, K., 1990. Chromosome number and meiotic behavior of cultivated chia, *Salvia hispanica* (Lamiaceae). Hort Sci. 25, 1646–1647.

Freeman, M.P., Hibbeln, J.R., Wisner, K.L., Davis, J.M., Mischoulon, D., Peet, M., Keck, P.E., Marangell, L.B., Richardson, A.J., Lake, J., Stoll, A.L., 2006. Omega-3 fatty acids: evidence basis for treatment and future research in psychiatry. J. Clin. Psychiat. 67, 1954–1967.

Guevara-Cruz, M., Tovar, A.R., Aguilar-Salinas, C.A., Medina-Vera, I., Gil-Zenteno, L., Hernández-Viveros, I., López-Romero, P., Ordaz-Nava, G., Canizales-Quinteros, S., Guillen Pineda, L.E., Torres, N., 2012. A dietary pattern including nopal, chia seed, soy protein, and oat reduces serum triglycerides and glucose intolerance in patients with metabolic syndrome. J. Nutr. 142, 64–69.

Hedge, I., 1970. Observation on the mucilage of Salvia fruits. Roy. Bot. Gard. Edinburgh 30, 79–95.

Illian, T.G., Casey, J.C., Bishop, P.A., 2011. Omega 3 chia seed loading as a means of carbohydrate loading. J. Strength Cond. Res. 25, 61–65.

Ixtaina, V.Y., Martinez, M.L., Spotorno, V., Mateo, C.M., Maestri, D.M., Diehl, B.W.K., Nolasco, S.M., Tomas, M.C., 2011. Characterization of chia seed oils obtained by pressing and solvent extraction. J. Food Comp. Anal. 24, 166–174.

Ixtaina, V.Y., Nolasco, S.M., Tomas, M.C., 2008. Physical properties of chia (*Salvia hispanica* L.) seeds. Ind. Crops Prod. 28, 286–293.

Jamboonsri, W., Phillips, T., Geneve, R., Cahill, J., Hildebrand, D., 2012. Extending the range of an ancient crop, *Salvia hispanica* L.—a new ω3 source. Gen. Resourc. Crop Evol. 59, 171–178.

Jeong, S.K., Park, H.J., Park, B.D., Kim, I.H., 2010. Effectiveness of topical chia seed oil on pruritus of end-stage renal disease (ESRD) patients and healthy volunteers. Ann. Dermatol. 22, 143–148.

Kamal-Eldin, A., Appelqvist, L.-Å., 1996. The chemistry and antioxidant properties of tocopherols and tocotrienols. Lipids 31, 671–701.

Komprda, T., Zornikova, G., Rozikova, V., Borkovcova, M., Przywarova, A., 2013. The effect of dietary *Salvia hispanica* seed on the content of n-3 long-chain polyunsaturated fatty acids in tissues of selected animal species, including edible insects. J. Food Comp. Anal. 32, 36–43.

Lin, K.Y., Daniel, J.R., Whistler, R.L., 1994. Structure of chia seed polysaccharide exudate. Carbohydr. Polym. 23, 13–18.

Lu, Y., Foo, L.Y., 2002. Polyphenolics of Salvia–a review. Phytochemistry 59, 117–140.

Marineli, R.D.S., Moraes, É.A., Lenquiste, S.A., Godoy, A.T., Eberlin, M.N., Maróstica Jr., M.R., 2014. Chemical characterization and antioxidant potential of Chilean chia seeds and oil (*Salvia hispanica* L.). LWT-Food Sci. Technol. 59, 1304–1310.

Medic, J., Atkinson, C., Hurburgh Jr., C.R., 2014. Current knowledge in soybean composition. J. Am. Oil Chem. Soc. 91, 363–384.

Nieman, D., Gillitt, N., Meaney, M., Dew, D., 2015. No positive influence of ingesting chia seed oil on human running performance. Nutrients 7, 3666.

Nieman, D.C., Cayea, E.J., Austin, M.D., Henson, D.A., Mcanulty, S.R., Jin, F.X., 2009. Chia seed does not promote weight loss or alter disease risk factors in overweight adults. Nutr. Res. 29, 414–418.

Nieman, D.C., Gillitt, N., Jin, F., Henson, D.A., Kennerly, K., Shanely, R.A., Ore, B., Su, M., Schwartz, S., 2012. Chia seed supplementation and disease risk factors in overweight women: a metabolomics investigation. J. Altern. Complement. Med. 18, 700–708.

Nijveldt, R.J., Van Nood, E., Van Hoorn, D.E., Boelens, P.G., Van Norren, K., Van Leeuwen, P.A., 2001. Flavonoids: a review of probable mechanisms of action and potential applications. Am. J. Clin. Nutr. 74, 418–425.

Oliva, M.E., Ferreira, M.R., Chicco, A., Lombardo, Y.B., 2013. Dietary Salba (*Salvia hispanica* L.) seed rich in alpha-linolenic acid improves adipose tissue dysfunction and the altered skeletal muscle glucose and lipid metabolism in dyslipidemic insulin-resistant rats. Prostaglandins Leukot. Essent. Fatty Acids 89, 279–289.

Prasad, K., 2009. Flaxseed and cardiovascular health. J. Cardiovasc. Pharmacol. 54, 369–377.

Rao, S., Abdel-Reheem, M., Bhella, R., Mccracken, C., Hildebrand, D., 2008. Characteristics of high α-linolenic acid accumulation in seed oils. Lipids 43, 749–755.

Reyes-Caudillo, E., Tecante, A., Valdivia-López, M.A., 2008. Dietary fibre content and antioxidant activity of phenolic compounds present in Mexican chia (*Salvia hispanica* L.) seeds. Food Chem. 107, 656–663.

Rossi, A.S., Oliva, M.E., Ferreira, M.R., Chicco, A., Lombardo, Y.B., 2013. Dietary chia seed induced changes in hepatic transcription factors and their target lipogenic and oxidative enzyme activities in dyslipidaemic insulin-resistant rats. Brit. J. Nutr. 109, 1617–1627.

Sargi, S.C., Silva, B.C., Santos, H.M.C., Montanher, P.F., Boeing, J.S., Junior, O.O.S., Souza, N.E., Visentainer, J.V., 2013. Antioxidant capacity and chemical composition in seeds rich in omega-3: chia, flax, and perilla. Cienca Tecnol. Alimen. 33, 541–548.

Silva, B.E., Dos Santos, H., Montanher, P., Boeing, J., Almeida, V.D., Visentainer, J., 2014. Incorporation of Omega-3 fatty acids in Nile Tilapia (*Oreochromis niloticus*) fed chia (*Salvia hispanica* L.). Bran. J. Am. Oil Chem. Soc. 91, 429–437.

Simopoulos, A., De Meester, F., 2009. A balanced omega-6/omega-3 fatty acid ratio, cholesterol and coronary heart disease. World Rev. Nutr. Diet. 100, 110–121.

Stones, W., Campbell, S., Krishnan, K., 2012. The role of vitamin E in prostate cancer. Ocidative stress in cancer biology and therapy. In: Spitz, D.R., Dornfeld, K., Krishnan, K., Gius, D. (Eds.), Ocidative Stress in Applied Basic Research and Clinical Practice. Humana Press, pp. 333–354.

Taylor, C.G., Noto, A.D., Stringer, D.M., Froese, S., Malcolmson, L., 2010. Dietary milled flaxseed and flaxseed oil improve N-3 fatty acid status and do not affect glycemic control in individuals with well-controlled type 2 diabetes. J. Am. Coll. Nutr. 29, 72–80.

Ulbricht, C., Chao, W., Nummy, K., Rusie, E., Tanguay-Colucci, S., Iannuzzi, C., Plammoottil, J., Varghese, M., Weissner, W., 2009. Chia (*Salvia hispanica*): a systematic review by the natural standard research collaboration. Rev. Recent Clin. Trials 4, 168–174.

Vázquez-Ovando, A., Rosado-Rubio, G., Chel-Guerrero, L., Betancur-Ancona, D., 2009. Physicochemical properties of a fibrous fraction from chia (*Salvia hispanica* L.). LWT-Food Sci. Technol. 42, 168–173.

Vuksan, V., Jenkins, A.L., Dias, A.G., Lee, A.S., Jovanovski, E., Rogovik, A.L., Hanna, A., 2010. Reduction in postprandial glucose excursion and prolongation of satiety: possible explanation of the long-term effects of whole grain Salba (*Salvia hispanica* L.). Eur. J. Clin. Nutr. 64, 436–438.

Vuksan, V., Jovanovski, E., Dias, A., Lee, A., Rogovik, A., Jenkins, A., 2009. Comparable dose-response glucose-lowering effect with whole versus finely ground novel omega-3-Rich grain salba (*Salvia hispanica* L.) baked into white bread. Pharmaceut. Biol. 47, S13.

Vuksan, V., Whitham, D., Sievenpiper, J.-L., Jenkins, A.-L., Rogovik, A.-L., Bazinet, R.-P., Vidgen, E., Hanna, A., 2007. Supplementation of conventional therapy with the novel grain salba (*Salvia hispanica* L.) improves major and emerging cardiovascular risk factors in type 2 diabetes: results of a randomized controlled trial. Diabetes Care 30, 2804–2810.

Wagner, K.-H., Kamal-Eldin, A., Elmadfa, I., 2004. Gamma-tocopherol–An underestimated vitamin? Ann. Nutr. Metabol. 48, 169–188.

Zingg, J.-M., Azzi, A., 2004. Non-antioxidant activities of vitamin E. Curr. Med. Chem. 11, 1113–1133.

Chapter 11.2

Cuphea (*Cuphea* spp.)

Thomas A. McKeon
United States Department of Agriculture, Agricultural Research Service, Western Regional Research Center, Albany, CA, United States

INTRODUCTION

Cuphea species were identified as part of the USDA-ARS New Crops Initiative for seed oils with industrial uses (Miller et al., 1964; Graham and Knapp, 1989). Most of the 265 *Cuphea* species (Lythraceae) produce oils containing medium-chain fatty acids (MCFAs) that are saturated, principally caprylic (C8:0), capric (C10:0), and lauric (C12:0) (Fig. 11.2.1) (Gesch et al., 2006; Phippen et al., 2006). These fatty acids (FAs) are of commercial value in producing soaps and detergents, lubricants, and plasticizers (Miller et al., 1964). *Cuphea* oils rich in C10:0 are valued in production of biodiesel as they have a suitable cetane number, low pour point and cloud point, and excellent lubricity properties (Knothe et al., 2009, and see chapter: Biodiesel and Its Properties of this volume), even improving the lubricity of ultralow sulfur diesel derived from petroleum. Thus, *Cuphea* oils are useful additives for improving the properties of biodiesel produced from widely available seed oils containing mainly C16 and C18 saturated and unsaturated FAs such as soy or canola (Knothe, 2014). The current commercial sources of MCFAs are coconut oil and palm kernel oil, both imported from a limited number of producing countries in the tropics, and about half are derived from chemical modification of petroleum (Phippen et al., 2006).

In addition to these industrial uses, *Cuphea* oils and the FAs derived from *Cuphea* oils have considerable food value. The MCFAs are metabolized rapidly for energy compared to the long-chain FAs. Substitution of MCFA oils in the diet results in reduced deposition of fat in adipose tissue and has been implicated in reduction of plasma cholesterol. As a result, these oils and FAs are of value incorporated in such food products as energy bars, in diets for patients with certain metabolic disorders, and for those suffering cachexia (Graham and Knapp, 1989; Takeuchi et al., 2008).

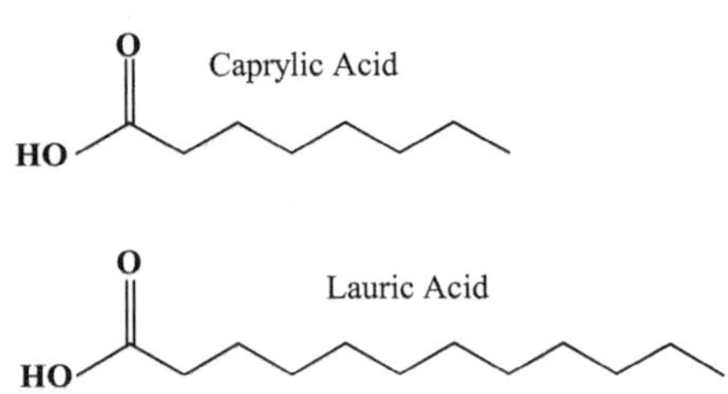

FIGURE 11.2.1 Structures of two major medium chain fatty acids (MCFA) present in *Cuphea* seed oils, caprylic acid, and lauric acid.

CUPHEA CROP RESEARCH

One goal of oilseed research in both North America and Europe has been development of a temperate crop source of MCFAs, so breeding programs to improve agronomic attributes of *Cuphea* sp. have been a part of agricultural research programs in both continents (Graham and Knapp, 1989; Zanetti et al., 2013). This research has identified *Cuphea* species that grow well in temperate climates and are insensitive to photoperiod, allowing cultivation in much of North America and Europe. The time from planting to seed maturation can be as short as 3 months. Table 11.2.1 depicts the MCFA composition of several *Cuphea* species that have been of interest for cultivation in temperate areas. A number of these selections are stored in the USDA Germplasm Resources Information Network (GRIN).

The general characteristics of *Cuphea* species that pose agronomic problems include seed shattering, seed dormancy, indeterminate inflorescences, and a shallow root system resulting in low tolerance to drought (Kim et al., 2011; Gesch et al., 2006). Such traits have been successfully bred out of soybean (Graham and Knapp, 1989), suggesting a continuing effort in breeding *Cuphea* would result in a more readily cultivated crop plant.

Although not yet grown as a commercialized crop some cultivars have been planted on a field scale. The cultivar PSR23, an interspecific hybrid of *C. viscosissima* and *C. lanceolata*, is considered the most promising cultivar from a commercial standpoint and has been grown in various locations by two different research groups (Gesch et al., 2006; Kim et al., 2011). Yields for the crop ranged from 283 to 1155 kg/ha with oil content ranging from 27% to 33%. Yields were dependent on location and seed shatter. The agronomic study carried out by Gesch et al. (2006) indicated a breakeven price for *Cuphea* would be

TABLE 11.2.1 Medium Chain Fatty Acid Content of *Cuphea* Species

Cuphea Species	Oil (%)	C8:0 (%)	C10:0 (%)	C12:0 (%)	C14:0 (%)
C. lanceolata[a]	20–33	1–2	78–91	1–4	1–5
C. calophylla[a]	26–34	0.4–4	19–33	58–72	2–7
C. tolucana[a]	30–38	0.4–0.7	21–42	46–65	1–5
C. procumbens[a]	19–35	1	81–89	1–2	1–2
C. viscosissima[a]	30	17	72	3	1
C. hookeriana[b]	16	65	24	0.1	0.2
PSR23[c]	27–33	1	77–84	2–3	2–4

[a]*Phippen et al. (2006).*
[b]*Miller et al. (1964).*
[c]*Kim et al. (2011).*

approximately \$1200 per metric ton. In contrast, soybean would have a break-even price of about \$240 per metric ton, putting *Cuphea* cultivation at a severe commercial disadvantage as a commodity oil.

There has been considerable interest in using genes from *Cuphea* to engineer other crop plants to produce oils containing high levels of MCFA. The *FatB2* genes encode MCFA-specific acyl-ACP thioesterases in *Cuphea* species. These thiesterases are responsible for the production of MCFA by "interrupting" fatty acid biosynthesis, releasing the MCFAcyl group from the growing acyl-ACP chain before achieving full length of C16:0 or C18:0. (See chapter: Introduction to Industrial Oil Crops for a discussion of fatty acid biosynthesis in plants.) The MCFA released by the thioesterase is subsequently incorporated into *Cuphea* oil (Dehesh et al., 1996). This research group cloned the *FatB2* gene from *C. hookeriana* and, on insertion in canola, obtained an oil enriched in C8:0 and C10:0, approximately 40 mol%. Although this technology is promising, it has not yet resulted in a commercial crop.

The MCFAs laurate and myristate available from *Cuphea* species would fill commercial niches that are key to production of surfactants, plasticizers, biodiesel improvement, and a food source of nutriceutical and pharmaceutical value. Although *Cuphea* shows some promise as a temperate crop producing MCFA, it has not yet reached commercial potential. *Cuphea* will require additional breeding efforts to achieve acceptance as an economically viable crop.

REFERENCES

Dehesh, K., Jones, A., Knutzon, D.S., Voelker, T.A., 1996. Production of high levels of 8:0 and 10:0 fatty acids in transgenic canola by overexpression of *Ch FATB2*, a thioesterase cDNAfrom *Cuphea hookeriana*. Plant J. 9, 167–172.

Gesch, R., Forcella, F., Olness, A., Archer, D., Hebard, A., 2006. Agricultural management of cuphea and potential for commercial production in the Northern corn belt. Ind. Crops Prod. 24, 300–306.

Graham, S.A., Knapp, S.J., 1989. *Cuphea*: a new plant source of medium–chain fatty acids. Crit. Rev. Food Sci. Nutr. 28 (2), 139–173.

Kim, K.-I., Gesch, R.W., Cermak, S.C., Phippen, W.B., Berti, M.T., Johnson, B.L., Marek, L., 2011. Cuphea growth, yield, and oil characteristics as influenced by climate and soil environments across the upper Midwest USA. Ind. Crops Prod. 33, 99–107.

Knothe, G., 2014. Cuphea oil as a potential biodiesel feedstock to improve fuel properties. J. Energ. Eng. 140, A5014001.

Knothe, G., Cermak, S.C., Evangelista, R.L., 2009. Cuphea oil as source of biodiesel with improved fuel properties caused by high content of methyl decanoate. Energ. Fuel 23, 1743–1747.

Miller, R.W., Earle, F.R., Wolff, I.A., 1964. Search for new industrial oils. IX. *Cuphea*, a versatile source of fatty acids. J. Am. Oil Chem. Soc. 41, 279–280.

Phippen, W.B., Isbell, T.A., Phippen, M.E., 2006. Total seed oil and fatty acid methyl ester contents of Cuphea accessions. Ind. Crops Prod. 24, 5259.

Takeuchi, H., Sekine, S., Kojima, K., Aoyama, T., 2008. The application of medium-chain fatty acids: edible oil with a suppressing effect on body fat accumulation. Asia Pac. J. Clin. Nutr. 17 (S1), 320–323.

Zanetti, F., Monti, A., Berti, M.T., 2013. Challenges and opportunities for new industrial oilseed crops in EU-27: a review. Ind. Crops Prod. 50, 580–595.

Chapter 11.3

Hemp (*Cannabis sativa* L.)

Hirotada Fukushige, David F. Hildebrand
Department of Plant and Soil Sciences, University of Kentucky, Lexington, KY, United States

INTRODUCTION

Cannabis species are annual herbaceous plants that have been cultivated for many centuries throughout the world for its fiber, as well as a source of food and medicine (Yanushevich, 1989; Pringle, 1997; Okazaki et al., 2011). Hemp (*Cannabis sativa* L. and a broad-leaf hemp biotype of *C. indica* L.) has a low content of Δ9-tetrahydrocannabinol (THC), a psychoactive compound, and has been cultivated for many centuries throughout the world for its fiber and seeds (Clarke and Merlin, 2013). However, the invention of the cotton threshing machine in the late 18th century and arrival of the steam engine in the early 19th century caused the decline of hemp production (Allegret, 2013). Subsequently, the outlawing of marijuana, a subgroup of *C. indica* L. with high THC content, in 1937 eventually led to the cessation of the hemp production in the United States. Recently, with increasing demand for sustainable food and chemical sources, interest in hemp production has been renewed. In 1998, production of industrial hemp (*C. sativa* L.) with a THC content of <0.3% was legalized in Canada and hemp acreage has increased annually to 58,000 ha in 2012 mainly in response to increasing demand for hempseed in foods in the United States (Leson, 2013). In 2014, the U.S. Agricultural Act of 2014 allowed any state to legalize industrial hemp production for research purposes. Cultivars of industrial hemp with a THC content of <0.2% and registered in the E.U. Approved Common Catalog of Cultivars (Regulation EC No 1251/99 and subsequent amendments) have been permitted to grow in Europe since 1997.

SEED COMPOSITION: OVERVIEW

Hempseeds, technically achenes, contain 19–38% oil and about 20–38% protein (Kriese et al., 2004; Iványi, 2006; Da Porto et al., 2014; Vonapartis et al., 2015). Other components include about 27–36% fiber and 4–6% minerals (House et al., 2010; Vonapartis et al., 2015) as well as tocopherols, carotenoids, and phenolic compounds (Oomah et al., 2002; Kriese et al., 2004; Chen et al., 2010; Vonapartis et al., 2015). Much of the fiber resides in the seed hull (House et al., 2010). For this reason, shelled/dehulled hemp seeds contain higher levels of oil (28–53%) and protein (20–39%) with lower levels of antinutritional compounds (House et al., 2010; Chen et al., 2010).

HEMPSEED OIL

Cold-pressed hempseed oil varies from off-yellow to dark green due to the presence of carotenoids and chlorophylls at different levels (Deferne and Pate, 1996; Oomah et al., 2002). While the presence of carotenoids and tocopherols increases the oxidative stability of oil, the presence of chlorophylls increases the risk of oxidation. With high unsaturation levels (see later and Table 11.1.1), virgin hempseed oil was reported to have <1 h of oxidative stability based on the Rancimat test at 120°C (Matthäus and Brühl, 2008), 1.32 h at 110°C (Uluata and Özdemir, 2012), and 6.4–7.6 h at 100°C (Dimić et al., 2009). A slightly better value of 1.35–1.72 h at 120°C was also reported (Anwar et al., 2006). Therefore, great care must be taken to prevent oil oxidation.

Hempseed oil is composed of linoleic acid ($18{:}2^{\Delta 9,12}$: LA) (50–70%), α-linolenic acid ($18{:}3^{\Delta 9,12,15}$: ALA) (12–25%), oleic acid ($18{:}1^{\Delta 9}$) (8–17%), palmitic acid (16:0) (4–8%), stearic acid (18:0) (2–3%), and a few minor fatty acids (FAs), which include γ-linolenic acid ($18{:}3^{\Delta 6,9,12}$: GLA) (0–6.8%) and stearidonic acid ($18{:}4^{\Delta 6,9,12,15}$: SDA) (0–2.5%) (Mölleken and Theimer, 1997; Kriese et al., 2004; Iványi, 2006; Dimić et al., 2009; Chen et al., 2010; Da Porto et al., 2014; Vonapartis et al., 2015) (Table 11.3.1).

As mentioned, hempseed oil consists of about 90% polyunsaturated fatty acids (PUFAs) and has an ω-6/ω-3 FA ratio between 2:1 and 3:1 and closer to the recommended ratio beneficial to human health than many other oils consumed generally (soybean oil is 7:1) (Callaway, 2004).

ESSENTIAL FAs

Both ω-3 and ω-6 FAs are essential FAs to higher animals including humans since all mammals must obtain them through their diets (Simopoulos, 2010). A target ω-6/ω-3 FA ratio of 1:1 to 2:1 is recommended for human nutrition since the balance between ω-6 and ω-3 FA metabolites in the human brain is close to 1:1. Unfortunately, current Western diets are deficient in ω-3 FAs and the ω-6/ω-3 FA ratio is 10–20:1. Therefore, more food ingredients rich in ω-3 FA such as fish, flax, chia, walnuts, etc. are recommended. However, the conversion of ALA, the major ω-3 FA in plant oils, to eicosapentaenoic ($20{:}5^{\Delta 5,8,11,14,17}$: EPA) and docosahexaenoic acids ($22{:}6^{\Delta 4,7,10,13,16,19}$: DHA), well-known essential FAs found in fish oils (Fig. 11.3.1), is rather poor in many humans (about 8% conversion from ALA to EPA) (James et al., 2003; Hussein et al., 2005; Burdge and Calder, 2005; Arterburn et al., 2006). Hence, the American Heart Association still recommends the inclusion of fatty fish in diets (twice a week) with an additional intake from ALA-rich sources for healthy individuals and the consumption of 1 g/day of EPA plus DHA for patients with coronary heart disease (Kris-Etherton et al., 2003). On the other hand, SDA, a precursor of EPA and DHA, is reported to be converted to EPA more efficiently than ALA (James et al., 2003; Maki and Rains, 2012; Walker et al., 2013) and shares similar biological properties to these FAs (Whelan, 2009), but more studies are needed.

TABLE 11.3.1 Oil Content and Fatty Acid Composition of Hempseed Cultivars

Cultivar	Year	Field Location	Oil%	Palmitic Acid 16:0	Stearic Acid 18:0	Oleic Acid 18:1 ω-9	Linoleic Acid 18:2 ω-6	α-Linolenic Acid 18:3 ω-3	γ-Linolenic acid 18:3 ω-6	References
Felina 34	2004	Hungary	32.2	7.1	2.9	14.0	56.0	15.1	2.8	Ivanyi (2006)
Futura 75	2004	Hungary	28.3	6.8	3.3	12.7	56.4	17.2	2.0	Ivanyi (2006)
Fibranova	2004	Hungary	27.5	7.3	3.0	13.8	57.4	15.5	1.3	Ivanyi (2006)
Ferimon	2004	Hungary	27.2	7.8	2.7	14.8	56.8	19.1	1.3	Ivanyi (2006)
Kompolii	2004	Hungary	21.6	7.0	3.2	12.1	55.6	19.3	1.2	Ivanyi (2006)
Finola	2003	U.K.	24.6	6.4	2.5	9.7	55.6	18.9	4.2	Ivanyi (2006)
Finola	2005	Hungary	22.9	7.1	2.4	11.1	55.6	16.5	4.6	Ivanyi (2006)
Beniko	2007?	Serbia	N.D.	7.0	2.7	13.7	55.7	13.0	1.9	Dimic et al. (2009)
Futura 75	2007?	Serbia	N.D.	7.6	3.3	16.4	51.9	12.4	1.7	Dimic et al. (2009)
Carmagnola S.	2007?	Serbia	N.D.	7.2	2.5	14.2	54.8	15.4	1.4	Dimic et al. (2009)
Yunma No. 1	2009?	Yunma, China	N.D.	5.2	2.6	8.5	56.7	25.2	0.9	Chen et al. (2010)
Beian	2009?	Beian, China	N.D.	4.2	2.3	11.0	61.6	17.3	2.4	Chen et al. (2010)
Baotou	2009?	Baotou, China	N.D.	4.6	2.5	14.7	62.9	13.5	0.5	Chen et al. (2010)
Chamaeleon	2011	Italy	N.D.	5.5	2.0	11.1	58.8	18.3	3.3	Da Porto et al. (2014)

Continued

TABLE 11.3.1 Oil Content and Fatty Acid Composition of Hempseed Cultivars—cont'd

Cultivar	Year	Field Location	Oil%	Palmitic Acid 16:0	Stearic Acid 18:0	Oleic Acid 18:1 ω-9	Linoleic Acid 18:2 ω-6	α-Linolenic Acid 18:3 ω-3	γ-Linolenic acid 18:3 ω-6	References
Chamaeleon	2012	Italy	N.D.	6.3	1.8	9.8	59.5	18.6	3.1	Da Porto et al. (2014)
Uso31	2011	Italy	N.D.	2.8	2.4	12.5	57.3	17.3	6.5	Da Porto et al. (2014)
Uso31	2012	Italy	N.D.	2.9	2.3	12.1	57.6	17.4	6.4	Da Porto et al. (2014)
Finola	2011	Italy	N.D.	5.7	1.9	9.2	55.9	19.4	6.0	Da Porto et al. (2014)
Finola	2012	Italy	N.D.	6.7	1.7	9.3	56.1	20.0	5.6	Da Porto et al. (2014)
Finola	2012	Canada	30.6	6.9	2.1	9.4	56.2	17.3	4.5	Vonapartis et al. (2015)
Delores	2012	Canada	26.9	6.7	2.7	12.8	56.2	14.7	2.9	Vonapartis et al. (2015)
Jutta	2012	Canada	27.6	6.8	2.7	13.0	55.6	15.0	2.9	Vonapartis et al. (2015)

N.D., Not determined.

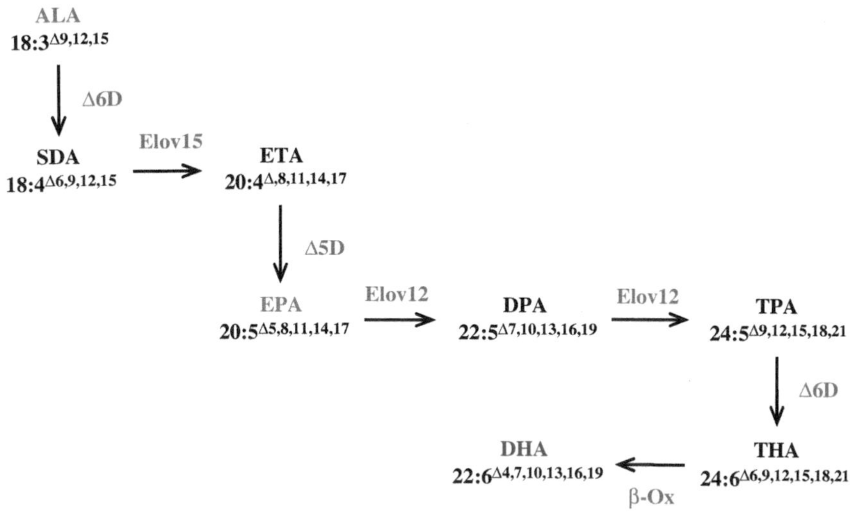

FIGURE 11.3.1 ω-3 Long-chain polyunsaturated fatty acid biosynthesis in mammals. *ALA*, α-linolenic acid; *β-Ox*, β-oxidation; *Δ5D*, Δ5-desaturase; *Δ6D*, Δ6-desaturase; *DHA*, docosahexaenoic acid; *DPA*, docosapentaenoic acid; *Elovl2*, elongase-2; *Elovl5*, elongase-5; *EPA*, eicosapentaenoic acid; *ETA*, eicosatetraenoic acid; *LA*, linoleic acid; *SDA*, stearidonic acid; *THA*, tetracosahexaenoic acid; *TPA*, tetracosapentaenoic acid.

ANTIOXIDANTS IN HEMPSEED OIL

As mentioned earlier, antioxidants such as carotenoids, tocopherols, and phenolic compounds exist in cold-pressed hempseed oil as well as in crude solvent-extracted oil.

Tocopherols and tocotrienols are vitamin E compounds and consist of four forms: α-, β-, γ-, and δ-tocopherol (Kamal-Eldin and Appelqvist, 1996). In general, their antioxidant activity are in order of α>β>γ>δ, although actual potency in vivo differs depending on other physical parameters. α-Tocopherol has been studied extensively for its antioxidant activity mainly due to its high uptake (Brigelius-Flohé and Traber, 1999; Zingg and Azzi, 2004; Brigelius-Flohé, 2006) and predominant vitamin E form in foods. Especially in the United States, γ-tocopherol has been reported to play an important role in reducing the risk of cardiovascular disease and cancer (Wagner et al., 2004; Stone et al., 2012). In hempseed oil, like many other vegetable oils, γ-tocopherol is the major tocopherol followed by α- and δ-tocopherols, while β-tocopherol is present in very small amounts. Their concentrations vary among the varieties and regions of production, from 174 to 791 mg/kg of γ-tocopherol, though the extraction methods affect levels significantly (Oomah et al., 2002; Anwar et al., 2006; Chen et al., 2010). When compared with flaxseed oil and canola oil, cold-pressed hempseed oil contained more γ-tocopherol as well as phenolic compounds (Teh and Birch, 2013).

Phenolic compounds such as hydroxylated derivatives of caffeic, cinnamic, and benzoic acids as well as flavonoids occur as natural hydrophilic antioxidants in plants (Silva et al., 2000; Balasundram et al., 2006). In hempseeds grown in Canada, a 15–52 g/kg gallic acid equivalent of phenolic compounds was detected (Vonapartis et al., 2015), while in cold-pressed oil, the total phenolic acids were found to be 1.9 g/kg, which is a fraction of the levels found in seeds, but still higher than other vegetable oils (Teh and Birch, 2013).

Carotenoid contents were reported to be 20–53 mg/kg oil by Oomah et al. (2002) with lutein 5 mg/kg but no β-carotene was detected by Rovellini et al. (2013).

HEMPSEED MEAL

Hemp protein isolates are mostly composed of edestin, a member of the legumin family of seed storage proteins, at ~82% (Tang et al., 2006; Wang et al., 2008). A methionine- and cysteine-rich protein was also isolated from hempseeds (Odani and Odani, 1998). Hempseed meal and its derived products contain all essential amino acids required by humans, although lysine is the most limiting amino acid based on amino acid scores (Tang et al., 2006; Wang et al., 2008; House et al., 2010). These results suggest that the hemp protein isolates are a good source of protein for human nutrition. With growing demand for health foods in the United States, whole seeds, hulled seeds, as well as hempseed meal/cake are being incorporated into protein bars, energy bars, granola, and other food products (House et al., 2010; Leson, 2006).

Hempseeds have been a major component of cage bird seeds. In 2002, >85% of an estimated 5300 t of hempseeds harvested in Europe were sold for animal feed, mainly as bird seed (Karus, 2004). Broiler chicks fed with feed supplemented with whole hempseed flour had higher body weight gain and feed conversion rate with lower feed intake (Khan et al., 2010). Feeding laying hens with hempseed and hempseed meal resulted in enrichment of ω-3 FAs in eggs without significant effects on egg production, feed consumption, or body weight change (Silversides and Lefrançois, 2005). Gakhar et al. (2012) observed some negative effects on laying hens related to body weight and feed intake from hempseed oil feeding but saw no effect on egg production as well as higher egg weight and ω-3 FA enrichment.

There was no improvement in body weight, feed intake, or feed conversion rate seen in Japanese quail fed with hempseed (Konca et al., 2014a). However, feeding Japanese quail with hempseed resulted in ω-3 FA enrichment in eggs and increased serum antioxidant activity (Konca et al., 2014b).

Mustafa et al. (1999) showed the higher levels of rumen undegraded proteins (RUP) that would be available for digestion in the intestine in ruminally fistulated cows fed hemp meal compared to regular canola meal. Total tract dry matter digestibility with sheep was similar between hemp and canola meal, suggesting hemp meal can replace canola meal in isonitrogenous diets for ruminants. However, Karlsson and Martinsson (2011) found the hempseed cake supplement to a barley-based diet did not improve the growth performance of lambs, while peas and rapeseed cake supplementation significantly improved it. The authors suggested this is due to the high content of fiber and low digestibility of RUP in hempseed cake but disagrees with the result from Mustafa et al. (1999). Later, Karlsson et al. (2012) showed decreased ruminal degradability of total amino acids and increased intestinal digestibility of moist heat–treated hempseed cake. Growing steers fed with cold-pressed hempseed cake compared to soybean meal showed similar weight gain and carcass traits due to improved rumen function from higher fiber content, suggesting cold-pressed hempseed cake as an alternative protein feed for growing cattle (Hessle et al., 2008).

Modern Nonfood Use

Hempseed oil is rich in minerals, vitamins, and EFAs and considered to be a perfect ingredient for light body oils (Kowalska et al., 2015). Other products such as soap, hair care products, and lip balm listed hempseed oil as an ingredient on a Canadian Hemp Trade Alliance Website (http://www.hemptrade.ca/products.php).

SUMMARY

There is renewed interest in hempseed oil from industrial hemp due to the increased interest in healthy foods in the United States, although it is one of the oldest crops. Hempseed oil has high unsaturation levels (~90%), making it one of the best natural drying oils. The ω-3/ω-6 fatty acid ratio of hempseed oil is close to the ideal ratio for human consumption. With the additional higher levels of GLA than most plant oils as well as good antioxidant amounts, hempseed oil can be of value in many health products.

ACKNOWLEDGMENTS

Our hempseed research is supported by the KY Science and Engineering Foundation and the KY Agricultural Experiment Station.

REFERENCES

Allegret, S., 2013. The histoy of hemp. In: Bouloc, P., Allegret, S., Arnaud, L. (Eds.), Hemp: Industrial Production and Uses.

Anwar, F., Latif, S., Ashraf, M., 2006. Analytical characterization of hemp (*Cannabis sativa*) seed oil from different agro-ecological zones of Pakistan. J. Am. Oil Chem. Soc. 83, 323–329.

Arterburn, L.M., Hall, E.B., Oken, H., 2006. Distribution, interconversion, and dose response of n−3 fatty acids in humans. Am. J. Clin. Nutr. 83, S1467–S1476.

Balasundram, N., Sundram, K., Samman, S., 2006. Phenolic compounds in plants and agri-industrial by-products: antioxidant activity, occurrence, and potential uses. Food Chem. 99, 191–203.

Brigelius-Flohé, R., 2006. Bioactivity of vitamin E. Nutr. Res. Rev. 19, 174–186.

Brigelius-Flohé, R., Traber, M.G., 1999. Vitamin E: function and metabolism. FASEB J. 13, 1145–1155.

Burdge, G.C., Calder, P.C., 2005. Conversion of α-linolenic acid to longer-chain polyunsaturated fatty acids in human adults. Reprod. Nutr. Dev. 45, 581–597.

Callaway, J.C., 2004. Hempseed as a nutritional resource: an overview. Euphytica 140, 65–72.

Chen, T., He, J., Zhang, J., Zhang, H., Qian, P., Hao, J., Li, L., 2010. Analytical characterization of hempseed (seed of *Cannabis sativa* L.) oil from eight regions in China. J. Diet. Suppl. 7, 117–129.

Clarke, R., Merlin, M., 2013. Cannabis: Evolution and Ethnobotany. University of California Press, Berkeley, CA, USA.

Da Porto, C., Decorti, D., Natolino, A., 2014. Potential oil yield, fatty acid composition, and oxidation stability of the hempseed oil from four *Cannabis sativa* L. cultivars. J. Diet. Suppl. 12, 1–10.

Deferne, J.-L., Pate, D.W., 1996. Hemp seed oil: a source of valuable essential fatty acids. J. Int. Hemp Assoc. 3, 4–7.

Dimić, E., Romanić, R., Vujasinović, V., 2009. Essential fatty acids, nutritive value and oxidative stability of cold pressed hempseed (*Cannabis sativa* L.) oil from different varieties. Acta Aliment. 38, 229–236.

Gakhar, N., Goldberg, E., Jing, M., Gibson, R., House, J.D., 2012. Effect of feeding hemp seed and hemp seed oil on laying hen performance and egg yolk fatty acid content: evidence of their safety and efficacy for laying hen diets. Poult. Sci. 91, 701–711.

Hessle, A., Eriksson, M., Nadeau, E., Turner, T., Johansson, B., 2008. Cold-pressed hempseed cake as a protein feed for growing cattle. Acta Agri. Scand. A Anim. Sci. 58, 136–145.

House, J.D., Neufeld, J., Leson, G., 2010. Evaluating the quality of protein from hemp seed (*Cannabis sativa* L.) products through the use of the protein digestibility-corrected amino acid score method. J. Agri. Food Chem. 58, 11801–11807.

Hussein, N., Ah-Sing, E., Wilkinson, P., Leach, C., Griffin, B.A., Millward, D.J., 2005. Long-chain conversion of [^{13}C]linoleic acid and α-linolenic acid in response to marked changes in their dietary intake in men. J. Lipid Res. 46, 269–280.

Iványi, I., 2006. Effect of genotype and environmental on the oil content and fatty acid composition of hempseed. Cereal Res. Commun. 34, 493–496.

James, M.J., Ursin, V.M., Cleland, L.G., 2003. Metabolism of stearidonic acid in human subjects: comparison with the metabolism of other n–3 fatty acids. Am. J. Clin. Nutr. 77, 1140–1145.

Kamal-Eldin, A., Appelqvist, L.-Å., 1996. The chemistry and antioxidant properties of tocopherols and tocotrienols. Lipids 31, 671–701.

Karlsson, L., Martinsson, K., 2011. Growth performance of lambs fed different protein supplements in barley-based diets. Livestock Sci. 138, 125–131.

Karlsson, L., Ruiz-Moreno, M., Stern, M.D., Martinsson, K., 2012. Effects of temperature during moist heat treatment on ruminal degradability and intestinal digestibility of protein and amino acids in hempseed cake. Asian-Australas. J. Anim. Sci. 25, 1559–1567.

Karus, M., 2004. European hemp Industry 2002. J. Ind. Hemp 9, 93–101.

Khan, R.U., Durrani, F.R., Chand, N., Anwar, H., 2010. Influence of feed supplementation with *Cannabis sativa* on quality of broilers carcass. Pak. Vet. J. 30, 34–38.

Konca, Y., Cimen, B., Yalcin, H., Kaliber, M., Beyzi, S.B., 2014a. Effect of hempseed (*Cannabis sativa* sp.) inclusion to the diet on performance, carcass and antioxidative activity in Japanese quail (*Coturnix coturnix japonica*). Korean J. Food Sci. Anim. Resourc. 34, 141–150.

Konca, Y., Yalcin, H., Karabacak, M., Kaliber, M., Durmuscelebi, F.Z., 2014b. Effect of hempseed (*Cannabis sativa* L.) on performance, egg traits and blood biochemical parameters and antioxidant activity in laying Japanese quail (*Coturnix coturnix japonica*). Brit. Poult. Sci. 55, 785–794.

Kowalska, M., Ziomek, M., Żbikowska, A., 2015. Stability of cosmetic emulsion containing different amount of hemp oil. Int. J. Cosmetic Sci. 37, 408–416.

Kriese, U., Schumann, E., Weber, W.E., Beyer, M., Brühl, L., Matthäus, B., 2004. Oil content, tocopherol composition and fatty acid patterns of the seeds of 51 *Cannabis sativa* L. genotypes. Euphytica 137, 339–351.

Kris-Etherton, P.M., Harris, W.S., Appel, L.J., 2003. Fish consumption, fish oil, omega-3 fatty acids, and cardiovascular disease. Arterioscler. Thromb. Vasc. Biol. 23, e20–e30.

Leson, G., 2006. Hemp foods in North America. J. Ind. Hemp 11, 87–93.

Leson, G., 2013. Hemp seed for nutrition. In: Bouloc, P., Allegret, S., Arnaud, L. (Eds.), Hemp: Industrial Production and Uses.

Maki, K.C., Rains, T.M., 2012. Stearidonic acid raises red blood cell membrane eicosapentaenoic acid. J. Nutr. 142, 626S–629S.

Matthäus, B., Brühl, L., 2008. Virgin hemp seed oil: an interesting niche product. Eur. J. Lipid Sci. Technol. 110, 655–661.

Mölleken, H., Theimer, R.R., 1997. Survey of minor fatty acids in *Cannabis sativa* L. fruits of various origins. J. Int. Hemp Assoc. 4, 13–17.

Mustafa, A.F., Mckinnon, J.J., Christensen, D.A., 1999. The nutritive value of hemp meal for ruminants. Can. J. Anim. Sci. 79, 91–95.

Odani, S., Odani, S., 1998. Isolation and primary structure of a methionine- and cystine-rich seed protein of *Cannabis sativa*. Biosci. Biotechnol. Biochem. 62, 650–654.

Okazaki, H., Kobayashi, M., Momohara, A., Eguchi, S.-I., Okamoto, T., Yanagisawa, S.-I., Okubo, S., Kiyonaga, J., 2011. Early Holocene coastal environment change inferred from deposits at Okinoshima archeological site, Boso Peninsula, central Japan. Quat. Int. 230, 87–94.

Oomah, B.D., Busson, M., Godfrey, D.V., Drover, J.C.G., 2002. Characteristics of hemp (*Cannabis sativa* L.) seed oil. Food Chem. 76, 33–43.

Pringle, H., 1997. Ice age communities may be earliest known net hunters. Science 277, 1203–1204.

Rovellini, P., Folegatti, L., Baglio, D., De Cesarei, S., Fusari, P., Venturini, S., Cavalieri, A., 2013. Chemical characterization of oil obtained by the cold pressing of *Cannabis sativa* L. seeds. Riv. Ital. Sostanze Gr. 90, 139–152.

Silva, F.A.M., Borges, F., Guimarães, C., Lima, J.L.F.C., Matos, C., Reis, S., 2000. Phenolic acids and derivatives: studies on the relationship among structure, radical scavenging activity, and physicochemical parameters. J. Agri. Food Chem. 48, 2122–2126.

Silversides, F.G., Lefrançois, M.R., 2005. The effect of feeding hemp seed meal to laying hens. Brit. Poult. Sci. 46, 231–235.

Simopoulos, A.P., 2010. The omega-6/omega-3 fatty acid ratio: health implications. Ol. Corps Gras Lipides 17, 267–275.

Stone, W.L., Campbell, S.E., Krishnan, K., 2012. The role of vitamin E in prostate cancer. In: Spitz, D.R., Dornfeld, K.J., Krishnan, K., Gius, D. (Eds.), Oxidative Stress in Cancer Biology and Therapy. Humana Press.

Tang, C.-H., Ten, Z., Wang, X.-S., Yang, X.-Q., 2006. Physicochemical and functional properties of hemp (*Cannabis sativa* L.) protein isolate. J. Agri. Food Chem. 54, 8945–8950.

Teh, S.-S., Birch, J., 2013. Physicochemical and quality characteristics of cold-pressed hemp, flax and canola seed oils. J. Food Comp. Anal. 30, 26–31.

Uluata, S., Özdemir, N., 2012. Antioxidant activities and oxidative stabilities of some unconventional oilseeds. J. Am. Oil Chem. Soc. 89, 551–559.

Vonapartis, E., Aubin, M.-P., Seguin, P., Mustafa, A.F., Charron, J.-B., 2015. Seed composition of ten industrial hemp cultivars approved for production in Canada. J. Food Comp. Anal. 39, 8–12.

Wagner, K.-H., Kamal-Eldin, A., Elmadfa, I., 2004. Gamma-tocopherol - an underestimated vitamin? Ann. Nutr. Metabol. 48, 169–188.

Walker, C.G., Jebb, S.A., Calder, P.C., 2013. Stearidonic acid as a supplemental source of ω-3 polyunsaturated fatty acids to enhance status for improved human health. Nutrition 29, 363–369.

Wang, X.-S., Tang, C.-H., Yang, X.-Q., Gao, W.-R., 2008. Characterization, amino acid composition and in vitro digestibility of hemp (*Cannabis sativa* L.) proteins. Food Chem. 107, 11–18.

Whelan, J., 2009. Dietary stearidonic acid is a long chain (n-3) polyunsaturated fatty acid with potential health benefits. J. Nutr. 139, 5–10.

Yanushevich, Z.V., 1989. Agricultural evolution north of the Black Sea from the Neolithic to the iron Age. In: Harris, D.R., Hillman, G.C. (Eds.), Foraging and Farming: The Evolution of Plant Exploitation. Unwin Hyman Ltd, London, UK.

Zingg, J.-M., Azzi, A., 2004. Non-antioxidant activities of vitamin E. Curr. Med. Chem. 11, 1113–1133.

Chapter 11.4

Jatropha (*Jatropha curcas* L.)

Guanqun Chen
Department of Biological Sciences, University of Manitoba, Winnipeg, Manitoba, Canada

INTRODUCTION

Jatropha (*Jatropha curcas*) is a plant that originated in Mexico and now grows widespread in tropical and subtropical areas in Latin America, Asia, and Africa (Dias et al., 2012). Jatropha seeds generally contain toxic components but produce 27–40% oil, rich in palmitic acid (C16:0, 13.4–15.3%), oleic acid (C18:1, 34.3–45.8%), and linoleic acid (C18:2, 29.0–44.2%) (Meher et al., 2013). This oil-bearing plant is widely recognized as a potential renewable energy source because it can grow and fruit on marginal or nonagricultural areas and has attractive peculiar features such as drought tolerance, pest resistance, rapid growth, easy propagation, and small gestation period. Due to the toxic nature of the plant and the extracted oil, the oil is considered unfit for food use and therefore of value in biodiesel, especially in countries that ban use of food crops for fuel use. Unlike other potential oil crops, jatropha is so attractive that it has been planted in large scale (>10 million ha globally) even without the agronomic improvement and evaluation needed for adaptation as a useful crop (Singh et al., 2014). At the same time, jatropha crop research has been a hot topic recently: about 1500 articles including >180 reviews have been published in the past 10 years (Edrisi et al., 2015).

Studies with large-scale plantations of jatropha indicated, however, that the growth performance and oil yield of this plant were far below than expected. The anticipated seed yields range from 2 to 12 t/ha but the actual ones reported from various countries are between 0.5 and 2 t/ha (Edrisi et al., 2015; Singh et al., 2014). The oil yield of jatropha is much lower than other oil-producing plants such as oil palm (eg, *Elaeis guineensis*), rapeseed (*Brassica napus*), and coconut (*Cocos nucifera*) (Yue et al., 2013). The commonly believed conceptions that jatropha is resistant to diseases and can grow well with high oil yield on marginal lands were proved to be wrong (Sanderson, 2009; Singh et al., 2014). Crude jatropha oil contains up to 15% of free fatty acids and the total fatty acids contain 30–50% of polyunsaturated fatty acids (mainly linoleic acid), which are not good for the production of high-quality biodiesel (Meher et al., 2013; Yue et al., 2013). Therefore, in order for jatropha to meet the demands as a renewable energy resource, this plant needs to be well domesticated.

J. CURCAS TOXICITY

While *Jatropha* spp. are distant relatives of castor (*Ricinus communis* L.) and both plants are considered toxic, nonfood plants, the nature of their toxicity differs. Both produce seeds containing a ribosome-inactivating protein (RIP), but curcin from *J. curcas* is a type 1 RIP, while ricin from castor is a type 2 RIP, and it is the type 2 RIP that is primarily responsible for castor seed toxicity (see chapter: *Brassica* spp. Oils, *Ricinus communis* L., the Castor Oil Plant). Curcin has the same *N*-glycosidase activity that the ricin A-chain has, but curcin lacks the strong lectin subunit of the type 2 RIPs that results in cell binding and endocytosis to deliver the *N*-glycosidase that inactivates the ribosome. On the other hand, most *J. curcas* seed contains phorbol esters, diterpenes that are irritants and promote tumors (Nakao et al., 2015). Seeds of toxic *Jatropha* cultivars contain 0.01–0.05% phorbol ester (He et al., 2011), which will be concentrated in the oil derived from the seed, making the oil inedible. Moreover, in addition to residual phorbol esters, the seed cake also contains phytate, trypsin inhibitors, and other antinutritional factors (Pradhan et al., 2012). In contrast to castor, there are edible varieties of *J. curcas*. These lack the phorbol esters in the seed, while the curcin is still present (He et al., 2011), although what toxic activity it does have may be alleviated by roasting.

JATROPHA CROP RESEARCH

Jatropha curcas L. belongs to the Euphorbiaceae family. Concerted efforts to domesticate jatropha have only begun within the past decade owing to the renewed impetus to develop biodiesel (Carels, 2009; Edrisi et al., 2015). Both conventional breeding and molecular breeding approaches have been used to increase the oil yield and solve other agronomic problems such as seed yield, oil content and composition, female-to-male flower ratio, synchronous flowering and fruiting, oil quality, and branch number, resulting in some improvements. Conventional domestication of jatropha is ongoing in several countries. Comparison of jatropha germplasm collected from different regions was conducted by several research groups, and the results showed that *J. curcas* had a low genetic variation (Basha and Sujatha, 2007). Several selective breeding methods including mass selection, recurrent selection, hybrid breeding, and induced mutation breeding have been used to improve trait performance (Divakara et al., 2010). About 1850 candidate trees have been selected from 5000 accessions collected in India for further examination (Yue et al., 2013). In another research project, over 800 accessions were collected and 100 accessions were identified with high oil yield (Yang et al., 2012). One selected line had an oil yield of 1566 kg/ha (Yang et al., 2010). Several commercial companies have reported elite varieties with oil yield of 2000 kg/ha/year under good management or in small-scale tests. While these results are promising, the results need to be tested by large-scale field examination. Overall, the reported best oil yield of the elite traits selected by conventional breeding approaches is about 2000 kg/ha/year, which is higher than that of rapeseed, flaxseed (*Linum usitatissimum*), or soybean (*Glycine max*) but lower than that of palm and coconut (Yue et al., 2013).

Genetic engineering has also been applied for rapid domestication of jatropha. Genomic resources of *J. curcas* such as molecular markers, linkage maps, transcriptome, and genome sequences are available now. Several traditional DNA markers including random amplified polymorphic DNA, amplified fragment length polymorphism, and intersimple sequence repeat markers, as well as two new types of DNA markers, microsatellites and single nucleotide polymorphism, have been developed (Yue et al., 2013). These DNA markers can be directly used for germplasm characterization and molecular breeding. A first-generation linkage map of 96 individuals from two families, *J. curcas* and *J. integerrima*, has been constructed, which can be used as a framework for genome-wide identification of associations between DNA markers and jatropha traits (Doerge, 2002; Wang et al., 2011). As for the transcriptome of *J. curcas*, over 100,000 expressed sequence tag sequences (ESTs) have been deposited in Genbank. These can be used for metabolic engineering of *J. curcas* to increase oil content, to modify fatty acid composition, and to improve agronomic performance. Recently, the draft genome sequence of *J. curcas* was published, which will serve as a valuable resource for research on improving *J. curcas* (Hirakawa et al., 2012; Sato et al. 2011).

The genomic resources mentioned here have been applied in molecular breeding of *J. curcas*. Examples of these applications are: (1) DNA markers have been used in analyzing genetic diversity of *J. curcas* (Gupta et al., 2012), (2) genetic barcoding system has been developed to identify and protect elite *J. curcas* trees (Yue et al., 2013), (3) genes related to important traits including fatty acid synthesis and stress resistance have been isolated and modified to increase the quality of biodiesel from *J. curcas* (Gu et al., 2012), and (4) quantitative trait loci (QTL, gene clusters or chromosomal regions influencing the expression of a quantitative trait) mapping has been conducted for oil yield and quality of *J. curcas* (Liu et al., 2011).

Transgenic technologies have been used to increase seed oil content and optimize the fatty acid composition of jatropha oil. Effective transformation systems were established for the genetic transformation of *J. curcas* (Li et al., 2008; Qu et al., 2012). As described in the Introduction, jatropha oil is not ideal for direct use in producing biodiesel due to its high percentage of linoleic acid. A recent study, however, showed that downregulation of three *fad2* genes encoding $\Delta 12$-desaturase under a seed-specific promoter increased oleic acid content to about 80% and decreased polyunsaturated fatty acids to <3% in seed oil without negative impact on other jatropha agronomic traits (Qu et al., 2012). Although results obtained from transgenic technology are promising, no transgenic *J. curcas* plants have been generated as a commercial crop so far.

Genes from jatropha have also been used in other plants to increase seed oil levels and alter oil quality. For example, when the *JcDGAT1* gene encoding diacylglycerol acyltransferase 1 was overexpressed in *Arabidopsis*, the transgenic *Arabidopsis* lines had a 30–40% higher content of total oil in seeds (Misra et al., 2013). The level of oleic acid significantly decreased and linolenic acid content increased (Misra et al., 2013).

CONCLUSION

In summary, the unique characteristics of jatropha differentiate it from many other potential oilseed plants, but additional efforts in domestication of jatropha and improvement in agronomic practices are crucial to make the production of biodiesel from *J. curcas* sustainable. Since molecular breeding hold the greater promise than conventional breeding in accelerating genetic improvement for sustainable production of plant oils, further development and application of *Jatropha* genomic resources are essential for making *J. curcas* an oilseed crop.

REFERENCES

Basha, S.D., Sujatha, M., 2007. Inter and intra-population variability of *Jatropha curcas* (L.) characterized by RAPD and ISSR markers and development of population-specific SCAR markers. Euphytica 156, 375–386.

Carels, N., 2009. *Jatropha curcas*: a review. In: Kader, J.C., Delseny, M. (Eds.), Advances in Botanical Research, Vol 50, pp. 39–86.

Dias, L.A.S., Missio, R.F., Dias, D.C.F.S., 2012. Antiquity, botany, origin and domestication of *Jatropha curcas* (Euphorbiaceae), a plant species with potential for biodiesel production. Gen. Mol. Res. 11, 2719–2728.

Divakara, B.N., Upadhyaya, H.D., Wani, S.P., Gowda, C.L.L., 2010. Biology and genetic improvement of *Jatropha curcas* L.: a review. Appl. Energ. 87, 732–742.

Doerge, R.W., 2002. Mapping and analysis of quantitative trait loci in experimental populations. Nat. Rev. Genet. 3, 43–52.

Edrisi, S.A., Dubey, R.K., Tripathi, V., Bakshi, M., Srivastava, P., Jamil, S., Singh, H.B., Singh, N., Abhilash, P.C., 2015. *Jatropha curcas* L.: a crucified plant waiting for resurgence. Renew. Sust. Energ. Rev. 41, 855–862.

Gu, K., Yi, C., Tian, D., Sangha, J.S., Hong, Y., Yin, Z., 2012. Expression of fatty acid and lipid biosynthetic genes in developing endosperm of *Jatropha curcas*. Biotechnol. Biofuels 5, 47.

Gupta, P., Idris, A., Mantri, S., Asif, M., Yadav, H., Roy, J., Tuli, R., Mohanty, C., Sawant, S., 2012. Discovery and use of single nucleotide polymorphic (SNP) markers in *Jatropha curcas* L. Mol. Breed. 30, 1325–1335.

He, W., King, A.J., Khan, M.A., Cuevas, J.A., Ramiaramanana, D., Graham, I.A., 2011. Analysis of seed phorbol-ester and curcin content together with genetic diversity in multiple provenances of *Jatropha curcas* L. from Madagascar and Mexico. Plant Physiol. Biochem. 49, 1183–1190.

Hirakawa, H., Tsuchimoto, S., Sakai, H., Nakayama, S., Fujishiro, T., Kishida, Y., Kohara, M., Watanabe, A., Yamada, M., Aizu, T., Toyoda, A., Fujiyama, A., Tabata, S., Fukui, K., Sato, S., 2012. Upgraded genomic information of *Jatropha curcas* L. Plant Biotechnol. 29, 123–130.

Li, M., Li, H., Jiang, H., Pan, X., Wu, G., 2008. Establishment of an Agrobacteriuim-mediated cotyledon disc transformation method for *Jatropha curcas*. Plant Cell Tissue Organ Cult. 92, 173–181.

Liu, P., Wang, C.M., Li, L., Sun, F., Yue, G.H., 2011. Mapping QTLs for oil traits and eQTLs for oleosin genes in jatropha. BMC Plant Biol. 11, 132.

Meher, L.C., Churamani, C.P., Arif, M., Ahmed, Z., Naik, S.N., 2013. *Jatropha curcas* as a renewable source for bio-fuels-A review. Renew. Sust. Energ. Rev. 26, 397–407.

Misra, A., Khan, K., Niranjan, A., Nath, P., Sane, V.A., 2013. Over-expression of *JcDGAT1* from *Jatropha curcas* increases seed oil levels and alters oil quality in transgenic *Arabidopsis thaliana*. Phytochemistry 96, 37–45.

Nakao, M., Hasegawa, G., Yasuhara, T., Ishihara, Y., 2015. Degradation of *Jatropha curcas* phorbol esters derived from Jatropha oil cake and their tumor-promoting activity. Ecotoxicol. Environ. Saf. 114, 357–364.

Pradhan, S., Naik, S.N., Khan, M.A.I., Sahoo, P.K., 2012. Experimental assessment of toxic phytochemicals in *Jatropha curcas*: oil, cake, bio-diesel and glycerol. J. Sci. Food Agric 92, 511–519.

Qu, J., Mao, H.-Z., Chen, W., Gao, S.-Q., Bai, Y.-N., Sun, Y.-W., Geng, Y.-F., Ye, J., 2012. Development of marker-free transgenic Jatropha plants with increased levels of seed oleic acid. Biotechnol. Biofuels 5.

Sanderson, K., 2009. Wonder weed plans fail to flourish. Nature 461, 328–329.

Sato, S., Hirakawa, H., Isobe, S., Fukai, E., Watanabe, A., Kato, M., Kawashima, K., Minami, C., Muraki, A., Nakazaki, N., Takahashi, C., Nakayama, S., Kishida, Y., Kohara, M., Yamada, M., Tsuruoka, H., Sasamoto, S., Tabata, S., Aizu, T., Toyoda, A., Shin-i, T., Minakuchi, Y., Kohara, Y., Fujiyama, A., Tsuchimoto, S., Kajiyama, S., Makigano, E., Ohmido, N., Shibagaki, N., Cartagena, J.A., Wada, N., Kohinata, T., Atefeh, A., Yuasa, S., Matsunaga, S., Fukui, K., 2011. Sequence analysis of the genome of an oil-bearing tree, *Jatropha curcas* L. DNA Res. 18, 65–76.

Singh, K., Verma, S.K., Patra, D.D., Singh, B., 2014. *Jatropha curcas*: a ten year story from hope to despair. Renew. Sust. Energ. Rev. 35, 356–360.

Wang, C.M., Liu, P., Yi, C., Gu, K., Sun, F., Li, L., Lo, L.C., Liu, X., Feng, F., Lin, G., Cao, S., Hong, Y., Yin, Z., Yue, G.H., 2011. A first generation microsatellite- and SNP-based linkage map of Jatropha. PLoS One 6, e23632.

Yang, C-y, Deng, X., Fang, Z., Peng, D.-P., 2010. Selection of high-oil-yield seed sources of *Jatropha curcas* L. for biodiesel production. Biofuels 1, 705–717.

Yang, C.-Y., Fang, Z., Li, B., Long, Y-f, 2012. Review and prospects of Jatropha biodiesel industry in China. Renew. Sust. Energ. Rev. 16, 2178–2190.

Yue, G.H., Sun, F., Liu, P., 2013. Status of molecular breeding for improving *Jatropha curcas* and biodiesel. Renew. Sust. Energ. Rev. 26, 332–343.

Chapter 11.5

Jojoba (*Simmondsia chinensis*)

Thomas A. McKeon
United States Department of Agriculture, Agricultural Research Service, Western Regional Research Center, Albany, CA, United States

Jojoba, *Simmondsia chinensis* (Link) C.K. Schneider, is a desert shrub native to the Sonoran Desert in north–central Mexico and the southwestern United States (Gentry, 1958). The seed of this plant has been used for centuries by native tribes who extracted the seed by grinding and boiling with water to obtain a waxy paste for treating wounds and burns (Kleiman and Dwyer, 2006). Interest in jojoba as a source of industrial materials grew when it was determined that instead of a triacylglycerol-based seed oil, the jojoba seed contained a liquid wax composed of very-long-chain fatty alcohols esterified to very-long-chain fatty acids (Fig. 11.5.1) (Green et al., 1936). The only other known biological source of liquid wax at that time was sperm whale oil, which was required for high-pressure lubricants and imported by the United States. With the inclusion of the whale on the endangered species list in the 1970s, imports of all whale products were forbidden and jojoba represented the best agricultural alternative to the 40–50 million pounds of sperm whale oil imported by the United States at that time (Wisniak, 1994). The seed contains from 45% to 55% liquid wax (Gunstone, 1990) and the composition of the esters are generally from 38 carbons up to 44 carbons, with minor amounts of shorter (C-36) and longer (C-46, C-48, C-50) esters (Table 11.5.1) (Miwa, 1984). Numerous industrial products have been proposed for jojoba oil, but production has never achieved the supply and cost effectiveness required for industrial utilization. As a result, the principal uses of jojoba oil have been in cosmetics and hair care products, where high cost is of lesser importance.

AGRONOMY AND PRODUCTION

The jojoba plant, *Simmondsia chinensis* (Link) Schneider, is a monotypic species of the family Simmondsiaceae in the order Caryophyllales (Crawley and Hilu, 2012). It is native to the Sonoran Desert in North America and, despite the name, jojoba is not native to China. The perennial evergreen shrub generally grows

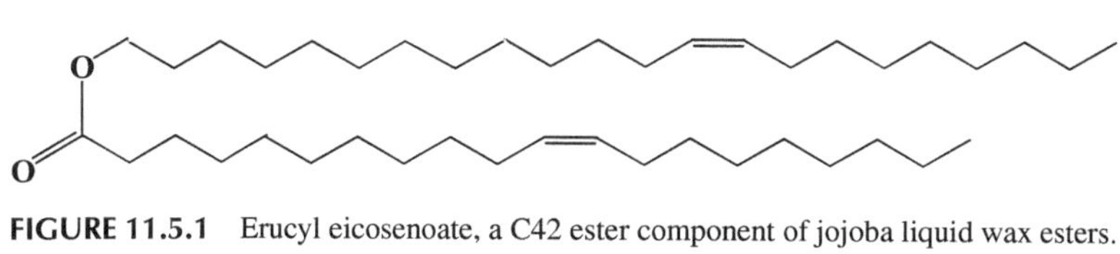

FIGURE 11.5.1 Erucyl eicosenoate, a C42 ester component of jojoba liquid wax esters.

TABLE 11.5.1 Distribution of Carbon Chain Lengths for Jojoba Oil Esters[a]

Ester Chain Length	Composition (%)
C34	0.1
C36	1.9
C38	6.8
C40	30.3
C42	50
C44	10
C46	0.8
C48	0.1
C50	0.02

[a]*Miwa (1984).*

to a height up to 5 m in semiarid conditions and can endure wide temperature fluctuations, although it is sensitive to temperatures below −6 to −17 °C, depending on water balance (Gentry, 1958; Wisniak, 1977). Jojoba is dioecious and the seed is produced from pollinated female ovules as a single seed enclosed in a (usually) dehiscent capsule. Although it is a desert plant, the jojoba seed requires water at critical times in development when it flowers in late winter/early spring and also early in seed development. The seed takes 4–5 months to mature from the pollinated ovule (Benzioni et al., 2007). After a stand of jojoba is well established, requiring 2–3 years, it is less dependent on water, growing well when receiving 30 cm rainfall per year on average and as little as 13 cm annually. As a perennial, it is not commercially productive until 5 years after planting and reaches maximum production after 10–15 years (Benzioni et al., 1999). Jojoba stands are productive for 100–200 years (Gentry, 1958).

Upon germination, the seed produces a taproot of 30–45 cm prior to emergence of the cotyledon, and the established plant has several taproots extending >3 m into the soil (Gentry, 1958). The plant grows well with limited agricultural inputs, and can withstand saline conditions (Benzioni et al., 1999). Jojoba is planted in semiarid locations in all six nonpolar continents, with area estimated to be 60,000 ha in 1990 (Gunstone, 1990). Planted area is believed to be higher in 2015, because plantings in India and the Middle East have expanded in recent years, although current reported yield does not reflect such widespread plantings (vide infra).

Jojoba is dioecious and open-pollinated, with both parental genotypes affecting wax yield (Benzioni and Vaknin, 2002). It is therefore important to control the parentage of a planting to ensure consistent wax production, quality, and yield. As a result, seed propagation of plants is not desirable, especially since male progeny exceed female progeny by factors up to fivefold (Sharma et al., 2008), and it takes 2 years after planting before the sex of the plant can be determined on the basis of flowering (Benzioni et al., 1999). Molecular methods to identify gender of seedlings have been developed. Inter-simple sequence repeats (ISSR) have been developed that distinguish male and female jojoba plants (Sharma et al., 2008). Additional markers have also been developed to enhance the reliability of male detection (Ince et al., 2010). While earlier approaches used rooting of known plants, methods for micropropagation of jojoba plantlets were later developed (Mills et al., 2009). Castor plantations are planted at a female-to-male ratio of 9:1 to maximize production from the females without diminishing pollination (Naqvi and Ting, 1990).

While there are insect, fungal, and even small animal pests associated with the jojoba plant (Gentry, 1958), these are generally less of a hindrance than local climate and availability of water in winter and their effect on wax yield. The expense of registering pesticides for what is still a minor crop does not make economic sense, and microbial damage does not appear to be significant (Naqvi and Ting, 1990).

The main emphases of jojoba breeding programs are yield, wax content, quality and composition. Jojoba has not yet lived up to what some scientists consider its potential ability to produce yields of up to 8 t of seeds/ha (Naqvi and Ting, 1990). Yields in experimental plots of >3 kg seed/plant have been reported (Benzioni et al., 1999), and plantings of 2000 plants/ha are considered reasonable (Naqvi and Ting, 1990); therefore, the forecasted yield appears to be achievable. Part of the reason that higher yields have not been achieved is that the plant is open-pollinated, and, with such a long time elapsing between planting and evaluation of long-term yield, traditional breeding approaches proceed slowly. However, characteristics such as third year height, flower index, relative fruit set, and seed weight are useful predictors of long-term yield (Benzioni et al., 1999). Global production of jojoba in 2014 was estimated at 7000 metric tons, yielding 3200 metric tons of cold pressed oil selling at $15–30/kg [personal communication, Jonathan Regev, Jojoba Desert (A.C.S) Ltd., D.N. Hanagev, Israel].

TOXICITY AND MEDICINAL USES

The seed contains the compound simmondsin [2-(cyanomethylene)-3-hydroxy-4,5-dimethoxycyclohexyl β-D-glucoside] as well as its ferulate ester (Elliger et al., 1974). Levels of 6–33% total simmondsins in defatted seed have been reported (Cappillino et al., 2003; Al-Soqeer et al., 2012). The presence of simmondsins makes the seed meal of jojoba toxic to animals, and analytical procedures for determining the level of simmondsin and its esters in the meal have been developed (Van Boven et al., 1996). Although considered to aid in appetite suppression according to folklore, chronic exposure to 0.5% simmondsin in the diet of Sprague–Dawley rats causes significant weight loss and death with some kidney, liver, and heart enlargement as well as severe anemia (York et al., 2000). The meal can be detoxified by physical, chemical, or fermentative means (Wisniak, 1994).

Jojoba wax has been used in folk medicine by native tribes indigenous to the Sonoran Desert area for centuries, first documented in written literature in 1701 by a Spanish missionary (Kleiman and Dwyer, 2006). The oil was used to treat wounds and burns, ease childbirth, and for skin and hair care, with the latter two applications serving as the major uses of jojoba oil in the present day. The oil contains antioxidants including tocopherols and phenolics that maintain stability for prolonged periods if stored sealed and protected from light and are lost upon bleaching or solvent stripping that will eliminate the antioxidants (Kampf et al., 1986). In one study with cold-pressed jojoba oil, it was found that topical application of jojoba oil was effective in reducing swelling caused by application of a skin irritant, and intravenous application jojoba oil reduced the levels of several biochemical indicators of inflammatory response (Habashy et al., 2005). Jojoba oil was shown to have very low toxicity in vitro to keratinocytes and dermal fibroblasts. In addition to stimulating signaling pathways involved in healing, it also stimulated production of collagen in fibroblasts, a critical factor in wound healing (Ranzato et al., 2011).

CHEMISTRY, BIOCHEMISTRY, AND MOLECULAR BIOLOGY

The waxy nature of jojoba was described in 1789 (Knoepfler and Vix, 1958). It is unique among widely known plants for storing lipid in the form of a liquid wax ester instead of triacylglycerol. The fatty acid and fatty alcohol components of the esters are almost exclusively ω-9 monounsaturates derived from oleate, with chain lengths of 18–22 and 20–24 carbons for the fatty acid and alcohol, respectively (Miwa, 1984). Table 11.5.1 provides an example of the distribution of overall chain length for wax esters in jojoba oil. Factors affecting the distributions of ester chain length include growing conditions and genotype of the male and female parental strains (Benzioni and Vaknin, 2002).

Because the goal for developing jojoba as a major crop has not been achieved, there is considerable interest in developing transgenic sources of an equivalent wax to meet industrial needs (Iven et al., 2015). Early approaches characterized biosynthesis of wax using a cell-free homogenate from developing jojoba seeds (Pollard et al., 1979). Several biochemical reactions relevant to wax biosynthesis were identified: elongation of long-chain acyl-CoA, reduction of long-chain acyl-CoA to the corresponding alcohol, and esterification of the long-chain acyl-CoA and long-chain alcohol to form a wax ester (Fig. 11.5.2) (Pollard et al., 1979). Both the long-chain fatty acyl-CoA reductase (Metz et al., 2000) and the wax synthase (Lardizabal et al., 2000) were purified from jojoba embryo microsomal membranes. Based on the

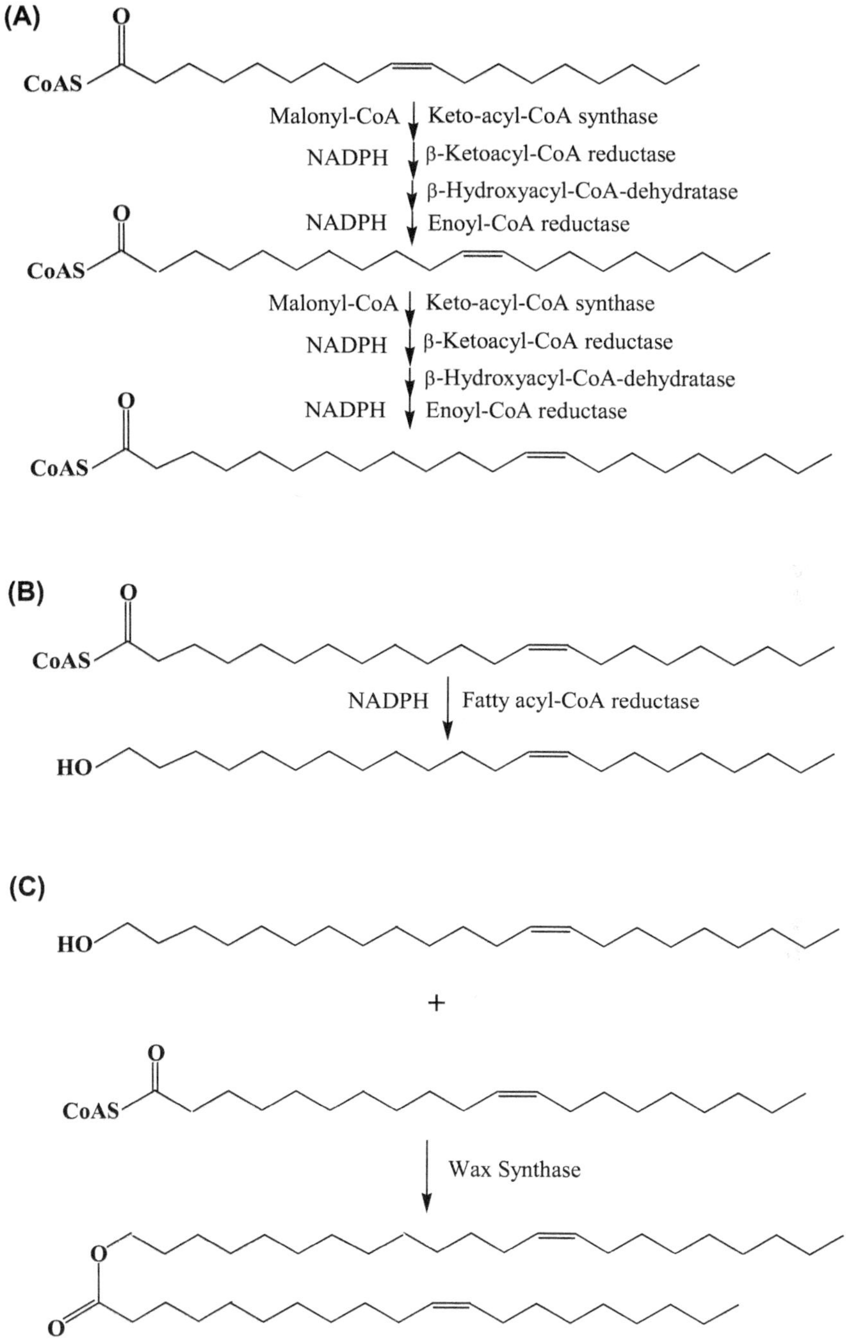

FIGURE 11.5.2 Jojoba wax biosynthesis: (A) Long-chain acyl-CoA elongation; (B) Long-chain acyl-CoA reductase; (C) Wax synthase.

N-terminal sequence of each protein, corresponding DNA primers were used to clone the cDNA for each enzyme (Metz et al., 2000; Lardizabal et al., 2000). These cDNAs, along with a cDNA for a component of a very-long-chain fatty acid elongase from *Lunaria annua,* were inserted in the model plant *Arabidopsis thaliana* and the seeds of plants derived from the transformants contained up to 70% of the seed oil as wax ester (Lardizabal et al., 2000). The cited study generated great interest in producing wax esters transgenically. Both rapeseed (see chapter: *Brassica* spp. Oils) and crambe (see chapter: Crambe (*Crambe abyssinica*)) produce oils containing >50% erucic acid; therefore, the two well-studied industrial oil–bearing crops, both of which reside in temperate climates, are logical choices to serve as hosts for expressing the reductase and wax synthase, for production of a liquid wax similar to jojoba oil.

INDUSTRIAL APPLICATIONS

Since the late 1700s, when jojoba oil was identified as a source of liquid wax similar in properties to sperm whale oil, numerous applications have been proposed for the oil. Jojoba oil production has never achieved the levels needed to match the demand for sperm whale oil, for example, 18–22 million kg imported annually by the United States until the 1970s (Wisniak, 1994). At the present time, nearly all the jojoba oil produced is used in cosmetic and hair treatments, niche applications in which it can command a high price for perceived health and cosmetic benefits (Kleiman and Dwyer, 2006).

As described, jojoba-like oils can potentially be produced in larger volumes from transgenic oil–bearing plants such as rapeseed and crambe, thereby allowing for the oil's potential use in many industrial applications that traditionally used sperm whale oil. The principal use for sperm whale wax was production of factice, a high-pressure lubricant additive derived from sulfurization of the wax. Compared to the whale oil factice, factice derived from jojoba oil had similar or superior performance characteristics as "an extreme-pressure additive in motor oils, gear lubricants and automatic transmission fluids" (Miwa and Rothfus, 1979). Another use for jojoba oil is as a solid wax. Hydrogenation of jojoba gives a hard wax with a melting point of 67°C, intermediate between beeswax and carnauba wax, important waxes for production of candles, coatings for fruit, wood coatings, and polish (Wisniak, 1994).

As an oxidatively stable lipid, jojoba oil could help to meet the demand for biodiesel fuel. Jojoba oil itself or in a 50:50 blend with petroleum was effective as diesel fuel under appropriate conditions (Al-Widyan and Al-Muhtaseb, 2010). The methanolysis of jojoba oil was optimized at 25°C using KOH as catalyst, resulting in a yield of 83.5% methyl ester from starting material. As a mixture of mostly C-20 and C-22 monounsaturated methyl esters, these would be well suited for biodiesel (Bouaid et al., 2007). Jojoba oil can also generate fuel components from thermal cracking, and there is an advantage to using jojoba oil versus triacylglycerols (TGs) in thermal cracking—jojoba oil does not generate the reactive small molecules and radicals that form from glycerol when cracking TG (Kozliak et al., 2013). Principal identified products obtained upon heating at 450°C in N_2 under pressure were short- to medium-chain length saturated hydrocarbons, fatty acids, and alkenes, although >50% of the products were unidentified. Cross-metathesis of jojoba oil with ruthenium-based Grubbs catalysts (see Polymers section of chapter: Castor (*Ricinus communis* L.)) also produced hydrocarbons for biofuel as well as oligoesters (Butilkov and Lemcoff, 2014). While each of these approaches to using jojoba oil as a fuel has potential utility, given the high price of jojoba oil compared to fossil fuels, the demand for jojoba oil by the cosmetic and hair care industry, and the oil's limited supply, the idea of using jojoba oil to generate fuel is currently impractical.

SUMMARY

Jojoba wax is a unique natural plant product that has numerous uses in both consumer and industrial products. It has high yield potential and is adapted to semiarid conditions, providing a useful crop for areas not otherwise well suited for agricultural production.

REFERENCES

Al-Soqeer, A., Motawei, M.I., Al-Dakhil, M., El-Mergawi, R., Al-Khalifah, N., 2012. Genetic variation and chemical traits of selected new jojoba (*Simmondsia chinensis* (Link) Schneider) genotypes. J. Am. Oil Chem. Soc.

Al-Widyan, M.I., Al-Muhtaseb, M.A., 2010. Experimental investigation of jojoba as a renewable energy source. Energ. Conv. Manage. 51, 1702–1707.

Benzioni, A., Shiloh, E., Ventura, M., 1999. Yield parameters in young jojoba plants and their relation to actual yield in later years. Ind. Crops Prod. 10, 85–95.

Benzioni, A., Vaknin, Y., 2002. Effect of female and male genotypes and environment on wax composition oin jojoba. J. Am. Oil Chem. Soc. 79, 297–302.

Benzioni, A., Van Boven, M., Ramamoorthy, S., Mills, D., 2007. Dynamics of fruit growth, accumulation of wax esters, simmondsins, proteins and carbohydrates in jojoba. Ind. Crops Prod. 26, 337–344.

Bouaid, A., Bajo, L., Martinez, M., Aracil, J., 2007. Optimization of biodiesel production from jojoba oil. Process Safe Env. Protec. 85, 378–382.

Butilikov, D., Lemcoff, N.G., 2014. Jojoba olefin methathesis: a valuable source for bio-renewable materials. Green Chem. 16, 4728–4733.

Cappillino, P., Kleiman, R., Botti, C., 2003. Composition of Chilean jojoba seeds. Ind. Crops Prod. 17, 177–182.

Crawley, S.S., Hilu, K.W., 2012. Caryophyllales: evaluating phylogenetic signal in trnK intron versus matK. J. Syst. Evol. 50, 387–410.

Elliger, C.A., Waiss, A.C., Lundin, R.E., 1974. Cyanomethylenecyclohexyl glucosides from *Simmondsia californica*. Phytochemistry 13, 2319–2320.

Gentry, H.P., 1958. The natural history of Jojoba (*Simmondsia chinensis*) and its cultural aspects. Econ. Bot. 12, 261–295.

Green, T.G., Hilditch, T.P., Stainsby, W.J., 1936. The seed wax of *Simmondsia californica*. J. Chem. Soc. 1750–1755.

Gunstone, F.D., 1990. Jojoba oil. Endeavour. New Ser. 14, 40–43.

Habashy, R.R., Abdel-Naim, A.B., Khalifa, A.E., Al-Azizi, M.M., 2005. Anti-inflammatory effects of jojoba liquid wax in experimental models. Pharmacol. Res. 51, 95–105.

Ince, A.G., Karaca, M., Onus, A.N., 2010. A reliable gender diagnostic PCR assay for jojoba (*Simmondsia chinensis* (Link) Schneider). Genet. Resour. Crop Evol. 57, 773–779.

Iven, T., Hornung, E., Heilmann, M., Feussner, I., 2015. Synthesis of oleyl oleate wax esters in *Arabidopsis thaliana* and *Camelina sativa* seed oil. Plant Biotech. J. http://dx.doi.org/10.1111/pbi.12379.

Kampf, A., Grinberg, S., Galun, A., 1986. The oxidative stability of jojoba wax. J. Am. Oil Chem. Soc. 63, 246–248.

Kleiman, R., Dwyer, K., 2006. A natural alternative. INFORM 17, 346–347.

Knoepfler, N.B., Vix, H.L.E., 1958. Review of chemistry and research potential of *Simmondsia chinensis* (Jojoba) oil. J. Agric. Food Chem. 6, 118–121.

Kozliak, E., et al., 2013. Non-catalytic cracking of jojoba oil to produce fuel and chemical by-products. Ind. Crops Prod. 43, 386–392.

Lardizabal, K.D., Metz, J.G., Sakamoto, T., Hutton, W.C., Pollard, M.R., Lassner, M.W., 2000. Purification of a jojoba embryo wax synthase, cloning of its cDNA, and production of high levels of wax in seeds of transgenic Arabidopsis. Plant Physiol. 122, 645–655.

Metz, J.G., Pollard, M.R., Anderson, L., Hayes, T.R., Lassner, M.W., 2000. Purification of a jojoba embryo fatty acyl-coenzyme A reductase and expression of its cDNA in high erucic acid rapeseed. Plant Physiol. 122, 635–644.

Mills, D., Yangqing, Z., Benzioni, A., 2009. Effect of substrate, medium composition, irradiance and ventilation on jojoba plantlets at the rooting stage of micropropagation. Sci. Hortic. 121, 113–118.

Miwa, T.K., Rothfus, J.A., 1979. Extreme-pressure lubricant tests on jojoba and sperm whale oils. J. Am. Oil Chem. Soc. 56, 765–770.

Miwa, T.K., 1984. Structural determination and uses of jojoba oil. J. Am. Oil Chem. Soc. 61, 407–410.

Naqvi, H.H., Ting, I.P., 1990. Jojoba: a unique liquid wax producer from the American desert. In: Janick, J., Simon, J.E. (Eds.), Advances in New Crops. Timber Press, Portland, OR, USA, pp. 247–251 (article updated in 1997) https://www.hort.purdue.edu/newcrop/proceedings1990/V1-247.html.

Pollard, M.R., McKeon, T., Gupta, L.M., Stumpf, P.K., 1979. Studies on biosynthesis of waxes by developing jojoba seed II. The demonstration of wax biosynthesis by cell-free homogenates. Lipids 14, 651–662.

Ranzato, E., Martinotti, S., Burlando, B., 2011. Wound healing properties of jojoba liquid wax: an in vitro study. J. Ethnopharm. 134, 443–449.

Sharma, K., Agrawal, V., Gupta, S., Kumar, R., Prasad, M., 2008. ISSR marker-assisted selection of male and female plants in a promising dioecious crop jojoba (*Simmondsia chinensis*). Plant Biotechnol. Rep. 2, 239–243.

Van Boven, M., Daenens, P., Tytgat, J., 1996. Determination of simmondsins and simmondsin ferulates in jojoba meal and feed by high performance chromatography. J. Agric. Food Chem. 44, 2239–2243.

Wisniak, J., 1977. Jojoba oil and its derivatives. Prog. Chem. Fats Lipids 15, 167–218.

Wisniak, J., 1994. Potential uses of jojoba oil and meal – a review. Ind. Crops Prod. 3, 43–68.

York, D.A., Singer, L., Oliver, J., Abbott, T.P., Bray, G.A., 2000. The detrimental effect of simmondsin on food intake and body weight of rats. Ind. Crops Prod. 12, 183–192.

Chapter 11.6

Lesquerella (*Physaria* spp.)

Guanqun Chen
Department of Biological Sciences, University of Manitoba, Winnipeg, Manitoba, Canada

INTRODUCTION

Hydroxylated fatty acids (HFAs) have valuable applications in the production of industrial materials (for more information, see chapter: Castor (*Ricinus communis* L.)). Currently, the only commercially available HFA is ricinoleic acid (C18:1-OH) from castor bean (*Ricinus communis*). Large-scale agricultural production of castor bean, however, is limited by the presence of the highly toxic protein ricin in the seeds (Lee et al., 2015). *Physaria* species (also known as *Lesquerella* species) can accumulate high amounts of HFA, including lesquerolic acid (C20:1-OH), in seeds, and are considered as a safe source of HFA (Fig. 11.6.1A). Lesquerolic acid is two carbons longer than ricinoleic acid, and lesquerolic oil has more favorable fuel properties than castor oil (Goodrum and Geller, 2005; Knothe et al., 2012). In *Physaria fendleri*, lesquerolic acid is synthesized from ricinoleic acid (Fig. 11.6.1B) (Reed et al., 1997). Briefly, oleic acid (C18:1) at *sn*-2 position of phosphatidylcholine (PC) is hydroxylated by a Δ12-hydroxylase (FAH12) to produce ricinoleic acid (C18:1-OH). Ricinoleic acid is then released from PC and elongated by a fatty acid–condensing enzyme (KCS) to synthesize lesquerolic acid (C20:1-OH).

Physaria belongs to the mustard family (Brassicaceae). These plants can grow well in areas with 250–400 mm of rainfall and thus are ideal for the semi-arid regions of North America (Dierig et al., 2011). Among *Physaria* species, *P. fendleri* produces 24–36% of oil in seeds with the most agronomically important characteristics (Dierig and Ray, 2009; Dierig et al., 2011). In addition to the high content of HFA in seed oil, *Physaria* seed meal has high lysine content and is

(A)

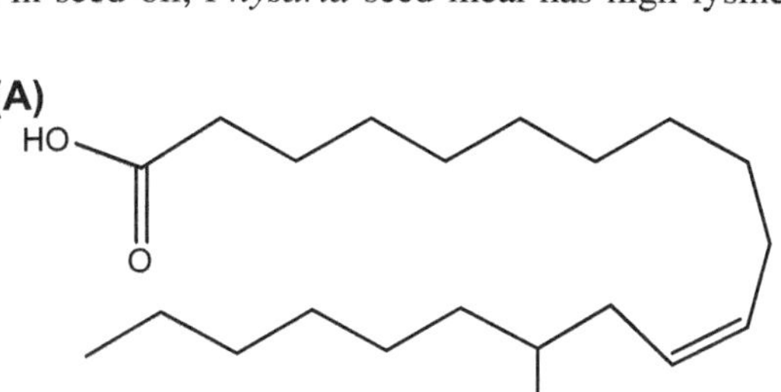

(B)

FIGURE 11.6.1 (A) Structure and (B) biosynthesis of lesquerolic acid (C20:1-OH). *FAH12*, Δ12 hydroxylase; *KCS*, fatty acid-condensing enzyme.

suitable for livestock and poultry feed (Carlson et al., 1990). Although *P. fendleri* is a short-lived perennial, it can tolerate freezing temperatures and thus can be planted in fall and harvest in June throughout the southwestern United States similar to other small grain crops. Moreover, farm equipment generally requires only slight modification for cultivating *Physaria*. *Physaria* will not compete with current commodity crops but can be placed in rotation with them (Dierig et al., 2011).

PHYSARIA CROP RESEARCH

Physaria is a genus native to the Americas, with the greatest concentration of taxa identified in the southwestern United States. More than 230 *Physaria* accessions are publicly available in the National Plant Germplasm System (NPGS) and over 10,000 germplasm lines are recorded in the USDA database, providing a rich genetic resource for *Physaria* crop research. Because *Physaria* has valuable oil composition and high crop potential but no real biological barriers for domestication, it has attracted researchers' interests to start breeding programs as early as 1950s (Dierig et al., 2011).

Several significant advances have been achieved through *Physaria* breeding. Oil content of *P. fendleri* has been improved from 24% to >30%. Several factors including soil temperature, moisture, depth of planting, planting dates, planting methods, and nitrogen fertilizer have been studied for their contribution to seed yields (Cruz et al., 2012, 2013a,b, 2014; Dierig and Crafts-Brandner, 2011; Dierig et al., 2011, 2012; Liu et al., 2014; Pastor–Pastor et al., 2015; Windauer et al., 2013). The current seed yield is ~1800 kg/ha. The evaluation of the *Physaria* species in NPGS including seed oil characteristics, plant architecture, autofertility, etc. also provided diverse traits for breeding programs. For instance, seeds of 195 accessions (32 species) harvested in various years of germplasm regeneration (1995–2004) were systemically evaluated for fatty acid profiles (Jenderek et al., 2009). The results indicated that HFA content was highly variable among *Physaria* species. The highest HFA content of most *P. fendleri* is ~67%, which is consistent with the fact that only *sn*-1 and *sn*-3 of triglyceride are occupied with HFA (Hayes et al., 1995; Jenderek et al., 2009). A few lines of *P. pallida* and *P. lindheiimeri* have HFA content >80%, which indicated that the *sn*-2 position of triglycerides is also filled with HFA in these lines. Interspecific hybrids between *P. fendleri* and *P. pallida* or *P. lindheiimeri* improved lesquerolic acid content to >75% (Dierig et al., 2004). Interesting genes specifically involved in placing HFA at the *sn*-2 position of triglyceride may be identified from *P. pallida* and *P. lindheiimeri* and could be used to generate transgenic *P. fendleri* lines with higher HFA content via metabolic engineering approach.

Physaria breeding is still ongoing, focusing on the improvement of several traits required for commercialization. These efforts include improving lesquerolic acid and total seed oil contents. The average oil content of a *P. fendleri* is in the range of 24–35% and a *P. fendleri* line with 45% of seed oil content was reported in one study (Dierig and Ray, 2009). Although the lesquerolic acid content of 67% appears to be the upper limit in *P. fendleri*, likely limited by HFA only being installed at the *sn*-1 and *sn*-3 position of triacylglycerol, it can be further improved by enabling lesquerolic acid to occupy the *sn*-2 position of triacylglycerol, either via traditional hybridization or genetic engineering. A line with >80% lesquerolic acid in seed oil would satisfy most market demands currently met by castor bean. Moreover, *P. fendleri* has a potential of yielding 2500–3000 kg/ha. This seed yield potential may be achieved through breeding based on harvest index and for specific environments and through improved agronomic practices such as good plant spacing, fine irrigation management, and efficient harvest (Dierig and Ray, 2009). Furthermore, the development of autofertile lines would be attractive as it can eliminate the requirement of pollinators to increase seed yield (Dierig and Ray, 2009). A few *Physaria* species are autofertile, and some hybrid *P. fendleri* lines with autofertility have been generated for further investigation (Dierig et al., 2011). Development of traits with herbicide resistance via genetic or hybrid approaches would also significantly improve weed management and reduce production costs.

The rapid development in the fields of molecular biology and metabolic engineering has contributed to *Physaria* crop research. New biotechnologies have been developed for *Physaria* breeding. A plastid transformation protocol in *P. fendleri* was reported a decade ago. In that study, transplastomic *Physaria* plants were fertile and produced seed (Skarjinskaia et al., 2003). Recently, an effective transformation system was established for the stable genetic transformation of *P. fendleri* (Chen, 2011). A *P. fendleri* seed transcriptome has been established for discovering genes involved in the synthesis of triacylglycerols, which provided a useful resource for the biotechnological production of HFA (Kim and Chen, 2015).

Genes from *Physaria* have been used to engineering other plants to produce oils containing high levels of HFA (Lee et al., 2015). For instance, the overexpression of *PfFAH12* in *Arabidopsis* and canola resulted in 16.5% and 9.9% of HFA in their seed oil, respectively (Broun et al., 1998). When a *PfKCS* gene and a castor *FAH12* gene were co-expressed in *Camelina*, the seeds accumulated 20% of total fatty acids as HFA, whereas the overexpression of castor *FAH12* only resulted in an HFA production of 15% (Snapp et al., 2014). Interestingly, in *Camelina* lines harboring both *PfKCS* and *RcFAH12*, but not the ones harboring only *RcFAH12*, the oil content and germination rate were restored to levels similar to those of nontransgenic lines, which was valuable from a breeding perspective.

In summary, *Physaria* is a potential new candidate crop for the production of HFA. Considerable breeding efforts have been made and significant progress has been accomplished in the domestication of *Physaria*. Additional breeding endeavors with the combination of agronomics, genetic engineering, and breeding approaches could make this species an additional HFA source along with castor.

REFERENCES

Broun, P., Boddupalli, S., Somerville, C., 1998. A bifunctional oleate 12-hydroxylase: desaturase from *Lesquerella fendleri*. Plant J. 13, 201–210.

Carlson, K., Chaudhry, A., Bagby, M., 1990. Analysis of oil and meal from *Lesquerella fendleri* seed. J. Am. Oil Chem. Soc. 67, 438–442.

Chen, G.Q., 2011. Effective reduction of chimeric tissue in transgenics for the stable genetic transformation of *Lesquerella fendleri*. Hort Sci. 46, 86–90.

Cruz, V.M.V., Comas, L.H., Dierig, D.A., 2014. Root phenotypic characterization of lesquerella genetic resources. Ind. Crop Prod. 62, 130–139.

Cruz, V.M.V., Kilian, A., Dierig, D.A., 2013a. Development of DArT marker platforms and genetic diversity assessment of the US collection of the new oilseed crop *Lesquerella* and related species. Plos One 8.

Cruz, V.M.V., Romano, G., Dierig, D.A., 2012. Effects of after-ripening and storage regimens on seed-germination behavior of seven species of *Physaria*. Ind. Crop Prod. 35, 185–191.

Cruz, V.M.V., Walters, C.T., Dierig, D.A., 2013b. Dormancy and after-ripening response of seeds from natural populations and conserved *Physaria* (syn. *Lesquerella*) germplasm and their association with environmental and plant parameters. Ind. Crop Prod. 45, 191–199.

Dierig, D., Ray, D.T., 2009. New crops breeding: *Lesquerella*. In: Vollmann, J., Rajcan, I. (Eds.), Oil Crops, pp. 507–516.

Dierig, D.A., Crafts-Brandner, S.J., 2011. The relationship of temperature to plant morphology of *Lesquerella*. Crop Sci. 51, 2165–2173.

Dierig, D.A., Tomasi, P.M., Salywon, A.M., Ray, D.T., 2004. Improvement in hydroxy fatty acid seed oil content and other traits from interspecific hybrids of three *Lesquerella* species: *Lesquerella fendleri, L. pallida*, and *L. lindheimeri*. Euphytica 139, 199–206.

Dierig, D.A., Wang, G., McCloskey, W.B., Thorp, K.R., Isbell, T.A., Ray, D.T., Foster, M.A., 2011. *Lesquerella*: new crop development and commercialization in the US. Ind. Crop Prod. 34, 1381–1385.

Dierig, D.A., Wang, G.S., Crafts-Brandner, S.J., 2012. Dynamics of reproductive growth of lesquerella (*Physaria fendleri*) over different planting dates. Ind. Crop Prod. 35, 146–153.

Goodrum, J.W., Geller, D.P., 2005. Influence of fatty acid methyl esters from hydroxylated vegetable oils on diesel fuel lubricity. Bioresour. Technol. 96, 851–855.

Hayes, D.G., Kleiman, R., Phillips, B.S., 1995. The triglyceride composition, structure, and presence of estolides in the oils of *Lesquerella* and related species. J. Am. Oil Chem. Soc. 72, 559–569.

Jenderek, M.M., Dierig, D.A., Isbell, T.A., 2009. Fatty-acid profile of *Lesquerella* germplasm in the National Plant Germplasm System collection. Ind. Crop Prod. 29, 154–164.

Kim, H.U., Chen, G.Q., 2015. Identification of hydroxy fatty acid and triacylglycerol metabolism-related genes in lesquerella through seed transcriptome analysis. BMC Genom. 16.

Knothe, G., Cermak, S.C., Evangelista, R.L., 2012. Methyl esters from vegetable oils with hydroxy fatty acids: comparison of lesquerella and castor methyl esters. Fuel 96, 535–540.

Lee, K.-R., Chen, G.Q., Kim, H.U., 2015. Current progress towards the metabolic engineering of plant seed oil for hydroxy fatty acids production. Plant Cell Rep. 34, 603–615.

Liu, J., Bronson, K.F., Thorp, K.R., Mon, J., Badaruddin, M., McCloskey, W.B., Ray, D.T., Chu, Q., Wang, G., 2014. Lesquerella seed and oil yield response to split-applied N fertilizer. Ind. Crop Prod. 60, 273–279.

Pastor-Pastor, A., Gonzalez-Paleo, L., Vilela, A., Ravetta, D., 2015. Age-related changes in nitrogen resorption and use efficiency in the perennial new crop *Physaria mendocina* (Brassicaceae). Ind. Crop Prod. 65, 227–232.

Reed, D.W., Taylor, D.C., Covello, P.S., 1997. Metabolism of hydroxy fatty acids in developing seeds in the Genera *Lesquerella* (Brassicaceae) and *Linum* (Linaceae). Plant Physiol. 114, 63–68.

Skarjinskaia, M., Svab, Z., Maliga, P., 2003. Plastid transformation in *Lesquerella fendleri*, an oilseed Brassicacea. Transgenic Res. 12, 115–122.

Snapp, A.R., Kang, J., Qi, X., Lu, C., 2014. A fatty acid condensing enzyme from *Physaria fendleri* increases hydroxy fatty acid accumulation in transgenic oilseeds of *Camelina sativa*. Planta 240, 599–610.

Windauer, L.B., Ploschuk, E.L., Benech-Arnold, R.L., 2013. The growth rate modulates time to first bud appearance in *Physaria mendocina*. Ind. Crop Prod. 49, 188–195.

Chapter 11.7

Meadowfoam (*Limnanthes alba*)

Thomas A. McKeon
United States Department of Agriculture, Agricultural Research Service, Western Regional Research Center, Albany, CA, United States

INTRODUCTION

Meadowfoam (*Limnanthes alba* Hartweg ex. Benth) is a short herbaceous plant in the Limnanthaceae family adapted to growth in marshes and a cool climate (Steiner et al., 2006) that produces seed containing oil composed of >95% C20 and C22 unsaturated fatty acids, mostly monounsaturated (Miller et al., 1964). It is native to northern California, southern Oregon, and western Canada. In North America, it is well adapted to winter/spring production in northern California, Southern Oregon, Maryland, and the Pacific coast of Canada but only as a summer crop in Alaska (Higgins et al., 1971; Jenderek and Hannan, 2009). Since *Limnanthes douglasii* was adopted in northwestern Europe as an ornamental plant, that is also a suitable location for meadowfoam. The plant grows well between 4°C and 16°C, although it is sensitive to prolonged exposure to temperatures <5°C (Jenderek and Hannan, 2009).

Interest in meadowfoam arose as part of a program to identify new, industrially useful crops for the U.S. farm economy (Miller et al., 1964; Higgins et al., 1971). Meadowfoam seed oil was found to have >90% of its fatty acids with a chain length of C20 or C22 (Miller et al., 1964) (Table 11.7.1). The main components of the *Limnanthes alba* oil were identified as *cis*-5-eicosenoic

TABLE 11.7.1 Fatty Acid Composition of Meadowfoam

	Fatty Acid Content (%)							
Cultivar	<C_{18}	C_{18}	C20:0	C20:1 Δ^5	C22:1 Δ^5	C22:1 Δ^{13}	C22:2 $\Delta^{5,13}$	>C20
L. alba,[a] cv. alba	1	2.5	0.5	62.5	2.5	12	18	0.5
L. alba[b,c] cv. Benth	0.7	1.7	0.7	61		15[b,c]	20	0.6
L. alba[b,c] cv. versicolor	0.4	1.3	0	60		28[b,c]	10	0.3

[a]*Purdy and Craig (1987).*
[b]*Miller et al. (1964).*
[c]*Includes some cis-5.*

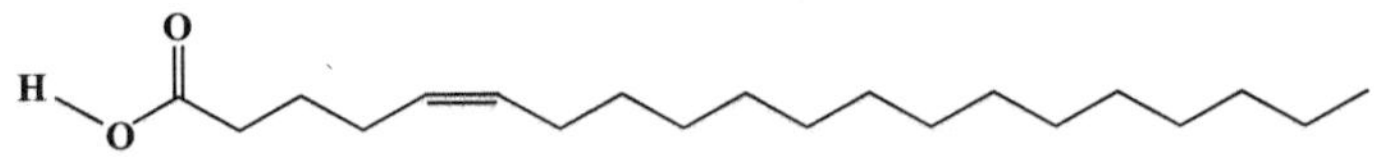

FIGURE 11.7.1 *cis*-5 Eicosenoic acid.

(Fig. 11.7.1), *cis*-5-docosenoic, and *cis*-5,13-docosadienoic acids as well as some erucic (*cis*-13-docosenoic) acid (Miller et al., 1964). The combination of high monounsaturate content (>75%) and chain length of C20 or C22 imparted high oxidative stability, as well as potential for use as a lubricant, plasticizer, or wax feedstock (Miller et al., 1964; Higgins et al., 1971). The unique fatty acid composition and resulting potential for industrial applications have garnered interest in meadowfoam as a crop in government-supported research programs in the United States and the European Union.

AGRONOMY AND PRODUCTION

Originally, chemists were interested in *L. douglasii*, an ornamental flowering plant widely grown in northwestern Europe and North America, that produces a seed oil containing >95% C20 and C22 fatty acids for a number of potential industrial applications (Smith et al., 1960). Seeds collected from seven species of *Limnanthes* were analyzed for oil content and fatty acid composition with all containing at least 70% total eicosenoic and docosenoic acid in the oil (Miller et al., 1964). *Limnanthes alba* was identified as being the best choice to serve as a meadowfoam crop, because it had a compact, erect habit, better seed retention, and a lower water requirement (Gentry and Miller, 1965; Higgins et al., 1971). The authors found seed yield was highest for *L. alba*, reaching 1550 kg/ha, but also that the yield was inconsistent and highly dependent on planting date, with spring planting in temperate climates producing very poor yields of seed, due to warm temperatures stopping vegetative growth prematurely (Higgins et al., 1971). *Limnanthes* spp. seed germinates optimally at temperatures from 5°C to 16°C, although there is a lag in germination at the lower temperatures (Toy and Willingham, 1966). Exposure to temperatures >16°C induces secondary dormancy (Toy and Willingham, 1967), a problem associated with early autumn or late spring planting. Some progress has been made in developing varieties and identifying growing conditions to reduce the impact of secondary dormancy (Joliff et al., 1994). While most meadowfoam cultivars are open-pollinated, some self-pollinated lines have been identified by direct selection from open-pollinated populations grown under reduced pollinator access (Knapp and Crane, 1997).

The erucic acid content of meadowfoam wild-type cultivars ranges from 8 to 24% (Gandhi et al., 2009). While oil containing erucic acid >3% is considered unfit for human consumption, the cosmetic and industrial applications of meadowfoam oil would most likely supersede its use for food. Nevertheless, chemical uses of meadowfoam oil, described later in this chapter, would benefit from more uniform Δ5-enoate composition and a reduced amount of the Δ5,13-dienoate content. A mutagenized population of meadowfoam cultivar Mermaid (9.7% erucic acid content) was screened for low erucic acid production, self-pollinated, and a fourth generation (M_4) of screened seeds was generated. Although the oil from seed of the selected M_4 plant produced only 3.2% erucic acid, the content of the *cis*-5,13-eicosadienoate increased from 19% to 33.6% (Gandhi et al., 2009). Achieving a high and more uniform content of *cis*-5-enoate remains a goal of meadowfoam breeding efforts. Hayes and Kleiman (1993) demonstrated that the selectivity of lipases could be used to isolate *cis*-5-enoate and *cis*-5,13-dienoate from erucic acid. They found that the *cis*-5 fatty acyl groups served as poor substrates for lipases relative to the erucyl, oleoyl, and linoleoyl acyl groups present in meadowfoam oil, presumably due to steric hindrance of the *cis*-5 acyl groups for penetration into the enzymes' active sites. The cited study demonstrated *Chromobacterium viscosum* lipase-catalyzed esterification of 1-butanol and free fatty acids generated from *Limnanthes alba* oil selectively esterified erucic, oleic, and linoleic acid, yielding a free fatty acid fraction containing >95% of the *cis*-5 acyl groups at >99% purity.

Meadowfoam is currently produced on 4000 acres (1600 ha) in Oregon, with yields of 1100–1500 kg seed/ha in commercial fields (Burden, 2012). This is an improvement over the 850 kg seed/ha described in 1990 (Oelke et al., 1990), although far from the 2300 kg seed/ha observed in some field trials (Oelke et al., 1990). Oil yields range from 27% to 35% (Purdy and Craig, 1987). Optimal solvent extraction of oil from flaked seed was obtained from very thin flakes heated to 94°C (Carlson et al., 1998). The heating of the seeds was necessary to inactivate endogenous enzymes that would release isothiocyanate compounds from glucosinolates into the oil. Following this procedure, the authors obtained a yield of 29.8% oil, approximately 95% yield of total oil contained in the seeds (Carlson et al., 1998).

CHEMISTRY, BIOCHEMISTRY, AND MOLECULAR BIOLOGY

The main fatty acid components of *L. douglasii* had been previously identified as *cis*-5-eicosenoic and *cis*-5-docosenoic acids as well as *cis*-5,13-docosadienoic and erucic acids (Smith et al., 1960; Bagby et al., 1961). Miller et al. (1964) confirmed that these fatty acids were present in a collection of seven *Limnanthes* spp. including seeds from wild isolates and some cultivated varieties, with oil content ranging from 20% to 33% and total C20 and C22 monoenoic acids ranging from 70% to 88% (Miller et al., 1964). Since the proportion of fatty acids in meadowfoam oil shorter than C20 is <3–5%, most of the triacylglycerol contains two C20 and one C22 or one C20 and two C22 acyl groups (Nikolova-Damyanova et al., 1990).

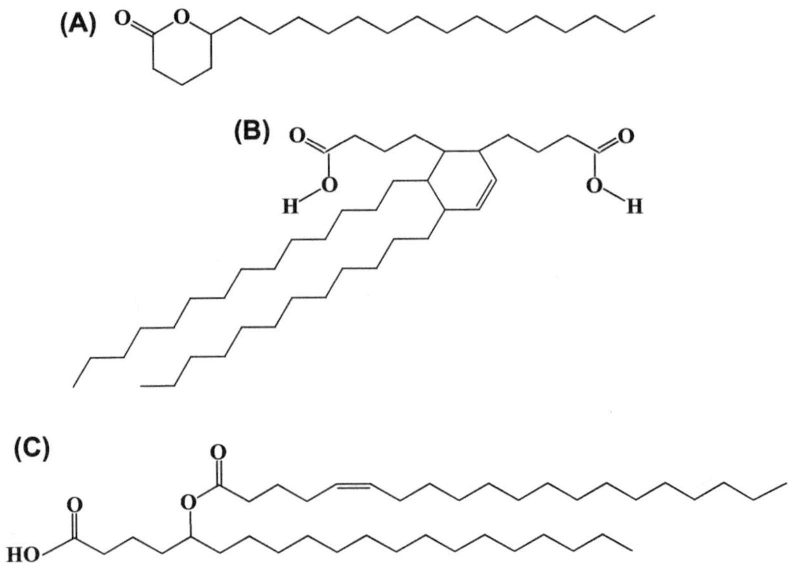

FIGURE 11.7.2 Chemicals derived from *cis*-5 eicosenoic acid: (A) δ-Lactone; (B) dimer acid; (C) estolide.

The position of the double bond in *cis*-5-eicosenoic and *cis*-5-docosenoic acids imparts some interesting chemical reactivity. In strong acid the Δ*cis*-5 mono- and di-unsaturated fatty acids are converted to the γ- or δ-lactone (Fig. 11.7.2a), which are readily converted to the 4-hydroxy or 5-hydroxy fatty acid, respectively (Isbell and Plattner, 1997). Conditions that maximized the production of the reactive δ-lactone were determined as well. Heating meadowfoam fatty acids in the presence of clay and water under pressure produces dimer acids (Burg and Kleiman, 1991). These are long chain dibasic and compounds composed of one six-membered ring system formed by the Diels–Alder reaction (Lee et al., 2015; described in chapter: Polymeric Products Derived from Industrial Oils for Paints, Coatings, and Other Applications). A reaction product derived from dimerization of *cis*-5-eicosenoic acid is depicted in Fig. 11.7.2b. The reaction conditions can also lead to formation of trimer acids (Burg and Kleiman, 1991). The dimer and trimer acids are of use in production of lubricants, adhesives, resins, and nontoxic plasticizers (Lee et al., 2015). By altering conditions in the reaction, such as addition of enough water to keep some in the liquid phase, monoestolides (Fig. 11.7.2c) can be formed with little or no dimer acid formation (Erhan et al., 1993). Estolides are useful as lubricant additives for viscosity adjustment, plasticizers, and in cosmetics (described in chapter: Polymeric Products Derived from Industrial Oils for Paints, Coatings, and Other Applications).

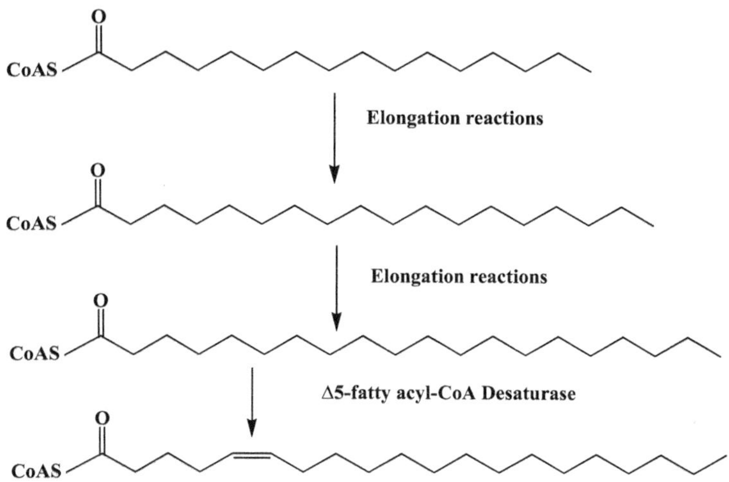

FIGURE 11.7.3 Pathway for biosynthesis of *cis*-5-eicosenoyl-CoA.

The production of the unusual fatty acids containing a double bond at the carbon 5 position of the long acyl chain results from the action of a *cis*-5-fatty acyl-CoA desaturase, an enzyme identified in cell-free extracts derived from cotyledons of meadowfoam (Moreau et al., 1981). In an attempt to further elucidate the biosynthesis of these fatty acids, DNA copies (cDNA) were prepared from RNA expressed in cotyledons during seed development. Over 1100 cDNA copies were cloned and sequenced and the sequence assembly searched for homology to DNA sequences encoding fatty acid desaturases and elongation enzymes. Homologues to fatty acyl-CoA desaturases found in non-plant eukaryotes as well as a long chain acyl-CoA elongase (β-ketoacyl-CoA synthase) were obtained from the search. When expressed together in somatic soybean embryo, *cis*-5-eicosenoic and *cis*-5-docosenoic acids were produced, comprising 18% of the fatty acids produced (Cahoon et al., 2000). The results obtained from these experiments confirmed the early observations of Pollard and Stumpf (1980), that the pathway to these fatty acids involves production of palmitoyl-CoA, which is elongated to eicosanoyl-CoA or docosanoyl-CoA and then converted to the *cis*-5-monounsaturaes by a *cis*-5-acyl-CoA desaturase (Fig. 11.7.3).

APPLICATIONS

The wholesale price of meadowfoam oil described by Burden (2012) is as high as $33/kg. It is thus best suited for use in cosmetics and hair care products due to its stability and ability to act as a carrier for other oils. In this way, it is very similar to jojoba. Numerous applications for meadowfoam oil have been proposed, including use as a lubricant additive, source of wax similar to sperm whale wax, factice (sulfur-treated) for use in rubber, lubricity additive, or precursor to chemicals with additional uses (Miller et al., 1964; Burden, 2012). Until meadowfoam oil production can be expanded, its main use is likely to remain in the consumer care and cosmetic industries.

MEADOWFOAM SEED MEAL

The glucosinolate content of meadowfoam seed meal reduces its use as an animal feed, as the glucosinolates and breakdown products are anti-feedants or toxic and the seed meal requires treatment for non-bovine feed (Oelke et al., 1990). However, the seed meal provides nitrogen for fertilizer and the breakdown products of glucosinolates generate isothiocyantes and thiocyanates that serve as herbicides and insect deterrents. In field trials with lettuce, application of meadowfoam seed meal treated with freshly crushed meadowfoam seed to provide active thioglucosidase to fields 7 days prior to planting resulted in 71% reduction of weeds and 8.5 times the N-content compared to lettuce planted in control fields (Intanon et al., 2015).

CONCLUSION

Oil derived from meadowfoam is unique in its very high content of C20 and C22 fatty acids. Even though it contains a significant amount of dienoic fatty acid, it is still very stable to oxidation, as the double bonds are separated by seven methylene groups. The position of the double bond in the 5 position imparts unique chemistry, allowing facile synthesis of 4- or 5-hydroxy fatty acids as well as estolides. The seed meal contains glucosinolates that make it useful as a preemergence organic herbicide. Further breeding improvement can enhance its production in hospitable conditions.

REFERENCES

Bagby, M.O., Smith, C.R., Miwa, T.K., Lohmar, R.L., Wolff, I.A., 1961. A unique fatty acid from *Limnanthes douglasii* seed oil: the C_{22} diene. J. Org. Chem. 26, 1261–1265.

Burden, D., 2012. Meadowfoam. Agricultural Marketing Resource Center (AgMRC). http://www.agmrc.org/commodities__products/grains__oilseeds/meadowfoam/ (30, 2015).

Burg, D.A., Kleiman, R., 1991. Preparation of dimer acids and dimer esters, and their use as lubricants. J. Am. Oil Chem. Soc. 68, 600–603.

Cahoon, E.B., Marillia, E.-F., Stecca, K.L., Hall, S.E., Taylor, D.C., Kinney, A.J., 2000. Production of fatty acid components of meadowfoam oil in somatic soybean embryos. Plant Physiol. 124, 243–251.

Carlson, K.D., Phillips, B.S., Isbell, T.A., Nelsen, T.C., 1998. Extraction of oil from meadowfoam flakes. J. Am. Oil Chem. Soc. 75, 1426–1436.

Erhan, S.M., Kleiman, R., Isbell, T.A., 1993. Estolides from meadowfoam fatty acids and other monounsaturated fatty acids. J. Am. Oil Chem. Soc. 70, 461–465.

Gandhi, S.D., Kishore, V.K., Crane, J.M., Slabaugh, M.B., Knapp, S.J., 2009. Selection for low erucic acid and genetic mapping of loci affecting the accumulation of very long-chain fatty acids in meadowfoam seed storage lipids. Genome 52, 547–556.

Gentry, H.S., Miller, R.W., 1965. The search for new industrial crops IV. Prospectus for *Limnanthes*. Econ. Bot. 19, 25–32.

Hayes, D.G., Kleiman, R., 1993. The isolation and recovery of fatty acids with Δ5 unsaturation from meadowfoam oil using lipase-catalyzed hydrolysis and esterification. J. Am. Oil Chem. Soc. 70, 555–560.

Higgins, J.J., Calhoun, W., Willingham, B.C., Dinkel, D.H., Raisler, W.L., White, G.A., 1971. Agronomic evaluation of prospective new crop species II. The American *Limnanthes*. Econ. Bot. 25, 44–54.

Intatnon, S., Hulting, A.G., Mallory-Smith, C.A., 2015. Field evaluation of meadowfoam (*Limnanthes alba*) seed meal for weed management. Weed Sci. 63, 302–311.

Isbell, T.A., Plattner, B.A., 1997. A highly regioselective synthesis of δ-lactones from meadowfoam fatty acids. J. Am. Oil Chem. Soc. 74, 153–158.

Jenderek, M.M., Hannan, R.M., 2009. Diversity in seed production characteristics within the USDA-ARS *Limnanthes alba* germplasm collection. Crop Sci. 49, 1387–1394.

Joliff, G.D., Seddigh, M., Franz, R.E., 1994. Seed germination and dormancy of new meadowfoam (*Limnanthes* spp.) phenotypes. Ind. Crops Prod. 2, 179–187.

Knapp, S.J., Crane, J.M., 1997. The development of self-pollinated inbred lines of meadowfoam by direct selection in open-pollinated populations. Crop Sci. 37, 1770–1775.

Lee, S., Ko, K.-H., Shin, J., Kim, N.-K., Kim, Y.-W., Kim, J.-S., 2015. Effects of the addition of dimer acid alkyl esters on the properties of ethyl cellulose. Carb. Polym. 121, 284–294.

Miller, R.W., Daxenbichler, M.E., Earle, F.R., Gentry, H.S., 1964. Search for new industrial oils. VIII. The genus *Limnanthes*. J. Am. Oil Chem. Soc. 41, 167–169.

Moreau, R.A., Pollard, M.R., Stumpf, P.K., 1981. Properties of a Δ5-fatty acyl-CoA desaturase in the cotyledons of developing *Limnanthes alba*. Arch. Biochem. Biophys. 209, 376–384.

Nikolova-Damyanova, B., Christie, W.W., Herslof, B., 1990. The structure of the triacylglycerols of meadowfoam oil. J. Am. Oil Chem. Soc. 67, 503–507.

Oelke, E.A., Oplinger, E.S., Hanson, C.V., Kelling, K.A., 1990. Meadowfoam. Alternative Field Crops Manual. http://www.hort.purdue.edu/newcrop/afcm/meadowfoam.html (August 30, 2025).

Pollard, M.R., Stumpf, P.K., 1980. Biosynthesis of C_{20} and C_{22} fatty acids by developing seeds of *Limnanthes alba*: chain elongation and D5 desaturation. Plant Physiol. 66, 649–655.

Purdy, R.H., Craig, C.D., 1987. Meadowfoam: new source of long-chain fatty acids. J. Am. Oil Chem. Soc. 64, 1493–1498.

Smith, C.R., Bagby, M.O., Miwa, T.K., Lohmar, R.L., Wolff, I.A., 1960. Unique fatty acids from *Limnanthes douglasii* seed oil: the C_{20}- and C_{22}-monoenes. J. Org. Chem. 25, 1770–1774.

Steiner, J.J., Mueller-Warrant, G.W., Griffith, S.M., Banowetz, G.M., Whittaker, G.W., 2006. Conservation practices in western Oregon perennial grass seed systems: II. Meadowfoam rotation crop management. Agron. J. 98, 1501–1509.

Toy, S.J., Willingham, B.C., 1966. Effect of temperature on seed germination of ten species and varieties of *Limnanthes*. Econ. Bot. 20, 71–75.

Toy, S.J., Willingham, B.C., 1967. Some studies on secondary dormancy in *Limnanthes* seed. Econ. Bot. 21, 363–366.

Chapter 11.8

Pennycress (*Thlapsi* spp.)

Guanqun Chen
Department of Biological Sciences, University of Manitoba, Winnipeg, Manitoba, Canada

INTRODUCTION

Pennycress (*Thlaspi* sp.) is a member of the Brassicaceae family with species growing on all continents except Antarctica. This plant has a short life cycle, cold tolerance, high productivity potential (up to 840 L/ha oil and 1470 kg/ha press-cake), and high seed oil content (up to 38%) (Sedbrook et al., 2014). Therefore, this plant could serve as a winter oilseed producing cover crop. The major fatty acids of pennycress seed oils are erucic acid (27.5–38.4%), linolenic acid (8.4–15.2%), and oleic acid (7.7–17.1%) (Phippen and Phippen, 2013). Pennycress has been recently identified as a potential oilseed crop for fuel and industrial products. Pennycress oil-derived biodiesel has a lower pour point (−18°C) and cloud point (−10°C) than that produced from camelina, soybean, and canola oil but unfavorable higher kinematic viscosity (5.24 mm^2/s at 40°C) (Moser, 2012). Reduced content of erucic acid is a target for pennycress breeding as it is expected to result in low kinematic viscosity of the pennycress-derived biodiesel. Removal of glucosinolate from the meal will aid in its acceptance as animal feed.

PENNYCRESS CROP RESEARCH

Initial interest in pennycress for cultivation as a possible substitute for rapeseed came in the 1940s (Clopton and Triebold, 1944). This plant was later considered as a potential novel industrial oilseed crop because of its high content of inedible erucic acid in seed oil (Carr, 1993). There are at least 164 accessions available in the United States, providing a rich genetic resource for pennycress research (Sedbrook et al., 2014). Recently, the effort in developing alternative energy sources initiated a breeding program for pennycress.

Some advances in the genetic improvement of pennycress have been achieved in traditional breeding programs in the past decade. Over 100 populations of pennycress have been systematically evaluated for key agronomic traits such as plant type, cold and heat tolerance, plant height, seed size and yield, seed oil content and fatty acid profile, etc. (Sedbrook et al., 2014). The results indicated that most of the wild populations display good cold tolerance and have seed weight ranging from 0.8 to 2.4 mg, yield ranging from 1100 to 2250 kg/ha, and seed oil content ranging from 24% to 39% (Phippen and Phippen, 2013; Sedbrook et al., 2014). Several other agronomic characters have been studied at Western Illinois University (Phippen et al., 2010a,b; Sedbrook et al., 2014). For example, the preliminary studies indicated that drilled pennycress plots resulted in higher seed oil yield per hectare than broadcast plots, but the average seed and oil yield between these two planting methods at three seeding rates were not significantly different (Phippen et al., 2010a). Compared to planting in October, planting in September resulted in a significant higher seed and total oil yield prior to planting soybeans (Phippen et al., 2010b). Moreover, spring lines had a short height and less seed yield than winter lines (Phippen et al., 2010a). Alachlor and pendimethalin based herbicides helped control weeds but had no significant impact on pennycress seed yield (Phippen et al., 2010b). In addition, seed oil constituents or protein profiles of soybeans planted following pennycress had no significant difference to those planted following fallow ground. All these studies indicate that pennycress is a tremendous potential oil crop.

Genetic engineering approaches have been used for rapid genetic improvement of pennycress. The *de novo* assembly of the comprehensive gene expression profile (transcriptome) in pennycress was recently reported (Dorn et al., 2013). With RNA extracted from representative plant tissues as templates, a draft transcriptome assembly consisting of 33,873 overlapping DNA sequences (contigs) was generated and functionally annotated, allowing for the identification of genes putatively controlling important agronomic traits. Moreover, a draft pennycress genome sequence was published recently (Dorn et al., 2015). This draft genome contains 27,390 predicted protein-coding genes that were annotated and covered almost all of the significant sequence similarity to *Arabidopsis* proteins. Many pennycress gene homologues involved in important pathways have high sequence similarity with other Brassicaceae species. Therefore, the knowledge gained from other Brassicaceae plants will benefit the improvements for pennycress.

The metabolite fingerprinting of pennycress embryos was conducted to assess active biochemical pathways during oil biosynthesis (Tsogtbaatar et al., 2015). The results indicated that glucose and glutamine would be the main sources of carbon and nitrogen, respectively, and oxidative pentose phosphate pathway, the tricarboxylic acid cycle, and the Calvin cycle were active in developing pennycress embryos (Tsogtbaatar et al., 2015).

This information about the transcriptome, genome, and metabolite fingerprinting can be used to direct molecular breeding of pennycress. Several effective genetic engineering approaches such as Targeting Induced Local Lesions IN Genomes (TILLING) and Targeted genome editing (Clustered, Regularly Interspaced, Short Palindromic Repeats (CRISPR)/Cac9 and Transcription Activator-Like Effector Nucleases (TALEN), respectively) have been successfully applied in crop breeding programs and will facilitate rapid improvement of pennycress.

In order to breed pennycress into a readily cultivated crop plant, several agronomic problems including seed dormancy, oil quality, seed glucosinolates, flowering time and maturation, pod shatter, seed size and oil content, should be improved. Such traits have been successfully bred out of other crops and the knowledge gained from those achievements will facilitate pennycress cultivation as a winter crop. Some important genes related to those traits have been identified in some crops and the homologues have been reported in pennycress (Sedbrook et al., 2014). These genes can be attractive targets to improve the traits in pennycress. For example, a gene named *Delay of Germinations 1* is a major quantitative trait locus specific for controlling seed dormancy in Brassica species (Graeber et al., 2010); *Fatty Acid Elongase 1* catalyzes the synthesis of unfavorable erucic acid in canola (Wu et al., 2008); *Flowering Locus C* and *FRIGIDA* genes are the primary determinants of flowering time in brassicas (Tadege et al., 2001; Yuan et al., 2009); *Shatterproof 1* and *Shatterproof 2* control pod shatter in Arabidopsis (Ferrandiz et al., 2000; Muhlhausen et al., 2013). Manipulation of those genes may significantly improve agronomic traits of pennycress.

In summary, although pennycress has not reached its commercial potential yet, it has great promise as a possible triple use seed oil/meal/cover temperate crop. Additional breeding efforts including both genetic engineering and improvements in agronomic practices will be key to bring this "weed" to a successful crop.

REFERENCES

Carr, P.M., 1993. Potential of fanweed and other weeds as novel industrial oilseed crops. In: Janick, J., Simon, J.E. (Eds.), New Crops. Wiley, New York, pp. 384–388.

Clopton, J.R., Triebold, H.O., 1944. Fanweed seed oil. Ind. Eng. Chem. 36, 218–219.

Dorn, K.M., Fankhauser, J.D., Wyse, D.L., Marks, M.D., 2013. De novo assembly of the pennycress (*Thlaspi arvense*) transcriptome provides tools for the development of a winter cover crop and biodiesel feedstock. Plant J. 75, 1028–1038.

Dorn, K.M., Fankhauser, J.D., Wyse, D.L., Marks, M.D., 2015. A draft genome of field pennycress (*Thlaspi arvense*) provides tools for the domestication of a new winter biofuel crop. DNA Res. 22, 121–131.

Ferrandiz, C., Liljegren, S.J., Yanofsky, M.F., 2000. Negative regulation of the *SHATTERPROOF* genes by *FRUITFULL* during Arabidopsis fruit development. Science 289, 436–438.

Graeber, K., Linkies, A., Muller, K., Wunchova, A., Rott, A., Leubner-Metzger, G., 2010. Cross-species approaches to seed dormancy and germination: conservation and biodiversity of ABA-regulated mechanisms and the Brassicaceae *DOG1* genes. Plant Mol. Biol. 73, 67–87.

Moser, B.R., 2012. Biodiesel from alternative oilseed feedstocks: camelina and field pennycress. Biofuels 3, 193–209.

Muhlhausen, A., Lenser, T., Mummenhoff, K., Theissen, G., 2013. Evidence that an evolutionary transition from dehiscent to indehiscent fruits in *Lepidium* (Brassicaceae) was caused by a change in the control of valve margin identity genes. Plant J. 73, 824–835.

Phippen, W., Gallant, J., Phippen, M., 2010a. Evaluation of planting method and seeding rates with field pennycress (*Thlaspi arvense* L.). In: Poster presentation at the 22nd Annual Meeting of the Association for the Advancement of Industrial Crops, Fort Collins, CO, USA, 54, pp. 743–748.

Phippen, W., John, B., Phippen, M., Isbell, T., 2010b. Planting date, herbicide, and soybean rotation studies with field pennycress (*Thlaspi arvense* L.). In: Poster Presentation at the 22nd Annual Meeting of the Association for the Advancement of Industrial Crops, Fort Collins, CO, USA, pp. 19–22.

Phippen, W.B., Phippen, M.E., 2013. Seed oil characteristics of wild field pennycress (*Thlaspi arvense* L.) populations and USDA accessions. In: Advancement of Industrial Crops Annual Meeting, Washington D.C (Poster).

Sedbrook, J.C., Phippen, W.B., Marks, M.D., 2014. New approaches to facilitate rapid domestication of a wild plant to an oilseed crop: example pennycress (*Thlaspi arvense* L.). Plant Sci. 227, 122–132.

Tadege, M., Sheldon, C.C., Helliwell, C.A., Stoutjesdijk, P., Dennis, E.S., Peacock, W.J., 2001. Control of flowering time by FLC orthologues in *Brassica napus*. Plant J. 28, 545–553.

Tsogtbaatar, E., Cocuron, J.C., Sonera, M.C., Alonso, A.P., 2015. Metabolite fingerprinting of pennycress (*Thlaspi arvense* L.) embryos to assess active pathways during oil synthesis. J. Exp. Bot. 66, 4267–4277.

Wu, G., Wu, Y., Xiao, L., Li, X., Lu, C., 2008. Zero erucic acid trait of rapeseed (*Brassica napus* L.) results from a deletion of four base pairs in the *fatty acid elongase 1* gene. Theor. Appl. Genet. 116, 491–499.

Yuan, Y.X., Wu, J., Sun, R.F., Zhang, X.W., Xu, D.H., Bonnema, G., Wang, X.W., 2009. A naturally occurring splicing site mutation in the *Brassica rapa* FLC1 gene is associated with variation in flowering time. J. Exp. Bot. 60, 1299–1308.

Chapter 11.9

Perilla (*Perilla frutescens*)

Douglas G. Hayes
Department of Biosystems Engineering and Soil Science, University of Tennessee, Knoxville, TN, United States

David F. Hildebrand
Department of Plant and Soil Sciences, University of Kentucky, Lexington, KY, United States

PERILLA

Perilla (*P. frutescens*) (L.) Britton, family Lamaiaceae (Labiatae), is a summer annual crop adapted to warm and humid climates, an herb, which is native to Eastern Asia, particularly China, Korea, Japan, India, and Thailand. It is also known as Bhanjir, beefsteak plant, Chinese basil, and purple mint. It typically grows to 1 m in height, and contains small flowers and brown fruits (Brenner, 1993). Seed color ranges from black to white and various shades of gray and brown. There are two major varieties of the plant, var. *frutescens*, for seed and oil production, and var *crispa*, a spicy herb often used for traditional medicine (Yu et al., 1997; Nitta et al., 2003). This mini-chapter will focus upon the former seed variety. The fruits contain small seeds (approximately 4 mg in weight) that contain 30–45% oil. The *frutescens* variety is a tetraploid when cultivated; however, Perilla species existing in the wild are diploid (Nitta et al., 2003). Seeds, which contain a soft exterior, are typically planted in the early spring, approximately 1 cm deep (Brenner, 1993). Its chromosome number is $2n = 40$ (Nitta et al., 2003). Seedlings can be grown in a greenhouse and then transplanted into the soil (Rouphael et al., 2015); but, plants can readily be established in the field with direct seeding. Perilla can grow on marginal land. Drip irrigation may be needed if the soil moisture becomes significantly low (Rouphael et al., 2015). The flowers self-pollinate, without involvement of insects (Brenner, 1993). Major obstacles to their cultivation include the poor storage stability of the seeds and the potential toxicity to livestock (Brenner, 1993). A major toxin in Perilla is perilla ketone (Muller-Waldeck et al., 2010). Perilla ketone-free Perilla genotypes are known, enabling breeders to readily select Perilla lines devoid of this toxin (Muller-Waldeck et al., 2010). Seeds must be in a postdormancy state for plants to grow; moreover, seeds will reside in a dormant stage where they cannot germinate, with the dormancy state being difficult to detect and monitor (Evergreen YH Enterprises, 2015). The best approach is to store seeds at refrigerator temperatures for several days prior to seeding. Transcriptomic information on Perilla species has recently been published (Tong et al., 2015). Germplasm collections exist in the United States (USDA/Iowa State University) and Japan (National Institute of Agrobiological Resources, Kannondai, Japan) (Brenner, 1993).

PERILLA OIL

The oil is known for its high content of α-linolenic acid (18:3–9c,12c,15c; 40–70%; Table 11.9.1), and is therefore of interest in nutrition and health (eg, as an edible oil, salad oil, and for treatment of coronary heart disease, hypertension, cancers, inflammation, arthritis, and asthma), due to its high degree of polyunsaturation (Asif, 2012; Joshi et al., 2015). For the same reason, perilla oil is of interest as inks, drying oils, and for other paints and coatings (see chapter: Polymeric Products Derived From Industrial Oils for Paints, Coatings, and Other Applications). The oil consists of 90% neutral lipids, of which 90% are triacylglycerols (TAG) (Shin and Kim, 1994; Yoon and Noh, 2011). Other prominent fatty acids include linoleic (18:2–9c,12c), oleic (18; 1–9c), palmitic (16:0), and stearic (18:0) acids (Table 11.9.1). The saturated fatty acids have a slight preference for the *sn*-1(3)-acylglycerol position compared to the 2-position; in contrast, the other fatty acids are randomly distributed between the acylglycerol positions (Yoon and Noh, 2011). The five most abundant TAG in perilla oil are LnLnLn (22.6%), OLnLn (18.1%), LLnLn (16.5%), OLLn (8.7%), and PLnLn (6.6%), where Ln, O, L, and P refer to linoleic, oleic, linoleic, and palmitic acyl groups, respectively (Ciftci et al., 2012). Perilla oil contains a high amount of sterols, 4606 ppm, with β-sitosterol being the most abundant (3377 ppm), and several others being abundant (100–250 ppm), including cycloartenol, campesterol, and δ5-avenasterol (Ciftci et al., 2012). Tocopherols are also abundant in perilla oil (691 ppm), with γ-tocopherol consisting of 94.3% of the tocopherol content (Shin and Kim, 1994; Ciftci et al., 2012). Phospholipids amounted to 2–3% of lipids, with phosphatidylethanolamine being the most abundant (50–57%) (Shin and Kim, 1994). Annual production of perilla oil is 40,000 metric tons (Shin and Kim, 1994).

RECOVERY OF PERILLA OIL FROM SEEDS

The most common approach for oil recovery is to first roast the seeds and then apply mechanical pressing (Jung et al., 2012). Roasting produces pyrazines

TABLE 11.9.1 Fatty Acid Composition of *Perilla fructescens* L. Oil From Various Asian Countries (Joshi et al., 2015, and References Therein)

Fatty Acid	India	China	Japan	Korea	Thailand
Palmitic (16:0)	9.9	7.2	7.7	7.4	6.9
Stearic (18:0)	2.3	2.9	3.8	3.6	0.6
Oleic (18:1)	0.1	20.8	10.2	9.5	10.5
Linoleic (18:2)	14.5	10.5	17.9	16.5	20.8
α-Linolenic (18:3)	68.6	52.6	60.4	63.0	61.2
Others	4.6	6.0	0.0	0.0	0.0

(via a Maillard reaction between reducing sugars and amino or amide groups), which produce a nutty flavor to the oil (Kwon et al., 2013). An effective approach is to use pressing of the seeds, followed by solvent extraction (eg, using diethyl ether or hexane) (Jung et al., 2012). Supercritical CO_2-based extraction has also been employed successfully (Jung et al., 2012). Extraction using compressed *n*-propane provided a slightly lower yield of neutral lipids and tocopherols than solvent extraction but enhanced the oxidative stability of the oil (Da Silva Marineli et al., 2015). A recent study compared the recovery of perilla oil from roasted seed using hexane and supercritical CO_2 as extractants and mechanical pressing. The study found that hexane and supercritical CO_2 extractions (the latter conducted at 420 bar) were equally superior for enhancing oil yield (40%) (Jung et al., 2012). However, the former solvent produced a higher tocopherol and sterol yield but also a lower oxidative stability (Jung et al., 2012).

PERILLA CO-PRODUCTS

The leaves of perilla are well known for use in traditional medicine as an herbal treatment for several different diseases, such as asthma, influenza, bronchitis, and digestive system dysfunction (Yu et al., 1997). The leaves contain an essential oil enriched in monoterpenes, aldehydes, and ketones, which are potentially applicable as antimicrobials, fungicides, antioxidants, and insecticide (Yu et al., 1997; Seo and Baek, 2009; Tian et al., 2014; Tabanca et al., 2015). Seeds also contain volatile oil. The pomace residue resulting from oil extraction of the seeds is rich in phenolic compounds such asrosmarinic acid that are potentially valuable as antioxidants (Guan et al., 2014). Typically, the pomace is used as animal and bird feed (Asif, 2012). In Korea, leaves are used for wrapping meats and raw fish, and other food products (Nitta et al., 2003). In Japan, Perilla leaves are a common ingredient in the popular dish tempura and are frequently added to pickled fruit known as umeboshi in rice balls (Sawabe et al., 2006).

ASSESSMENT AS A POTENTIAL NEW CROP

P. frutescens (L.) *Britt.* var. *frutescens* is an important oilseed in Korea, and on a smaller scale in Japan and other regions of eastern Asia (Negi et al., 2011; Tong et al., 2015). There has been very little research on improving perilla via breeding, even though perilla species can undergo cross-pollination (Tong et al., 2015) with most Perilla breeding reports from Korea (Park et al., 2002). Genetic engineering protocols have been developed for Perilla (Hossain et al., 2010; Lee et al., 2005). Although Perilla is a potentially valuable replacement for flaxseed (linseed) oil, due to its high linoleic acid content, it is viewed as a niche crop in the United States, for Asian food (Brenner, 1993).

REFERENCES

Asif, M., 2012. Nutritional importance of monounsaturated and polyunsaturated fatty acids of *Perilla* oil. Int. J. Phytopharm. 2, 154–161.

Brenner, D.M., 1993. Perilla: botany, uses, and genetic resources. In: Jankck, J., Simon, J.E. (Eds.), New Crops. John Wiley and Sons, New York, pp. 322–328.

Ciftci, O.N., Przybylski, R., Rudzinska, M., 2012. Lipid components of flax, perilla, and chia seeds. Eur. JU Lipid Sci. Technol. 114, 794–800.

Evergreen YH Enterprises, 2015. Asian Vegetable Seeds – Evergreen Seeds. Anaheim, CA USA http://www.evergreenseeds.com (accessed 02.07.15.).

Guan, Z., Li, S., Lin, Z., Yang, R., Xhao, Y., Liu, J., Yang, S., Chen, A., 2014. Identification and quantitation of phenolic compounds from the seed and pomace of *Perilla frutescens* using HPLC/PDA and HPLC-ESI/QTOF/MS/MS. Phytochem. Anal. 25, 508–513.

Hossain, H., Kim, Y.H., Lee, Y.S., 2010. Stable transformation of leaf Perilla using apical buds as explants. SABRO J. Breed. Genet. 42, 21–34.

Joshi, A., Sharma, A., Pandey, D.P., Bachheti, R.K., 2015. Physico-chemical properties of *Perilla fructescens* seeds. Der Pharma Chemica 7, 35–41.

Jung, D.M., Yoon, S.H., Jung, M.Y., 2012. Chemical properties and oxidative stability of perilla oils obtained from roasted perrilla seeds as affected by extraction methods. J. Food Sci. 77, C1249–C1255.

Kwon, T.Y., Park, J.S., Jung, M.Y., 2013. Headspace-solid phase microextraction-gas chromatography-tandem mass spectrometry (HS-SPME-GC-MS) method for the determination of pyrazines in perilla seed oils: impact of roasting on the pyrazines in perilla seed oils. J. Agric. Food Chem. 61, 8514–8516.

Lee, B.-K., Yu, S.-H., Kim, Y.-H., Ahn, B.-O., Hur, H.-S., Lee, S.-C., Zhang, Z., Lee, J.-Y., 2005. *Agrobacterium*-mediated transformation of Perilla (*Perilla frutescens*). Plant Cell Tissue Organ Cult 83, 51–58.

Muller-Waldeck, F., Sitzmann, J., Schnitzler, W.H., Grabmann, J., 2010. Determination of toxic perilla ketone, secondary plant metabolites, and antioxidative capacity in five *Perilla fructescens* L. varieties. Food. Chem. Tox 48, 264–270.

Negri, V.S., Rawar, L.S., Phondani, P.C., XChandra, A., 2011. *Perilla frutescens* in transition: a medicinal and oil yielding plant need instant conservation, a case study from central Himalaya. India Environ. We Int. J. Sci. Technol. 6, 193–200.

Nitta, M., Lee, J.K., Ohnishi, O., 2003. Asian *Perilla* crops and their weedy forms; their cultivation, utilization, and genetic relationships. Econ. Bot. 57, 245–253.

Park, C., Bang, J., Lee, B., Kim, J., Lee, B., Chung, M., 2002. A high-oil and high-yielding perilla variety "Daesildeulkkae" with dark brown seed coat. Kor. J. Breed 34, 78–79.

Rouphael, Y., Raimondi, G., Paduano, A., Sacchi, R., Barbieri, G., De Pascale, S., 2015. Influence of organic and conventional farming on seed yield, fatty acid composition and tocoperhols of Perilla. Aust. J. Crop Sci. 4, 303–308.

Sawabe, A., Satake, T., Aizawa, R., Sakatani, K., Nishimoto, K., Ozeki, C., Hamada, Y., Komemushi, S., 2006. Toward use of the leaves of *Perilla frutescens* (L.) Britton var. Acuta Kudo (red perilla) with Japanese dietary pickled plum (umeboshi). J. Oleo Sci. 55, 413–422.

Seo, W.H., Baek, H.H., 2009. Characteristic aroma-active compounds of Korean perilla (*Perilla frutescens* Britton) leaf. J. Agric. Food Chem. 57, 11537–11542.

Shin, H.S., Kim, S.W., 1994. Lipid composition of perilla seed. J. Am. Oil Chem. Soc. 71, 619–622.

Tabanca, N., Demirci, B., Ali, A., Ali, Z., Blythe, E.K., 2015. Essential oils of green and red *Perilla frutescens* as potential sources of compounds for mosquito management. Ind. Crops Prod. 65, 36–44.

Tian, J., Zeng, X., Zhang, S., Wang, Y., Zhang, P., Lu, A., Peng, X., 2014. Regional variation in components and antioxidants and antifungal activities of *Perilla frutescens* essential oils in China. Ind. Crops Prods 59, 69–79.

Tong, W., Kwon, S.J., Lee, J., Choi, I.Y., Park, Y.J., Choi, S.H., Sa, K.J., Kim, B.W., Ju, K.L., 2015. Gene set by de novo assembly of *Perilla* species and expression profiling between *P. frutescens* (L.) var. *frutescens* and var. *crispa*. Gene 559, 155–163.

Yoon, S.H., Noh, S., 2011. Positional distribution of fatty acids in perilla (*Perilla frutescens* L.) oil. J. Am. Oil Chem. Soc. 88, 157–158.

Yu, H.C., Kosuna, K., Haga, M., 1997. Perilla: The Genus Perilla. Harwood Academic Publishers, Amsterdam.

Chapter 11.10

Pili (*Canarium ovatum*)

Laura J. Pham
Oils and Fats Laboratory, National Institute of Molecular Biology and Biotechnology (BIOTECH), University of the Philippines at Los Baños, Laguna, Philippines

INTRODUCTION

Pili is perhaps the second most important fruit-bearing tree and also one of the most important oilseeds of commercial value in the Philippines. The Pili species (*Canarium ovatum* Engl) is the most important of the several fruit trees bearing edible nuts that have their center of diversity in the Philippines. Its geographic distribution in the country remains limited to areas relatively close to its center of origin (Coronel, 1991). Thus, the present production of the plant is confined to a limited area. The economic uses of pili make it a highly appreciated nut. It has great potential for development as a major export crop and would offer stiff competition to other nuts of the world, and was popular in the US over 100 years ago (Editor, 1914). Pili could well be the secondary "Tree of Life" due to its various products. To the people of the Bicol region, it is their "Tree of Hope." There are potential processing and utilization possibilities as food, feed, industrial uses, fuel and handicraft material from every part of the tree (Marcone et al., 2002). Such potential makes pili one of the primary flagship commodities of the Philippines (PCARRD-DOST, 2010).

THE PILI TREE

Pili belongs to the genus *Canarium*, comprising 78 species (Weeks, 2009) in the family *Burseraceae* which includes four genera. There are about 40 species of *Canarium* in the Philippines. Some authorities believe that there are three or four species of pili in the island, of which only two yield the highly priced nuts. These are *Canarim ovatum* Engler, which is mainly found in the Bicol region and known for its oil. It is a fairly large tree that reaches a height of about 20 m and a trunk diameter of about 40 cm. Leaves are alternate and compound with opposite smooth leaflets, which are rounded at the base and pointed at the tip. *Canarium luzonicum*, on the other hand, reaches a height of 35 m and a diameter of 1 m. Leaves are odd pinnate and three pairs of opposite leaflets that are smooth, pointed at apex and rounded at base. It is abundant in the Quezon Province and propagated for its resin (Arribas, 1994). The pili tree is one of the most resistant species to typhoons but a number of trees are destroyed by devastating typhoons. The greatest threat to genetic diversity is human, as they reduce the population of the trees in times of food insecurity. Thus there is a need to continue planting seedlings to ensure that genetic viability and diversity of the species is maintained.

FIGURE 11.8.1 Pili fruits.

PILI FRUIT

Pili is known for its fruits (Fig. 11.8.1). The fruit is botanically classified as a drupe but is popularly called pili nut. The fruits are about 4–6 cm long of variable shapes, namely elliptical, oblong, oval and obovate. Generally, the fruit has three parts: the skin, the pulp and the nut. The pulp, which turns from green to dark purple to almost black upon ripening, surrounds the thick shelled, edible nut. The skin (exocarp) is smooth, thin and shiny, and the pulp (mesocarp) is fibrous, fleshy, and greenish yellow in color. The hard shell (endocarp) within protects a normally dicotyledonous embryo. The basal end of the shell (endocarp) is pointed and the apical end is more or less blunt. Between the seed and the hard shell (endocarp) is a thin, brownish, fibrous seed coat developed from the inner layer of the endocarp. This thin coat usually adheres tightly to the shell and/or the seed. Much of the kernel weight is made up of the cotyledons, which are about 4.1–16.6% of the whole fruit. It is composed of approximately 8% carbohydrate, 11.5–13.9% protein and 70% fat. The kernel is described as tasting like macadamia or walnut, and is used in confections and other food preparations.

PILI OIL

Pili kernel contains about 70% oil and it is high in saturated fat, with composition varying from 44% to 60% oleate and 33–38% palmitate. The oil can be fractionated to yield a high melting oil similar to cocoa butter and a low melting oil similar to olive oil (Kakuda et al., 2000). This oil content however varies depending on several factors such as the extent of dryness of the kernel and species of the pili used. The oil has a sweet taste, which makes it suitable for culinary purposes. It also possesses an antiseptic effect and antineuritic vitamins. The oil is suitable for replacing olive oil for sardine manufacture, salad dressing and other food preparations. The high melting fat obtained by fractionation, pili butter, can replace the cocoa butter in the food, pharmaceutical and cosmetic industries. The composition of pili nut oil is palmitic (34.58%), stearic (9.97%), oleic (44.06) and linoleic (10.31%). In a study on pili kernel oil's triacylglycerol profile of the hard fraction, analyses showed the presence of POP, POS and SOS + SSO molecular species, similar to cocoa butter (Kakuda et al., 2000).

Pili pulp oil on the other hand contains 28–36% oil and has color which varies from yellow green to dark green depending on the type of extraction. The fatty acid composition of pili pulp oil is palmitic (23.89%), palmitoleic (4.65%), Stearic (2.63%) Oleic (60.76%) and linoleic acid (6.61%). The pili pulp oil has a higher oleic fatty acid value than pili nut oil, as reflected by its higher iodine number. The saturated fatty acids stearic and palmitic are higher in pili nut oil. Pili pulp oil has more unsaponifiable matter than Pili nut oil, and this material includes sterols and carotenoids. The sterol fraction is the highest minor component present, and it is higher in pili pulp than in pili nut oil (Pham et al., 2003). Pili pulp oil could be a source of lipid molecular species such as carotenoids, squalene, tocopherols and sterols are of increasing interest for nutritional, nutraceutical, cosmetic and pharmaceutical applications. With these components and a highly monounsaturate content, there are strong possibilities for expanding use of pili pulp oil to the global market. Preliminary evaluation of fatty acid methyl esters prepared from pili pulp oil comply with biodiesel standards and present a potential source and alternative to explore for commercial exploitation (Razon, 2008).

PILI OIL EXTRACTION

Pili oil production is an emerging technology. For the oil from the kernel, the methods of extracting from plant materials like coconut meat and peanuts are applied. An aqueous extraction method for pili pulp oil using enzymes has been developed by researchers at the University of the Philippines, Los Banos, and it is now being used in some commercial production (Pham and Pham, 2008). Modification of the technology of extraction had been carried out by the Department of Science and Technology in the Bicol region. The technology generated renewed interest among local growers and is paving the way for the establishment of the pili pulp oil production industry (Flores, 2008) that was previously discarded as waste after removing the kernel.

CONCLUSION

The pili tree is reminiscent of oil palm in terms of producing an fruit with oils of differing, valued composition in the kernel and the pulp. The pulp oil of pili is similar to olive oil with its high oleate content. The pili kernel oil can be fractionated into a high melting fat resembling cocoa butter and a low melting fraction similar to olive oil. These properties point to excellent possibilities for use of these oils in numerous applications.

REFERENCES

Arribas, L., 1994. Primer on Pili (*Canarium Ovatum* Engl.). The Demontano Foundation, Makati, Metro Manila.

Coronel, R.E., 1991. *Canarium ovatum* Engl. In: Verheij, E.W.M., Coronel, R.E. (Eds.), Plant Resources of South-East Asia, vol. 2, Edible Fruits and Nuts. Pudoc Wageningen, The Netherlands, pp. 105–108.

Editor, 1914. Three new nuts. J. Hered. 5, 179–184.

Flores, H., 2008. Coming Soon: Virgin Pili Nut Oil. PhilStar.

Kakuda, Y., Jahaniaval, F., Marcone, F., Montevirgen, L., Montevirgen, Q., Umali, J., 2000. Characterization of pili nut (*Canarium ovatum*) oil: fatty acid and triacylglycerol composition and physicochemical properties. JAOCS 77 (9), 991–997.

Marcone, M.R., Kakuda, Y., Jahaniaval, F., Yada, R.Y., Montevirgen, L.S., 2002. Characterization of the proteins of Pili nut (*Canarium ovatum*, Engl.). Plant Foods Hum. Nutr. 57, 107–120.

Pham, L., Revellame, E., Rasco, P., 2003. Studies on the Lilid molecular species and minor components of some Philippine seed and nut oils. In: Murata, N., et al. (Ed.), Advance Research on Plant Lipids. Kluwer Academic Publishers, Netherlands, pp. 23–26.

Pham, L., Pham, C., 2008. Pili pulp: potential value. INFORM 2 (19), 83–86.

PCARRD-DOST, 2010. Philippine Council for Agriculture, Forestry and Natural Resources Research and Development (PCARRD). PCARRD-DOST, Annual Report 2009. PCARRD-DOST, Los Baños, Laguna, pp. 42–43, 78 p.

Razon, L., 2008. Preliminary Evaluation of Biodiesel from *Canarium Ovatum* (Pili) Pulp Oil and Psophocarpus Tetragonolobus (Winged Bean) Seed Oil. De La Salle University Research Abstracts.

Weeks, A., 2009. Evolution of the Pili nut genus (*Canarium* L., *Burseraceae*) and its cultivated species. Genetic Resources and Crop Evolution 5, 765–781.

Chapter 11.11

Epilogue

Thomas A. McKeon
United States Department of Agriculture, Agricultural Research Service, Western Regional Research Center, Albany, CA, United States

The chemical functionality of the fatty acid(s) contained in an oil has served as the major criterion for identifying a new crop having potential for industrial applications. While this book includes a number of those crops, there are still others that produce fatty acids of interest, but for a variety of reasons have not yet achieved much success as cultivated crops for industrial oil production. These include coriander, *Lunaria annua*, *Vernonia*, *Euphorbia lagascae*, calendula, and *Dimorphotheca pluvialis* (Zanetti et al., 2013). These crops continue to represent a challenge for agronomists to bring them into commercial production, although they remain of interest (Cruz and Dierig, 2012; Zanetti et al., 2013). For a number of these crops, the genes responsible for biosynthesis of the fatty acids have been cloned and expressed in transgenic plants, but only limited amounts of the unusual fatty acid have been produced (vide infra).

Coriander (*Coriandrum sativum* L.) is an herbaceous crop widely grown for the herb cilantro and for coriander seed, a common spice (Evangelista et al., 2015). The seed contains oil consisting of 57–75% petroselinic acid (Fig. 11.11.1), 6-*cis*-octadecenoic acid (Isbell, 2009). Ozonolysis of petroselenic acid would produce lauric acid (C12:0) for use in surfactants and adipic acid for Nylon 6,6 (Millam et al., 1997). The seed produces 12.8–30.2% oil with yields averaging 900–1120 up to 2800 kg/ha but has yet to achieve success as an oilseed crop (Isbell, 2009). Attempts to engineer petroselinic acid production in transgenic crops have also proven problematic (Suh et al., 2002).

Lunaria annua L., the honesty plant, produces an oil containing very long-chain moneoneoic acids, with ~50% erucic acid and 25% nervonic acid (Fig. 11.11.2) and seed oil of 30–40% (Princen, 1983). Seed yields of 2–2.5

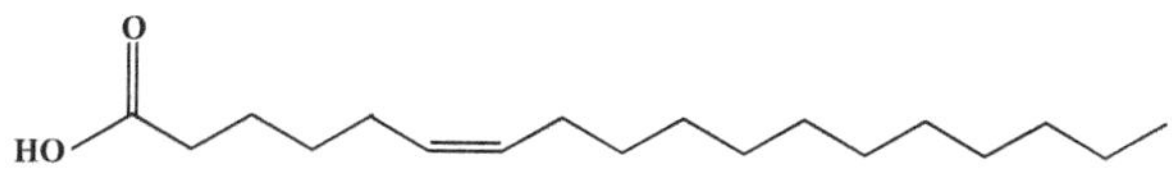

FIGURE 11.11.1 Petroselinic acid.

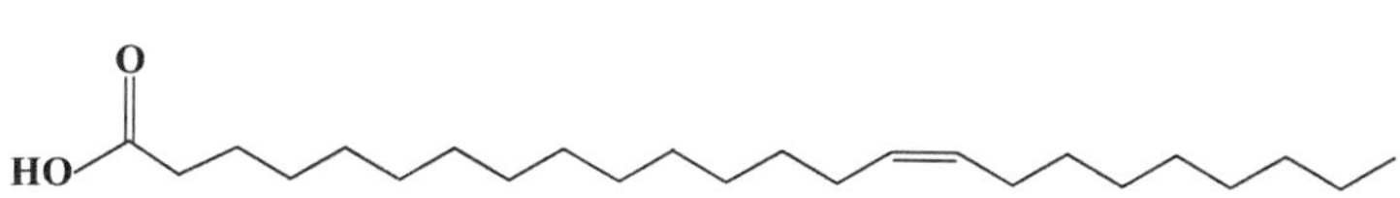

FIGURE 11.11.2 Nervonic acid.

t/ha have been described and the oil is useful for biodiesel and as a lubricating basestock (Dodos et al., in press). In addition to being a biannual crop, the other agronomic problems for Honesty include a long vernalization and limitations on mechanical harvesting (Walker et al., 2003; Zanetti et al., 2013). The fatty acid elongase from *L. annua* has been cloned and was used in experiments to duplicate jojoba wax ester in a transgenic plant (Lardizabal et al., 2000).

The epoxy fatty acid vernolic acid (Fig. 11.11.3) is a component of oil from several plants including *Centrapalus pauciflorus* var. *ethiopica* (formerly *Vernonia galamensis*) and *Euphorbia lagascae* (Shimelis et al., 2013; Zanetti et al., 2013). Epoxy oils would be useful in applications including plasticizers, adhesives, epoxy resins, paints, and coatings, as well as polyurethanes after epoxy ring opening (Shimelis et al., 2013). Currently, epoxidized oils are produced by chemical epoxidation of soybean oil or petrochemicals, while vernonia oil is a natural product from a renewable resource (Li et al., 2012). Vernonia (*C. pauciflora*) seed contains 20–39% oil with 70–78% veronilc acid content (Shimelis et al., 2013). However, limitations on cultivation of *Vernonia* include small, seeds, low yields, and limited geographical growing area due to photoperiod sensitivity (Li et al., 2012; Zanetti et al., 2013). *Euphorbia lagascae* also produces epoxy oil containing 58–69% vernolic acid in a seed containing 42–50% oil (Pascual-Villalobos et al., 1993). The plant has a potential yield of 1600 kg/ha but presents a challenge for achieving the yield in cultivation due to high seed dehiscence and poor germination (Pascual-Villalobos et al., 1993; Zanetti et al., 2013). The plant itself is of biomedical interest as terpenoids derived from the aerial portions of the plant may have use in cancer chemotherapy (Duarte et al., 2009). Attempts to produce vernolic acid in transgenic plants have had some success as soybean transformed with an epoxygenase from *Stokesia laevis* and DGAT from Vernonia produce normal levels of oil for soybean (~20%) and contain up to 26% vernolic acid (Li et al., 2012).

The pot marigold *Calendula officinalis* L., calendula, produces an oil containing 59–65% calendic acid (Fig. 11.11.4) (Eberle et al., 2014). As an oil containing a high proportion of conjugated fatty acid, it can be used in place

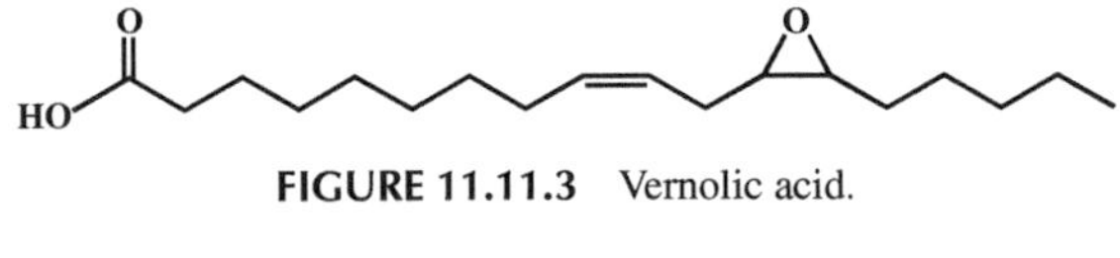

FIGURE 11.11.3 Vernolic acid.

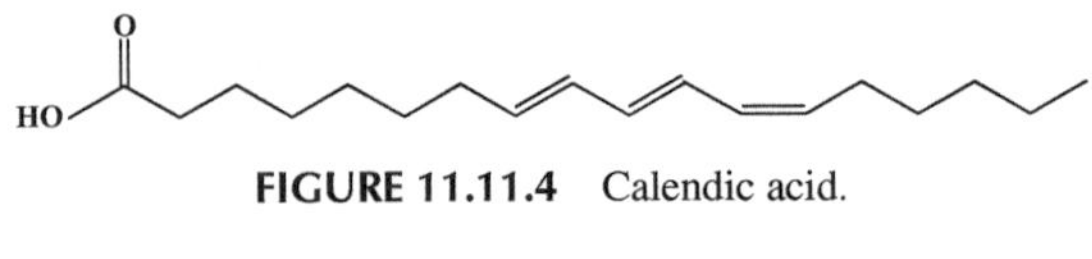

FIGURE 11.11.4 Calendic acid.

of tung oil as well as in applications suited for epoxy oils. It can thus replace high–volatile organic carbon (VOC) compounds derived from petroleum with a low-VOC oil (Gesch, 2013). As a crop, calendula is well adapted to temperate climates and cultivars suitable for including in crop rotation cylces have been developed (Gesch, 2013). The crop can be managed and harvested with conventional field equipment, and despite some reports of high seed shatter, the selected cultivars do not display that trait (Gesch, 2013). These cultivars produce seed with 18–20.5% oil with yields averaging up to 1800kg/ha but as high as 2400kg/ha. Seed germination is sensitive to temperature, with soil temperatures >35°C reducing germination >50% (Eberle et al., 2014). Transgenic plants modified to express the desaturase gene involved in producing calendic acid (Fritsche et al., 1999) have generated oils containing up to 22% calendic acid (Mietkiewska et al., 2014).

Dimorphotheca pluvialis L. Moench, the Cape mariold or weather prophet, is an annual decorative flowering plant. The seed contains 28–39% with 26–62% dimorphecolic acid, a conjugated hydroxy fatty acid (Fig. 11.11.5) (Knowles et al., 1965; Muuse et al., 1992). The unique structure suggest numerous possible uses in paints, coatings, or polymers (Hof and Dolstra, 1999); however, it is heat labile and very unstable to oxidation, with <1% of the stability of meadowfoam oil, a high monoenoic acid oil (Muuse et al., 1992). This susceptibility to heat and oxidation makes recovery of the oil or fatty acid difficult. It can be extracted by supercritical CO_2 or lipase action, but it cannot be vacuum distilled and retain its structure, so it is an expensive process to recover pure oil or dimorphecolic acid (Muuse et al., 1992; Derksen et al., 1993). The plant itself requires breeding and selection for improvement (Hof and Dolstra, 1999); early work on harvesting and yield indicated a high degree of dehiscence and poor yield of seed, approximately 450kg/ha (Breemhaar and Bouman, 1995). As might be expected for such an unuisual fatty acid, it is the product of two divergent oleate desaturases (Cahoon and Kinney, 2004). Transgenic plant production of an oil rich in dimorphecolic acid is likely to be an even greater challenge than production of a simple hydroxy fatty acid.

These emerging oil crops may eventually be brought into agricultural production or merely serve as genetic sources to be mined by plant scientists for developing transgenic plants (Taylor et al., 2011; Vanhercke et al., 2013). Yet these plants continue to pose a challenge to understand how and perhaps why they make these fatty acids.

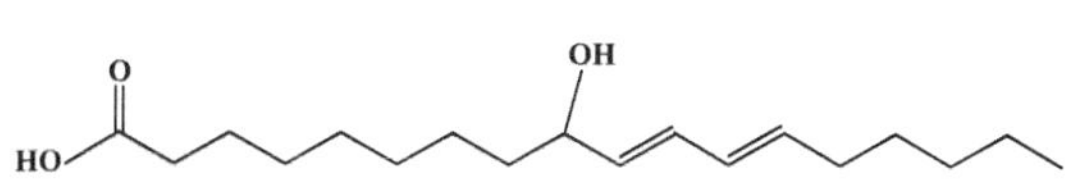

FIGURE 11.11.5 Dimorphecolic acid.

REFERENCES

Breemhaar, H.G., Bouman, A., 1995. Mechanical harvesting and cleaning of *Calendula officinalis* and *Dimorphotheca pluvialis*. Ind. Crops Prod. 3, 281–284.

Cahoon, E.B., Kinney, A.J., 2004. Dimorphecolic acid is synthesized by the coordinate activities of two divergent D12-oleic acid desaturases. J. Biol. Chem. 279, 12495–12502.

Cruz, V.M.V., Dierig, D.A., 2012. Trends in literature on new oilseed crops and related species: seeking evidence of increasing or waning interest. Ind. Crops Prod. 37, 141–148.

Derksen, J.T.P., Muuse, B.G., Cuperus, F.P., van Gelder, W.M.J., 1993. New seed oils for oleochemical industry: evaluation and enzyme-bioreactor mediated processing. Ind. Crops Prod. 1, 133–139.

Dodos, G.S., Karonis, D., Zannikos, F., Lois, E., 2015. Renewable fuels and lubricants from *Lunaria annua* L. Ind. Crops Prod 75, 43–50. http://dx.doi.org/10.1016/i.indcrop.2015.05.046.

Duarte, N., Ramalhete, C., Varga, A., Molnar, J., Ferreira, M.-J.U., 2009. Multidrug resistanmce modulation and apoptosis induction of cancer cells by terpenic compounds isolated from *Euphorbia* species. Anticancer Res. 29, 4467–4472.

Eberle, C.A., Forcella, F., Gesch, R., Peterson, D., Eklund, J., 2014. Seed germination of calendula in response to temperature. Ind. Crops Prod. 52, 199–204.

Evangelista, R.L., Hojilla-Evangelista, M.P., Cermak, S.C., Isbell, T.A., 2015. Dehulling of coriander fruit before oil extraction. Ind. Crops Prod. 69, 378–384.

Li, R., Yu, K., Wi, Y., Tateno, M., Hatanaka, T., Hildebrand, D.F., 2012. Vernonia DGATs can complement the disrupted oil and protein metabolism in epoxygenase-expressing soybean seeds. Metab. Eng. 14, 29–38.

Fritsche, K., Hornung, E., Peitzsch, N., Renz, A., Feussner, I., 1999. Isolation and characterization of a calendic acid producing (8, 11)-linoleoyl desaturase. FEBS Lett. 462, 249–253.

Gesch, R., 2013. Growth and yield response of calendula (Calendula officinalis) to sowing date in the northern U.S. Ind. Crops Prod. 45, 248–252.

Hof, L., Dolstra, O., 1999. Response to divergent mass selection for plant architecture and earliness and its effect on seed yield in *Dimorphotheca pluvialis*. Ind. Crops Prod. 10, 145–155.

Isbell, T., 2009. U.S. effort in the development of new crops (Lesquerella, Pennycress Coriander and Cuphea). OCL 16, 205–210.

Knowles, R.E., Goldblatt, L.A., Kohler, G.A., Toy, S.J., Haun, J.R., 1965. Oil seed composition of two species of Dimorphotheca grown in five locations in the United States. Econ. Bot. 19.

Lardizabal, K.D., Metz, J.G., Sakamoto, T., Hutton, W.C., Pollard, M.R., Lassner, M.W., 2000. Purification of a jojoba embryo wax synthase, cloning of its cDNA, and production of high levels of wax in seeds of transgenic Arabidopsis. Plant Physiol. 122, 645–655.

Mietkewska, E., Lin, Y., Weselake, R.J., 2014. Engineering production of C18 conjugated fatty acids in developing seeds of oil crops. Biocat Agric. Biotech. 3, 44–48.

Millam, S., Mitchell, S., Craig, A., Paoli, M., Moscheni, E., Angelini, L., 1997. In vitro manipulation as a means for accelerating improvement of some new potential oil crops. Ind. Crops Prod. 6, 213–219.

Muuse, B.G., Cuperus, F.P., Derksen, J.T.P., 1992. Composition and physical properties of oils from new oilseed crops. Ind. Crops Prod. 1, 57–65.

Pascual-Villalobos, M.J., Robbelen, G., Correal, E., Witzke, E-von, 1993. Performance test of Euphorbia lagascae Spreng., an oilseed species rich in vernolic acid, in southeast Spain. Ind. Crops Prod. 1, 185–190.

Princen, L.H., 1983. New oilseeds on the horizon. Econ. Bot. 37, 478–492.

Shimelis, H.A., Mashela, P.W., Hugo, A., 2013. Principal agronomic and seed oil traits in the industrial oil crop vernonia (*Centrapalus pauciflora* var. *ethiopica*). South Africa J. Plant Soil 30, 131–137.

Suh, M.C., Schultz, D.J., Ohlrogge, J.B., 2002. What limits production of unusual monoenoic fatty acids in transgenic plants? Planta 215, 584–595.

Taylor, D.C., Smith, M.A., Fobert, P., Mietkiewska, E., Weselake, R.J., 2011. Metabolic engineering of higher plants to produce bio-industrial oils. In: Moo-Young, M. (Ed.), Comprehensive Biotechnology, vol. 4. second ed. Elsevier, Amsterdam, pp. 67–85.

Vanhercke, T., Wood, C.C., Stymne, S., Singh, S.P., Green, A.G., 2013. Metabolic engineering of plant oils and waxes for use as industrial feedstocks. Plant Biotech. J. 11, 197–210.

Walker, R.L., Walker, K.C., Booth, E.J., 2003. Adaptation potential of the novel oilseed crop, Honesty, (*Lunaria annua* L.), to the Scottish climate. Ind. Crops Prod. 18, 7–15.

Zanetti, F., Monti, A., Berti, M.T., 2013. Challenges and opportunities for new industrial oilseed crops in EU-27: a review. Ind. Crops Prod. 50, 580–595.

Chapter 12

Successful Commercialization of Industrial Oil Crops

Andrew Hebard
Technology Crops International, Winston-Salem, NC, United States

INTRODUCTION

Based in the United States, Technology Crops International (TCI) is a business with global reach focusing entirely on specialty oilseed crop commercialization and supply chain management, "from soil to oil." Our mission is to remove the challenges associated with the commercial usage of specialty arable crops and develop robust, sustainable, and enduring supply chains for our stakeholders, which start with farmers and end with consumers. Ultimately, we are about creating value through the use of specialty crops, and this has become our main point of reference as we go through the process of developing commercialization strategies or building supply chains. We prioritize and rigorously test the question, "What value and what challenges does it present; and to whom?" I am the founder and CEO of TCI and have spent most of my career in commercial and executive roles relating to specialty arable crops.

From "soil to oil" covers a lot of ground, literally; we have found that planning or modeling a commercial supply chain is most effectively done when we assemble the components into either "Grow, Make or Sell" functions. Simply put, we *Grow* specialty crops, from which we *Make* products for industry, and we *Sell* what we make. Our business and this chapter looks exclusively at specialty arable crops, and excludes fruit, vegetables and plantation crops, as these have a different set of production and market dynamics than those crops that form part of rotational (mostly annual) arable crops.

As all good chefs know, before you embark on making a rabbit pie, first, catch your rabbit, and with that in mind, it would be sensible to qualify both "commercialization" and "industrial" as they relate to the subject of specialty oilseed crops.

Commercialization is the process or cycle of introducing a new product or method of production into the market where its success is measured by the extent to which the market adopts it. *Industrial* relates to those products or services that are used in, or by, industry. Therefore, the focus of this chapter is

Industrial Oil Crops. http://dx.doi.org/10.1016/B978-1-893997-98-1.00012-9

about how to maximize the chances of success when bringing a new oilseed crop to market. It is based on our own experiences with many different crops in many different regions, serving multiple end markets. To use the pie analogy, we have learned to catch, prepare, and cook the occasional rabbit.

GROW

Operationally, growing the crop is where it all begins. Bearing in mind that somewhere in the region of 50–75% of the final cost of the goods, which in this case is a fully refined vegetable seed oil, will be incurred in the crop production stage, it greatly impacts the overall economic viability of the supply chain. Naturally, growers or farmers are the predominant stakeholders in this link of the supply chain, and it is absolutely critical to understand the production dynamics they face and their thought processes as they ruminate on their cropping options season by season. They are acutely aware of the risk–reward dynamics of their business, and with agriculture as much as any weather-dependent industry, risk mitigation is often at the front of their minds. The diagram (Fig. 12.1), while not exhaustive, summarizes many of the risks that growers perceive are associated with moving away from tried and tested commodity crops to new or specialty crops. Of course, these risks are all over and above the day-to-day risks commensurate with commodity crop production.

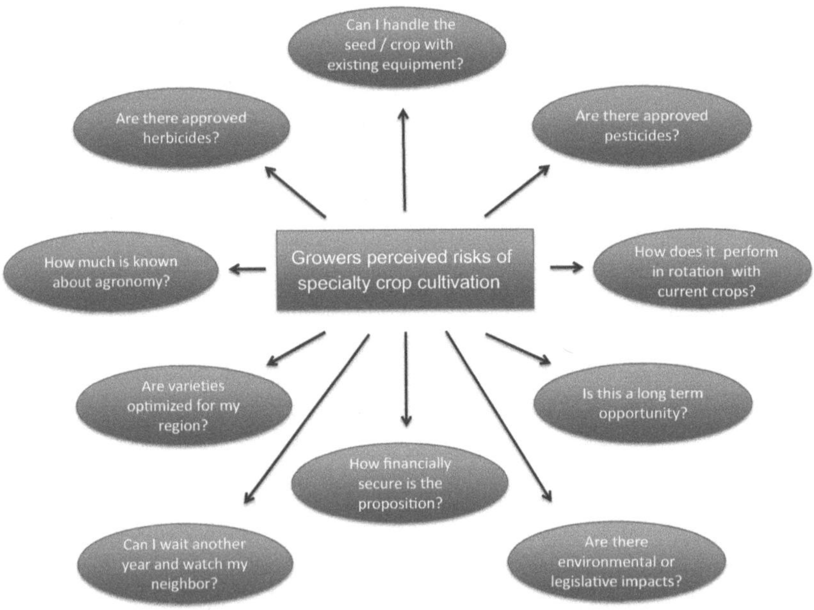

FIGURE 12.1 Growers' perceived risks of specialty crop cultivation.

When asked about growing specialty crops, the two overriding factors seen as high risk in the mind of a grower are (1) the degree of confidence they have in the commercial partner(s) that are supporting the crop's development (in other words, "are they credible?"), and (2) whether some form of income protection or crop insurance is available. As an example, in the United States, Federal Crop Insurance (FCI) is a specific and most widely applied form of subsidy and is available to growers of certain insurable crops. Outside of the United States, other forms of subsidy can provide income protection, yet through a different mechanism such as the Single Farm Payment (SFP) system that operates in most parts of the European Union. Access to and eligibility for different subsidy regimens is important to understand. Many new or specialty crops fall outside of the eligibility criteria for crop insurance, thereby increasing the perceived risk to the grower. The SFP effectively "decouples" subsidies from specific crops and instead provides an area-based subsidy payable to growers, requiring them to comply with various environmental stewardship practices aimed in part, to enhance the countryside. For specialty crops, one of the consequences of this is policy is that it creates the somewhat utopian "level playing field" such that there is no impartiality toward the biodiversity on a farm. For example, a grower that contemplates producing meadowfoam, borage, or camelina sativa in the European Union will broadly receive the same income protection as if he or she grew wheat, barley, or canola. In the United States, the former three specialty crops are outside of the FCI program that supports wheat, barley, and canola. The corollary of this is that greater arable agricultural biodiversity and specialty crop production takes place where economic factors do not handicap producers.

There is perhaps a compelling case that argues how preferentially subsidizing commodity crops such as corn and beans greatly disadvantages specialty crops, as these are the ones that do not attract the investment in breeding and agronomy that helps reduce risk and improve productivity. Conversely, there is the persuasion that specialty crops should stand up on their own commercial merits sans subsidization.

Not surprisingly, the higher the perceived risk a crop has to a grower, the disproportionately higher is the price required by the grower, which translates directly into higher oil costs. Therefore, the ability to mitigate risk is in part driven by the country of production and the agricultural policies it has in place. Navigating these polices and leveraging them can make the difference between viability or unattainable success.

A traditional farmer and a traditional arable rotation?

Let us take a basic cropping scenario for a grower cultivating 1000 acres comprising approximate equal areas of four commodity crops (soybeans, corn, canola, and winter-seeded wheat). They will likely have a good history of growing these crops in a well-established rotation and know the risks and benefits that preceding crops have on following crops. The grower could also have access to some or all of the following:

- World-class genetics from world-class seed companies, adapted to their region
- Genetically modified seeds with bespoke input and or output traits (eg, herbicide tolerance or high oleic oil composition)
- Tried and tested crop protection tools for controlling weeds, insects, and diseases
- Knowledge of optimal crop nutrition regimens (eg, nitrogen, phosphate, and potash requirements) as well as minor trace element requirements (such as selenium, boron, and sulfur)
- Years of experience and investment built up in optimizing the output of these crops on their own farm (soil type, climate, and topography)
- Market price data and price risk management tools (including local elevator delivery prices, futures, and options contracts)
- A farm mechanization and resource structure that has evolved to suit these four crops (including seeders, combines, grain drier, and silo storage)
- Subsidy support in the form of crop insurance or other income protection mechanisms
- Peer pressure from neighbors (not to be underestimated!)
- Advisors, textbooks, and internet resources to answer or give an opinion on nearly everything to do with these crops
- Having most of their entire annual income tied up in these crops on this 1000 acres

Each year, new and enhanced varieties of the same crop species are available to the grower to further advance his productivity and profitability. With all that in mind, moving away from the tried and tested could seem a particularly hard, possibly rash, decision to make. This is the hurdle to overcome with introducing specialty crops, and in our opinion presents the biggest challenge in building the supply chain.

Let us now contemplate introducing a new specialty oilseed crop, for example, one of the many new and minor oilseeds that have been explored for commercialization into this theoretical rotation. This could include meadowfoam, cuphea, camelina sativa, lesquerella, or echium; the options are many and while each one has its own unique situation, they all fall a long way short of the credentials possessed by the four commodity crops they are competing against. Without some risk management tools, it would be a remarkably bold decision to replace one of the four commodity crops in its entirety with a specialty crop, so now becomes the added challenge of having a fifth crop to integrate and manage on the farm.

It is therefore impactful to quantify the risk–reward model in the eyes of a grower as it puts some parameters around the level of income the grower puts at risk (opportunity cost) by moving away from the familiar commodity crops. Most growers and their cropping advisors have a strong grasp on what constitutes realistic revenue from their crops, and this is the measuring stick that a new crop is invariably held up to.

While no two situations are identical, our experience suggests that:

- If all other things are equal (which of course they rarely are) most growers will require a realistic potential gross margin (crop revenue minus direct production costs) benefit or upside of at least 25% to be persuaded away from a well-established "commodity" crop, and
- For each 1% increase in perceived risk, they look for a 2% increase in perceived reward up to a certain point, beyond this the relationship widens to the point at which income assurances or guarantees become required.

Finally, outside of the risk–reward consideration is the third leg of the stool, which is a little more abstract, and that is mindset. Some growers, albeit relatively few, fall into the "early adopter" camp and really enjoy or are motivated by the challenge and prospect of pioneering a new crop. Conversely, well-established, tried-and-tested corn + bean + corn + bean rotation growers would likely be some of the hardest to convince to change, and probably rightly so.

So to conclude these comments on grow, we have found that this is invariably the hardest link in the supply chain to get optimized, and the one that is most often underestimated in terms of cost and time. Therefore, it is also one that greatly benefits from good early stage planning and modeling by means of desktop feasibility studies. These can help identify risks, accelerate the commercialization timeline, reduce or avoid certain costs, or change the ingoing business plan prior to commencement. Selecting the best region and the most appropriate growers are the two biggest factors influencing success in the "grow" link of the chain.

MAKE

In the context of specialty oilseed crops, the make component of the supply chain is about taking a whole oilseed from the farm gate and converting it into a useable ingredient or feedstock for industry. Consequently, there are actually three separate products that require consideration in terms of both handling and processing, and these are the seeds, the meal, and the oil.

Directly following harvest, preserving the quality of the seed is a priority, and it involves ensuring the seed is dry (typically <10% moisture content), clean (typically <2% foreign material), cool (typically <60°F), and free of any deleterious items including insects and mold. Achieving these parameters enhances the storability of the crop and stabilizes the quality. Going back to the 1000-acre farm scenario, it is likely that little, if any, of the drying, cleaning, ventilating, storing, and handling equipment and capacity are available for a new specialty oilseed, and these initially need to be outsourced with a specialist seed handling business, incurring more cost and logistics issues to overcome. Some seeds are relatively inert when clean and dry and can be stored for many years with relatively little deterioration, while others are prone to rapid degradation and require processing into oil shortly after harvest. A *Camelina sativa* seed is as different in its handling requirements to corn, as a Ping-Pong ball is to a basketball, so knowing what physical parameters you are dealing with becomes necessary before you design your piece of the supply chain.

Seeds contain a great deal of biologically active material, some of which can be harmful to humans and animals, which introduces a different set of considerations. While these attributes will not be covered in detail here, they are a highly important factor in the commercialization process. The impacts of these materials can vary from mild irritancy to acute toxicity, and may be experienced during cleaning (eg, through dust or microscopic hairs), during crushing (eg, through the release of certain gases), or in the meal (eg, the exposure to certain proteins). A well-documented example of a seed with known toxicity and allergenicity issues is castor bean, which in turn has stimulated considerable interest in breeding varieties with low allergen/low ricin traits that can be cultivated in the European Union or United States.

Most oilseeds contain between 20% and 50% oil, and therefore the majority (between 50% and 80%) of their biomass is generally a lower value co-product, typically in the form of a protein-rich meal (exceptions may include soybean and canola, which have high-quality, high-protein meals that are in demand for animal feed). Therefore, in most cases removing as much of the oil as possible while maintaining the quality of the oil and meal is commercially imperative. Not having access to suitable oil recovery and subsequent refining technology, capability, and capacity can become rate-limiting steps in the supply chain. Scale and scalability also play a meaningful role.

Although this is somewhat generalized, here are the dynamics we have often encountered with oil recovery and refining:

- Mechanical pressing/expelling of oilseeds is relatively widely available, is low cost, but results in lower oil recovery (leaving >10% residual oil in meal), typically commanding crush fees of $50–150/mt.
- Small-scale solvent extraction (<50 mt/day) is relatively limited in capacity and region, is expensive, but results in high oil recovery (leaving <2% residual oil in meal), typically commanding extraction fees of $400–700/mt.
- Large-scale solvent extraction (>500 mt/day) is relatively accessible in terms of capacity and region, results in high oil recovery, but requires a significant minimum quantity, typically commanding extraction fees of $70–150/mt and minimum quantities to run the facility continuously for at least a few days.
- Small-scale refining is relatively limited in capacity and region, processing losses are generally high (>10%), typically commanding refining fees of $400–600/mt and batch sizes of 1.0–10.0 mt.
- Large-scale refining is relatively accessible in capacity and region, and processing losses are generally low (<6%), typically commanding refining fees of $100–200/mt and batch sizes of >25 mt.

During the "*Grow*" stage, maintaining a specialty crops "identity preservation" is generally done through ensuring good husbandry practices, avoiding situations such as potential pollen drift between the same species but with different traits and carry-over or volunteer seeds in a rotation. During the "*Make*" stage, maintaining identity preservation becomes harder as more variables are introduced and the seeds become mobile as they leave the field and, depending on the degree of either segregation or isolation required, can place significant restrictions or burdens on the operational components of the supply chain. Typically, the market will dictate the degree of identity preservation required, and in most cases this relates to what is determined as the acceptable level, absence or containment of a certain biological marker, or the exclusion of genetically modified material.

Numerous books, research papers, and studies have been published in the area of oil recovery and refining, so this is no more than a cursory overview of the considerations of "*Make*" when it comes to specialty oil crops. However, the importance of planning and avoiding unpleasant surprises can perhaps be illustrated and put into some economic context as follows.

Consider the example of the specialty oilseed crop *Echium plantageneum*, grown as a source of stearidonic acid for dietary supplements.

- It has low seed yields and considered high risk/high inconvenience by growers;
- It commands crop prices of ~\$3.00/lb of seed paid to the grower;
- It has a low oil content of 20%;
- Being high in pyrrolizidine alkaloids (PAs), the meal is not suitable for animal feed and the oil needs fully refining to remove any residual PAs; and
- Total processing and logistics costs are \$0.50/lb.

Therefore, the minimum cost of oil is \$17.50/lb [((\$3.00/lb + \$0.50/lb)/0.2)]. If during oil recovery (using mechanical recovery), 10% of the meal is residual oil, each 1.0 t of meal contains 200 lb of oil, with a potential cost (or loss of revenue) of \$3500. Typically this meal is placed into landfill or incinerated. Refining losses of 10% add a further debit to the profit and loss statement. So to look at it another way, if you purchase 1.0 t of seed at \$6,000, spend \$1000 to expel it into 1800 lb of meal and 200 lb of oil, the lost value of oil that remains in the meal is 50% of the price paid for the seed and processing it in the first place.

So, distilling this down into some conclusions, the following points are crucial for a successful "*Make*" component of the supply chain:

- Ensure you have a comprehensive understanding of the full economic impact of each variable,
- Assess technology, capacity, and capability for oil recovery and refining,
- Understand the importance of co-product values and what drives them,
- Understand the scalability requirements and how this impacts time, costs, and logistics,
- Understand any regulatory issues or show stoppers that are present, and
- Increasing scale drives revenue, increasing revenue drives investment, increasing investment drives performance, and increasing performance drives value, which in turn drives scale.

SELL

At the outset of this chapter, the question "What is the value and what challenges does it bring and to whom?" was posed as one of the most important items not to lose sight of as the prospects of commercializing a new crop sweep you along in a swath of excitement and opportunity. As in the early stages of many ventures, the size of the opportunity is often inflated, the true cost often underestimated and the timeline often unrealistically shortened. When taking the product or project to market, being prepared for the "so what?" question is essential planning. Typically the question is being asked from a buyer's perspective of "is it cheaper and or better than what I am currently using?" Ideally it is both, or can be, given time.

The question "is it cheaper or better?" can be the graveyard of specialty crops, if a compelling case cannot be made for switching from the tried and tested. Equally valuable to understand in the development of a new crop is whether you are in the "demand fulfillment" or "demand generation" business of selling, as they require a different approach and have completely different dynamics. Demand fulfillment is typically an industry pull strategy, where there is an existing market for your product and your raison d'être is to be better or cheaper than what, or who, is currently out there satisfying that demand. Demand creation is when you are looking to supply something entirely new, something that the consumer possibly did not even know they need and you are promoting and pushing it into industry. Here the questions of cheaper or better become less precise to answer, and the "so what?" question becomes the foundation on which to build your stage.

Case Study/Examples

To put some context and granularity to this, below are examples of a demand pull and a supply push scenario for two different specialty crops.

Example 1: Crambe abyssinica—*Demand Pull and Demand Fulfillment*

Erucic acid (EA; C22:1ω9) is a long-chain, monounsaturated fatty acid (FA) found in many garden ornamental flowers such as nasturtiums and wallflowers, as well as members of the *Brassica* family, a genus of plants from the mustard family (see chapter: *Brassica* spp. Oils). It was widely abundant in rapeseed oil in the pre-1970s, at which time it was effectively eliminated through targeted breeding efforts, creating a new quality of edible oil termed "canola" (also known as double low or "00" rapeseed oil in the European Union). The impetus behind this initiative was the reported deleterious health effects that EA exhibited in rat-feeding studies.

EA is a well-established, highly functional and essential component of many nonfood consumer products, most notably as an oleochemical feedstock where it is converted into erucamide and behenyl alcohol, both key building blocks for further manufacturing processes. Erucamide is used as a process aid in the manufacture of polyethylene film, while behenyl alcohol is an emulsifier, viscosity regulator, or emollient in many cosmetic products. Pre-1970s, it was readily available from commodity rapeseed oil. Now, however, with the distinction between canola and rapeseed, the EA crop is grown and processed under tightly controlled, identity-preserved conditions, both to protect the integrity of the canola industry and to maintain the quality of EA production. The crop is known by its acronym HEAR (high–EA rapeseed) and contains a minimum of 45% EA, whereas canola or double low rapeseed has a maximum of 2% EA.

Canola, double low rapeseed, and HEAR are virtually identical visually, agronomically and physiologically other than the two very different levels of EA they have in the seed oil. Being the same species, they also cross-pollinate, and volunteers (self-seeded progeny) can be difficult to control in crop rotations. HEAR and canola are often produced by the same growers, handled at the same elevators, and crushed/stored at the same manufacturing site, so strict identity-preservation protocols are employed at each link of the supply chain to prevent co-mingling of seeds, oil, and meal. Upon delivery to a processing site, testing each individual load for its EA composition is often impractical due to time, cost, and resource.

It is therefore obvious what some of the benefits are that could be achieved through finding an alternative source of EA that is cheaper and or better than HEAR. Wallflowers and nasturtiums clearly are not suitable candidate species, and this was the situation against which *Crambe abyssinica* (see also chapter: Crambe (*Crambe abyssinica*)) was explored and successfully commercialized as an alternative EA source.

Crambe is a somewhat drought-tolerant, low-input member of the Brassica family that contains 56–60% EA (~10% more than HEAR) and demonstrated agronomic traits that allowed it to be cultivated by growers with existing equipment and integrated into traditional crop rotations. It does not cross-pollinate with canola, and the seeds and plants are visibly dissimilar. The seeds are suitable for mechanical pressing and oil extraction using standard hexane technology, and the oil can be refined and split to separate and then recover the various FAs in the same way that HEAR oil can. So, all in all, crambe is a possible "drop in" for HEAR and hence became what would seemingly be a demand fulfillment exercise, notwithstanding the cheaper or better requirement. The species also benefitted from government-funded research and breeding activity, such that growers had access to field-ready genetics.

Now here is what happened in actuality. The significantly higher levels of EA created an immediate appeal for the crop, and its production increased steadily through until the mid-1990s, reaching an estimated 75,000 acres and delivering ~20,000 t of oil, a significant percentage of global usage. It could be grown relatively cheaply and benefitted from an FCI program that was successfully implemented, and the relatively low profitability of competing crops was such that there was only a small opportunity cost to producing it.

The low costs of crop and oil production initially more than offset the negative attributes of crambe oil, but over a relatively short period of time, the positive attributes that related to the "better" question became eclipsed by the negative attributes to the "cheaper" question. To quantify this, while the higher EA levels commanded a price premium of ~15% over HEAR, the lower oil content of crambe (33% vs 44% in HEAR), the high levels of sulfur in crambe oil (100 ppm vs 25 ppm in HEAR), and the high levels of glucosinolates in the meal (150 ppm vs 20 ppm in HEAR) all financially and operationally became increasingly disadvantageous to the crop. Counterintuitively, though, all of these features were present when crambe was first explored as an alternative to HEAR, so why did the economics move against crambe over the next two or so decades?

It was almost entirely driven by the economics of competing crops. Breeding advances greatly improved not just the productivity of existing crops in the crambe growing region but also provided accessibility to crops that previously were not well adapted to that region (eg, corn and soybeans) due to shortness of growing season. During the late 1980s and early 1990s when crambe production peaked in North Dakota (ND), there was virtually no corn and only limited soybean production in the state, and farm gross margins were close to break even for most cash crops. Further, as global commodity prices rose, so did cash crop gross margins, consequently increasing the opportunity cost of not growing a commodity crop. Today, cash crop gross margins are closer to $500/acre, and soybeans and corn are the number two and number three acreage crops in ND, respectively. NDSU reports that from 1989 to 2012 the annual average farm income in ND (for the top 20% of producers) rose from ~$75,000 to >$900,000 (Swenson, 2014). This significant increase in financial performance highlights how the risk-and-reward horizon has changed. As a consequence, ND has witnessed a dramatic reduction in the cropping diversity of the state and specialty crops command markedly reduced acres.

Compounding the situation further, as commodity prices rose, so did oil and meal values of the commodity crops. The limitations on usage, the higher meal content of the seed, and the lower nutritional value of crambe meal resulted in its price competitiveness falling significantly behind other protein sources, which effectively created a bigger hole in the overall economic model.

An important lesson from this is that introducing a new crop into an agricultural environment of marginal productivity and marginal profitability is significantly easier than into an environment of high margin, high profit, and high opportunity cost. Or, to put more simply, if there is little downside to the grower, what does he have to lose by trying something different? Conversely, when the stakes are high, human nature is to take fewer chances. Today, crambe is cultivated on <2000 acres globally and has established itself as a unique ingredient in some personal care formulations that can sustain the markedly higher oil production costs. At the time of writing, HEAR oil is \$0.60/lb and crambe oil is >\$3.00/lb.

In summary, crambe went through to successful commercialization when the opportunity cost of alternative crops was low, when the relative profitability of alternative crops was at parity, and when the regional agricultural industry was favorable to niche crops. Access to newly adapted and elite corn and soybean genetics, coupled with a surge in global commodity prices, dramatically changed the cropping options and corresponding decision making of growers. This is a salient point to consider with specialty crops. The annual investment in the breeding of commodity crops is justified by the acreages they command. Conversely, minor crops cannot sustain major annual investments in genetic improvement and year on year the productivity gap widens, therefore impacting the risk-and-reward equation.

Example 2: Ahiflower™ Demand Creation/Supply Push

The search for sustainable, cost-competitive, and biologically effective sources of omega-3 (ω-3) FAs that can provide meaningful consumer value has attracted a great deal of commercial interest and is largely a market-driven phenomenon, implying an imbalance between emerging demand and existing supply. Broadly speaking, there are three major commercial feedstock origins: algae, marine animals, and terrestrial plants. Each source has unique advantages and disadvantages, and increasing biological efficacy often comes at an elevated cost accompanied by increased concerns over provenance and sustainability (which for this exercise will be defined as Value replaced ≥ (value derived − value depleted)).

The ω-3 FAs eicosapentaenoic acid (EPA; C20:5 ω-3) and docosahexaenoic acid (DHA; C22:6 ω-3) are known to deliver greater health benefit than the shorter, less-unsaturated precursor molecule α-linolenic acid (ALA; C18:3 ω-3). EPA and DHA are traditionally derived from algal and marine sources, whereas ALA is typically found in oils from various terrestrial plants, notably flax, chia, and canola, which contain between 15% and 65% ALA in their oil fractions. ALA is the widely consumed plant-derived ω-3 in human diets, with a reported efficiency of conversion to EPA of ~2–8%, depending on total PUFA in the diet (Gibson et al., 2013) (Fig. 12.2).

Scientific studies generally concur that the major rate-limiting step for converting ALA to EPA in humans is that of initially introducing a fourth double bond, effectively converting ALA (C18:3 ω-3) into stearidonic acid (SDA; C18:4 ω-3) (Walker et al., 2013). Whereas the conversion of ALA to EPA in humans is relatively inefficient, SDA is converted to EPA with an efficiency of up to 30%. There is, therefore, a significant advantage to humans in having higher levels of SDA in ingested ω-3–rich oil to increase the EPA content in human tissue while at much lower intake levels compared to ALA.

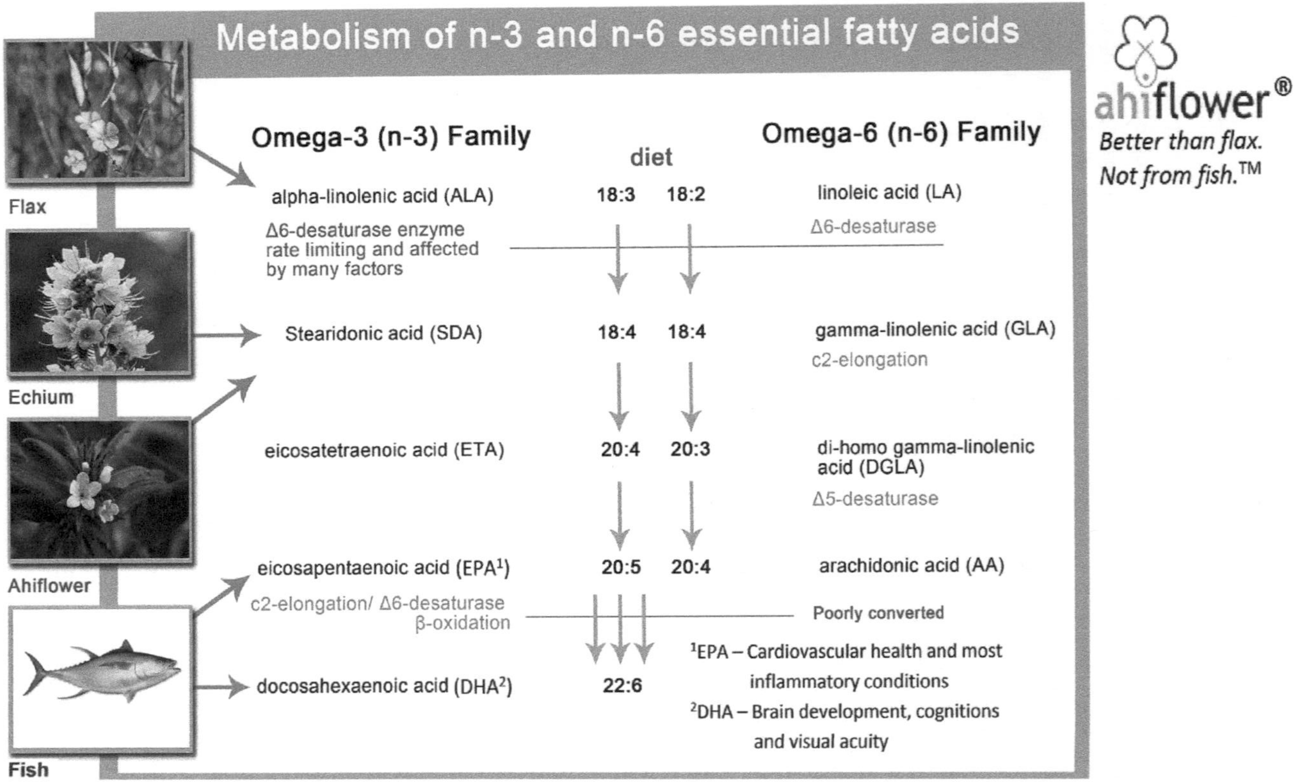

FIGURE 12.2 Metabolism of ω-3 fatty acids. The major plant-derived ω-3 PUFAs are ALA (linolenic acid, 18:3n-3) and SDA (stearidonic acid, 18:4n-3). ALA has important health benefits of its own; however, metabolism to longer chain ω-3's is relatively poor in the human body. The Δ-6-desaturase enzyme limits uptake. Since conversion of SDA is much higher than ALA, about 30% versus 5%, SDA is a much more effective source for longer chain ω-3's such as EPA.

Although some microorganisms can synthesize FA with more than four double bonds, rarely is this seen at meaningful levels in terrestrial plant oils. Even where some plants synthesize very long-chain FA such as EA (C22:1 ω-9) and nervonic acid (C24:1 ω-9), those having more than two double bonds are unusual, those having more than three double bonds are rare, and those having more than four double bonds are virtually unheard of.

This was the background against which our company sought to find a plant species that could produce high levels of ω-3 polyunsaturated FA (ALA or SDA) in an oil-rich seed suitable for commercial-scale growing and oil recovery. We also needed a seed that would let us use traditional oilseed processing and oil-refining techniques and equipment. Additionally, it had to meet all of our supply chain criteria for safe, scalable, and sustainable production. Basically after favorably being able to address the "what value and what challenges does it present; and to whom?" question, it had to be capable of passing the *Grow, Make*, and *Sell* test.

Many literature sources have identified the FA composition of oil-containing plant seeds, which helped us to narrow down the search to a handful of candidate plant families that (1) can be grown under temperate conditions, (2) appear to be malleable in terms of possible domestication, thus removing reliance on wild crafting (the gathering or collecting of uncultivated material from the wild), and (3) meet the commercial proposition intended.

Initially, hundreds of what on paper appeared to be promising species, mostly belonging to the Boraginaceae family, were screened, ranging from *Myositis* (the common forget-me-not) to the many different members of the *Echium* species (such as *Echium plantagineum*, also known unaffectionately in Australia as Paterson's curse or salvation Jane). Also considered were other currently commercially available oils that have ALA and SDA content, most notably hemp and black currant seed; however, none met the criterion of scalability or having sufficiently elevated levels of these desirable FA. In the late 1990s, commercialization of *E. plantagineum* as the preferred candidate crop to deliver ALA and SDA commenced, containing ~32% and ~13%, respectively. This species had another benefit, namely, modest levels of γ-linoleic acid (GLA; C18:3 ω-6), present at ~5% of total FA. From the outset, however, *Echium* production presented major challenges in each of the *Grow*, *Make*, and *Sell* categories for the reasons outlined earlier in the chapter.

Therefore, the search continued and focus shifted to the species *Buggloss-soidies arvensis* (also a member of the Boraginaceae family), a somewhat unprolific wild species native to many regions of the world. While the *Grow* characteristics did not initially meet our criteria, intuitively it was felt they might, if a concerted investment were made to understand its physiology and how to encourage it to adapt to modern row crop or arable cultivation. After a global germplasm search, selection, and evaluation program that is in its sixth year, there are now registered varieties under Plant Variety Protection (PVP) protocols, as well trademarks for this new crop, which we call Ahiflower™.

These are just two important examples of intellectual property creation and protection that are often overlooked with specialty crops.

These varieties represent the most prolific or physiologically suited wild types that were found. Ongoing identification and evaluation form a primary strategy as we look to assemble the world's largest repository of genetics of the species.

Understanding a plant's evolutionary mechanisms and then working with them to produce stable, reliable yields when cultivated is both a humbling and rewarding process. The very traits that enable its survival and competition as part of a biodiversity-rich wild environment are often the same ones that make it wholly unsuitable for cropping. Dormancy, low vigor, and poor uniformity as well as tolerance to climatic and environmental stimuli all need researching and understanding. These hurdles are inherent to the strategy of choosing biodiversity over biotechnology to deliver your target output. We chose biodiversity, largely driven by strong customer preference for a natural product, and our core competence within the life sciences is in bringing specialty crops to market, rather than being a genetics and plant breeding business.

After various setbacks that were more than offset by advancements, the crop is now being commercially cultivated under strict grower licensing agreements in both the Northern and Southern Hemispheres, to deliver two annual crop cycles. Field performance is highly respectable, with seed yields of 900 lb per acre being achieved, delivering ~250 lb of oil per acre. Our strategy is to work with best-in-class growers, those that want to become part of a fully integrated supply chain and to be innovative with their agronomic and business practices, and those that have a real passion for producing value-enhancing specialty crops. Interestingly, many of these same growers produce meadowfoam, borage, and crambe oils under contract also and are willing to try other new species as the opportunity arises.

Regarding the make link in the chain, this captures everything from when the crop is harvested through the manufacturing of the finished oil. Similar to other oilseeds, the process of oil recovery and refining must be modified to the characteristics of the seed and the oil itself. Protocols for seed cleaning and drying, mechanical pressing (expelling), solvent extraction, and oil refining—including neutralization, bleaching, and deodorizing—had to be developed. Once manufactured, the development of analytical testing methodologies, oil stabilization strategies (to prevent deterioration through oxidation), and product formulation or oil delivery systems (such as soft gel capsules and functional food ingredients) are necessary so that a high-quality, safe, efficacious, and consumer-acceptable product could be delivered. Under correct storage conditions Ahiflower™ seed can be held for 12 months before crushing and oil stored for 6 months after refining by using suitable protective antioxidant measures.

As defined earlier, the success of taking this oil to market is dependent upon demand creation, and if commercialization is a three-legged stool (composed of grow, make, sell), this sell leg of the stool needs to be in complete balance with the grow and the make legs. However, unlike these other two legs, which have their own unique hurdles to overcome for specialty crops, selling is almost universal in its challenge and not specific to specialty crops. The criteria of better and cheaper are relevant, as is the "so what" question.

To take Ahiflower™ to market, a consumer value proposition was developed that attempted to talk to these three considerations. Through market research we recognized that success in selling the product would be impacted by our ability to communicate some or all of the following attributes:

- Safety
- Authenticity or traceability of supply
- Efficacy
- Natural
- Sustainable
- Affordable

Each of these presents its own further set of accomplishments to undertake. For example, safety needs to be both proven scientifically through the analysis for any harmful components and regulatory through the gaining of approvals for human consumption in the target markets. Authenticity is about the products provenance and the ability to trace production and processing back to source. For this we use a proprietary crop stewardship and safety platform called Crop Assured 365™ enabling a full "soil to oil" traceability audit to be carried out.

Efficacy requires the proof of action such that the consumer understands what benefit can be expected from using the product and this too needs to be supported by empirical data. In the case of Ahiflower™, this was achieved through investing in human feeding studies. *Natural* is the preferred descriptor for many dietary supplements and health foods (the main target market for this oil) and basically shows a predilection for non-genetically or nonchemically modified products.

Sustainable means different things to different people, so we looked at both the sustainability of our product and the sustainability of products that we would be likely competing against. In the case of ω-3 supplements, these are predominantly fish oil and krill oil, so making a case that an annual oilseed crop is likely to be more sustainable than wild harvesting of marine organisms is straight forward. Finally, *affordability* must be considered. Will the consumer value proposition be strong enough to elicit purchasing habits such that it returns enough cash back to all stakeholders to franchise them into becoming long-term partners in the supply chain? This, of course, is what has to underpin all of the investment and resource utilization taken to bring the product to market.

While there are still challenges to overcome, all indications are that Ahiflower meets all the requirements for a successful new specialty oilseed crop, and will be a successful venture for all partners in the supply chain.

CONCLUSION

There is a lot of wisdom in recognizing that you or your company cannot be all things to all people, and in the commercialization of specialty crops, TCI recognized that our strengths lay in managing the grow, make, and sell functions for products where we looked to meet an existing demand. Creating demand for an entirely new product requires expertise in marketing, branding, consumer awareness, and many other skills that we do not possess. So in the case of Ahiflower™, our modus operandi has been to partner with the best-in-class businesses that can take the product to the consumer in partnership with ourselves through a business-to-business (B2B) shared risk, shared reward model.

To conclude, the successful commercialization of specialty arable crops requires:

- A deep understanding of the grow, make, and sell aspects of the supply chain at the operational, technical, and financial levels,
- The recognition and engagement with the best partners, including growers, processors, and marketers, and
- Understanding the bigger picture and the stakeholders it contains such as regulatory bodies, government policies on subsidies, and, of course, the investors and executives who will be responsible for the successes of the initiative.

REFERENCES

Gibson, R.A., Neumann, M.A., Lien, E.L., Boyd, K.A., Tu, W.C., 2013. Docosahexaenoic acid synthesis from alpha linolenic acid is inhibited by diets high in polyunsaturated fatty acids. Prostaglandins Leukot. Essent. Fatty Acids 88, 139–146.

Swenson, A., 2014. North Dakota Farm Financial Performance. https://www.ag.ndsu.edu/farmmanagement/documents/aglendnonrrv (August 31, 2015).

Walker, C.G., Jebb, S.A., Calder, P.C., 2013. Stearidonic acid as a suplemental source of ω-3 polyunsaturated fatty acids to enhance status for improved human health. Nutrition 29, 363–369.

Chapter 13

Applications of Doubled Haploidy for Improving Industrial Oilseeds

Alison M.R. Ferrie, Karen L. Caswell
National Research Council, Saskatoon, SK, Canada

INTRODUCTION

Crop improvement is essential to meet the demands of a changing world (eg, increased population, climate change, and decreasing land base). Plant breeding programs are continuously striving to increase crop yield, enhance crop quality, and improve crop tolerance to diseases and pests. It takes approximately 10 years to develop a new crop variety, although this depends on the species. The traditional plant breeding methods involve the time-consuming process of self-fertilization or back-crossing in order to produce homozygous lines. Doubled haploid (DH) production is the quickest method for developing homozygous breeding lines (as reviewed in Ferrie and Möllers, 2011). The production of DH lines from microspores or ovules makes it possible to obtain stable homozygous lines in a single step. This chapter provides an overview of in vitro DH methods for improving oilseeds intended for industrial purposes. The focus will be on *Brassica*, other cruciferous, Apiaceae, and Compositae species, because little or no DH work has been done on other potential industrial oilseeds.

DH METHODOLOGY

In vitro DH methods that use the culture of anthers, isolated microspores, or ovules, for the improvement of plant species have been reviewed recently (Dunwell, 2010; Ferrie and Caswell, 2011a). Androgenesis, the culture of the male gametophyte, is the preferred method in most of the oilseeds for which there are DH protocols (Table 13.1). This is not new technology but was first reported in 1964 (Guha and Maheshwari, 1964). The early work was mostly with the Solanaceae family but has now been extended to over 250 species (Maluszynski et al., 2003).

Industrial Oil Crops. http://dx.doi.org/10.1016/B978-1-893997-98-1.00013-0

TABLE 13.1 Doubled Haploid Protocols and Response for Selected Oilseed Species

Oilseed Species	DH Method	Response	Plant Regeneration	References
Cruciferae				
B. carinata	IMC[a]	Embryos	Haploid/DH	Barro and Martin (1999)
B. juncea	IMC	Embryos	Haploid/DH	Ferrie (2003)
B. napus	IMC	Embryos	Haploid/DH	Ferrie (2003)
B. nigra	IMC	Embryos	Haploid/DH	Ferrie and Keller, (2007)
B. rapa	IMC	Embryos	Haploid/DH	Ferrie (2003)
Camelina sativa	IMC	Embryos	Haploid/DH	Ferrie and Bethune (2011)
Crambe abyssinica	IMC	Embryos	Haploid/DH	Ferrie and Keller (2007)
Eruca sativa	IMC	Embryos	Diploid	Leskovšek et al. (2008)
Lesquerella sp.	IMC	Swollen cells	No plants	Tomasi and Ferrie (2003)
Sinapis alba	IMC	Embryos	Haploid	Ferrie and Keller (2007), Mithila and Hall (2007)
Apiaceae				
Caraway	Anther culture	Callus	Haploid/DH	Smykalova et al. (2009)
	IMC	Embryos	Haploid/DH	Ferrie et al. (2011b)
Dill	IMC	Embryos	Haploid/DH	Ferrie et al. (2011b)
Fennel	IMC	Embryos	Haploid/DH	Ferrie et al. (2011b)

Compositae				
Calendula	IMC	Callus	Haploid	Ferrie and Caswell (2011b)
Safflower	Anther culture	Callus	Haploid	Prasad et al. (1990, 1991) Caswell and Ferrie (unpublished)
	IMC	Embryos	No plants	Caswell and Ferrie (unpublished)
	Ovary culture	Callus	Shoots only	Caswell and Ferrie (unpublished)
Sunflower	Anther culture	Callus, embryos	Haploid	Gurel (1991a, 1991b), Thengane et al. (1994) Saji and Sujatha (1998), Mix (1985) Caswell and Ferrie, (unpublished)
	IMC	Callus	No plants	Coumans and Zhong (1995)
	Ovary culture	Callus, embryos	Haploid	Mix (1985), Gelebart and San (1987)
Other				
Flax	Anther culture	Callus	Haploid	Reviewed by Obert et al. (2009)
	IMC	Callus	Haploid	Reviewed by Obert et al. (2009)
	Ovary culture	Callus	DH	Burbulis et al. (2011), reviewed by Obert et al. (2009)

[a]*IMC, isolated microspore culture.*

Microspore embryogenesis can be influenced by genotype, donor plant growth conditions, stage of microspore development, composition of the culture medium, and environmental conditions during culture (Ferrie and Caswell, 2011a). The frequency with which microspore-derived embryos are produced will depend on whether these factors are favorable. Embryo conversion to plants is dependent on genotype (in some species) and can be influenced by media composition and culture conditions. In many cases, these plants are haploid but spontaneous doubling of the chromosomes can sometimes occur resulting in a homozygous DH plant. For plants that remain haploid, treatment with colchicine or other doubling agents will create a DH plant.

There are many advantages to using haploidy technology for practical applications (eg, varietal development, mutagenesis, and transformation) and basic research (eg, genomics, biochemical, and physiological studies), and this has been reviewed recently (Dunwell, 2010; Ferrie and Möllers, 2011). Using DH methodology in a breeding program can accelerate the development of new cultivars by 3–4 years, depending on the species (Ulrich et al., 1984). Other advantages include easier identification of recessive traits, more efficient selection, and being able to work with smaller population sizes. Furthermore, marker-assisted breeding is more efficient in haploid embryos and/or plants as only a single allele is present at any given locus.

Microspores, since they are single cells and haploid, make ideal material for inducing mutation, transformation, and the newer methods of gene editing. One can make changes at the single cell level by exposing the cells to a mutagen or by transformation and subsequently regenerate a plant from the cell. Traits can be fixed by chromosome doubling to achieve homozygosity. Other advantages of using microspores in mutagenesis or transformation are the large number of uniform cells that can be treated in a relatively small space and the ease of selection by applying the appropriate selection pressure.

DH methods can be used for biochemical and physiological studies. The developmental sequence of a microspore-derived haploid embryo is comparable to its zygotic counterpart, and many of the biochemical pathways are similar. Many studies have been conducted in the *Brassica* species looking at fatty acid accumulation and composition (Pomeroy et al., 1991; Taylor et al., 1990; Weselake et al., 1991; Wiberg et al., 1991).

BRASSICA SPECIES

There have been many studies on *Brassica* tissue culture and regeneration, and these protocols have been widely used for the improvement of the oilseed *Brassica* crops (Mollika et al., 2011; Rani et al., 2013). Efficient isolated microspore culture protocols are available for most of the *Brassica* species: *Brassica napus*, *Brassica rapa*, *Brassica juncea* (Ferrie, 2003), and *Brassica carinata* (Barro and Martín, 1999; Ferrie, 2003). However improvements in these protocols are continually being made as there are significant genotype differences. Additives to the medium, such as 2-(*N*-morpholino)ethansulfonic acid (MES), gum arabic, bleomycin, *p*-chlorophenoxyisobutyric acid (PCIB), polyethylene glycol, and brassinosteroids (Ahmadi et al., 2012; Ferrie et al., 2005; Ferrie and Keller, 2007; Yuan et al., 2012; Zhang et al., 2011), as well as precise bud selection, can enhance embryogenesis in some of the more recalcitrant genotypes (Bhowmik et al., 2011b).

Although DH techniques are routine for many of the *Brassica* species, there have been very few reports of microspore culture in *Brassica nigra* (Ferrie et al., 2005; Ferrie and Keller, 2007). While *B. nigra* is not a major oilseed crop, there is considerable interest in the species as it is one of the three progenitor species in the genus *Brassica*, contributing the B genome to *B. juncea* (AABB) and *B. carinata* (BBCC). There is significant effort in the oilseed industry to promote *B. juncea* as a drought- and heat-tolerant species (Potts et al., 1999); however, this effort is limited by the lack of suitable genetic variation within the B genome. *B. juncea* varieties developed to date have been derived through interspecific hybridization with *B. rapa* (A genome) and *B. napus* (A and C genomes) for B genome introgression of appropriate quality trait genes through nonhomologous recombination. Development of novel variation within the B genome would allow for more rapid development of a broad *B. juncea* germplasm base that would contribute directly to development of varieties. Five *B. nigra* lines were evaluated for microspore culture response including Ni100, a previously developed DH line (Ferrie, unpublished). The microspore culture protocol was based on the *Brassica* protocol described by Ferrie and Keller (1995) and Ferrie et al. (2008). A number of factors were evaluated in order to enhance microspore embryogenesis in *B. nigra*. Microspore-derived embryos were observed for all five genotypes evaluated, with differences in embryogenic response between genotypes, as to be expected. Ni100 was the most embryogenic line with a frequency of 665 embryos/100 buds. In most cases, plants grown at 20/15°C (day/night temperature) were the most embryogenic, in contrast to *B. napus* and many of the other *Brassica* species that require a colder temperature (10/5°C). NLN culture medium (Lichter, 1982) supplemented with 25% PEG 4000 (polyethylene glycol; molecular weight of 4000) or 13% sucrose induced embryo development (Ferrie and Keller, 2007).

As shown for the *Brassica* species, the frequency of embryogenesis is sufficient for this system to be used for practical applications such as mutagenesis studies (Table 13.2). Microspore mutagenesis studies have been carried out in the *Brassica* species using a variety of mutagenic agents, such as EMS (ethyl methanesulfonate), ENU (ethyl nitrosourea), NaN_3 (sodium azide), or γ- and ultraviolet (UV) irradiation. Different traits of value have been produced and identified including disease resistance (Ahmad et al., 1991; Zhang and Takahata, 1999), cold tolerance (McClinchey and Kott, 2008), herbicide tolerance (Swanson et al., 1988), and fatty acid modification (Barro et al., 2001, 2002; Beaith et al., 2005; Ferrie et al., 2008; Sonntag and Rudloff, 2004).

TABLE 13.2 Microspore Mutagenesis for Oil Quality Characteristics

Species	Mutagen	Characteristics Evaluated	Response
Brassica carinata	EMS	Modification in erucic acid levels	Barro et al. (2001)
Brassica carinata	UV	Modification in glucosinolate levels	Barro et al. (2002)
Brassica juncea	EMS	Alterations in FA profile	Ferrie et al. (2008)
Brassica napus	UV	Changes in saturated fatty acid levels	Beaith et al. (2005)
Brassica napus	UV	Changes in saturated fatty acid levels	Sonntag and Rudloff (2004)
Brassica napus	EMS	Alterations in FA profile	Ferrie et al. (2008)
Brassica napus	ENU	Alterations in FA profile	Wong and Swanson (1991)
Brassica nigra	EMS	Alterations in FA profile	Ferrie (unpublished)
Brassica rapa	EMS	Alterations in FA profile	Ferrie et al. (2008)

A large microspore mutagenesis project generated more than 90,000 haploid/DH plants of *B. napus*, *B. rapa*, and *B. juncea* (Ferrie et al., 2008) and to a smaller extent *B. carinata* and *B. nigra* (Ferrie, unpublished). The mutagen used was EMS, and a killing curve determined the concentration of EMS to apply in the mutagenesis experiments. EMS concentrations ranging from 0.1% to 0.4% were evaluated. At 0.1% EMS, there was a 20–50% reduction in embryogenesis, whereas for 0.4% EMS the reduction in embryogenesis was much higher (>95%), although this varied depending on the species. For the mutagenesis experiments, the microspores were incubated with EMS (0.1% or 0.2%) for 90 min and then cultured as per the *Brassica* microspore culture protocol (Ferrie and Keller, 1995; Ferrie et al., 2008). The microspores were also subjected to an in vitro colchicine treatment to double the chromosome number. The embryos that survived both the EMS and colchicine treatments were regenerated to plants. Fatty acid analysis was conducted on 7000 *B. rapa* DH plants grown in the greenhouse, with differences observed among the DH lines (Ferrie et al., 2008). From the 7000 DH lines, 197 lines had oleic acid (C18:1) levels greater than 55% (parental line 50.3%) and 69 lines had less than 8% α-linolenic acid (C18:3) (parental line 12.3%). For total saturates (C12–C24), the levels ranged from 3.8% to 45.0%, with 157 lines exhibiting less than 5% (parental line 6.9%). Other interesting fatty acid profiles were identified;

for example, high myristic acid (C14:0), high stearic acid (C18:0), high arachidonic acid (C20:0), high eicosanoic acid (C20:1c11), high hexacosanoic acid (C26:0), or low palmitic acid (C16:0) compared with the parental line. Field trials were conducted with the *B. napus* and *B. juncea* DH lines. Again, differences were observed among the fatty acid profiles of the DH lines. For *B. napus*, the saturate levels ranged from 5.0% to 7.7%. One line had 45.7% erucic acid, 15.1% oleic acid, and 14.2% α-linolenic acid. All other lines were of canola quality similar to the parental line. For the 3630 *B. juncea* lines that were evaluated, there were differences among the DH lines; however, these differences were within the range normally seen in *B. juncea* germplasm. Differences among the DH lines were also observed in terms of agronomic response, such as plant height and time to flowering (Ferrie et al., 2008).

A similar microspore mutagenesis project was carried out with two genotypes of *B. nigra*, including Ni100. Most of the microspore mutagenesis experiments used 0.1% and 0.2% EMS, which had previously been used in *B. napus*, *B. juncea*, and *B. rapa* experiments. At 0.2% EMS, there was a reduction in embryogenic frequency in the range of 36–91%, depending on the experiment (Ferrie, unpublished). A total of 129 mutagenesis experiments were conducted with *B. nigra*, which resulted in 870 DH lines with seed. There were differences between the parental line and the DH lines for levels of fatty acids; however, further analyses must be completed to determine if the differences are significant.

Other mutagenic agents such as shortwave UV light have also been used to generate alterations in the fatty acid profile of *B. napus* (Beaith et al., 2005). Freshly isolated microspores of four *B. napus* genotypes were exposed to UV light for 90 s. The developing embryos were subsequently subjected to a heat stress, which facilitated the identification of lines with lower saturate levels. Differences were found among the four genotypes in terms of embryogenic response. Cotyledonary tissue from developing embryos was sampled for fatty acid analysis, and the embryos were then regenerated to DH plants. Seed from 211 DH lines was analyzed and compared with the appropriate control. Reduced palmitic acid content was observed in 76% of the DH lines and 68% had reduced stearic acid content.

Brassica carinata microspores have been mutagenized using UV light (Barro et al., 2002) or EMS (Barro et al., 2001). DH lines with modifications in glucosinolate and erucic acid contents were identified. The normal range for erucic acid content in *B. carinata* genotypes is between 35% and 45% of the total fatty acid composition (Mnzava and Olsson, 1990). From 400 DH lines derived from EMS mutagenized microspores, the authors identified six lines with erucic acid greater than 52% and three lines with less than 25% erucic acid (Barro et al., 2001). In the experiments using UV for mutagenesis, DH lines with low or high glucosinolates were identified (Barro et al., 2002). Glucosinolates ranged from 37.5 to 99.2 μmol/g seed. The untreated control had a mean of 80 μmol/g seed. Higher erucic acid levels were also identified with an increase from 42.8% in the untreated control to 48.8% in the mutagenized lines.

Microspores, microspore-derived embryos, and DH plants have been used for transformation studies and have been reviewed recently (Ferrie and Möllers, 2011). As mentioned previously, because of the haploid single cell nature of the microspores and the ability to regenerate these cells to plants with high efficiency, this is an ideal system for transformation. Different methods of transformation have been used with *Brassica* microspores including electroporation (Jardinaud et al., 1993), particle bombardment (Fukuoka et al., 1998), microinjection (Jones-Villeneuve et al., 1995), *Agrobacterium tumefaciens* (Dormann et al., 1995, 2001; Pechan, 1989), and the newer methods of macroinjection with nanocarriers (Bhowmik et al., 2011a). Using microspore transformation methods with the newer methods of gene editing for precise targeting of genes could enhance breeding for different oil profiles.

The *Brassica* species mentioned here are good examples of utilization of an efficient DH protocol for the improvement of the crops. For the other species in this chapter, the DH protocols are not well established and therefore have not been used in breeding programs or for mutagenesis or transformation.

OTHER CRUCIFEROUS SPECIES

Camelina sativa (L.) Crantz

Camelina sativa, a member of Brassicaeae, has a unique seed oil profile, with the majority of the fatty acids being linolenic (C18:3), oleic (C18:1), linoleic (C18:2), and eicosenoic (C20:1) (Gugel and Falk, 2006). There is interest in using this oil for industrial purposes, human nutrition, and pharmaceuticals. Compared with *B. napus*, camelina has a number of advantages as it is more tolerant to insects (Gugel and Falk, 2006), diseases (Conn et al., 1988; Salisbury, 1987), and drought (Gugel and Falk, 2006). Despite these beneficial qualities of camelina, improvements in seed yield, oil content, and seed size are required to increase commercial production of the crop. Very little tissue culture research has been published on camelina, even with the interest in this crop. Regeneration studies have looked at several different explants, with leaf explants being more responsive than hypocotyl explants (Tattersall and Millam, 1998). These studies were used as a basis for an in vitro transformation method that was developed and patented (Kuvshinov et al., 2011). However, most studies use the in planta transformation method of floral dip for camelina (Liu et al., 2012; Lu and Kang, 2008). Several protoplast fusion studies have been conducted between *C. sativa* and *Brassica* species; however, the objective of these studies has been to improve the *Brassica* partner (Hansen, 1998; Jiang et al., 2009; Narasimhula et al., 1994; Sigareva and Earle, 1999). These hybrid products produced shoots, but rooting was a problem, most likely due to the camelina parent.

A *Brassica* DH protocol with modifications was used to generate haploid and DH plants of camelina (Ferrie and Bethune, 2011). The culture medium used for camelina was NLN (Lichter, 1982) with 12.5% sucrose and 12.5% PEG 4000. The microspores were plated at a low density (10,000 microspores/mL) compared with *Brassica* species (50,000–100,000 microspores/mL). Camelina microspores were given a 72-h glutamine starvation similar to the *B. rapa* protocol (Ferrie, 2003). The frequency of embryogenesis was low compared with *B. napus*, with a frequency of approximately 38 embryos per 10^5 microspores (Ferrie and Bethune, 2011). Despite the low embryogenic response, over 100 DH plants were regenerated and preliminary fatty acid analysis of 68 of these DH lines indicated that there were differences among the DHs as well as between the DH lines and the parental genotype (Ferrie and Smith, unpublished).

Lesquerella Species

Lesquerella is a member of Brassicaceae and has an oil profile rich in hydroxyl fatty acids (HFA) that are used for making nylons, plastics, lubricants, and waxes (Dierig et al., 1993; Thompson et al., 1989). *Lesquerella fendleri* has 60% lesquerolic acid (C20:1OH) and 1.5% densipolic acid (C18:2OH) (Barclay et al., 1962). In some *Lesquerella* species, the HFA level can be as high as 90% (Rollins and Shaw, 1973). This profile is similar to that of castor (*Ricinus communis* L.) without the highly toxic component ricin, and therefore its oil could replace the role of castor oil for industrial purposes—hence, the interest in commercializing this plant. Crop improvements, such as increased seed yield, are required in order to make *Lesquerella* more appealing to the producer. In vitro methods developed for *Lesquerella* include protoplast fusion (Ovcharenko et al., 2011; Skarzhinskaya et al., 1996), embryo rescue (Du et al., 2008), ovule culture (Dierig et al., 2004), transformation (Chen, 2011; Wang et al., 2008), and cell suspensions (Kharenko et al., 2011).

Microspore culture experiments have been attempted in five *Lesquerella* species (i.e., *Lesquerella fendleri*, *Lesquerella pallida*, *Lesquerella auriculata*, *Lesquerella densipila*, and *Lesquerella lindheimeri*) (Tomasi and Ferrie, 2003). Different bud sizes, microspore density, and media compositions were evaluated; swollen microspores were produced, but there was no embryo development. Microspore developmental stages have been correlated to floral bud and petal length, which assists in bud selection (Tomasi et al., 1999). It was found that late uninucleate microspores (determined to be embryogenic in *Brassica* species by Dunwell et al., 1985) were observed in buds 3.7–4 mm in size. However, anther culture and microspore culture experiments using bud sizes from 3.13 to 4.05 mm (correlating to early uninucleate to vegetative/generative microspores) were unsuccessful in generating embryos (Tomasi et al., 1999).

Eruca sativa Mill

Eruca sativa, also known as Rocket or arugula, is usually used as a salad component, but there is interest in using it as a lubricant or biodiesel crop (Li et al., 2009). The seeds contain approximately 45% erucic acid and about 9% gadoleic acid (C20:1) (Lazzeri et al., 2004). Plantlet regeneration via somatic embryogenesis (Chen et al., 2011), protoplast culture (Sikdar et al., 1987), and

Agrobacterium-mediated transformation (Slater et al., 2011) have been reported for *Eruca sativa*.

In terms of DH, a microspore culture protocol has been developed for *E. sativa* based on the *B. napus* protocol (Leskovšek et al., 2008). Buds (4–5 mm in length) in the late uninucleate to early binucleate stage of microspore development were selected for the experiments. NLN basal medium (Lichter, 1982) with 13% sucrose was used as the wash medium and the culture medium. Activated charcoal was added to the NLN culture medium. A heat shock of 32°C for 24 h was beneficial. Embryos were observed 14–21 days after initial culture; however, conversion to plants was low most likely due to secondary embryogenesis. The majority of the resulting plants were diploid (Leskovšek et al., 2008).

Crambe abyssinica Hochst. ex R.E. Fries

Crambe abyssinica is an oilseed crop with high levels of erucic acid (55–60%) (Mandal et al., 2002). It is used as a lubricant or corrosion inhibitor, as well as an ingredient in synthetic rubbers, plastic films, and nylons. In vitro regeneration has been successful using hypocotyls (Li et al., 2011; Palmer and Keller, 2011), cotyledons, or leaves (Furmanek and Banaś, 2011) as explants. Hypocotyl explants have been used in *Agrobacterium* transformation of crambe resulting in transgenic plants (Chhikara et al., 2012; Li et al., 2010), which carried the wax synthase and fatty acid reductase genes (Li et al., 2010).

Crambe is considered a recalcitrant species in DH. However, the use of PEG 4000 was beneficial as embryos were produced using NLN medium (Lichter, 1982) supplemented with 25% PEG 4000. The addition of 3 mL of NLN medium with 13% sucrose 14 days after culture initiation was beneficial for continued development of the embryos (Ferrie and Keller, 2007). Fertile DH plants were regenerated from these microspore-derived embryos. However, the frequency of embryogenesis was still very low and insufficient for practical application.

APIACEAE SPECIES

The Apiaceae are a diverse group of plants that includes vegetables, herbs, and spices. With the exception of carrot (*Daucus carota* L.), there are very few crop improvement programs for these species. The Apiaceae are a source of secondary metabolites and compounds that have medicinal value as well as industrial applications. *Anethum graveolens* L. (dill) could be used as a biopesticide (Evergetis et al., 2013), whereas caraway (*Carum carvi* L.), which contains carvone, has been used in agriculture as a sprouting inhibitor (Bouwmeester and Kuijpers, 1993), antifungal agent, and insect repellent (Toxopeus and Bouwmeester, 1993).

Protocols for regeneration via embryogenesis or organogenesis have been reported in caraway (Ammirato, 1974; Furmanowa et al., 1984; Krens et al., 1997) and dill (Jana and Shekhawat, 2011). Somatic embryogenesis of the apiaceae species was reviewed by Ekiert (2000). These in vitro protocols are important for further improvement of this group of plants. These protocols would allow for rapid advancements in breeding and for studying important traits. DH plants have been generated from caraway via anther culture (Smykalová et al., 2009).

An isolated microspore culture protocol has been developed for some of the Apiaceae species (Ferrie et al., 2011b). Twenty Apiaceae species were evaluated for microspore culture response. Of those 20 species, 11 of them were induced to develop microspore-derived embryos and DH plants were generated from the majority of those species that produced embryos. For 10 of the species (eg, fennel (*Foeniculum vulgare* Mill.), dill, and anise (*Pimpinella anisum* L.)), NLN basal medium (Lichter, 1982) was beneficial, whereas for caraway, AT-3 (Touraev and Heberle-Bors, 1999) was advantageous for embryo production (Ferrie et al., 2011b). The presence of elevated concentrations of sucrose (20–25%) resulted in higher embryo yields for some species, as did the use of brassinosteroids. DH lines of dill and caraway were grown under field conditions for two to five seasons (Ferrie et al., 2011a). Differences were observed among the DH lines and between the DH lines and the parental donor material for height, seed yield, essential oil concentration, and percentages of carvone and limonene. Several of the dill DH lines were short and early maturing, which are desired characteristics for a breeding program. The essential oil content was similar or higher in the DH lines than in the parental material.

COMPOSITAE SPECIES

The Compositae or Asteraceae family comprises a diverse group of species which are valuable for human food, oil production, ornamentals, insecticides, medicinal uses, and industrial applications (reviewed in Dempewolf et al., 2008). Tissue culture techniques are available for some of the species; however, this family is considered recalcitrant regarding DH techniques. In the author's laboratory, experiments were conducted in an effort to develop an efficient and consistent protocol for the production of DHs in a few of the compositae species. Anther culture, isolated microspore culture, and ovary culture were explored by testing preculture factors (eg, growth conditions of donor plants, genotype, and pretreatments) and postculture factors (eg, media and incubation conditions) (Ferrie, 2007; Ferrie and Caswell, 2011a) with limited success.

Carthamus tinctorius L.

Safflower (*Carthamus tinctorius*) mainly produces an edible oil, as well as a drying oil for enamels, caulks, putties, and nonyellowing drying paints (reviewed in Ekin, 2005). In vitro regeneration and *Agrobacterium*-mediated transformation have been reported in safflower (Başalma et al., 2008; Motamedi et al., 2011; Sri Shilpa et al., 2010; Vijaya Kumar et al., 2008).

Prasad et al. (1990, 1991) were successful in producing haploid plants from safflower anthers. They evaluated five different basal media N6 (Chu et al., 1975), B5 (Gamborg et al., 1968), Chaleff's (Chaleff and Stolarz, 1981), LS (Linsmaier and Skoog, 1965), and MS (Murashige and Skoog, 1962) and determined that MS was the best for inducing callus production (Prasad et al., 1991). Shoot regeneration was observed on MS supplemented with 6-benzylaminopurine (BAP) and α-naphthaleneacetic acid (NAA), and roots were regenerated when shoots were transferred to half-strength MS with NAA. Of the plants regenerated, 64% were haploid.

We observed in our anther culture experiments that calli could be produced from five genotypes, using four basal media (D1 (Smykalová et al., 2009), M3 (Gurël et al., 1991a, 1991b), modified R92.25 (Theiler-Hedtrich and Hunter, 1996), MS + 2 mg/L BA + 0.5 mg/L NAA (Caswell and Ferrie, unpublished), and MSG-Cl (Becwar et al., 1990)) at two culture temperatures (24°C, 32°C). Multiple shoots were produced from calli when subcultured to a B5-based medium (Gamborg et al., 1968) with 2% sucrose supplemented with glutamine, serine, BAP, and NAA (Caswell and Ferrie, unpublished). In a few cases, the shoots produced flower buds in vitro but in most cases failed to produce roots.

Preliminary microspore culture experiments were conducted using eight cultivars of safflower (Caswell and Ferrie, unpublished). Calli developed from three genotypes. In all cases, the calli developed in a medium that contained growth regulators, and therefore the possibility exists that the calli developed from somatic hairs rather than from microspores. This phenomenon has been observed in other Compositae species (Coumans and Zhong, 1995; Ferrie and Caswell, 2011b). Somatic hairs can sometimes produce callus in microspore culture media, especially medium with growth regulators and in some cases, plants can be regenerated. Achieving a clean culture, free of somatic tissue is essential.

Further microspore culture experiments were conducted with two genotypes, using medium without growth regulators. A total of nine embryos/embryo-like structures were produced from these lines in NLN media (Lichter, 1982) with 10% or 13% sucrose and a heat shock treatment of 7–21 days at 32°C, prior to culture at 24°C. Embryos could not be consistently produced, even though the experiments were repeated multiple times. These embryo/embryo-like structures were not sufficiently developed for ploidy analysis.

The culture of safflower ovaries resulted in calli being produced with seven types of media, all of which contained growth regulators (eg, D1 (caraway medium, Smykalová et al., 2009), D1-MS, BR1, M3 (sunflower medium, Gurël et al., 1991a, 1991b), modified R92.25 (chicory medium, Theiler-Hedtrich and Hunter, 1996), MS + 3 mg/L BA (Caswell and Ferrie, unpublished)). Shoots were produced from calli initiated on D1 or D1-MS with a subsequent transfer to D1 medium, but again rooting was difficult (Caswell and Ferrie, unpublished).

Helianthus annuus L.

Sunflowers (*Helianthus annuus*) are grown for oil production, used as a confection, and for the bird feed market (National Sunflower Association of Canada, 2012). Although reports of sunflower anther culture (Gurël et al., 1991a, 1991b; Saji and Sujatha, 1998; Thengane et al., 1994; Zhong et al., 1995), isolated microspore culture (Coumans and Zhong, 1995), and ovary culture (Gelebart and San, 1987; Mix, 1985) appear in the literature, there has been limited success.

In our studies, experiments were conducted using 12 genotypes. Sunflower microspores appeared to reach the uninucleate stage, which is desirable for microspore culture, prior to developing a thick exine. No embryos were produced from any treatment using isolated microspores: calli were produced in several cases but no shoots were produced. For anther culture studies, callus developed, as did leafy structures, from 6 of the 12 sunflower cultivars and with seven different types of medium (Caswell and Ferrie, unpublished). These shoots were rooted and transplanted to soil. Flow cytometry analysis indicated that two of the five resulting plants were haploid, proving that they originated from microspores, not from diploid anther wall or filament tissue. Those that are diploid may have originated from somatic tissue, or may have originated from haploid tissue and spontaneously doubled (Caswell and Ferrie, unpublished). Zhong et al. (1995) conducted histological examination of cultured sunflower anthers and determined that embryos in the cultures were of somatic origin.

Calendula officinalis L.

Although calendula (*Calendula officinalis*) is predominantly used as an ornamental plant, the plants accumulate calendic acid (octadeca-8:10:12-trienoic acid), which can be used as a drying agent in paints, varnishes, and plastics (Beerentrup and Rőbbelen, 1987). In vitro regeneration (Çőçü et al., 2004; Leal et al., 2009; Victório et al., 2012) and cell suspension cultures (Grzelak and Janiszowska, 2002; Legha et al., 2012) have been reported for calendula.

For DH, we have generated haploid plants from isolated microspores of calendula (Ferrie and Caswell, 2011b). A number of factors influencing microspore embryogenesis were evaluated using two genotypes ("Erfurter Orangefarbigen" and "Pacific Beauty"). Callus was produced from both genotypes using R92.01 (Theiler-Hedtrich and Hunter, 1996) basal medium. The callus was transferred to a modified R92.25 regeneration medium (Theiler-Hedtrich and Hunter, 1996) and shoots were produced. Ten plants were analyzed by flow cytometry, and two of these were haploid. Anther culture experiments also resulted in callus production and plants; however, these plants were determined to be diploid.

OTHER SPECIES

The production of flax (*Linum usitatissimum* L.) DH plants has been reported in the literature and has been reviewed by Obert et al. (2009). Anther culture, isolated microspore culture, and ovary culture methods have yielded calli that have been regenerated to haploid or DH plants. However, the authors concluded that further improvements in the methods are required before the method is suitable for use in breeding programs. For other oilseed species, very few DH protocols are available (Table 13.1).

CONCLUSION

Although microspore-derived embryos and DH plants have been reported in a number of oilseeds, the frequency of embryogenesis is low, and therefore application of the protocol is not feasible at this time except in the *Brassica* species. Over the past few years, there have been improvements in the DH protocols and the frequency of embryogenesis of other oilseed crops, but further research is still required to be able to apply the protocols in breeding. Using a highly efficient embryogenic system in mutagenesis and transformation for the production of industrial oils has been demonstrated in the *Brassica* species. With advances in the basic microspore culture protocols and combining these methods with more precise methods of mutagenesis and transformation, further progress in improving industrial oils will be realized.

REFERENCES

Ahmad, I., Day, J.P., MacDonald, M.V., Ingram, D.S., 1991. Haploid culture and UV mutagenesis in rapid-cycling *Brassica napus* for the generation of resistance to chlorsulfuron and *Alternaria brassicolia*. Ann. Bot. 67, 521–525.

Ahmadi, B., Alizadeh, K., Teixeira da Silva, J.A., 2012. Enhanced regeneration of haploid plantlets from microspores of *Brassica napus* L. using bleomycin, PCIB, and phytohormones. Plant Cell Tissue Organ Cult. 109, 525–533.

Ammirato, P.V., 1974. The effects of abscisic acid on the development of somatic embryos from cells of caraway (*Carum carvi* L.). Bot. Gaz. 135, 328–337.

Barclay, A.S., Gentry, H.S., Jones, Q., 1962. The search for new industrial crops. II. *Lesquerella* (Cruciferae) as a source of new oilseeds. Econ. Bot. 16, 95–100.

Barro, F., Fernández-Escobar, J., De la Vega, M., Martin, A., 2001. Doubled haploid lines of *Brassica carinata* with modified erucic acid content through mutagenesis by EMS treatment of isolated microspores. Plant Breed. 120, 262–264.

Barro, F., Fernández-Escobar, J., De la Vega, M., Martín, A., 2002. Modification of glucosinolate and erucic acid contents in doubled haploid lines of *Brassica carinata* by UV treatment of isolated microspores. Euphytica 129, 1–6.

Barro, F., Martín, A., 1999. Response of different genotypes of *Brassica carinata* to microspore culture. Plant Breed. 118, 79–81.

Başalma, D., Uranbey, S., Mirici, S., Kolsarici, Ö., 2008. TDZ×IBA induced shoot regeneration from cotyledonary leaves and in vitro multiplication in safflower (*Carthamus tinctorius* L.). Afr. J. Biotechnol. 7, 960–966.

Beaith, M.E., Fletcher, R.S., Kott, L.S., 2005. Reduction of saturated fats by mutagenesis and heat selection in *Brassica napus* L. Euphytica 144, 1–9.

Becwar, M.R., Nagmani, R., Wann, S.R., 1990. Initiation of embryogenic cultures and somatic embryo development in loblolly pine (*Pinus taeda*). Can. J. For. Res. 20, 810–817.

Beerentrup, H.M., Rőbbelen, G., 1987. Calendula and coriandrum–new potential oil crops for industrial uses. Fat Sci. Technol. 89, 227–230.

Bhowmik, P., Dirpaul, J., Ferrie, A.M.R., Polowick, P.L., 2011a. Microspore Transformation Methodology by Macroinjection. PCT/CA2011/001326.

Bhowmik, P., Dirpaul, J., Polowick, P., Ferrie, A.M.R., 2011b. A high throughput *Brassica napus* microspore culture system: influence of percoll gradient separation and bud selection on embryogenesis. Plant Cell Tissue Organ Cult. 106, 359–362.

Bouwmeester, H.J., Kuijpers, A.M., 1993. Relationship between assimilate supply and essential oil accumulation in annual and biennial caraway (*Carum carvi* L.). J. Essent. Oil Res. 5, 143–152.

Burbulis, N., Blinstrubiene, A., Kupriene, R., 2011. Effect of genotype and medium composition on linseed (*Linum usitatissimum*) ovary culture. Biologia 66, 465–469.

Chaleff, R.S., Stolarz, A., 1981. Factors influencing the frequency of callus formation among cultured rice (*Oryza sativa*) anthers. Physiol. Plant. 51, 201–206.

Chen, G.Q., 2011. Effective reduction of chimeric tissue in transgenics for the stable genetic transformation of *Lesquerella fendleri*. HortScience 46, 86–90.

Chen, K., Zhang, X.B., Jiang, J.L., Wang, X.Y., 2011. Plantlet regeneration from cotyledon, cotyledon petiole, and hypocotyl explants via somatic embryogenesis pathway in roquette (*Eruca sativa* Mill.). Plant Biosyst. 145, 68–76.

Chhikara, S., Dutta, I., Paulose, B., Jaiwal, P.K., Dhankher, O.P., 2012. Development of an *Agrobacterium*-mediated stable transformation method for industrial oilseed crop *Crambe abyssinica* 'BelAnn'. Ind. Crops Prod. 37, 457–465.

Chu, C.C., Wang, C.-C., Sun, C.S., Hsu, C., Yin, K.-C., Chu, C.-Y., Bi, F.-Y., 1975. Establishment of an efficient medium for anther culture of rice through comparative experiments on the nitrogen sources. Sci. Sin. 18, 659–668.

Çöçü, S., Uranbey, S., Pek, A., Khawar, K.M., Sarihan, E.O., Kaya, M.D., Parmaksiz, I., Özcan, S., 2004. Adventitious shoot regeneration and micropropagation in *Calendula officinalis* L. Biol. Plant. 48, 449–451.

Conn, K.L., Tewari, J.P., Dahiya, J.S., 1988. Resistance to *Alternaria brassicae* and phytoalexin-elicitation in rapeseed and other crucifers. Plant Sci. 56, 21–25.

Coumans, M., Zhong, D., 1995. Doubled haploid sunflower (*Helianthus annuus*) plant production by androgenesis: fact or artifact? Part 2. In vitro isolated microspore culture. Plant Cell Tissue Organ Cult. 41, 203–209.

Dempewolf, H., Rieseberg, L.H., Cronk, Q.C., 2008. Crop domestication in the Compositae: a family-wide trait assessment. Genet. Resour. Crop Evol. 55, 1141–1157.

Dierig, D.A., Thompson, A.E., Nakayama, F.S., 1993. Lesquerella commercialization efforts in the United States. Ind. Crops Prod. 1, 289–293.

Dierig, D.A., Tomasi, P.M., Salywon, A.M., Ray, D.T., 2004. Improvement in hydroxyl fatty acid seed oil content and other traits from interspecific hybrids of three *Lesquerella* species: *Lesquerella fendleri, L. pallida*, and *L. lindheimeri*. Euphytica 139, 199–206.

Dormann, M., Wang, H.-M., Datla, N., Ferrie, A.M.R., Keller, W.A., Oelck, M.M., 1995. Transformation of Freshly Isolated Brassica Microspores and Regeneration to Fertile Homozygous Plants. 9th International Rapeseed Congress, Cambridge, 3, pp. 816–818.

Dormann, M., Wang, H.-M., Oelck, M., 2001. Transformed Embryogenic Microspores for the Generation of Fertile Homozygous Plants. USA No. 6316694.

Du, E.-Z., Ge, X.-H., Zhao, Z.-G., Li, Z.-Y., 2008. Chromosome elimination and fragment introgression and recombination producing intertribal partial hybrids from *Brassica napus* × *Lesquerella fendleri* crosses. Plant Cell Rep. 27, 261–271.

Dunwell, J.M., 2010. Haploids in flowering plants: origins and exploitation. Plant Biotechnol. J. 8, 377–424.

Dunwell, J.M., Cornish, M., De Courcel, A.G.L., 1985. Influence of genotype, plant growth temperature and anther incubation temperature on microspore embryo production in *Brassica napus* ssp. *oleifera*. J. Exp. Bot. 36, 679–689.

Ekiert, H., 2000. Medicinal plant biotechnology: the Apiaceae family as the example of rapid development. Pharmazie 55, 561–567.

Ekin, Z., 2005. Resurgence of safflower (*Carthamus tinctorius* L.) utilization: a global view. J. Agron. 4, 83–87.

Evergetis, E., Michaelakis, A., Haroutounian, S.A., 2013. Exploitation of Apiaceae family essential oils as potent biopesticides and rich source of phellandrenes. Ind. Crops Prod. 41, 365–370.

Ferrie, A.M.R., 2003. Microspore culture of *Brassica* species. In: Maluszynski, M., Kasha, K.J., Forster, B.P., Szarejko, I. (Eds.), Doubled Haploid Production in Crop Plants. Kluwer Academic Publishers, Dordrecht, pp. 205–215.

Ferrie, A.M.R., 2007. Doubled haploid production in nutraceutical species: a review. Euphytica 158, 347–357.

Ferrie, A.M.R., Bethune, T.D., 2011. A microspore embryogenesis protocol for *Camelina sativa*, a multi-use crop. Plant Cell Tissue Organ Cult. 106, 495–501.

Ferrie, A.M.R., Bethune, T.D., Arganosa, G.C., Waterer, D., 2011a. Field evaluation of doubled haploid plants in the Apiaceae: dill (*Anethum graveolens* L.), caraway (*Carum carvi* L.), and fennel (*Foeniculum vulgare* Mill.). Plant Cell Tissue Organ Cult. 104, 407–413.

Ferrie, A.M.R., Bethune, T.D., Mykytyshyn, M., 2011b. Microspore embryogenesis in Apiaceae. Plant Cell Tissue Organ Cult. 104, 399–406.

Ferrie, A.M.R., Caswell, K.L., 2011a. Isolated microspore culture techniques and recent progress for haploid and doubled haploid plant production. Plant Cell Tissue Cult. 104, 301–309.

Ferrie, A.M.R., Caswell, K.L., 2011b. Review of doubled haploidy methodologies in ornamental species. Propag. Ornam. Plants 11, 63–77.

Ferrie, A.M.R., Dirpaul, J., Krishna, P., Krochko, J., Keller, W.A., 2005. Effects of brassinosteroids on microspore embryogenesis in *Brassica* species. In Vitro Cell Dev. Biol. Plant 41, 742–745.

Ferrie, A.M.R., Keller, W.A., 1995. Microspore culture for haploid plant production. In: Gamborg, O.L., Phillips, G.C. (Eds.), Plant Cell, Tissue and Organ Culture. Fundamental Methods. Springer-Verlag, Berlin, pp. 155–164.

Ferrie, A.M.R., Keller, W.A., 2007. Optimization of methods for using polyethylene glycol as a non-permeating osmoticum for the induction of microspore embryogenesis in the *Brassicaceae*. In Vitro Cell Dev. Biol. Plant 43, 348–355.

Ferrie, A.M.R., Möllers, C., 2011. Haploids and doubled haploids in *Brassica* spp. for genetic and genomic research. Plant Cell Tissue Organ Cult. 104, 375–386.

Ferrie, A.M.R., Taylor, D.C., MacKenzie, S.L., Rakow, G., Raney, J.P., Keller, W.A., 2008. Microspore mutagenesis of Brassica species for fatty acid modifications: a preliminary evaluation. Plant Breed. 127, 501–506.

Fukuoka, H., Ogawa, T., Matsuoka, M., Ohkawa, Y., Yano, H., 1998. Direct gene delivery into isolated microspores of rapeseed (*Brassica napus* L.) and the production of fertile transgenic plants. Plant Cell Rep. 17, 323–328.

Furmanek, T., Banaś, W., 2011. Embryogenic callus formation by cotyledon and leaf explants of *Crambe abyssinica* seedlings. Biotechnologia 92, 209–213.

Furmanowa, M., Oledzka, H., Sowinska, D., 1984. Regeneration of plants by embryogenesis with callus cultures of *Carum carvi* L. J. Plant Physiol. 115, 209–210.

Gamborg, O.L., Miller, R.A., Ojima, K., 1968. Nutrient requirements of suspension cultures of soybean root cells. Exp. Cell Res. 50, 151–158.

Gelebart, P., San, L.H., 1987. Production of haploid plants of sunflower (*Helianthus annuus* L.) by in vitro culture of non-fertilized ovaries and ovules. Agronomie 7, 81–86.

Grzelak, A., Janiszowska, W., 2002. Initiation and growth characteristics of suspension cultures of *Calendula officinalis* cells: changes in the level of oleanolic acid during the cell growth cycle of the culture. Plant Cell Tissue Organ Cult. 71, 29–40.

Gugel, R.K., Falk, K.C., 2006. Agronomic and seed quality evaluation of *Camelina sativa* in western Canada. Can. J. Plant Sci. 86, 1047–1058.

Guha, S., Maheshwari, S.C., 1964. In vitro production of embryos from anthers of *Datura*. Nature 204, 497.

Gürel, A., Kontowski, S., Nichterlein, K., Friedt, W., 1991a. Embryogenesis in microspore culture of sunflower (*Helianthus annuus* L.). Helia 14, 123–128.

Gürel, A., Nichterlein, K., Friedt, W., 1991b. Shoot regeneration from anther culture of sunflower (*Helianthus annuus*) and some interspecific hybrids as affected by genotyped and culture procedure. Plant Breed. 106, 68–76.

Hansen, L.N., 1998. Intertribal somatic hybridization between rapid cycling *Brassica oleracea* L. and *Camelina sativa* (L.) Crantz. Euphytica 104, 173–179.

Jana, S., Shekhawat, G.S., 2011. Plant growth regulators, adenine sulfate and carbohydrates regulate organogenesis and in vitro flowering of *Anethum graveolens*. Acta Physiol. Plant. 33, 305–311.

Jardinaud, M.-F., Souvre, A., Alibert, G., 1993. Transient GUS gene expression in *Brassica napus* electroporated microspores. Plant Sci. 93, 177–184.

Jiang, J., Zhao, X., Tian, W., Li, T., Wang, Y., 2009. Intertribal somatic hybrids between *Brassica napus* and *Camelina sativa* with high linolenic acid content. Plant Cell Tissue Organ Cult. 99, 91–95.

Jones-Villeneuve, E., Huang, B., Prudhomme, I., Bird, S., Kemble, R., Hattori, J., Miki, B., 1995. Assessment of microinjection for introducing DNA into uninuclear microspores of rapeseed. Plant Cell Tissue Organ Cult. 40, 97–100.

Kharenko, O.A., Zaharia, L.I., Giblin, M., Čekić, V., Taylor, D.C., Palmer, D., Abrams, S.R., Loewen, M.C., 2011. Abscisic acid metabolism and lipid accumulation of a cell suspension culture of *Lesquerella fendleri*. Plant Cell Tissue Organ Cult. 105, 415–422.

Krens, F.A., Keizer, L.C.P., Capel, I.E.M., 1997. Transgenic caraway, *Carum carvi* L.: a model species for metabolic engineering. Plant Cell Rep. 17, 39–43.

Kuvshinov, V., Kanerva, A., Koivu, K., Pehu, E., Kuvshinova, S., 2011. A Transformation System in Camelina Sativa. (7910803).

Lazzeri, L., Errani, M., Leoni, O., Venturi, G., 2004. *Eruca sativa* spp. *oleifera*: a new non-food crop. Ind. Crops Prod. 20, 67–73.

Leal, F., Rodrigues, A., Fernandes, D., Nunes, F.M., Cipriano, J., Ramos, J., Teixeira, S., Vieira, S., Carvalho, L.M., Pinto-Carnide, O., 2009. In vitro multiplication of *Calendula arvensis* for secondary metabolites extraction. Acta Hortic. 812, 251–256.

Legha, M.R., Prasad, K.V., Singh, S.K., Kaur, C., Arora, A., Kumar, S., 2012. Induction of carotenoid pigments in callus cultures of *Calendula officinalis* L. in response to nitrogen and sucrose levels. In Vitro Cell Dev. Biol. Plant 48, 99–106.

Leskovšek, L., Jakše, M., Bohanec, B., 2008. Doubled haploid production in rocket (*Eruca sativa* Mill.) through isolated microspore culture. Plant Cell Tissue Organ Cult. 93, 181–189.

Li, S., Wang, Y., Dong, S., Chen, Y., Cao, F., Chai, F., Wang, X., 2009. Biodiesel production from Eruca Sativa Gars vegetable oil and motor, emissions properties. Renew. Energy 34, 1871–1876.

Li, X., Ahlman, A., Lindgren, H., Zhu, L.-H., 2011. Highly efficient in vitro regeneration of the industrial oilseed crop *Crambe abyssinica*. Ind. Crops Prod. 33, 170–175.

Li, X., Ahlman, A., Yan, X., Lindgren, H., Zhu, L.-H., 2010. Genetic transformation of the oilseed crop *Crambe abyssinica*. Plant Cell Tissue Organ Cult. 100, 149–156.

Lichter, R., 1982. Induction of haploid plants from isolated pollen of *Brassica napus*. Z. Pflanzenphysiol. 105, 427–434.

Linsmaier, E.M., Skoog, F., 1965. Organic growth factor requirements of tobacco tissue cultures. Physiol. Plant. 18, 100–127.

Liu, X., Brost, J., Hutcheon, C., Guilfoil, R., Wilson, A.K., Leung, S., Shewmaker, C.K., Rooke, S., Nguyen, T., Kiser, J., De Rocher, J., 2012. Transformation of the oilseed crop *Camelina sativa* by *Agrobacterium*-mediated floral dip and simple large-scale screening of transformants. In Vitro Cell Dev. Biol. Plant 48, 462–468.

Lu, C., Kang, J., 2008. Generation of transgenic plants of a potential oilseed crop *Camelina sativa* by *Agrobacterium*-mediated transformation. Plant Cell Rep. 27, 273–278.

Maluszynski, M., Kasha, K.J., Szarejko, I., 2003. Published doubled haploid protocols in plant species. In: Maluszynski, M., Kasha, K.J., Forster, B.P., Szarejko, I. (Eds.), Doubled Haploid Production in Crop Plants: A Manual. Kluwer Academic Publishers, The Netherlands, pp. 309–335.

Mandal, S., Yadav, S., Singh, R., Begum, G., Suneja, P., Singh, M., 2002. Correlation studies on oil content and fatty acid profile of some Cruciferous species. Genet. Resour. Crop Evol. 49, 551–556.

McClinchey, S.L., Kott, L.S., 2008. Production of mutants with high cold tolerance in spring canola. Euphytica 162, 51–67.

Mithila, J., Hall, J.C., 2007. Production of an auxinic herbicide-resistant microspore-derived doubled haploid wild mustard (*Sinapis arvensis* L.) plant. Crop Prot. 26, 357–362.

Mix, G., 1985. Antheren-und Ovarienkultur von sonnenblumen (*Helianthus annuus* L.). Landbauforsch. Volkenrode 3, 153–156.

Mnzava, N.A., Olsson, K., 1990. Studies on tropical vegetables. Part I. Seed amino, fatty acid and glucosinolate profile of Ethiopian mustards (*Brassica carinata* Braun). Food Chem. 35, 229–235.

Mollika, S.R., Sarker, R.H., Hoque, M.I., 2011. In vitro plant regeneration in *Brassica* spp. Plant Tissue Cult. Biotechnol. 21, 127–134.

Motamedi, J., Zebarjadi, A., Kahrizi, D., Salmanian, A.H., 2011. In vitro propagation and *Agrobacterium*-mediated transformation of safflower (*Carthamus tinctorius* L.) using bacterial mutated aroA gene. Aust. J. Crop Sci. 5, 479–486.

Murashige, T., Skoog, F., 1962. A revised medium for rapid growth and bioassays with tobacco tissue cultures. Physiol. Plant. 15, 473–497.

Narasimhula, S.B., Kirti, P.B., Bhatt, S.R., Prakash, S., Chopra, V.L., 1994. Intergeneric protoplast fusion between *Brassica carinata* and *Camelina sativa*. Plant Cell Rep. 13, 657–660.

National Sunflower Association of Canada, 2012. The Sunflower Production Guide. pp. 64.

Obert, B., Žáčková, Z., Šamaj, J., Pret'ová, A., 2009. Doubled haploid production in flax (*Linum usitatissium* L.). Biotechnol. Adv. 27, 371–375.

Ovcharenko, O., Momot, V., Cherep, N., Sheludko, Y., Komarnitsky, I., Rudas, V., Kuchuk, N., 2011. Transfer of transformed *Lesquerella fendleri* (Gray) Wats. chloroplasts into *Orychophragmus violaceus* (L.) O.E. Schulz by protoplast fusion. Plant Cell Tissue Organ Cult. 105, 21–27.

Palmer, D., Keller, W.A., 2011. Somatic embryogenesis in *Crambe abyssinica* Hochst. Ex R.E. Fries using seedling explants. Plant Cell Tissue Organ Cult. 104, 91–100.

Pechan, P.M., 1989. Successful cocultivation of *Brassica napus* microspores and proembryos with *Agrobacterium*. Plant Cell Rep. 8, 387–390.

Pomeroy, M.K., Kramer, J.K.G., Hunt, D.J., Keller, W.A., 1991. Fatty acid changes during development of zygotic and microspore-derived embryos of *Brassica napus*. Physiol. Plant. 81, 447–454.

Potts, D.A., Rakow, G.W., Males, D.R., 1999. Canola-quality *Brassica juncea*, a new oilseed crop for the Canadian prairies. In: Proceedings of the 10th International Rapeseed Congress, Canbera, Australia, 23–26 September 1999.

Prasad, B.R., Khadeer, M.A., Seeta, P., Anwar, S.Y., 1990. Influence of genotype and cold pretreatment on anther culture response in safflower, *Carthamus tinctorius* L. Indian J. Exp. Biol. 28, 924–927.

Prasad, B.R., Khadeer, M.A., Seeta, P., Anwar, S.Y., 1991. In vitro induction of androgenic haploids in Safflower (*Carthamus tinctorius* L.). Plant Cell Rep. 10, 48–51.

Rani, T., Yadav, R.C., Yadav, N.R., Rani, A., Singh, D., 2013. Genetic transformation in oilseed brassicas – a review. Indian J. Agric. Sci. 83, 367–373.

Rollins, R.C., Shaw, E., 1973. The Genus Lesquerella (Cruciferae) in North America. Harvard University Press, Cambridge, MA. pp. 288.

Saji, K.V., Sujatha, M., 1998. Embryogenesis and plant regeneration in anther culture of sunflower (*Helianthus annuus* L.). Euphytica 103, 1–7.

Salisbury, P.A., 1987. Blackleg resistance in weedy crucifers. Cruciferae News Lett. 12, 90.

Sigareva, M.A., Earle, E.D., 1999. Camalexin induction in intertribal somatic hybrids between *Camelina sativa* and rapid-cycling *Brassica oleracea*. Theor. Appl. Genet. 98, 164–170.

Sikdar, S.R., Chatterjee, G., Das, S., Sen, S.K., 1987. Regeneration of plants from mesophyll protoplasts of the wild crucifer *Eruca sativa* Lam. Plant Cell Rep. 6, 486–498.

Skarzhinskaya, M., Landgren, M., Glimelius, K., 1996. Production of intertribal somatic hybrids between *Brassica napus* L. and *Lesquerella fendleri* (Gray) wats. Theor. Appl. Gen. 93, 1243–1250.

Slater, S.M.H., Keller, W.A., Scoles, G., 2011. *Agrobacterium*-mediated transformation of *Eruca sativa*. Plant Cell Tissue Organ Cult. 106, 253–260.

Smykalová, I., Smirous Jr., P., Kubosiova, M., Gasmanova, N., Griga, M., 2009. Doubled haploid production via anther culture in annual, winter type of caraway (*Carum carvi* L.). Acta Physiol. Plant. 31, 21–31.

Sonntag, K., Rudloff, E., 2004. Microspore mutagenesis in transgenic oilseed rape for the modification of fatty-acid composition. Acta Univ. Latviensis Biol. 676, 227–230.

Sri Shilpa, K., Kumar, V.D., Sujatha, M., 2010. *Agrobacterium*-mediated genetic transformation of safflower (*Carthamus tinctorius* L.). Plant Cell Tissue Organ Cult. 103, 387–401.

Swanson, E.B., Coumans, M.P., Brown, G.L., Patel, J.D., Beversdorf, W.D., 1988. The characterization of herbicide tolerant plants in *Brassica napus* L. after in vitro selection of microspores and protoplasts. Plant Cell Rep. 7, 83–87.

Tattersall, A., Millam, S., 1998. Establishment and in vitro regeneration studies of the potential oil crop species *Camelina sativa*. Plant Cell Tissue Organ Cult. 55, 147–149.

Taylor, D.C., Weber, N., Underhill, E.W., Pomeroy, M.K., Keller, W.A., Scowcroft, W.R., Wilen, R.W., Moloney, M.M., Holbrook, L.A., 1990. Storage protein regulation and lipid accumulation of microspore embryos of *Brassica napus* L. Planta 181, 18–26.

Theiler-Hedtrich, R., Hunter, C.S., 1996. Microspore culture in chicory. In: Jain, S.M., Sopory, S.K., Veilleux, R.E. (Eds.), In Vitro Haploid Production in Higher Plants, vol. 3, pp. 99–113.

Thengane, S.R., Joshi, M.S., Khuspe, S.S., Mascarenhas, A.F., 1994. Anther culture in *Helianthus annuus* L., influence of genotype and culture conditions on embryo induction and plant regeneration. Plant Cell Rep. 13, 222–226.

Thompson, A.E., Dierig, D.A., Johnson, E.R., 1989. Yield potential of *Lesquerella fendleri* L. (gray) Wats., a desert plant resource for hydroxyl fatty acids. J. Arid. Environ. 16, 331–336.

Tomasi, P., Dierig, D.A., Backhaus, R.A., Pigg, K.B., 1999. Floral bud and mean petal length as morphological predictors of microspore cytological stage in *Lesquerella*. HortScience 34, 1269–1270.

Tomasi, P.M., Ferrie, A.M.R., 2003. Microspore culture in genus Lesquerella. In: Association for the Advancement of Industrial Crops Annual Meeting, October 12–15, 2003.

Touraev, A., Heberle-Bors, E., 1999. Microspore embryogenesis and in vitro pollen maturation in tobacco. In: Hall, R.D. (Ed.), Methods in Molecular Biology, vol. 111: Plant Cell Culture Protocols. Humana Press, Totowa, pp. 281–291.

Toxopeus, H., Bouwmeester, H.J., 1993. Improvement of caraway essential oil and carvone production in the Netherlands. Ind. Crops Prod. 1, 295–301.

Ulrich, A., Furtan, W.H., Downey, R.K., 1984. Biotechnology and rapeseed breeding: some economic considerations. Sci. Counc. Can. Rep. 67.

Victório, C.P., Lage, C.L.S., Sato, A., 2012. Tissue culture techniques in the proliferation of shoots and roots of *Calendula officinalis*. Revist. Cienc. Agron. 43, 539–545.

Vijaya Kumar, J., Ranjitha Kumari, B.D., Castaño, E., 2008. Cyclic somatic embryogenesis and efficient plant regeneration from callus of safflower. Biol. Plant. 52, 429–436.

Wang, W., Wang, C., Huang, B.-L., Huang, B., 2008. *Agrobacterium-tumefaciens*-mediated transformation of *Lesquerella fendleri* L., a potential new oil crop with rich lesquerolic acid. Plant Cell Tissue Organ Cult. 92, 165–171.

Weselake, R.J., Taylor, D.C., Pomeroy, M.K., Lewson, S.L., Underhill, E.W., 1991. Properties of diacylglycerol acyltransferase from microspore-derived embryos of *Brassica napus* L. Phytochemistry 30, 3533–3538.

Wiberg, E., Rahlen, L., Hellman, M., Tillberg, E., Glimelius, K., Stymne, S., 1991. The microspore-derived embryo of *Brassica napus* L. as a tool for studying embryo-specific lipid biogenesis and regulation of oil quality. Theor. Appl. Genet. 82, 515–520.

Wong, R.S.-C., Swanson, E., 1991. Genetic modification of canola oil: high oleic acid canola. In: Haberstroh, C., Morris, C.E. (Eds.), Fat and Cholesterol Reduced Foods: Technologies and Strategies. Portfolio Publ. Co, The Woodlands, TX, USA, pp. 153–164.

Yuan, S.-X., Su, Y.-B., Liu, Y.-M., Fang, Z.-Y., Yang, L.-M., Zhuang, M., Zhang, Y.-Y., Sun, P.-T., 2012. Effects of pH, MES, arabinogalactan-proteins on microspore cultures in white cabbage. Plant Cell Tissue Organ Cult. 110, 69–76.

Zhang, F.-L., Takahata, Y., 1999. Microspore mutagenesis and in vitro selection for resistance to soft rot disease in Chinese cabbage (*Brassica campestris* L. ssp. *pekinensis*). Breed. Sci. 49, 161–166.

Zhang, Y., Wang, A., Liu, Y., Wang, Y., Feng, H., 2011. Effects of the antiauxin PCIB on microspore embryogenesis and plant regeneration in *Brassica rapa*. Sci. Hortic. 130, 32–37.

Zhong, D., Michaux-Ferriére, N., Coumans, M., 1995. Assay for doubled haploid sunflower (*Helianthus annuus*) plant production by androgenesis: fact or artifact? Part 1. In vitro anther culture. Plant Cell Tissue Organ Cult. 41, 91–97.

Chapter 14

Genetic Transformation of Crops for Oil Production

Priti Maheshwari
Lethbridge Research Station, Lethbridge, AB, Canada

Igor Kovalchuk
University of Lethbridge, Lethbridge, AB, Canada

INTRODUCTION

Since the dawn of civilization, plant-derived oils have always played a key role in a man's life by providing essential nutrition. After thousands of years of domestication of oil plants, in today's modern world, they are approved for limited human consumption, and moreover, their application is expanding in many directions. Plant-derived oils are currently used not only as biolubricants but also as biofuels, which are cost-effective sustainable sources for renewable energy.

The challenge we face today is to meet the ever-increasing demand for food coupled with rising global food prices. In addition, the amount and type of dietary fat intake are directly related to various chronic cardiovascular and other diseases. Dietary regulations with a proper balance of saturated and polyunsaturated fats have positive health benefits. Hence, oil plants have to be modified to not only meet public health standards for food and animal feed requirements but also produce the desired spectrum of oils that would be able to meet industrial needs. This can be achieved by using the technique of genetic engineering in which a gene or genes of interest aimed to produce the desired trait can be introduced in a plant to obtain what is called a transgenic plant. An example of numerous applications of plant transgenesis is using reverse engineering to produce plants that, like fish, would accumulate very high levels of omega-3 very long-chain polyunsaturated fatty acids (ω-3 VLC PUFAs) (Napier et al., 2004).

Innovation has always been a vital key to the progress of humanity, and science is a major driving force in this process. From the perspective of agriculture, molecular genetics coupled with modern plant biotechnology has set the stage for the advancement of agricultural practices by delivering high-quality crops that will be able to meet the changing lifestyle and global food and health demands. Agrobiotechnology enables the manipulation of the plant genome,

Industrial Oil Crops. http://dx.doi.org/10.1016/B978-1-893997-98-1.00014-2

thus paving the way for plant genetic engineering technology (PGMT) and for the production of genetically modified (GM) plants by means of metabolic engineering and genomics. Tissue culture techniques, hybrid production, doubled haploids, various gene transfer methods, and many other techniques have made this possible (Paul, 2008). A better understanding of lipid synthesis metabolism and storage combined with transgenic technology going hand-in-hand with conventional plant-breeding approaches and the domestication of plants that synthesize useful fatty acids (FAs) will allow for a real breakthrough in the area of oil crops. The first transgenic plant was reported in 1983, and since then, the progress has been promising. By 2011, over half a billion acres of transgenic oil crops were grown in the United States alone. Many of these have been modified to contain traits that control oil characteristics alongside with herbicide, pesticide, and stress tolerance. Raising public awareness on GM crops and gaining public acceptance of GM food can have positive implications in the future when GM products will become common commodities that satisfy human food and health standards.

This chapter will focus on recent advances toward the development of transgenic oilseed crops with a predefined FA composition by metabolic and genetic engineering techniques. Here, current and future endeavors will be discussed aiming at producing oil crops that can serve as renewable energy sources in industries because of high levels of the engineered and more stable FAs. Various means to overcome technological bottlenecks and challenges will also be highlighted.

MULTIPURPOSE AND TECHNICAL OILS

FA composition of vegetable oils dictates their physical and chemical properties and henceforth their applications (Murphy, 1994). Five triglycerides predominate in a vegetable oil content (90%); they are saturated palmitic (C16:0) and stearic (C18:0) acids, the monounsaturated oleic acid (C18:1), and polyunsaturated linoleic (C18:2) and α-linolenic (C18:3) acids (Gunstone, 2004). Besides these, a few 100 other unusual FAs have been found in smaller quantities in different species that are potential sources for a wide array of important industrial applications. These FAs vary in the length of chains from 8 to 24 carbon atoms that have different numbers of double bonds and epoxy and hydroxyl functional groups (Metzger and Bornscheuer, 2006). Based upon oil quality, oils can be categorized as multipurpose oils and technical oils (Carlsson, 2009). The former are oils that are rich in mono- and diunsaturated oleic acids and linoleic acid (LA) and saturated palmitic acid and hence can be used for both food/feed and chemical feedstock purposes. All six major oil crops—soybean (*Glycine max*), rapeseed/canola (*Brassica napus*), palm (*Elaeis guineensis*), sunflower (*Helianthus annuus*), cottonseed (*Gossypium hirsutum*), and peanut (*Arachis hypogaea*)—are sources of multipurpose oils. Oils obtained from these plants are mostly used for various edible purposes such as cooking, frying, preparation of margarine, processed foods, and animal feed. Most of the world's vegetable oils (72%) come from oil palm, soybean, rapeseed, and sunflower, in that order. Some edible oils have a special composition that suits specific needs and command high prices. For instance, cocoa butter has unique characteristics that make it desirable for the confectionery industry. High oleic acid varieties of these major oil crops make them useful in various industrial applications, the main ones being biofuels, industrial feedstocks, paints, varnishes, textiles, lubricants, cosmetics, etc. On the other hand, technical oils have the special composition and properties that make them valuable only as industrial feedstocks. Major technical oil–containing plants include castor bean (*Ricinus communis*), tung-oil tree (*Aleurites fordii*), jojoba (*Simmondsia chinensis*), crambe (*Crambe abyssinica*), flax (*Linum usitatissimum*), Ethiopian mustard (*B. carinata*), camelina (*Camelina sativa*), jatropha (*Jatropha curcas*), and many others.

Though only five major FAs account for 95% of the world's oil production, an amazing diversity exists in the plant kingdom, and thousands of FAs with unique characteristics, including medicinal properties, have been identified over the years. Some of them have already drawn attention to themselves, and many others are waiting to be explored. The most important reason for the underutilization of these plants is the high costs associated with their low oil yields. A great alternative is the manipulation of commercial oil crops for producing varieties with different oil composition and unique qualities, thus opening new markets for various potential applications. Further advancement in our understanding of lipid biosynthetic pathways, coupled with new tools of genetic engineering and classic methods of biochemistry and plant breeding, holds great promise for tomorrow.

OILS—AN ESSENTIAL DIETARY REQUIREMENT FOR HUMANS

Linoleic acid (LA) and α-linolenic acid (ALA) present in plant oils are necessary dietary requirements for humans due to the lack of the Δ12- and Δ15-desaturase activity (López Alonsoa and García Marotob, 2000; Ruiz-López et al., 2009). LA and ALA act as precursors for the synthesis of VLC PUFAs, 20-carbon FAs that contain three or more double bonds in the *cis* position. Depending on the position of the first double bond, they can be either ω-6 (*n*-6) or omega-3 (*n*-3) FAs. ω-3 VLC PUFAs, especially eicosapentaenoic acid (EPA) and docosahexaenoic acid (DHA), modulate membrane fluidity and permeability and hence play an essential role in providing metabolism for human health benefits. Since humans have a limited capacity to synthesize ω-3 VLC PUFAs, their supplementation in diet is required for well-being. Marine fish is the main source of EPA and DHA in the human diet. Fish accumulate them by ingesting marine algae, which have the capacity to synthesize them de novo. Besides algae, some lower plants such as fungi, bacteria, and moss can synthesize them. The lack of VLC PUFAs in higher plants, a very quick decline of world fish stocks, and the health risks associated with eating fish contaminated by environmental pollutants have all led to the search of alternates for providing these essential oils for human consumption. After the identification of genes involved in FA synthesis over the past decade, the generation of transgenic oil plants capable of producing desired oil characteristics has been possible (Cahoon et al., 2007).

PLANT OILS AS AN ALTERNATE TO PETROLEUM—BIOFUELS

Ninety percent of the total petroleum consumption is used in transportation and electricity generation, and the rest is used as feedstock (raw material) in various industries for the production of multiple products such as paints, varnishes, plastics, cosmetics, rubber, detergents, solvents, lubricants, etc. The use of plant oils for the production of oleochemicals is a sustainable alternative to the world's fast-depleting nonrenewable oil deposits and has drawn great interest due to environmental benefits caused by reducing greenhouse gas emissions. Plant-derived oils can be used as raw materials for the production of biofuels and biodiesels and as an alternative to petroleum feedstocks in the chemical industry due to their similar combustion properties (Durret et al., 2008). According to the International Energy Agency's (IEA), the production of biofuels has increased several-fold in recent years (Tonkin, 2009). Biodiesel is a diesel-like biofuel made from renewable biological materials. Technically, its FA methyl esters are obtained via *trans*-esterification of vegetable oils with the addition of methanol in the presence of acid or alkali (Shahid and Jamal, 2011). Edible oils were the first oils used for the production of biodiesel, and hence biodiesels obtained from them are known as first-generation biodiesels. They account for more than 95% of the world's biodiesel (84% rapeseed, 13% sunflower oil, 1% palm oil, 2% soybean oil, and others) (Hass, 2005). First-generation biofuel crops have demonstrated their limitations such as low oil yields to be used for the purpose apart from their major use as food (Fairless, 2007). Due to rising food prices and concerns raised regarding food security and land management, other alternatives are being tested (Atabani et al., 2012). Nonedible oils are now used as second-generation biofuels as the probable source for reducing utilization of edible oils. These plants are also more economical and efficient, the most important being ratanjyote (*Jatropha curcas*), karanja (*Pongamia pinnata*), and jojoba (*Simmondsia chinensis*). Having emerged as third-generation biofuel, microalgae cultured in bioreactors are the highest-yielding oil feedstock that can produce up to 25 times more than oil palm and 250 times more than soybeans, with the limitation being high production costs (Vega-Sánchez and Ronal, 2010).The most promising approach for satisfying the future needs lies in genetic modifications of conventional and nonconventional oil crops by introducing desirable traits for the production of biodiesel from oil-producing microalgae and fungi (Moser, 2009).

OILSEED CROP PLATFORMS FOR ANIMAL FEEDSTOCK

Besides being used as animal feedstock, plant-derived oils are drawing attention as an alternative to marine oils to feed aquaculture and other farmed carnivores. Although plant oils are considered to be a cheap and renewable source for animal feed, there is still a need for modification to meet the nutritional requirements of fish and animal stocks (Thompson, 2000). Vegetable oils for fish should contain 18:2 *n*-6, 18:1 *n*-9, and *n*-3 PUFA. Rapeseed oil contains

16–25% 18:2 *n*-6 and is a promising substitute for fish oil in feed diets of Atlantic salmon. Enzyme systems for FA desaturation in seed can be altered by genetic engineering techniques coupled with plant-breeding methods using seed-specific promoters. A novel alternative pathway, a PKS gene cluster, can also be used to further increase the amount of *n*-3 PUFA and make way for novel agricultural cash crops.

CONVENTIONAL METHODS OF OIL CROP IMPROVEMENT

During the 10,000 years since the development of agriculture in human history, plants with desirable traits have been selected, and these improved characteristics have been passed on to future plant generations, thus contributing to the appearance of improved varieties. The advent of genetic principles and the implementation of this knowledge toward crop improvement allowed for the development of the newer plant-breeding methods. Plant breeding relies on natural diversity within plant species and closely related species to acquire desirable characteristics through hybridization techniques aided by the tools of grafting, tissue culture, micropropagation, anther culture, embryo rescue and assisted pollination, double haploid production, protoplast fusion and hybridization, marker-assisted (molecular-assisted) breeding, etc. These variations within a species appear as a result of mutations, growth conditions, and genetic dissimilarities, or they can even be induced (mutagenesis). Plant breeders constantly screen large populations, select plants with superior traits, and cross different genotypes producing the newer varieties in successive progenies. Newer breeding tools, such as zinc finger nuclease technology (ZFN-1, ZFN-2, and ZFN-3), oligonucleotide-directed mutagenesis (ODM), *cis*-genesis and intragenesis, RNA-dependent DNA methylation (RdDM), grafting onto GM rootstock, reverse breeding integrated with genomics, proteomics, and high-throughput genotyping, will enable large-scale screening of oilseed germplasms and identification of novel alleles.

The best examples of using plant breeding for the improvement of edible oil crops by Monsanto and Dupont include the development of edible canola oil (an acronym for Canada oil low acid), a double-zero cultivar with the low content of erucic acid and glucosinolates, from high-erucic acid rapeseed oil; high-oleic acid (75%) and <1% ALA safflower and sunflower oils; and low-LA soybean (>3%). These examples share common characteristics, namely, the levels of FAs in the plants were altered by reducing the activity of endogenous enzymes. Another classic example is the improvement of agronomic and economic traits of cultivated sunflower by broadening of the narrow genetic base by the infusion of genes from wild species. Similarly, a high-monounsaturated FA containing a sunflower mutant was obtained by chemical mutagenesis and backcrossing (Fernandez-Martinez et al., 1997). Oils that contain predominantly either oleic acid or stearic acid and are low in palmitic acid do not occur naturally in the major cultivated oilseed species. However, plant breeders have developed such FA profiles in a number of oilseed crops. A good example of this is the selection of high-oleic safflower and peanut oils.

On the other hand, breeding approaches have been less successful in manipulating FA profiles of edible oil crops for producing industrial oils. The reason is that most crop plants do not contain genes allowing them to accumulate such FA for industrial applications. Limited success has been seen with high-oleate/low-polyunsaturate lines of soybean and rapeseed/canola developed in the United States and Europe. These oils have already been used as biodegradable lubricating fluids due to a higher oxidative stability. Palm oil is also a very promising source for manipulations, but its genetic modification is challenging due to the slow growth of trees and a long generation life time, which make tissue culture manipulations a very time-consuming process. In addition, the tree requires 5–7 years to flower and produce oil. Over the past decade, two oil palm species, *Elaeis guineensis* and *Elaeis oleifera*, have been identified in Africa and South America. They can contain up to 65% oleate in the mesocarp oil. By using DNA-based molecular markers, it is possible to cross high-oleate lines into existing commercial lines for producing commercial amounts of high-oleic palm oil by around 2020. *Elaeis oleifera* has been crossed with *E. guineensis* to impart several desirable characteristics, including slow height increment, high unsaturation, and resistance to diseases (such as *Fusarium* wilt), to the economically important *E. guineensis*.

RECOMBINANT DNA TECHNOLOGY—METHODS AND TOOLS FOR OBTAINING TRANSGENIC PLANTS

Recombinant DNA technology allows the specific identification, isolation, manipulation, and reintroduction of desired gene(s) into a living organism to obtain a transgenic organism. This is one of the fundamental techniques used in plant molecular genetics for crop improvement and is replacing conventional plant-breeding methods at various levels so as to incorporate desirable traits, which was not possible to do using conventional methods. Also, FA biosynthesis is regulated by multiple genes, and it is difficult to obtain any type of Mendelian segregation of individual traits, which further limits plant-breeding methods.

Genetic engineering is a state-of-the-art technique that requires the following major steps for its successful implementation and use in crop improvement: a good understanding of genes and enzymes involved in metabolic pathways; gene construction (cloning the desired genes using the principle of overexpression, gene silencing, positional cloning); the availability of suitable *cis*-regulatory promoters to achieve gene expression in the desired tissue at the proper time; the availability of very efficient regeneration/transformation methods; the selection, regeneration, and analysis of putative transformants; and testing genetic stability of the trait.

Transgenic plants with increased level of activity for an enzyme of interest are usually obtained through overexpression of the transgene by placing it under the transcriptional control of a strong promoter. Similarly, gene silencing is used to reduce the activity of target enzyme. Gene silencing can be accomplished via various methods, namely, site-directed mutagenesis (SDM), transcriptional gene silencing (TGS), co-suppression, posttranscriptional gene silencing (PTGS), intron-spliced inverted repeats (ISIR), antisense RNA (RNAi), and direct inverted repeats (IR). Of these methods, antisense RNA is the most commonly used. Here, a modified gene of interest that produces RNA transcripts (hairpin RNA) complementary to the gene to be silenced is transferred into a host cell. On the other hand, co-suppression involves the transfer of a portion of the target-coding sequence in the sense orientation under the control of a strong promoter. The pHANNIBAL vector is proved to be very useful for the construction of novel transformation vectors with antisense, inverted repeats, and intron-spliced inverted repeats of the gene of interest (Bhore and Shah, 2012).

There is a wide range of methods available for the transfer of DNA into plant cells, each one having its merits and demerits. Some important techniques for obtaining transgenic plants followed by specific examples of success in oil crops will be discussed in the following sections.

Agrobacterium-Mediated Transformation

Agrobacterium, a naturally occurring gram-negative soil inhabitant, possesses the ability to transfer T-DNA (transferred DNA), a portion of DNA from its Ti (tumor-inducing) plasmid, to plant cells (Gelvin, 2003). T-DNA integration in the host plant genome followed by the expression of bacterial oncogenes that it encodes leads to the formation of crown gall disease in the infected plants. Bacterial genes in T-DNA can be replaced by desired gene(s) of interest, and this disarmed Ti-plasmid can then be used as a vehicle to deliver foreign DNA in plant cells. This method was used for plant transformation as early as in 1980s (Bevan, 1984). A modern *Agrobacterium*-mediated transformation technique utilizes a binary vector that consists of T-DNA and the vector backbone. One portion contains T-DNA, a segment delimited by the right border (RB) and the left border (LB), multiple cloning sites (mcs), a selectable marker gene for plants (*nptII*, *hpt*, or *bar*), a reporter gene (commonly *gus* or *gfp*), and gene(s) of interest. The most popular binary vectors are pBin19, pBI121, pPZP, pCAMBIA, and pFGC.

The *Agrobacterium*-mediated method is very efficient, and its success depends on the capability of bacteria to infect plant cells, to deliver T-DNA into cells and to incorporate it in the genome; it also depends on the successful regeneration of plants from transformed cells. Compared to monocots, dicotyledonous species are more responsive toward this method. A new category of vectors called superbinary vectors have been tested that have a better capability to transform monocots. They contain *vir*B, *vir*G, and *vir*C genes derived from pTiBo542, a plasmid responsible for imparting the supervirulence phenotype to the *Agrobacterium tumefaciens* strain A281. Significant improvements in plant tissue culture and regeneration techniques over the past two decades have made this natural genetic engineer harness its potential to transform a wide array of crop plants, including oil-yielding species. The main advantage of the method over all others is that it results in a greater proportion of less complex, single-copy stable transgene integration events.

Biolistic Gene Gun Method

Also known as the microprojectile bombardment method, a biolistic gene gun is a commonly used method for the transformation of those species that are recalcitrant to *Agrobacterium*-mediated transformation (mostly monocots such as cereals and legumes). Micron-sized gold or tungsten particles are coated with the DNA of interest. These particles are then accelerated at high speeds by helium/air pressure and are bombarded on plant tissue with the help of a specially designed instrument known as the "gene gun." The method was developed and tested first by John Sanford at Cornell University in 1987. Its advantages include high efficiency, species-independent transformation, and ability to exclude the plasmid backbone, to introduce polygenic traits, to target mitochondria and chloroplast DNA, and to introduce DNA fragments that are as large as yeast artificial chromosomes (YACs). The shortcomings of this method are the inability to systemically deliver DNA, a high copy number of integrated transgenes, the reduced regeneration capability of bombarded tissue due to cell damage, and the high cost of equipment. The advantages of the *Agrobacterium*-mediated method and biolistic transformation have been combined in a novel transformation technique termed agrolistic transformation. This approach makes it possible to co-transform the gene/s of interest with a vector that contains *Agrobacterium* virulence genes *vir*D1, *vir*D2, and *vir*E2 by using the particle bombardment method (Hansen and Chilton, 1996).

Nanoparticle-Mediated Gene Transfer

This approach has emerged as a very promising technique of genetic transformation of plants. Mesoporous silica nanoparticles (MSNs) serve as containers for gene delivery into isolated plant cells via mannose receptor–mediated endocytosis (Pérez-de-Luque and Rubiales, 2009). MSNs face the reduced challenges of transfer across cell walls due to their small size. A transfection agent, Lipofectamine, aids nanoparticle-mediated DNA delivery into target cells. Cell-penetrating peptides (CPPs) are a class of short peptides that possess the ability to translocate across the plasma membrane. They have been used as delivery vehicles to transport macromolecules, including DNA, into plant cells. Their low cytotoxicity and extreme versatility to undergo diverse modifications have made them a very attractive tool in plant genetic engineering. Though a precise mode of delivery still needs to be revealed, it is believed that they deliver cargo through the lipid bilayer by inverted micelle formation. CPPs such as Tat monomer (Tat) and dimer (Tat2) have been shown to deliver DNA into plant cells (Chugh et al., 2009).

Recently, we developed a novel method of transgene delivery into monocots (Ziemienowicz et al., 2012). The method relies on the use of an *in vitro* prepared nanocomplex consisting of T-DNA, VirD2, and RecA proteins. The preassembled complex was then delivered to triticale microspores with the help of a Tat2 cell-penetrating peptide. In a series of experiments, we showed that this approach (1) allowed for single transgene copy integration events and (2) prevented the degradation of delivered DNA, thus leading to the integration of intact copies of the transgene into the genome of triticale plants. Northern and Western blots analyses showed that the transgene was expressed in all transgenic plants regenerated from microspores transfected with the full T-DNA–protein complex. This approach can easily substitute for the bombardment technique that is currently used for monocots, thus becoming a highly valuable tool for plant biology and biotechnology. Disadvantages of this technique include primarily the lengthy and somewhat tricky protocol for the preparation of nano–T-DNA–complex particles and the fact that the protocol so far works only for microspores. We are currently developing protocols that allow the transformation of somatic cells.

Electroporation and Polyethyleneglycol-Mediated Transformation

An electric pulse of 10,000–100,000 V/cm (depending on cell size) is applied temporarily for a few microseconds to disturb the lipid bilayer of protoplasts, thus permitting the transport of DNA, RNA, proteins, drugs, and dyes through these electropores inside the cell (Fromm et al., 1985). Polyethyleneglycol (PEG) can also be used in place of the electric pulse for DNA delivery in protoplasts (Paszkowski et al., 1984). The technique is very versatile and requires a very small amount of DNA. The first transgenic plants were obtained by the transformation of protoplasts, but since the regeneration from protoplasts is a challenge, this system is mostly used for transient expression analysis.

Other, Less-Efficient Techniques

The pollen-tube pathway (PTP) method: Styles are cut after pollination, and the DNA that is applied over them reaches the ovule through the pollen tube. This method was first applied for rice transformation, and later it was also applied in wheat, soybean, and other species.

Liposomes: Liposome-based gene transfer is a nonviral gene delivery method that uses the combination of phospholipids, cholesterol, and some polymers for the formation of spherical vesicles. DNA is delivered into cells by liposomes via membrane fusion and endocytosis. This method has several advantages such as enhanced delivery, nontoxicity, the protection of nucleic acids from nuclease activity, and the ability to target DNA into a variety of cell types, including protoplasts, via plasmodesmata.

Microinjection: DNA is introduced in the nucleus or cytoplasm of a plant cell with the help of a glass microcapillary pipette. This method has limited applications in plant systems due to the presence of a well wall that acts as a barrier for glass microtools, but it has been demonstrated to transform certain plant species.

Silicon carbide whisker-mediated transformation: Silicon carbide crystals with a diameter of 0.6 μm and a length of 10–80 μm are mixed in liquid medium by vortexing in the presence of DNA and plant cells. The crystals pierce the cell walls and facilitate DNA transfer into plant cells (Kaeppler et al., 1990). This technique is considered to be one of the simplest techniques of DNA transfer to plant cells; it is easy and quick but has a limited use due to low transformation efficiency, damage to cells that affects their regenerative capability, and the health hazards associated working with asbestos and silicon fibers. This technique has been successfully demonstrated in species such as maize, wheat, tobacco, etc.

Gene delivery vehicles: The DNA chain is collapsed into a compact globule through neutralization of its negative charge by cationic polymers such as polyamines and multivalent cations such as spermidine and cobalt hexamine [$Co(NH_3)6^{3+}$] (Wagdy, 2001).

Beads: Micrometer-sized calcium alginate beads that were named "bio-beads" encapsulate vectors carrying genes of interest; they have been successfully tested in tobacco cells (Sone et al., 2002) (Figure 14.1).

GENETIC TRANSFORMATION OF OIL CROPS FOR SUSTAINABLE DEVELOPMENT

The composition of FAs stored in plant seeds is highly variable among different species and is amenable to changes, making them a target for genetic manipulations without disturbing plant physiology. One of the earliest implementation of biotechnology was for oil crop improvement. Rapeseed (the family Brassicaceae) was one of the first oilseed crops that has benefited from the technology due to its relatedness to the *Arabidopsis* genome. Rapeseed with increased levels of lauric acid was the first GM crop approved for commercialization in the year 1996 in the United States (Murphy, 1996). Most GM crops in market today are herbicide- and pest-resistant varieties, and increasingly, new cultivars are being tested that have valuable traits aimed at satisfying the demands of today's lifestyle and health concerns. There is an increased interest not only in improving the quality of oil yield in major oil crops but also in transferring novel traits from nonedible oil plants to soybean and rapeseed using the recombinant DNA technology (Table 14.1).

Considerable progress has been made over the past decades in characterizing the lipid synthesis pathways in many oil plants; genes for making various enzymes have been identified and cloned. Along with genome maps now available for major oil crops and advances in genomics, proteomics, and

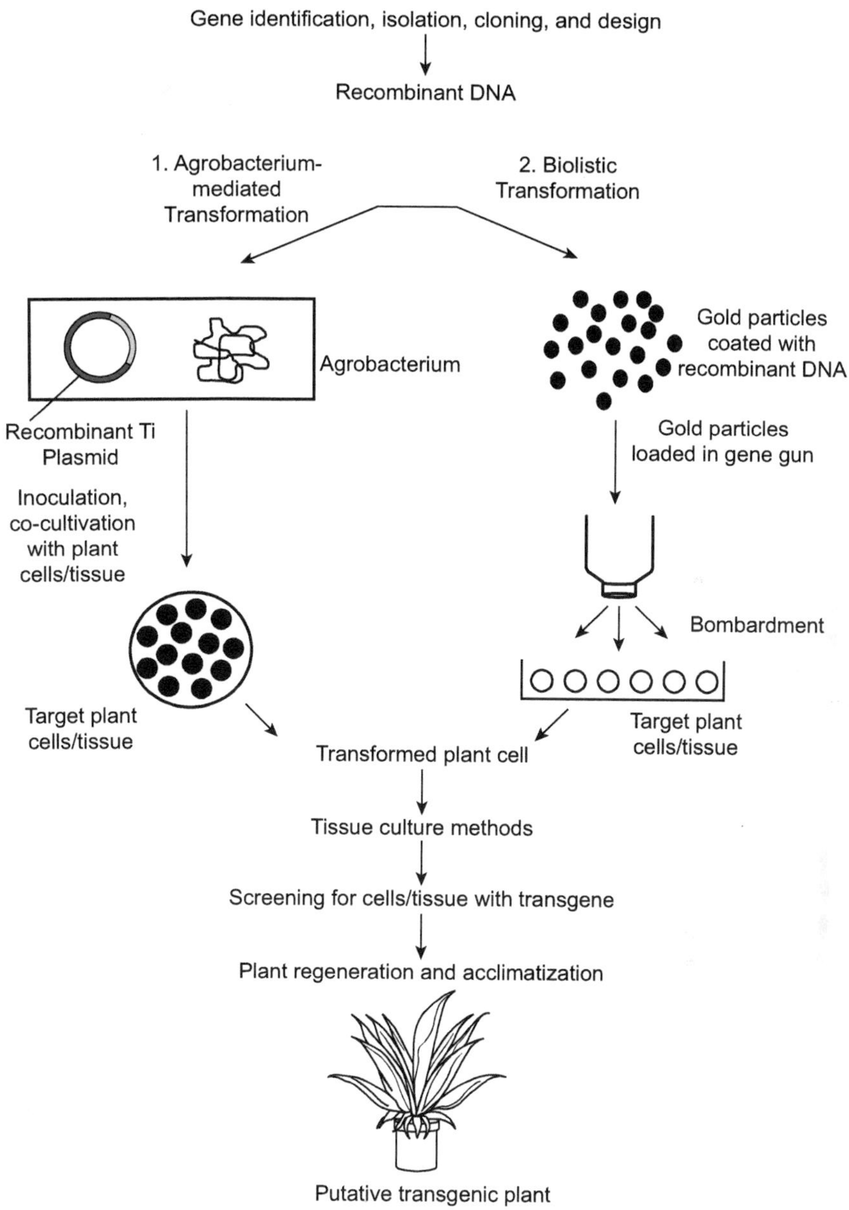

FIGURE 14.1 The *Agrobacterium tumefaciens*–mediated and biolistic transformation methods in plants.

TABLE 14.1 List of Selected Examples of Oil Modifications in Transgenic Oilseed Crops

Genus	Fatty Acid Modification	Target Enzyme Engineered	Gene Source	Gene Regulation	References
Soybean	Stearic acid	Stearoyl-ACP-Δ9- desaturase and oleoyl-Δ12-desaturase	Soybean	Downregulation	Kinney (1996)
	Oleic acid	Oleoyl-Δ12-desaturase	Soybean	Downregulation	Kinney (1996)
	Eleostearic acid	Conjugase	*Momordica*	Overexpression	Cahoon et al. (1999)
	Δ5-Eicosenoic acid	β-Ketoacyl-CoA synthase and acyl-CoA desaturase	Meadowfoam	Overexpression	Cahoon et al. (1999)
	Oleic acid	Δ12-Fatty acid desaturase	Soybean	Overexpression	Buhr et al. (2002)
	Palmitic acid	Palmitoyl-thioesterase	Soybean	Down regulation	Buhr et al. (2002)
	ϒ-Linolenic acid and stearidonic acid	Δ6-Desaturase	*Arabidopsis thaliana*	Overexpression	Sato et al. (2004)
	Arachidonic acid	Δ5-Desaturase, Δ6-desaturase, GLELO elongase, Δ15-desaturase	*Mortierella alpina* IS-4 and soybean	Overexpression	Chen et al. (2006)
	Stearidonic acid	Δ6-Desaturase, Δ15-desaturase	*B. officinalis* and *A. thaliana*	Overexpression	Eckert et al. (2006)
	α-Linolenic acids (18:3)	ω-3 Fatty acid desaturase	Soybean	Down regulation	Flores et al. (2008)
	C20-LC PUFAs	Δ6-Desaturase, Δ6-elongase, Δ5-desaturase	*Marchantia polymorpha*	Overexpression	Kajikawa et al. (2008)
	Higher oil content	Diacylglycerol acyltransferase 2A	*Umbelopsis ramanniana*	Overexpression	Lardizabal et al. (2008)
	Oleic acid	Δ12-Fatty acid desaturase	Soybean	Overexpression	Wang and Xu (2008)
	Higher oil content	Sphingolipid compensation	*Saccharomyces cerevisiae*		Rao and Hildebrand (2009)

	Improved oil yield and composition	Fatty acid ω-3 desaturase 2, acyl–acyl carrier protein thioesterase, diacylglycerol acyl-transferase, dihydrodipicolinate synthase, high-lysine protein, truncated cysteine synthase	Soybean, *Yarrowia lipolytica*, *Corynebacterium glutamicum*, barely	Overexpression	Li et al. (2010)
	Epoxy fatty acid	Epoxygenase, diacylglycerol acyl-transferase	*Stokesia laevis*, *Vernonia galamensis*	Overexpression	Li et al. (2010)
Brassica napus	Stearic acid	Stearoyl-acyl carrier protein (ACP)-Δ9- desaturase	*B. napus*	Downregulation	Knutzon et al. (1992)
	Palmitic acid	Acyl-ACP thioesterase	*Cuphea*	Overexpression	Jones et al. (1995)
	Lauric acid	Acyl-ACP thioesterase	California Bay	Overexpression	Voelker et al. (1996)
	Caprylic acid	Acyl-ACP thioesterase	*Cuphea*	Overexpression	Dehesh et al. (1996)
	Stearic acid	Acyl-ACP thioesterase	Mangosteen	Overexpression	Hawkins and Kridl (1998)
		Laurate-specific lysophosphatidic acid acyltransferase (LPAAT), acyl-ACP thioesterase	Coconut, California Bay	Overexpression	Knutzon et al. (1999)
	Oleic acid	Oleoyl-Δ12-desaturase	*B. napus*	Downregulation	Stoutjesdijk et al. (2000)
	γ-Linolenic acid (18:3 ω-6)	Oleoyl-Δ6-desaturase and Oleoyl-Δ12-desaturase	*Mortierella alpina*	Overexpression	Ursin et al. (2000)
	Saturated fatty acids	Palmitoyl-ACP-desaturase	Cat's claw (*Doxantha unguis-cati*)	Downregulation	Bondaruk et al. (2007)
	Oil rich in ricinoleic acid	Fatty acid hydroxylase	castor	Overexpression	van de Loo et al. (1995)
Cottonseed	Stearic acid	Stearoyl-ACP-Δ9- desaturase	Cottonseed	Downregulation	Liu et al. (2000)
	Oleic acid	Oleoyl-Δ12-desaturase	Cottonseed	Downregulation	Liu et al. (2000)
B. juncea	Oleic acid	Oleoyl-Δ12-desaturase	*Brassica*	Downregulation	Stoutjesdijk et al. (2000)
Flax	C18 ω-3, stearidonic acid	ω-3 desaturase	Primula	Overexpression	Ruiz-López et al. (2009)

metabolomics methods, biotechnology is now a realistic tool that has the full potential toward sustainable development by producing novel crop varieties with desirable traits required to meet the demands of future generations without a significant impact on the environment.

A wide range of goals exist for plant biotechnologists that are based on various applications of plant oils:

- Increasing oil accumulation and yield: One of the prerequisites for the commercial viability of a GM oil crop is that it should accumulate desired FAs at the highest possible yield and purity (ideally, up to 80–90%). This is theoretically feasible as plants such as the castor bean accumulate as much as 80% of a single FA. The current oil yield ranges from 1 metric ton (MT)/ha in temperate crops to about 4–10 MT/ha in tropical oil palms. An increase in the oil production by 50% will be sufficient to meet food demands of the world's growing population and to get feed stocks that are reasonably priced for industries.
- Tailoring oil composition for improving the quality of oil: by reducing the amount of bad FAs and incorporating and increasing the content of healthier FAs; oil seed crops need to be engineered to meet global health and nutritional requirements. Oils that can reduce risk of heart diseases, cancer, obesity, diabetes, etc. are the targets.
- Increasing oil stability: The development of high-oleic oils in order to provide high-stability cooking/frying oils. These oils can replace the most commonly and widely used saturated fats and hydrogenated oils that significantly increase the risk of cardiovascular disease due to the presence of saturated and *trans*-FAs that elevate the LDL cholesterol level in the blood. Similarly, oils with an increased content of stearic acid would enable the production of solid fats without the need for hydrogenation. These oils with high oxidative and thermal stability will open up markets for their use as feedstocks for industrial biodiesel production.
- Rendering plants tolerant to stress and resistant to insects, pests, herbicides, and heat- and drought-tolerant excessive weeds, thus reducing plants competition for sunlight, nutrition and hence affecting crop yield. Herbicide-resistant crops make possible the use of nonselective herbicides to get rid of all weeds in one single application. Targets for pest resistance include *Bacillus thuringiensis* Cry proteins, chitinases, glucanases, lectins, alpha amylase inhibitors, protease inhibitors, cystatins that activate plant innate immune responses to produce phytoalexins, etc. Viruses and fungi are responsible for a great number of diseases, such as bunt, rust, gray mold, powdery mildew, downy mildew, etc. The approaches to virus resistance include the introduction of genes encoding viral coat proteins.
- Engineering major oil crops that will be able to produce useful oils generally produced only by wild and nonedible oil plants.
- Designing oilseeds to replace petroleum products.

- Imparting other valuable traits such as an increased vitamin A and E contents.
- Producing VLC PUFAs in transgenic plants: oilseeds producing high levels of EPA and DHA along with reduced amounts of ω-6 FAs will not only be useful for medicinal and nutraceutical applications in human health, they are also a good source as fish feedstock.

SOYBEAN

Soybean (*Glycine max* L. Merrill), a subtropical species, is the largest source of edible oil worldwide. Soybean seeds contain about 40% protein and about 20% oil. In the world market, the United States ranks number one in soybean production (33%), followed by Brazil (27%), Argentina (21%), and China (7%); India and Paraguay are also substantial producers (FAOSTAT, 2011). Soybean oil is used in the food, pharmaceutical, and feed industries, as well as for biodiesel purposes. Among major food crops, soybean production has increased rapidly in the past 20 years, but it is still not enough to meet all requirements. In 2010, 81% of soybean cultivated in the world was transgenic, and in 2011, the percentage was >99% in Argentina and >94% in the United States. Nonselective herbicide-resistant (Roundup Ready) soybean that was resistant to glyphosate(*N*-phosphonomethylglycine) was the first transgenic variety introduced for the commercial production in 1996 (Padgette et al., 1995). Currently, >90% of soybean traded worldwide is herbicide tolerant.

Soybean oil is composed of palmitic, stearic, oleic, linoleic, and linolenic acids, which made it a promising source of biodiesel production. In Brazil, soybean accounts for 80% of all biodiesel production; in Argentina, for 100%; in the United States, for ~74%; and in Europe, for 16%. But the presence of large amounts of unsaturated esters makes this oil unstable to oxidation. Soybean has been genetically engineered to decrease the amount of PUFAs through downregulation of FA desaturases by ribozyme termination of RNA transcripts and RNAi (Kasai and Kanazawa, 2012). The synthesis of other FAs, such as γ-linolenic, stearidonic, arachidonic, eicosapentaenoic, and vernolic acids, has led to further improvement of oil quality. In addition, an increased accumulation of α-tocopherol (up to 95%) led to an increase in an antioxidant activity, which enhanced health benefits. Many newer traits have been genetically incorporated, but some challenges continue (Table 14.1).

The expressed sequence tag (EST) databases containing full-length cDNAs and cDNA microarrays have been useful tools for engineering the soybean genome (Stacey et al., 2004; Umezawa et al., 2008). Recently, the soybean genomics database has also been developed using the whole genome sequence (Schmutz et al., 2010). However, being a complicated process, soybean regeneration has remained recalcitrant to routine transformation methods.

There are only two successful and repeatable methods: *Agrobacterium*-mediated transformation of cotyledonary node explants from young seedlings or imbibed seeds and particle-bombardment transformation of somatic embryos

(Donaldson and Simmonds, 2000). Cotyledonary nodes are prewounded by a scalpel, a needle, or even a stainless steel microbrush, and after *Agrobacterium* co-cultivation, they are cultured with cytokinins, thus allowing shoot proliferation. The use of a superbinary vector and acetosyringone as reducing agents in the co-cultivation medium has increased successful transformation. In particle-bombardment-mediated transformation via somatic embryogenesis, somatic embryos are induced from cotyledonary explants under the influence of the auxin 2,4-dichlorophenoxyacetic acid (2,4-D). The use of linear transgene constructs lacking vector backbone sequences has been shown to regenerate transgenic soybean plants with a low transgene copy number. Somatic embryos can be cultured as suspension cultures that can retain their regeneration potential for over a year. They can also be used as models of zygotic embryos for studying oil accumulation and composition as well as for the assessment of transgenic seed traits even before regeneration of whole plants. However, somatic embryogenesis is genotype specific—hence, its application for the production of desired genotypes is limited. The cultivar Jack has been the most promising one and is used widely for soybean transformation (Tomlin et al., 2002). The competence for somatic embryogenesis has been incorporated in other cultivars through breeding methods (Kita et al., 2007, 2010), thus allowing to broaden the scope of genotypes to be used. Also, newer improved protocols for regeneration hold a promise for more genotypes amenable to transformation (Ishimoto et al., 2010; Takagi et al., 2011).

OIL PALM

Oil palm (*Elaeis guineensis* Jacq.) is believed to be indigenous to West Africa. More than 80% of the world's palm oil comes from Southeast Asia, mainly Malaysia (50%) and Indonesia. Another species, such as American oil palm (*Elaeis oleifera*), are also cultivated for vegetable oil. One hectare of oil palm plantation produces 4.5 t of oil per year which is three times the yield of coconut and more than 10 times that of soybean. Oil palms accumulate ~90% of their oil in the mesocarp of fruits (palm oil contains mostly palmitic and oleic acid, together ~80%:10% LA), and the remaining comes from seed kernels (they contain >60% of medium-chain FAs, mainly, lauric acid). Palm oil is mostly used in food products, as well as an oleochemical feedstock and a feedstock for biodiesel production. Palm kernel oil is a most important feedstock used in the production of detergents and other applications of medium-chain FAs.

A palm tree has a long regeneration time, a narrow gene pool, and the open pollination behavior, along with the high commercial implication tagged to the crop, together makes its improvement through either conventional breeding methods or genetic engineering alone very limited (Rajanaidu and Jalani, 1995). Basta-resistant oil palm has been developed using genetic transformation through biolistics (Parveez, 2000). Both *Cowpea trypsin inhibitor* (*CpTI*) and *Bacillus thuringiensis* (*Bt*) crystal insecticidal protein genes and *chitinase* were shown to be successful to address problems related to basal stem rot (Sharma et al., 2002). The Malaysian

Palm Oil Board is the major body involved in genetic engineering of oil palm, and its major goals include developing high-oleate and high-stearate transgenic palms (Parveezet al., 1999; Parveez, 2003). Large amounts of oleate and linoleate in palm oil have been shown to lower total serum cholesterol by reducing the level of LDL, which makes it nutritionally beneficial as well as useful as a good industrial feedstock. A further increase in oleate content will also be advantageous for oleochemical industries. While high-stearate palm oil can be used as a cocoa butter substitute, high–palmitoleic acid oil finds nutraceutic applications (Sambanthamurthi et al., 2002). Since palm fruits have two storage tissues (mesocarp and kernel) that can be targeted for accumulating GM products, the scope is unlimited. The characterization of mesocarp- and kernel-specific promoters will help in this invention. Until now, the biolistic method has been more successful than the *Agrobacterium*-mediated approach (Parveez, 2000). Recent improvements in the biolistic techniques (Parveez and Bahariah, 2012) and research on *Agrobacterium*-mediated transformation, namely, the use of immature embryos and embryogenic calli derived from young leaves as starting materials, using acetosyringone, exposure to a plasmolysis medium, and physical injury via biolistics, hold great promise for future advancements (Izawati et al., 2012).

RAPESEED

Rapeseed (*Brassica napus* L.) is the third largest oil crop after soybean and oil palm, with 12% of the world oil market and the main production areas in Europe (35%), China (26%), India (14%), and Canada (8%). Canola oil (double zero rapeseed) developed through conventional plant breeding from rapeseed is used for the production of margarine, spreads, dairy blends, animal feed, emulsifiers, vitamin E, healthy cooking oils, etc.

Rapeseed is very closely related phylogenetically to *Arabidopsis*; therefore, the results of genome analysis of *Arabidopsis* can be directly applied to canola. *Arabidopsis* genes can be directly used for downregulating rapeseed genes. The initial GM rapeseed varieties obtained carried improved herbicide-tolerance traits, as well as fungus- and insect-resistant traits (Stewart et al., 1996). The most significant development made by Calgene in 1993–1 994 came in the form of rapeseed containing 40% stearic and 40% lauric acid. Normally, rapeseed contains 1–2% stearic acid and <0.1% lauric acid. This lauric rapeseed variety was approved for unrestricted commercial cultivation in the United States in 1995. It contains a lauroyl-ACP thioesterase gene from the California Bay tree (*Umbellularia californica*) that leads to premature termination of the growing FA chain, thus resulting in the accumulation of lauric acid. Oils rich in lauric acid are mainly used for the production of soaps, detergents, confectionery, milk formulas, etc. Another good example is the high-stearic variety that contains an antisense copy of a *Brassica* stearate desaturase gene. This inhibits the function of the indigenous gene, resulting in the accumulation of high-stearic oils, rather than their desaturation to high-oleic oils. The resulting high-stearic oil has advantages over the normal rapeseed oil for the production of solid fats such as margarines. The FA hydroxylase gene responsible for the synthesis of ricinoleic acid in castor bean has also been transformed in canola.

Rapeseed is very easy to manipulate under the conditions in tissue culture, and it has a high transformation efficiency which makes it the most desirable plant for genetic modifications. Various explants have been tested for regeneration potential in canola, namely, hypocotyls, cotyledonary leaves, stems, true leaves, roots, microspores, protoplasts, etc. Among these hypocotyls, the explants are the most promising choice. The *Agrobacterium*-mediated method is the most preferred method of choice for canola transformation (Cardoza and Stewart, 2004).The existing protocols for transformation and regeneration focus mostly on the cultivar Westar, which is not agronomically desirable, and the protocols established for it are often not directly applicable to commercial cultivars (De Block et al., 1989). To address this, we have conducted an elaborate investigation of the conditions for regenerating shoots from hypocotyl explants from four genetic lines: Invigor 5020, Westar, Topas, as well as a microspore culture-derived line of Topas (Line 4079). A detailed analysis of the effect of hormonal combinations on regeneration medium, donor plant age, and explant type on the regeneration capacity of these cultivars indicated that a careful selection of age of donor plants and slight changes in hormonal combinations can drastically improve regeneration frequencies (Maheshwari et al., 2011). Significant progress has also been made in microspore culture, thus paving the way for further efficient transformation systems (Wen et al., 2012).

CASTOR

Ricinus communis is a monotypic species containing up to 60% unique oil, 90% of it being ricinoleic acid (12-C hydroxyoleic acid). Some special characteristics of this acid, such as a high molecular weight, a low melting point (5°C), a very low solidification point (−12°C to −18°C), and the highest and most stable viscosity, render it extremely useful for industrial purposes. Vegetable oils that are rich in ricinoleic acid have properties that are desirable in the production of nylon, lubricants, soaps, and resins. The main suppliers of oil worldwide are India (the world's leader with 60% of the total production), Brazil, and China. Its cultivation faces some challenges due to the presence of the toxic ricin, an alkaloid ricinine, and an allergenic protein polysaccharide CB-1A in castor beans. Ricin inhibits protein synthesis by inactivating ribosomes, and it is lethal when taken in any way. Besides engineering castor plants for producing novel oils, such as epoxy acids for the production of premium oil–based paints without the liberation of volatile pollutants, there is a need to eliminate toxic compounds and render it safe to handle. Ricin-free castor meal will have significant value for the production of animal feeds due to its high nutritional content. However, its manipulation has faced challenges because these plant species are highly recalcitrant to transformation. Embryo axes and shoot tips are the only promising explants that can be used for transformation (Sujatha and Sailaja, 2005, 2007). Vacuum infiltration of seedlings and wounded flower buds and the use of other materials such as hypocotyls and cotyledons have shown limited success. Though insecticide-resistant castor plants have been reported, efforts for knocking out the genes responsible for the production of ricin and other toxins are not yet fully successful (Malathi et al., 2006).

SUNFLOWER

Sunflower (*Helianthus annuus* L.) is adapted well to both subtropical and temperate areas and is drought tolerant. Its main oil storage organ is an achene with 50% oil content. Sunflower oil accounts for 9% of the world plant oil production, the main producers being Russia (20%), Europe (19%), Ukraine (15%), and Argentina (13%). Triacylglycerols, the main constituents of sunflower oil, are liquid at room temperature and have a low melting point. It produces high-quality edible oil. Besides food, sunflower oil is the main feedstock for the oleochemical and cosmetics industries, for the synthesis of polyester films, modified resins, lacquers, copolymers, plasticizers, etc. Also, the high-oleic acid sunflower variety is used for biodiesel production. Three main types of oils are available—a traditional variety composed of palmitic (6%), oleic (24%), and linoleic (65%) acids; a high-oleic variety (80% oleic acid); and a high-palmitic variety (28% palmitic acid). To date, the improvement of sunflower has been restricted to conventional breeding, and no GM cultivar has been released. The reason was that genetic transformation of sunflower has been very challenging due to its very high recalcitrance to in vitro culture and poor transmission of the transgenes to the progeny. *Agrobacterium*-mediated and direct gene transfer into protoplasts have been demonstrated, but with limited success. Regeneration of sunflower can be achieved through organogenesis and embryogenesis, but since embryogenesis is highly genotype dependent, the implementation of this technique in different genotypes is limited. Innovative strategies, such as wounding the explants with glass beads during co-cultivation, microprojectile bombardment of half shoot apices in the combination with *Agrobacterium* co-cultivation, have been tested to increase the efficiency of transformation (Rao and Rohini, 1999; Liu et al., 2011).

COTTON

Cotton (*Gossypium hirsutum*) is mainly grown for its fiber, and its seeds are collected as byproducts. Cottonseed oil content is low (20%), but due to large amounts of cotton production, cottonseed oil accounts for 5% of the total oil production in the world. Its oil is rich in ALA (56%) and palmitic and oleic acids (23% and 17%, respectively). Crude cottonseed contains gossypol, a phenolic compound, and cyclopropene acids, which render seed products not usable for food/feed without refining and hydrogenation, making them more desirable for industrial applications. Refined oil is the fifth most widely consumed vegetable oil in the world (termed Heart Oil) and is enriched in vitamin E (tocopherol) and is a very good frying oil due to high levels of palmitic acid, which provides stability. Interspecific hybridization has been difficult in cotton, whereas in vitro regeneration techniques such as ovule culture, protoplast culture, somatic embryogenesis, and genetic transformation using *Agrobacterium* and particle bombardment methods are well established. Cotton is the second most widely used GM crop on the market. GM traits used for cultivation are herbicide and insecticide (Bt) resistance (the *cry1Ab* gene for the resistance against *Heliothis armigera*) (Tohidfar et al., 2008). Currently, there are no examples of commercial GE traits engineered for oil quality, but high-oleic and high-stearic acid varieties have already been developed through genetic engineering.

FLAX

Linseed (*Linum usitatissimum* L.) is an annual oil crop that is grown either for its fiber (fiber flax) or for its oil (oilseed flax). The area under annual cultivation is 3 million hectares, accounting for 1% of the total world production of plant oil. The five major flax-producing countries are Canada (the world leader), China, India, the United States, and the European Union. Oilseed flax oil has a very high content of LA (40–60%) and is used as a fast-drying oil for technical applications such as linoleum flooring, paints, varnishes, and other coatings. Due to its high content of LA, oilseed flax oil has recently been increasingly used for human consumption. Its daily consumption aids in the prevention of atherosclerotic cardiovascular disease due to the presence of ALA, lignan, phytoestrogen, and soluble fiber, which makes it suitable as a source of drugs and therapeutics. Flax can be transformed by *Agrobacterium*-mediated and biolistic methods; its regeneration system was established from protoplasts, hypocotyls, cotyledons, and leaves (Burbulis et al., 2009). Flax varieties containing single gene modifications conferring resistance to insects and herbicides (glyphosate, glufosinate, and sulfonylurea) have been commercialized (McHughen, 2002). The only GE flax cultivar in the world, "CDC Triffid," resistant to sulfonylurea herbicide, was considered for commercial release in Canada in 1998. Being a minor crop, flax has not fully benefited from genetic engineering approaches. Oil with >70% ALA would increase its drying quality. Zero percent ALA cultivars of flax can be obtained by reducing the activity of Δ15-desaturase. Being the richest plant source of ω-3 FA, flax has tremendous potential for its further accumulation, which makes it a potential vegetarian alternative to fish oil. Therefore, it might be used not only as nutraceuticals but also as a valuable feedstock for fish and poultry (eggs containing high levels of ω-3 FA).

MAIZE

GM maize (*Zea mays*) resistant to corn borer was introduced for the first time in 1997 in the United States and Canada, and now it is the only commercially grown crop in Europe. Approximately 65% of maize grain is used for feeding animals, making it the most important feedstock for the production of meat, eggs, milk, and fuel in the world. High-oil maize shows a greater feed efficiency than normal-oil maize due to a higher caloric content of oil. Also, corn oil is the most valuable co-product of industrial processing of its grain for human consumption. Compared with soybean oil, which contains 6.8% linolenic acid, maize oil is more stable because it contains less than 1.0% linolenic acid. Biolistics-mediated transformations have been more successful in maize, but *Agrobacterium*-mediated methods are fully developed and promising (Takavar et al., 2010).

JATROPHA

Jatropha curcas L. is a tropical and subtropical drought-resistant shrub that has been getting a lot of attention lately as a potential biodiesel plant. It has long been used as lamp oil and for producing soap. Its oil content is 34%; it is rich in oleic (42%), linoleic (35%), palmitic (14%), and stearic (6%) acids. The plant contains several toxic compounds such as lectin, saponin, carcinogenic phorbol, and a trypsin inhibitor. The plant requires the elaborative breeding and genetic engineering techniques to improve its oil quality for increasing yield, stability, and reducing levels of toxic compounds. Limitations for improving traits by breeding include a lack of knowledge of the genetic base, poor yield, low genetic diversity, and vulnerability to a wide array of insects and diseases. Efficient methods of in vitro regeneration have been established that use direct and callus-mediated shoot regeneration, somatic embryogenesis from endosperm, seedling tissues, leaves, cotyledon discs, petioles and hypocotyls, and axillary nodes and provide a potential for exploitation of Jatropha through applications of biotechnological tools (Jha et al., 2007; Misra and Misra, 2010). A good example can be silencing the Δ9- or Δ12-desaturase genes for increasing the accumulation of stearic and oleic acids (King et al., 2009).

CAMELINA

Camelina (*Camelina sativa* L.) is an underused oilseed crop adapted to a wide range of climatic conditions. Its oil content is 38–47%, and oil is highly polyunsaturated (46% linolenic, 21% linoleic and 17% oleic FAs, 3% erucic acid). It can be stably transformed by simple *Agrobacterium*-based methods such as vacuum infiltration and floral dip methods.

TRANSGENIC OIL CROPS—MEANS AND CHALLENGES

Genetic engineering is the ability to incorporate genes from distant species that is not possible through conventional plant breeding. Though an efficient technique, it comes with many challenges. The technique is still not very precise; insertions of multiple copies at random loci can pose a threat to the stability of the plant genome, thus affecting the physiology and biochemistry of the host plant.

Despite more than two decades of research on genetic modifications of oil crops, success has been limited. In recent years, genes have been cloned for all the major enzymes that control FA biosynthesis in oilseeds, but the complex mechanisms underlying oil synthesis, regulation and FA flux remain partially understood. Further progress in the area is fully dependent on the identification of target genes and a complete understanding of their genetic and biochemical control mechanisms. Major challenges facing the modification of oil profiles include, but are not limited to, the following: the transgenes sometimes are poorly or not expressed at all in seeds; oil stored in seeds is necessary for the germination and the modified oil content might interfere with the process of germination; at times, it is difficult for the modified plant to metabolize a new oil profile; structural and storage lipids share some common steps in biosynthetic pathways; there is a likelihood that a novel FA might get incorporated in the cell membrane; and constitutive expression of transgenes can be detrimental.

Seed oil content in plants is controlled by multiple steps in the oil biosynthetic pathway, and hence, the strategy of single-enzyme alterations does not always result in an increased accumulation of desired FAs. Theoretically, it is feasible to express almost any FA in plant oil by genetic engineering. However, despite decades of research, most novel FAs still fail to accumulate at desirable levels. At times, this complex metabolic pathway demands the insertion of additional transgenes such as acyltransferases and co-expression of desaturases and elongases for the production of VLC PUFAs. In addition, in crops such as canola, a portion of novel acyl groups is shown to integrate in storage lipids instead of being channeled to seeds (Broun et al., 1999).

Increased levels of FA accumulation for the improvement of nutritional values should be achieved without compromising the functionality or any other agronomic characteristic of the crop. It must be ensured that the transgene is expressed in correct tissues at correct time. The simplest means of modifying the oil synthesis pathway is via single gene modifications. This can be achieved by downregulation of target genes to reduce the levels of unwanted FAs or by overexpression of target genes to increase the content of desired FAs. Some examples of this are: silencing linoleate-desaturase genes for obtaining low levels of α-linolenate and high-oleate transgenic lines of canola, soybean, and cottonseed; increasing the content of stearic acid in rapeseed by introducing stearoyl-ACP desaturase cDNA in the antisense orientation; lowering the content of a palmitoyl-acyl carrier protein thioesterase (*PATE*), one of the key genes involved in FA biosynthesis in plastids, by posttranscriptional *PATE* gene silencing (Bhore and Shah, 2012); and decreasing the amounts of ALA and LA by antisense expression of ω-3 and ω-6 desaturase in rapeseed (Voelker et al., 1992) and soybean (Topfer et al., 1995), etc.

Newer approaches and tools need to be developed to meet the aforementioned challenges. The use of seed-specific promoters, antisense RNA technology, reengineering enzymes to create novel products, enzyme crystallization to study their structures in details, the use of global transcription factors that control many genes of the pathway, the use of genes expressed only in seeds, and the use of Δ12-desaturase that shares common motifs with regulatory elements provide some good alternatives for achieving high oil yields. Overexpression of a transcription factor, maize ZmWRI1, has been shown to increase the content of seed oil without affecting seed germination (Grotewold, 2008). ESTs generated by rapid throughput automated cDNA sequencing can be fully exploited to facilitate gene isolation in the oil biosynthesis pathway. The FA synthesis pathways are well characterized in canola and soybean and can be implicated in other oil crops. Overexpression of the yeast glycerol 3-phosphate dehydrogenase gene (*ghpd1*) in canola seeds has led to a 40% increase in the flux of lipids due to an increased supply of glycerol 3-phosphate, a substrate for TAG synthesis (Vigeolas et al., 2007).

After new traits have been genetically engineered, plant breeding comes into play to help in the selection of lines with improved yield, oil content and quality, as well as other traits that later can be combined with the original characteristics. Recent developments in plant genomics, microarray technology, proteomics, and metabolomics will soon make it possible to understand the complex relationship between genes and will provide a clear understanding of their function and regulation, making way for a successful future and progress of GM crops. This will finally depend on the abilities to engineer predictable stable expression of multiple transgenic traits coupled with the prevention of transfer of undesirable material to nontransgenic crops and other species, thus rendering the technique safe and sustainable.

DEVELOPMENT OF CLEAN MARKER-FREE TRANSGENIC PLANTS

While genetically transforming a plant with a gene of interest, marker gene(s) are also introduced to enable the selection of transformed cells/tissue/plants and determine the successful stable integration and expression of the gene of interest introduced. Over 50 different selectable marker genes (SMGs) have been used to date, the most common being antibiotic (*nptII* and *hpt* genes for resistance to the aminoglycoside antibiotics kanamycin and hygromycin) and herbicide resistance genes (*bar* gene for resistance to herbicide phosphinothricin). The cultivation of GM crops might lead to the transfer of these SMGs to bacteria in the GI tract or into soil and render them resistant to antibiotics. Similarly, weeds might incorporate the herbicide-resistance gene and render weeding difficult. Besides SMGs, a screenable marker reporter gene is introduced. The products of these genes allow a visible estimation of levels of foreign gene expression in transgenic tissues. Some main examples include β-glucuronidase (*GUS*), luciferase (*LUC*), β-galactosidase, and green florescence protein (*GFP*). These markers facilitate screening using simple biochemical, histochemical or fluorimetric assays. The presence of a selection marker in the transformation system initially is a prerequisite and is crucial. However, it becomes redundant once the transgenic plants have been selected, and its removal ensures biosafety of the technology. Recent research in this area has provided two different approaches for producing marker-free plants: the use of biosafe SMGs, and the elimination of the selectable marker after selection. The removal of SMG also enables transgene stacking by sequential transformations and avoiding gene silencing by multiplication of promoters and poly(A) signals. Besides SMG elimination, the use of marker-free binary vectors is successfully demonstrated in peanut (Bhatnagar et al., 2010).

Biosafe Markers

Biosafe markers are categorized into four groups: positive, visual, negative, and pathogen based (Wei et al., 2012).

A positive selection system: Transformed cells can be identified without injury or death of nontransformed cells. It imparts to the transformed cells the ability to metabolize a special compound rendering them to have a growth advantage over nontransformed cells. The addition of this new compound in the culture medium allows growth and differentiation of transformed cells, while nontransformed cells fail to metabolize it and hence are not able to regenerate, such as the *uidA* gene (the β-glucuronidase enzyme), the *manA* gene (the phosphomannose isomerase enzyme), *xylA* (xylose isomerase), *ipt* (isopentenyl transferase), *pmi* (phosphomannose isomerase), *AK* (aspartate kinase), *DHPS* (dihydrodipicolinate synthase), *lyr* (lysine recemase), and *TSB1* (tryptophan synthase beta 1).

A visual selection system: A visual identification enables the selection of transformed cells, such as *Gus*, *lacZ* (β-galactosidase), *luc* and *lux* (luciferase, firefly *Photinus pyralis*, and *Vibrio harveyi*, respectively), *gfp* (jellyfish *Aequorea victoria*), and AtMYB12 (the transcriptional activator in *Arabidopsis thaliana* that is indispensable for the biosynthesis of flavonols).

Negative selection: It allows survival of only transformed cells, such as *galT* (galactose-1-phosphate uridylyltransferase), *badh* (betaine aldehyde dehydrogenase), DOGR1 (2-deoxyglucose-6-phosphate phosphatase), and codA (cytosine deaminase).

Elimination of Marker Genes

Tuteja et al. (2012) described the progress in the technique. The most important approaches are as follows:

Co-transformation—A very simple and most successful approach based on the co-transformation of target cells with two plasmids that target the insertion at two different loci in the plant genome. One carries the desired gene and the other one—SMG. These two are inserted into separate T-DNAs, which then segregate independently in a Mendelian fashion after sexual reproduction, thus enabling the selection of SMG-free transgenic plants. Since the frequency of co-transformation often exceeds 50%, at least in Arabidopsis, the method is highly efficient. It has been tested in rapeseed (Daley et al., 1998) and soybean (Xing et al., 2000).

The multi-autotransformation (MAT) vector system: This system uses an oncogene *ipt* (isopentenyl transferase) of *Agrobacterium tumefaciens* for the selection of transgenic plants and the site-specific recombination (SSR) system (R/Rs), and the gene of interest is cloned adjacent to it. The expression of the *ipt* gene causes abnormal shoot morphology. Recombinase (R) catalyzes recombination between two directly oriented recognition sites (Rs) and removes a "hit and run" cassette from the plant genome, which reverts them into normal shoots. The recombinase (R) gene is cloned under a chemically inducible promoter (IP) in order to avoid early removal of the *ipt* gene. Several improved versions have been designed and tested in various plant species.

Site-specific recombination: The two best examples include the *Cre*/*lox*P and FLP/FRT systems. The *Cre*/*lox*P system consists of two components: the first one contains two *lox*P sites, each consisting of 34-bp inverted repeats cloned in direct orientation flanking a DNA sequence. The second one is the *cre* gene that encodes a 38-kDa recombinase protein that specifically binds to the *loxP* sites. It excises the intervening sequence along with one of the *loxP* sites (Zuo et al., 2001). The system has been shown to work efficiently in soybean (Li et al., 2007) and rapeseed (Kopertekh et al., 2009). The FLP/FRT system comprises the 2-μm plasmid of *Saccharomyces cerevisiae* where the FLP enzyme catalyzes recombination between two directly repeated FLP recombination target (*frt*) sites, eliminating the sequence between them (Lyznik et al., 1996).

TARGETED GENOME ENGINEERING FOR LOCUS-SPECIFIC MODIFICATIONS

Current methods of gene transfer involve transgene incorporation at random sites, which leads to unpredictable gene expression because of the chromosomal position effect. If the transgene integrates in the region of transcriptionally inactive heterochromatin of the host genome, it may be very poorly expressed or even silenced. Targeted genome modification in plants helps overcome the challenges associated with conventional genetic engineering methods and enables specific site-directed genome modifications required in reverse genetics. Gene targeting (GT) allows the precise integration of a transgene into a predefined site in the host genome that has predefined expression characteristics (Day et al., 2000). GT is based on either SSR or homologous recombination (HR). SSR requires a site-specific recombinase that catalyzes the recombination between highly specific target recognition sites located on both the donor and target DNA.

While in HR, the donor DNA that is homologous to the target loci in the host genome is transferred to cells where genetic information is exchanged between the homologues. HR-based GT has been difficult in plants because a foreign

DNA delivered by *Agrobacterium*-mediated genetic transformation integrates in the host genome by the nonhomologous end joining (NHEJ) repair pathway. HR can be enhanced by the introduction of a specific double-strand break (DSB) into the target loci by restriction enzymes. One such class of enzymes consists of ZFNs—synthetic restriction enzymes that consist of three or four zinc finger DNA-binding motifs and a nonspecific nuclease domain derived from the *FokI* restriction enzyme (Porteus and Carroll, 2005). Each zinc finger is a stretch of ~30 amino acids stabilized by a zinc ion that binds to a particular three-base DNA sequence. The binding motif can be customized to design zinc fingers capable of recognizing a desired genomic locus. Though ZFNs have proved to be very versatile and efficient for introducing DSB in DNA at user-specified locations, its use in crop improvement is still at the initial stages, and in coming years, extensive research is required to more precisely establish its use in modifying the plant genome (Weintahl et al., 2010).

TALENs represent a new tool that allows gene-specific modifications, including gene inactivation through point mutations or deletions. TALEN™ is a modified version of TALE in which the TALE DNA binding domain that provides sequence-specific recognition of the DNA is fused to an endonuclease catalytic domain that enables creating DSBs in the DNA. This allows TALEN to recognize a specific sequence with high precision and cut it. DSB repair often (in >50% of cases) results in point mutations or small deletions leading to an inactivation of the gene. TALENs can be designed to target multiple genes at a time, thus being able to inactivate an entire pathway (Reyon et al., 2012). TALENs have been used for targeting animal and plant cells. Disease-resistant rice varieties have been generated using TALENs (Li et al., 2012). TALENs can also dramatically increase the frequency of GT, thus allowing for more efficient gene replacement by HR (Li et al., 2011). TALENs have a huge potential for targeted gene inactivation and gene replacement in oil crops.

THE DEBATE ON GM CROPS

In 2010, GM crops occupied >10% of the world's arable land (~1,525 million ha), and >30,000 products on the world market contain GM material. Although numerous transgenic varieties have already been tested and are waiting for being cultivated on a large scale, the future of GM crops is very uncertain. In many parts of the world, they are not widely accepted and their use has been very controversial. Fear of unknown dangers associated with the technique and underlying risk factors has been a strong concern not only among common men but in the scientific community as well. The key reasons of controversy are: food safety, horizontal gene transfer (HGT, the transfer of genetic material directly into a living cell or an organism among unrelated species such as between plants and microbes), vertical gene transfer (VGT, the transfer of herbicide resistance genes to weeds), the effects on ecosystems, gene flow in non-GM crops through pollen- or seed-mediated flow, some moral/religious/ethical concerns, as well as concerns related to corporate control of the food and seed supply (food crops are not free from patent applications).

The risks associated with GM crop depend on the biology of the species and the function of transgenic traits. The transfer of antibiotic resistance marker gene from GM plants to bacteria is creating a threat of the appearance of antibiotic-resistant pathogenic bacteria. Herbicide-tolerant transgenic crops represent the majority of GM crops across the world. In 2010 alone, of the 148 million ha of herbicide-tolerant GM crops cultivated, 83% were often associated with insect tolerance. All of these have led to higher applications of glyphosate, risking the development of glyphosate-resistant weeds. Other dangers include potential allergic reactions, the susceptibility to insect pests, the integration of free DNA in microbes, the probability of hybrid production with related species, and the probable effect on food chain.

Agronomical concerns include premature pod shattering, low seed content, poor germination rates, the reduced diversity at the farm level, and the use of cultivars that can cross-pollinate with nontransgenic varieties. These challenges are being addressed and worked upon.

Genetic engineering is still an imprecise technology, and the random incorporation of genes might lead to unpredictable disturbances in the plant genome, physiology, and biochemistry. Since GMPs cannot be tested for unknown health hazards, the only probability to safeguard their use is to develop a battery of tests for the analysis of potential toxicity, mutagenicity and allergenicity of GM crops. Until now, there is no evidence that GM crops are harmful. Thorough risk assessments of GMP involving alongside careful monitoring and management of cultivation areas at the ecological and molecular levels and the development of marker free plants are some of the solutions to address the above-mentioned issues. Although the number of countries growing GM crops has increased to 29 in recent years, there is still much to be learned and understood.

OIL CROPS OF THE FUTURE

With the increasing knowledge of the regulation of various biochemical pathways at both the genetic and protein level, a new era of oil crops seems to be not very far away. Besides the six traditional major oil crops, other edible crops, such as Sorgum (*Sorghum bicolor*), potato (*Solanum tuberosum*), cassava (*Manihot esculenta*), sugarcane (*Saccharum*), etc., that produce small amounts of oil are potential sources of the new generation of oil crops that can be engineered to produce more oil than starch and sugar (products that they normally produce). Such modifications will require a complete understanding of the mechanisms by which the carbon flow is directed toward oil accumulation. However, important information about the biosynthetic pathways can be obtained by investigating plants that accumulate high amounts of oil (10–20%) alongside starch in their seeds, such as oat (*Avena sativa*), corn (*Zea mays*) or quinoa (*Chenopodium quinoa*). Another good model plant could be the cultivated form of yellow nutsedge (*Cyperus esculentus*) which accumulates more than 25% of oil and 30% of starch in its tubers. There is also a significant promise in the transfer of traits for hydrocarbon synthesis from cyanobacteria (the third generation of biodiesel producers), the enhancement of light and carbon uptake and cold tolerance, and the improvement of photosynthesis capacity in oilseed crops.

Other examples of potential progress in plant oil industry include modifications of oil producers, such as *Camelina sativa*, for producing desirable oils and terpenes; producing plants that are engineered to store oils in stems and leaves not just in seeds, generating oil seed crops with increased tolerance to drought and temperature; modifications of oil content in nutty plants, such as almond, walnut, etc., for increasing their shelf life and preventing foul odor production while stored. It is also of a great interest to engineer plants that accumulate unusual FAs synthesized in various plants but not in common oilseeds, such as palmitoleic acid. Palmitoleate increases cell membrane fluidity, inhibits oncogenesis, reduces inflammation associated with diabetes and heart disease, and there is an increasing demand for it in the food and nutraceutical industries for health purposes. Besides, it can serve as a renewable source of 1-octene, a chemical feedstock used in the production of polyethylene (Wua et al., 2012).

CONCLUSION

Worldwide investigations exploring the innumerable possibilities of genetic manipulations in a wide variety of plant species alongside numerous studies analyzing the complexity of lipid metabolism—they all are set to put forward innovative strategies for developing a rationally designed platform for developing genetically engineered plants that can accumulate the desired characteristics of end products. Major breakthroughs in a complete understanding of the lipid metabolism pathway are still awaited, but significant progress has already been made, implying a high promise toward achieving the goal. In recent years, success has been significant, and plant biotechnology has proved to be the key to the successful engineering of new cost-effective oil crop plants. Seeing the importance of plant oils for both human health and nutrition as well as its role as a sustainable replacement for fast depleting petrochemicals, such progress should be globally accepted.

REFERENCES

Atabania, A.E., Silitonga, A.S., Badruddina, I.A., Mahliaa, T.M.I., Masjukia, H.H., Mekhilefd, S., 2012. A comprehensive review on biodiesel as an alternative energy resource and its characteristics. Renew. Sust. Energ. Rev. 16, 2070–2093.

Bevan, M., 1984. Binary Agrobacterium vectors for plant transformation. Nucleic Acids Res. 12, 8711–8721.

Bhatnagar, M., Prasad, K., Bhatnagar-Mathur, P., Narasu, M.L., Waliyar, F., Sharma, K.K., 2010. An efficient method for the production of marker-free transgenic plants of peanut *Arachis hypogaea* L. Plant Cell Rep. 29, 495–502.

Bhore, S.J., Shah, F.H., 2012. Genetic transformation of the American oil palm (*Elaeis oleifera*) immature zygotic embryos with antisense Palmitoyl-acyl carrier protein thioesterase (PATE) gene. World Appl. Sci. J. 16 (3), 362–369.

Bondaruk, M., Johnson, S., Degafu, A., Boora, P., Bilodeau, P., Morris, J., et al., 2007. Expression of a cDNA encoding palmitoyl-acyl carrier protein desaturase from cat's claw (*Doxantha unguis-cati* L.) in *Arabidopsis thaliana* and *Brassica napus* leads to accumulation of unusual unsaturated fatty acids and increased stearic acid content in the seed oil. Plant Breed. 126, 186–194.

Broun, P., Gettner, S., Somerville, C., 1999. Genetic engineering of plant lipids. Annu. Rev. Nutr. 19, 197–216.

Buhr, T., Sato, S., Ebrahim, F., Xing, A., Zhou, Y., Mathiesen, M., Schweiger, B., Kinney, A., Staswick, P., Clemente, T., 2002. Ribozyme termination of RNA transcripts down-regulate seed fatty acid genes in transgenic soybean. Plant J. 30, 155–163.

Burbulis, N., Blinstrubiene, A., Kupriene, R., 2009. Regeneration of adventitious shoots of linseed (*Linum usitatissimum* L.) from hypocotyl explants. Zemdirbyste Agric. 96 (3), 168–175.

De Block, M., De Brouwer, D., Tenning, P., 1989. Transformation of *Brassica napus* and *Brassica oleracea* using *Agrobacterium tumefaciens* and the expression of the bar and neo genes in the transgenic plants. Plant Physiol. 91, 694–701.

Cahoon, E.B., Carlson, T.J., Ripp, K.G., Schweiger, B.J., Cook, G.A., Hall, S.E., Kinney, A.J., 1999. Biosynthetic origin of conjugated double bonds: production of fatty acid components of high-value drying oils in transgenic soybean embryos. Proc. Natl. Acad. Sci. U.S.A. 96, 12935–12940.

Cahoon, E.B., Shockey, J.M., Dietrich, C.R., Gidda, S.K., Mullen, R.T., Dyer, J.M., 2007. Engineering oilseeds for sustainable production of industrial and nutritional feedstocks: solving bottlenecks in fatty acid flux. Curr. Opin. Plant. Biol. 10, 236–244.

Cardoza, V., Stewart Jr., 2004. Agrobacterium Mediated Transformation of Canola. In: Curtis, I.S. (Ed.), Transgenic Crops of the World—Essential Protocols. Kluwer Academic Publishers, Netherlands, pp. 379–387.

Carlsson, A.S., 2009. Plant oils as feedstock alternatives to petroleum—A short survey of potential oil crop platforms. Biochimie 91, 665–670.

Chen, R., Matsui, K., Ogawa, M., Oe, M., Ochiai, M., Kawashima, H., Sakuradani, E., Shimizu, S., Ishimoto, M., Hayashi, M., et al., 2006. Expression of Δ6, Δ5 desaturase and GLELO elongase genes from *Mortierella alpina* for production of arachidonic acid in soybean [*Glycine max* (L.) Merrill] seeds. Plant Sci. 170, 399–406.

Chugh, A., Amundsen, E., Eudes, F., 2009. Translocation of cell-penetrating peptides and delivery of their cargoes in triticale microspores. Plant Cell Rep. 28 (5), 801–810.

Daley, M., Knauf, V.C., Summerfelt, K.R., Turner, J.C., 1998. Co-transformation with one *Agrobacterium tumefaciens* strain containing two binary plasmids as a method for producing marker-free transgenic plants. Plant Cell Rep. 19, 489–496.

Day, C.D., Lee, E., Kobayashi, J., Holappa, L.D., Albert, H., Ow, D.W., 2000. Transgene integration into the same chromosome location can produce alleles that express at a predictable level, or alleles that are differentially silenced. Genes Dev. 14, 2869–2880.

Dehesh, K., Jones, A., Knutzon, D.S., Voelker, T.A., 1996. Production of high levels of 8:0 and 10:0 fatty acids in transgenic canola by overexpression of Ch FatB2, a thioesterase cDNA from *Cuphea hookeriana*. Plant J. 9, 167–172.

Donaldson, P.A., Simmonds, D.H., 2000. Susceptibility to *Agrobacterium tumefaciens* and cotyledonary node transformation in short-season soybean. Plant Cell Rep. 19, 478–484.

Durrett, T.P., Benning, C., Ohlrogge, J., 2008. Plant triacylglycerols as feedstocks for the production of biofuels. Plant J. 54, 593–607.

Eckert, H., Vallee, B.L., Schweiger, B.J., Kinney, A.J., Cahoon, E.B., Clemente, T., 2006. Co-expression of the borage Δ6 desaturase and the Arabidopsis Δ15 desaturase results in high accumulation of steridonic acid in the seeds of transgenic soybean. Planta 224, 1050–1057.

Fairless, D., 2007. Biofuel: the little shrub that could—maybe. Nature 449, 652–655.

FAOSTAT—Production Crops: <faostat.fao.org>; 2011.

Fernandez-Martinez, J.M., Mancha, M., Osorio, J., Garces, R., 1997. Sunflower mutant containing high levels of palmitic acid in high oleic background. Euphytica 97, 113–116.

Flores, T., Karpova, O., Su, x, Zeng, P., Bilyeu, K., Sleper, D.A., Nguyen, H.T., Zhang, Z.J., 2008. Silencing of GmFAD3 gene by siRNA leads to low α-linolenic acids (18:3) of fad3-mutant phenotype in soybean [*Glycine max* (Merr.)]. Transgenic Res. 17, 839–850.

Fromm, M., Taylor, L.P., Walbot, V., 1985. Expression of genes transferred into monocot and dicot plant cells by electroporation. Proc. Natl. Acad. Sci. U S A. 82 (17), 5824–5828.

Gelvin, B.S., 2003. *Agrobacterium*-mediated plant transformation: the biology behind the "Gene-Jockeying" tool. Microbiol. Mol. Biol. Rev. 67 (1), 16–37.

Grotewold, E., 2008. Transcription factors for predictive plant metabolic engineering: are we there yet? Curr. Opin. Biotechnol. 19, 138–144.

Gunstone, F.D., 2004. The Chemistry of Fats and Oils—Sources, Composition, Properties and Uses. CRC Press, New York, USA.

Haas, M.J., 2005. Improving the economics of biodiesel production through the use of low value lipids as feedstocks: vegetable oil soapstock. Fuel Process. Technol. 86, 1087–1096.

Hansen, G., Chilton, M.D., 1996. 'Agrolistic' transformation of plant cells: integration of T-strand generated in planta. Proc. Natl. Acad. Sci. U.S.A. 93, 14978–14983.

Hawkins, D.J., Kridl, J.C., 1998. Characterization of acyl-ACP thioesterases of mangosteen (*Garcinia mangostana*) seed and high levels of stearate production in transgenic canola. Plant J. 13, 743–752.

Ishimoto, M., Rahman, S.M., Hanafy, M.S., Khalafalla, M.M., El-Shemy, H.A., Nakamoto, Y., Kita, Y., Takanashi, K., Matsuda, F., Murano, Y., et al., 2010. Evaluation of amino acid content and nutritional quality of transgenic soybean seeds with high-level tryptophan accumulation. Mol. Breed. 25, 313–326.

Izawati, A.M.D., Parveez, G.K.A., Masani, M.Y.A., 2012. Transformation of oil Palm using *Agrobacterium tumefaciens*. Methods Mol. Biol. 847, 177–188.

Jha, T.B., Mukherjee, P., Datta, M.M., 2007. Somatic embryogenesis in JatrophacurcasLinn., an important biofuel plant. Plant Biotechnol. Rep. 1, 135–140.

Jones, A., Davies, H.M., Voelker, T.A., 1995. Palmitoyl-acyl carrier protein (ACP) thioesterase and the evolutionary origin of plant acyl-ACP thioesterases. Plant Cell 7, 359–371.

Kaeppler, H.F., Gu, W., Somers, D.A., Rines, H.W., Cockburn, A.F., 1990. Silicon carbide fiber mediated DNA delivery into plant cells. Plant Cell Rep. 9, 415–418.

Kajikawa, M., Matsui, K., Ochiai, M., Tanaka, Y., Kita, Y., Ishimoto, M., Kohzu, Y., Shoji, S., Yamato, K.T., Ohyama, K., et al., 2008. Production of arachidonic and eicosapentaenoic acids in plants using bryophyte fatty acid Δ6-desaturase, Δ6-elongase, and Δ5-desaturase genes. Biosci. Biotechnol. Biochem. 72, 435–444.

Kasai, M., Kanazawa, A., 2012. RNA silencing as a tool to uncover gene function and engineer novel traits in soybean. Breed. Sci. 61, 468–479.

King, A.J., He, W., Cuevas, J.A., Freudenberger, M., Ramiaramanana, D., Graham, I.A., 2009. Potential of *Jatropha curcas* as a source of renewable oil and animal feed. J. Exp. Bot. 60, 2897–2905.

Kinney, A.J., 1996. Development of genetically engineered soybean oils for food applications. J. Food Lipids 3, 273–292.

Kita, Y., Nakamoto, Y., Takahashi, M., Kitamura, K., Wakasa, K., Ishimoto, M., 2010. Manipulation of amino acid composition in soybean seeds by the combination of deregulated tryptophan biosynthesis and storage protein deficiency. Plant Cell Rep. 29, 87–95.

Kita, Y., Nishizawa, K., Takahashi, M., Kitayama, M., Ishimoto, M., 2007. Genetic improvement of the somatic embryogenesis and regeneration in soybean and transformation of the improved breeding lines. Plant Cell Rep. 26, 439–447.

Knutzon, D.S., Hayes, T.R., Wyrick, A., Xiong, H., Davies, H.M., Voelker, T.A., 1999. Lysophosphatidic acid acyltransferase from coconut endosperm mediates the insertion of laurate at the *sn*-2 position of triacylglycerols in lauric rapeseed oil and can increase total laurate levels. Plant Physiol. 120, 739–746.

Knutzon, D.S., Thompson, G.A., Radke, S.E., Johnson, W.B., 1992. Modification of *Brassica* seed oil by antisense expression of a stearoyl-acyl carrier protein desaturase gene. Proc. Natl. Acad. Sci. U.S.A. 89, 2624–2628.

Kopertekh, L., Broer, I., Schiemann, J., 2009. Developmentally regulated site-specific marker gene excision in transgenic *B. napus* plants. Plant Cell Rep. 28, 1075–1083.

Lardizabal, K., Effertz, R., Levering, C., Mai, J., Pedroso, M.C., Jury, T., Aasen, E., Gruys, K., Bennett, K., 2008. Expression of *Umbelopsis ramanniana* DGAT2A in seed increases oil in soybean. Plant Physiol. 148, 89–96.

Li, Z., Xing, A., Moon, B.P., Burgoyne, S.A., Guida, A.D., Liang, H., Lee, C., Caster, C.S., et al., 2007. A Cre/loxPmediated self-activating gene excision system to produce marker gene free transgenic soybean plants. Plant Mol. Biol. 65, 329–341.

Li, R., Yu, K., Hatanaka, T., Hildebrand, D.F., 2010. *Vernonia* DGATs increase accumulation of epoxy fatty acids in oil. Plant Biotechnol. J. 8, 184–195.

Li, T., Liu, B., Spalding, M.H., Weeks, D.P., Yang, B., 2012. High-efficiency TALEN-based gene-editing produces disease-resistant rice. Nat. Biotechnol. 30 (5), 390–392.

Liu, H., Xie, X., Sun, S., Zhu, W., Ji, J., Wang, G., 2011. Optimization of *Agrobacterium*-mediated transformation of sunflower (*Helianthus annuus* L.) immature embryos. AJCS 5 (12), 1616–1621.

Liu, Q., Singh, S., Green, A., 2000. Genetic modification of cotton seed oil using inverted-repeat gene-silencing techniques. Biochem. Soc. Trans. 28 (6), 927–929.

López Alonsoa, D., García Marotob, F., 2000. Plants as 'chemical factories' for the production of polyunsaturated fatty acids. Biotechnol. Adv. 18, 481–497.

Lyznik, L.A., Rao, K.V., Hodges, T.K., 1996. FLP-mediated recombination of FRT sites in maize genome. Nucleic Acids Res. 24, 3784–3789.

van de Loo, F.J., Broun, P., Turner, S., Somerville, C., 1995. An oleate 12-hydroxylase from Ricinus-communis L. is a fatty acyl desaturase homolog. Proc. Natl. Acad. Sci. U.S.A. 92, 6743–6747.

Maheshwari, P., Selvaraj, G., Kovalchuk, I., 2011. Optimization of *Brassica napus* (Canola) explants regeneration for genetic transformation. New Biotechnol. 29 (1), 144–155.

Malathi, B., Ramesh, S., Venkateswara, K., Reddy, D.V., 2006. *Agrobacterium*-mediated genetic transformation and production of semilooper resistant transgenic castor (*Ricinus communis* L.). Euphytica 147 (3), 441–449.

McHughen, A., 2002. In: Khachatorians, G.G., McHughen, A., Scorza, R., Nip, W.K., Hui, Y.H. (Eds.), Transgenic Plants and Crops Transgenic Linseed Flax. Marcel Dekker, Inc, New York, p. 899.

Metzger, J.O., Bornscheuer, U., 2006. Lipids as renewable resources: current state of chemical and biotechnological conversion and diversification. Appl. Microbiol. Biotechnol. 71, 13–22.

Misra, M., Misra, A.N., 2010. Jatropha: the biodiesel plant biology, tissue culture and genetic Transformation—A review. Int. J. Pure Appl. Sci. Technol. 1 (1), 11–24.

Moser, B.R., 2009. Biodiesel production, properties, and feedstocks. In Vitro Cell Dev. Biol. Plant 45, 229–266.

Murphy, D.J., 1994. Biogenesis, function and biotechnology of plant storage lipids. Prog. Lipid Res. 33 (112), 71–85.

Murphy, D.J., 1996. Engineering oil production in rapeseed and other oil crops. Trends Biotechnol. 14, 206–213.

Napier, J.A., Beaudoin, F., Michaelson, L.V., Sayanova, O., 2004. The production of long chain polyunsaturated fatty acids in transgenic plants by reverse engineering. Biochimie 86, 785–792.

Padgette, S.R., Kolacz, K.H., Delannay, X., Re, D.B., LaValee, B.J., Tinius, C.N., Rhodes, W.K., Otero, Y.I., Barry, G.F., Eichholtz, D.A., Peschke, V.M., Nida, D.L., Taylor, N.B., Kishore, G.M., 1995. Development, identification, and characterization of a glyphosate-tolerant soybean line. Crop Sci. 35, 1451–1461.

Parveez, G.K.A., 2003. Novel products from transgenic oil palm. AgBiotechNet 5, 1–8.

Paul, K., 2008. Risks from GMOs due to horizontal gene transfer. Environ. Biosafety Res. 7, 123–149.

Parveez, A.G.K., Bahariah, B., 2012. Biolistic-mediated production of transgenic oil Palm. Methods Mol. Biol. 847, 163–175.

Parveez, G.K.A., 2000. Production of transgenic oil palm (ElaeisguineensisJacq.) using biolistic techniques. In: Jain, S.M., Minocha, S.C. (Eds.), Molecular Biology of Woody Plants, vol. 2. Kluwer Academic Publishers, pp. 327–350.

Parveez, G.K.A., Ravigadevi, S., Abdullah, S.N.A., Othman, A., Ramli, U.S., Rasid, O., Masri, M.M., Cheah, S.C., 1999. Production of transgenic oil palm—current success and future considerations. In: Proceedings of PORIM International Palm Oil Congress. 1–6 February, Kuala Lumpur, Malaysia, pp. 283–297.

Paszkowski, J., Shillito, R.D., Saul, M., Mandak, V., Hohn, T., Hohn, B., 1984. Direct gene transfer to plants. EMBO J. 3, 2717–2722.

Pérez-de-Luque, A., Rubiales, D., 2009. Nanotechnology for parasitic plant control. Pest Manage. Sci. 65, 540–545.

Porteus, M.H., Carroll, D., 2005. Gene targeting using zinc finger nucleases. Nat. Biotechnol. 23, 967–973.

Rajanaidu, N., Jailani, B.S., 1995. World-wide performance of DXP planting material and future prospects. In: Proceedings of 1995 PORIM National Oil Palm Conference, pp. 1–29.

Rao, K.S., Rohini, V.K., 1999. *Agrobacterium*-mediated transformation of Sunflower (*Helianthus annuus* L.): a simple protocol. Ann. Bot. 83, 347–354.

Rao, S.S., Hildebrand, D., 2009. Changes in oil content of transgenic soybeans expressing the yeast SLC1 gene. Lipids 44, 945–951.

Reyon, D., Tsai, S.Q., Khayter, C., Foden, J.A., Sander, J.D., Joung, J.K., 2012. FLASH assembly of TALENs for high-throughput genome editing. Nat. Biotechnol. 30 (5), 460–465.

Ruiz-López, N., Haslam, R.P., Venegas-Calerón, M., Larson, T.R., Graham, I.A., Napier, J.A., Sayanova, O., 2009. The synthesis and accumulation of stearidonic acid in transgenic plants: a novel source of 'heart-healthy' omega-3 fatty acids. Plant Biotechnol. J. 7, 704–716.

Sambanthamurthi, R., SitiNorAkmar, A., Ahmad Parveez, G.K., 2002. Genetic manipulation of the oil palm—challenges and prospects. The Planter, Kuala Lumpur 78 (919), 547–562.

Sato, S., Xing, A., Ye, X., Schweiger, B., Kinney, A., Graef, G., Clemente, T., 2004. Production of γ-linolenic acid and stearidonic acid in seeds of marker-free transgenic soybean. Crop Sci. 44, 646–652.

Schmutz, J., Cannon, S.B., Schlueter, J., Ma, J., Mitros, T., Nelson, W., Hyten, D.L., Song, Q., et al., 2010. Genome sequence of the palaeopolyploid soybean. Nature 463, 178–183.

Sharma, H.C., Crouch, J.H., Sharma, K.K., Seetharama, N., Hash, C.T., 2002. Applications of biotechnology for crop improvement: prospects and constraints. Plant Sci. 163 (3), 381–395.

Shahid, E.M., Jamal, Y., 2011. Production of biodiesel: a technical review. Renew. Sust. Energ. Rev. 15, 4732–4745.

Sone, T., Nagamori, E., Ikeuchi, T., 2002. A novel gene delivery system in plants with calciumalginate micro-beads. J. Biosci. Bioeng. 94 (1), 87–91.

Stacey, G., Vodkin, L., Parrott, W.A., Shoemaker, R.C., 2004. National science foundation-sponsored workshop report: draft plan for soybean genomics. Plant Physiol. 135, 59–70.

Stewart Jr., N., Adang, M.J., All, J.N., Raymer, P.L., Ramachandran, S., Parrott, W.A., 1996. Insect control and dosage effects in transgenic canola containing a synthetic *Bacillus thuringiensis* cry 1Ac gene. Plant Physiol. 112, 115–120.

Stoutjesdijk, P.A., Hurlestone, C., Singh, S.P., Green, A.G., 2000. High-oleic acid Australian *Brassica napus* and *B. juncea* varieties produced by co-suppression of endogenous delta-12 desaturases. Biochem.Soc. Trans. 28, 938–940.

Sujatha, M., Sailaja, M., 2005. Stable genetic transformation of castor (*Ricinus communis* L.) via *Agrobacterium tumefaciens*-mediated gene transfer using embryo axes from mature seeds. Plant Cell Rep. 23 (12), 803–810.

Sujatha, M., Sailaja, M., 2007. Development of transgenic castor for insect resistance. In: Extended Summaries National Seminar Changing Global Vegetables Oils Scenario: Issues and Challenges Before India, Hyderabad, India; January 29–31, pp. 7–8.

Takagi, K., Nishizawa, K., Hirose, A., Kita, A., Ishimoto, M., 2011. Manipulation of saponin biosynthesis by RNA interference mediated silencing of β-amyrin synthase gene expression in soybean. Plant Cell Rep. 30, 1835–1846.

Takavar, S., Rahnama, H., Rahimian, H., Kazemitabar, K., 2010. *Agrobacterium* mediated transformation of Maize (*Zea mays* L.). J. Sci. Islam. Repub. Iran 21 (1), 21–29.

Thompson, R.D., 2000. Turning fields into grains. Nature 408, 39–41.

Tohidfar, M., Ghareyazie, B., Mosavi, M., Yazdani, S., Golabchian, R., 2008. *Agrobacterium*-mediated transformation of cotton (*Gossypium hirsutum*) using a synthetic *cry1Ab* gene for enhanced resistance against *Heliothis armigera*. Iranian J. Biotechnol. 6 (3), 164–173.

Tomlin, E.S., Branch, S.R., Chamberlain, D., Gabe, H., Wright, M.S., Stewart, C.N., 2002. Screening of soybean, *Glycine max* (L.) Merrill, lines for somatic embryo induction and maturation capability from immature cotyledons. In Vitro Cell Dev. Biol. Plant 38, 543–548.

Tonkin, I., 2009. Global Biodiesel Production Reaches Record High. http://www.roadtransport.com.

Topfer, R., Martini, N., Schell, J., 1995. Modification of plant lipid synthesis. Science 268, 681–686.

Tuteja, N., Verma, S., Sahoo, R.K., Raveendar, S., Reddy, I.N.B.L., 2012. Recent advances in development of marker-free transgenic plants: regulation and biosafety concern. J. Biosci. 37 (1), 167–197.

Umezawa, T., Sakurai, T., et al., 2008. Sequencing and analysis of approximately 40 000 soybean cDNA clones from a full-length-enriched cDNA library. DNA Res. 15 (6), 333–346.

Ursin, V., Knutzon, D., Radke, S., Thornton, J., Knauf, V., 2000. Production of beneficial dietary omega-3 and omega-6 fatty acids intransgenic canola. In: Abstracts of the 14th International Symposium on Plant Lipids, p. 13.

Vega-Sánchez, M.E., Ronald, P.C., 2010. Genetic and biotechnological approaches for biofuel crop Improvement. Curr. Opin. Biotechnol. 21, 218–224.

Vigeolas, H., Waldeck, P., Zank, T., Geigenberger, P., 2007. Increasing seed oil content in oil-seed rape (*Brassica napus* L.) by overexpression of a yeast glycerol-3-phosphate dehydrogenase under the control of a seed-specific promoter. Plant Biotechnol. J. 5, 431–441.

Voelker, T.A., Worrell, A.C., Anderson, L., Bleibaum, J., Fan, C., Hawkins, D.J., Radke, S.E., Davies, H.M., 1992. Fatty acid biosynthesis redirected to medium chains in transgenic oilseed plants. Science 257, 72–74.

Voelker, T.A., Hayes, T.R., Cranmer, A.M., Turner, J.C., Davies, H.M., 1996. Genetic engineering of a quantitative trait: metabolic and genetic parameters influencing the accumulation of laurate in rapeseed. Plant J. 9, 229–241.

Wagdy, A.S., 2001. Stable genetic transformation of cotton plants using polybrene-spermidine treatment. Plant Mol. Biol. Rep. 19, 377.

Wang, G., Xu, Y., 2008. Hypocotyl-based Agrobacterium-mediated transformation of soybean (*Glycine max*) and application for RNA interference. Plant Cell Rep. 27, 1177–1184.

Wei, Z., Wang, X., Xing, S., 2012. Current progress of biosafe selectable markers in plant transformation. J. Plant Breed. Crop Sci. 4 (1), 1–8.

Weinthal, D., Tovkach, A., Zeevi, V., Tzfira, T., 2010. Genome editing in plant cells by zinc finger nucleases. Trends Plant Sci. 15, 308–321.

Wen, Y., Fu, T., Wen, J., Tu, J., Ma, C., Shen, J., Zhang, S., Bin, Y., 2012. Profound improvements of isolated microspores culture techniques in winter *Brassica napus* L. Afr. J. Biotechnol. 11 (7), 1617–1623.

Wua, Y., Li, R., Hildebrand, D.F., 2012. Biosynthesis and metabolic engineering of palmitoleate production, an important contributor to human health and sustainable industry. Prog. Lipid Res. 51, 340–349.

Xing, A., Zhang, Z., Sato, S., Staswick, P., Clement, T., 2000. The use of two T-DNA binary system to derive marker-free transgenic soybeans. In Vitro Cell Dev. Biol. Plant 36, 456–463.

Zuo, J., Niu, Q.W., Moller, S.G., Chua, N.H., 2001. Chemical regulated site-specific DNA excision in transgenic plants. Nat. Biotechnol. 19, 157–161.

Ziemienowicz, A., Shim, Y.S., Matsuoka, A., Eudes, F., Kovalchuk, I., 2012. A novel method of transgene delivery into triticale plants using the Agrobacterium transferred DNA-derived nanocomplex. Plant Physiol. 158 (4), 1503–1513.

Chapter 15

Engineering Oil Accumulation in Vegetative Tissue

Randall J. Weselake
Department of Agricultural, Food and Nutritional Sciences, University of Alberta, Edmonton, AB, Canada

INTRODUCTION

Within the last decade, substantial research has been directed toward metabolic engineering of vegetative tissue to enhance oil production (Chapman et al., 2013; Vanhercke et al., 2014a,b). Unlike developing seeds of oil crops and mesocarp, such as that of palm (eg, *Elaeis guineensis*), leaves are known to produce relatively low levels of triacylglycerol (TAG) of up to about 0.5% (dry weight) (Lin and Oliver, 2008). There are, however, a few examples of nonseed tissues capable of producing considerably larger amounts of TAG. Wang et al. (2007) reported that "oil firewood" (*Tetraena mongolica* Maxim) can produce 9% (dry weight) TAG in the phloem of the stems of this species, whereas yellow nutsedge (*Cyperus esculentus*) can accumulate 25% oil in its tubers (Carlsson, 2009). Reprogramming vegetative tissue to produce increased amounts of TAG represents a potential way of substantially increasing the global supply of usable liquid fuel energy, especially if this could be achieved in high-biomass crops such as *Miscanthus* spp. and switchgrass (*Panicum virgatum*) (Ohlrogge and Chapman, 2011; Troncoso-Ponce et al., 2013). Indeed, there are concerns about using traditional oil crops as a source of oil for biodiesel production as the growing global population increases its demand for food oil (Carlsson, 2009).

Metabolic engineering of vegetative tissue to produce increased TAG content is still at the proof-of-concept stage. Advances in boosting the TAG content of leaves, however, have benefited from the knowledge gained from in-depth investigations of seed oil formation and biotechnological interventions to increase seed oil content (Weselake et al., 2009; Baud and Lepiniec, 2010; Chapman and Ohlrogge, 2012; Singer et al., 2013). Biotechnological modifications to encourage production of TAG in vegetative tissue have included: (1) increasing the supply of building blocks for TAG assembly, (2) enhancing the process of TAG assembly, and (3) reducing TAG turnover (Vanhercke et al., 2014a). Increases in TAG content in vegetative tissue have been achieved

Industrial Oil Crops. http://dx.doi.org/10.1016/B978-1-893997-98-1.00015-4

through the introduction of transcription factors and upregulation and/or down-regulation of specific enzyme-catalyzed reactions in primary metabolism (Durrett et al., 2008; Chapman et al., 2013; Troncoso-Ponce et al., 2013; Vanhercke et al., 2014a). Transcription factors can effect the expression of many different genes encoding enzymes and proteins involved in metabolic pathways in primary metabolism (Weselake et al., 2009). Both single-gene and multigene approaches were used to increase the TAG content of vegetative tissue, but the highest levels of TAG reached have involved multigene approaches. A simplified overview of biochemical processes leading to TAG production and turnover in vegetative tissue is shown in Fig. 15.1. Examples of biotechnological interventions to promote increased TAG accumulation are also indicated. Prior to discussing these strategies in more detail, a comparative overview of acyl-lipid metabolism in developing seeds and leaves is provided.

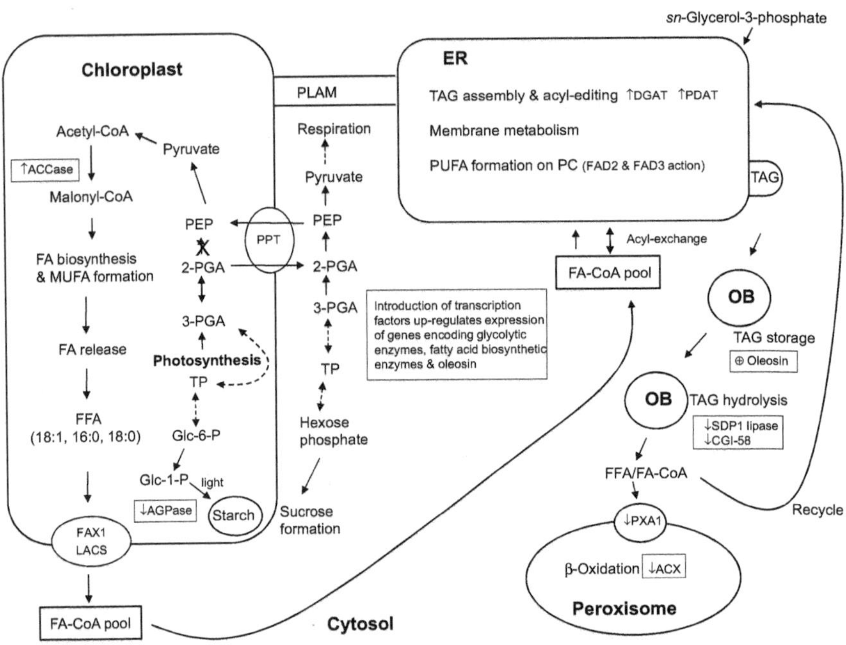

FIGURE 15.1 Simplified overview of metabolic events leading to triacylglycerol (TAG) formation and turnover in C3 vegetative tissue, and examples of metabolic engineering strategies for encouraging TAG accumulation in the cytoplasm. Formation of 3-phosphoglycerate (3-PGA) in the Calvin Benson cycle of photosynthesis can lead to the formation of sucrose in the cytosol and starch in the chloroplast. Although sufficient phosphoglyceromutase activity appears to be present to catalyze conversion of 3-PGA to 2-phosphoglycerate (2-PGA) in the chloroplast, enolase activity to catalyze conversion of 2-PGA to phosphoenolpyruvate (PEP) is lacking in this subcellular compartment. Instead, plastidial 2-PGA is exported to the cytosol in exchange for PEP, produced through cytosolic glycolysis, via the plastidial PEP/phosphate translocator (PPT). Plastidial pyruvate is converted to acetyl-CoA by the catalytic action of the pyruvate dehydrogenase complex. Malonyl-CoA is formed from acetyl-CoA, bicarbonate and ATP via the catalytic action of acetyl-CoA carboxylase

OVERVIEW OF ACYL-LIPID METABOLISM IN DEVELOPING SEEDS AND LEAVES

Developing seeds of traditional oil crops such as soybean (*Glycine max*) and oilseed rape (*Brassica napus*) are sucrose-receiving "sink" organs which are dedicated to TAG accumulation during seed maturation. The resulting TAG serves as an energy store to support germination and early seedling growth. Fatty acid and TAG biosynthesis in developing seeds of oleaginous plants has been the subject of numerous reviews (eg, Baud and Lepiniec, 2010; Chapman and Ohlrogge, 2012; Bates et al., 2013; Ohlrogge and Browse, 1995). Fatty acids are formed in plastids through sequential addition of two carbon fragments donated from malonyl-CoA (coenzyme A) through the action of the fatty acid synthase (FAS) complex. Malonyl-CoA is produced through the catalytic action of acetyl-CoA carboxylase (ACCase) which uses acetyl-CoA, ATP and bicarbonate as substrates. The fatty acyl chain grows while attached to acyl carrier protein (ACP). In temperate crops such as oilseed rape, fatty acid synthesis proceeds until 16-carbon and 18-carbon chains are formed. Oleoyl (18:1Δ^{cis9})-ACP is the major product of plastidial fatty acid synthesis in the seed plastids in plants such as *Arabidopsis thaliana* (Arabidopsis), oilseed rape and soybean. Free oleic acid, stearic acid (18:0) and palmitic acid (16:0) can be released from acyl-ACP through the catalytic action of plastidial thioesterase (acyl-ACP hydrolase). Some acyl groups formed by the action of FAS, however, can be used to form plastidial lipids in the prokaryotic pathway (Ohlrogge and Browse,

(ACCase). Malonyl-CoA provides 2-carbon fragments for elongation of the fatty acyl (FA) chain attached to acyl carrier protein (ACP) of the fatty acid synthase complex. A soluble stearoyl-ACP desaturase catalyzes the conversion of stearoyl-ACP to oleoyl-ACP with some FA remaining saturated. Thioesterase catalyzes the release of free fatty acid (FFA) from FA-ACP. Fatty acid export 1 (FAX1) may transport FFA across the inner plastidial envelope and activation of FFA to FA-CoA is catalyzed by long-chain acyl-CoA synthetase (LACS) on the outside of the plastid. FA-CoA serves as an FA donor to support glycerolipid metabolism in the endoplasmic reticulum (ER). *sn*-Glycerol-3-phosphate derived from dihydroxyacetone phosphate serves as glycerol backbone in for TAG assembly and membrane metabolism. Polyunsaturated fatty acids (PUFA) are formed at the level of phosphatidylcholine (PC) via the catalytic action of fatty acid desaturase (FAD) in forms FAD2 and FAD3. Acyl-CoA:diacylglycerol acyltransferase (DGAT) catalyzes the FA-CoA-dependent acylation of *sn*-1,2-diacylglycerol to produce TAG. In contrast, phospholipid:diacylglycerol acyltransferase (PDAT) uses PC as an FA donor to form TAG. Various acyl-editing pathways, including PDAT action, can channel PUFA into TAG. Oil bodies (OB) containing TAG pinch off the outer leaflets of the ER and can undergo hydrolysis catalyzed by sugar-dependent 1 (SDP1) lipase. Liberated FA chains can move into the peroxisome to undergo β-oxidation or be recycled for use in membrane biosynthesis. ⊕, addition of; ↓, downregulation; ↑, upregulation. Abbreviations: *ACX*, acyl-CoA oxidase; *AGPase*, ADP-glucose pyrophosphorylase; CGI-58, comparative gene identification-58; *Glc-1-P*, glucose-1-phosphate; *Glc-6-P*, glucose-6-phosphate; *MUFA*, monounsaturated fatty acid; *PXA1*, peroxisomal ABC-transporter one; *PLAM*, plastid-associated endoplasmic reticulum membrane; *TP*, triose phosphate. *This Figure was developed based on information from Geigenberger et al. (2005), Slocombe et al. (2009), Prabhakar et al. (2010), Flügge et al. (2011), Chapman and Ohlrogge (2012), Chapman et al. (2013) and Vanhercke et al. (2014a).*

1995). A newly discovered integral membrane protein known as fatty acid export 1 (FAX1) appears to be responsible for transporting liberated free fatty acids through the inner envelope of the plastid (Li et al., 2015). Free fatty acids are then converted to acyl-CoA on the cytosolic side of the plastid through the catalytic action of long-chain acyl-CoA synthetase (LACS). In some plants, the long-chain acyl-CoA can be further elongated in endoplasmic reticulum (ER) by the catalytic action of elongases. The extraplastidial acyl-CoA pool supports the acyl-CoA-dependent biosynthesis of TAG catalyzed by the acyltransferases of the *sn*-glycerol-3-phosphate (G3P) or Kennedy pathway, which also supports the formation of nitrogenous membrane lipids by providing *sn*-1,2,-diacylglycerol (DAG) as a precursor. In developing seeds enriched in polyunsaturated fatty acids (PUFAs) or unusual fatty acids such as ricinoleic acid (12-OH $18{:}1\Delta^{9cis}$), DAG enriched in 18:1 is converted to phosphatidylcholine (PC), which serves as substrate for fatty acid desaturases or hydroxylase, respectively, to produce PC enriched in PUFA or ricinoleic acid. Several different acyl-editing mechanisms have been identified for channeling PUFA or unusual fatty acids, such as ricinoleic acid, into TAG (Bates and Browse, 2012; Chapman and Ohlrogge, 2012; Wang et al., 2012; Pan et al., 2013, 2015). In the Kennedy pathway, acyl-CoA:diacylglycerol acyltransferase (DGAT) catalyzes the acyl-CoA-dependent acylation of DAG to form TAG (Liu et al., 2012). TAG can also be formed through an acyl-CoA-independent process. Phospholipid:diacylglycerol acyltransferase (PDAT) catalyzes the acylation of DAG using fatty acyl groups derived from the *sn*-2 position of PC (Liu et al., 2012). DGAT and PDAT seem to operate in a redundant fashion to catalyze the formation of TAG in developing Arabidopsis embryos (Zhang et al., 2009). TAG droplets form in the outer leaflets of the ER and eventually pinch off to resulting in oil bodies (OB) of 0.2 to 2 μm in the cytosol. Proteins known as oleosins coat the OB, preventing them from fusing (Huang, 1996).

During seed germination, a sugar-dependent 1 (SDP1) lipase catalyzes the hydrolysis of TAG (Quetteir and Eastmond, 2009). Recently, it was shown that Arabidopsis SDP1 is first localized on peroxisomes, and then during seedling growth eventually moves to the surface of OB (Thazar-Poulot et al., 2015). The translocation process involves the action of peroxisomal extensions and a multiprotein complex, known as a functional retromer, which participates in protein trafficking. Free fatty acids released through hydrolysis of TAG are transported into peroxisomes and are converted to 4-carbon compounds that are eventually converted to soluble sugars to support seedling growth (Graham, 2008). The process involves β-oxidation, the glyoxylate cycle and gluconeogenesis. The growth regulator abscisic acid (ABA) and sugars inhibit the mobilization of TAG.

During the day, leaves capture and reduce carbon dioxide through photosynthesis, converting it to sucrose for delivery to "sink" organs such as developing seeds. During the day, sugars can also be stored in chloroplasts in the form of starch for conversion back to sucrose during the night to support plant growth

(Stitt and Zeeman, 2012). ADP-glucose pyrophosphorylase (AGPase) catalyzes the formation of ADP-glucose and pyrophosphate from glucose-1-phosphate and ATP (see Fig. 15.1) (Geigenberger et al., 2005). Both light and sugars contribute to redox activation of AGPase.

Fatty acid synthesis also occurs in leaf chloroplasts, but the resulting fatty acyl chains are not used to produce TAG for carbon/energy storage. Instead, fatty acyl chains are used to support membrane lipid production during the active growth of leaves (Chapman et al., 2013; Vanhercke et al., 2014a). Fatty acids produced in the chloroplast can be used in the assembly of thylakoid lipids at the envelope membranes or sent to the ER to support extraplastidial lipid assembly (Wang and Benning, 2012). In addition, some lipid precursors can also return to the chloroplast from the ER. Evidence is accumulating in support of continuity between the chloroplasts and the ER (eg, Xu et al., 2008; Mehrshahi et al., 2013), which could potentially facilitate the movement of lipids between these organelles. In vegetative tissue, small amounts of cytoplasmic TAG may serve as a transient storage depot for free fatty acids produced during membrane lipid alteration (Chapman et al., 2013). TAG assembly in the ER of mesophyll leaf cells appears to be a similar process to that of oil-forming cells of developing seeds. He et al. (2006) found microsomal DGAT activity and immunochemically detected DGAT1 polypeptide in true leaves of castor (*Ricinus communis*) seedlings at 17 and 21 days after imbibition. Recently, pulse-chase radiolabeling experiments were used to demonstrate that young Arabidopsis leaves possess substantial TAG-biosynthetic capacity despite accumulating low levels of TAG (Tjellström et al., 2015). The results of these experiments also suggested that ER DGAT1 was the main enzyme involved in cytoplasmic TAG biosynthesis. Fan et al. (2013a), however, provided evidence that PDAT1 has a role in both diverting membrane lipids to TAG and enhancing fatty acid biosynthesis in Arabidopsis leaves.

TAG degradation in leaves also involves lipase action (Kelly et al., 2013), import of fatty acids into the peroxisome and β-oxidation in this organelle (Chapman et al., 2013) (see Fig. 15.1). Leaf-type peroxisomes specialize in the generation of citrate, but some plant species at the young seedling or senescence stage produce succinate via the glyoxylate cycle (Pracharoenwattana et al., 2005). The resulting citrate or succinate can be oxidized in the tricarboxylic acid cycle in the mitochondria or used in gluconeogenesis in the cytosol to support sucrose synthesis and its eventual export to "sink" organs (Graham, 2008; Chapman et al., 2013; Troncoso-Ponce et al., 2013).

A notable difference in primary carbon metabolism in the cells of leaves versus developing seed embryos is the absence of a complete glycolytic pathway in chloroplasts (Chapman et al., 2013). In developing seeds, pyruvate derived from stromal phosphoenolpyruvate (PEP) is a major precursor for fatty acid production (Andre et al., 2007; Baud et al., 2007a), whereas in leaf cells PEP must be imported into chloroplasts because of the absence of enolase activity required to convert PEP into pyruvate (Prabhakar et al., 2010; Flügge et al.,

2011) (see Fig. 15.1). The PEP/phosphate translocator, which transports PEP into the chloroplast, uses 2-phosphoglycerate as a counterexchange substrate (Flügge et al., 2011). It has been suggested that the absence of enolase activity in the chloroplast may represent an evolutionary adaptation which ensures that photosynthate captured in the chloroplast is preferentially used for export to "sink" tissues as opposed to supporting fatty acid biosynthesis in the chloroplast (Chapman et al., 2013).

Plastids have also been shown to produce TAG. An early study by Martin and Wilson (1984) indicated that spinach (*Spinacia oleracea*) leaf chloroplasts were capable of TAG-biosynthetic activity. Xue et al. (1997) later reported TAG production in pea (*Pisum sativum*) root plastids. In chloroplasts, abiotic stresses and senescence have been shown to promote TAG accumulation, possibly acting as a transient sink or buffer for free fatty acids released from galactolipids during stress or senescence-induced membrane degradation (Sakiki et al., 1990a,b; Kaup et al., 2002; Troncoso-Ponce et al., 2013). Similarly, phytol derived from degraded chlorophyll can lead to the transient formation of phytyl esters during abiotic stress or senescence (Lippold et al., 2012). Biosynthesis of phytyl esters in Arabidopsis chloroplasts is catalyzed by phytyl ester synthase (PSE) in the form of PSE1 and PSE2 (Lippold et al., 2012). The phytyl esters are deposited into plastoglobuli which also contain TAG. DGAT activity associated with these PSE enzymes may also account for the presence of this TAG although the nature of the acyl donor is not clear.

Recent studies have shown that heat stress can lead to increased accumulation of TAG enriched in PUFA in Arabidopsis vegetative tissue (Higashi et al., 2015; Mueller et al., 2015). Higashi et al. (2015) investigated the response to heat using lipidomic and transcriptomic analysis. Lipidomic analysis of molecular species of TAG indicated that 54:9 TAG increased in response to increasing temperature. Since 54:9 TAG is probably composed of three α-18:3 ($18{:}3\Delta^{9cis,12cis,15cis}$) chains, the data suggest that heat stress promotes the use of α-18:3 as a substrate in TAG biosynthesis. Complementary transcriptomic data indicated that genes encoding most of the Kennedy pathway enzymes along with PDAT increased in expression during heat stress. Although TAG accumulation increased under heat stress, there also appeared to be a potential for increased TAG degradation, since the expression of genes encoding leaf TAG lipases SDP1 and SDP1L was also increased during heat stress. In addition, the gene encoding PSE1, which may be involved in TAG biosynthesis in the chloroplast, also exhibited increased expression during heat stress. The expression of *FIB1a* and *FIB1b* was also increased. These transcripts encode fibrillins which are involved in the formation of plastoglobuli in chloroplasts (Youssef et al., 2010).

Yang et al. (2011) have demonstrated that ABA signaling is involved in governing the accumulation of TAG in Arabidopsis leaves and that AB14 is required for activation of *DGAT1* expression during nitrogen deficiency. AB14 encodes an activator protein-2/ethylene-responsive factor transcription factor that interacts with the CE1-like signature, CADDG, found in numerous ABA and sugar-responsive promoters (Finklestein et al., 1998).

TRANSCRIPTION FACTORS PROMOTE INCREASED TRIACYLGLYCEROL PRODUCTION IN VEGETATIVE TISSUE

Transcription factors are proteins which interact with DNA at specific sequences, thereby controlling the expression of target genes. The expression of many genes can be affected by a single type of transcription factor. Numerous transcription factors are known to be involved in regulating embryogenesis and seed maturation in Arabidopsis and oil crops (Santos Mendoza et al., 2008; Weselake et al., 2009; Baud and Lepiniec, 2010). Their effects are complex and interactive. *LEAFY COTYLEDON* genes (*LEC1* and *LEC2*) encode transcription factors which can be considered to be master regulators (Weselake et al., 2009; Baud and Lepiniec, 2010). The *LEC1* gene is necessary for early and late embryo maturation in Arabidopsis and encodes the CCAAT box-binding factor HAP3 (Lotan et al., 1998; Junker and Bäumlein, 2012). LEC2 contains a "B3" DNA-binding domain which is also shared with transcription factors FUSCA3 and INSENSITVE3 (Santos Mendoza et al., 2008). LEC1 was shown to also upregulate genes encoding ω-3 and ω-6 fatty acid desaturases, fatty acid elongase 1 and six oleosins (Mu et al., 2008). LEC2 was shown to interact with the promoter regions of genes encoding oleosin (S3/OLEO1) and steroleosin HSD1 (Braybrook et al., 2006; Baud et al., 2009). A recent study on a functional orthologue of Arabidopsis LEC1 in the lycophyte, *Selaginella moellendorffii*, suggests that LEC1 originated at least 30 million years before the first seed plants, and that it is involved in cellular processes in seedless plants similar to those occurring during seed development (Kirkbride et al., 2013).

Transcription factor WRINKLED1 (WRI1) operates one level below the master regulators LEC1 and LEC2, specifying their action on the fatty acid biosynthetic network (Baud et al., 2007b; Weselake et al., 2009; Baud and Lepiniec, 2010). Focks and Benning (1998) initially characterized the Arabidopsis mutant *wrinkled1* (*wri1*) showing an 80% decrease in seed oil content. *WRI1* encodes an APETALA2-ethylene-responsive element-binding protein (AP2/EREBP). Constitutive expression of the cDNA encoding AP2/EREBP in wild-type Arabidopsis resulted in transgenic plants exhibiting an increase in seed oil content of 10–20% more than for wild-type plants (Cernac and Benning, 2004). WRI1 is necessary for both germination and early seedling growth (Cernac et al., 2006). Transcriptomic analyses have revealed that WRI1 regulates a number of genes in glycolysis and fatty acid synthesis (Ruuska et al., 2002; Baud et al., 2007b). For a more in-depth introduction to the role of transcription factors in embryogenesis and seed development, the reader is encouraged to consult the minireview of P.R. Fobert (http://lipidlibrary.aocs.org/plantbio/transfactors/index.htm).

Seed transcription factors have proven useful in reprogramming vegetative tissue to increase TAG production. Lotan et al. (1998) provided early evidence that ectopic expression of *LEC1* in vegetative cells induced the expression of embryo-specific genes and initiated the production of embryo-like structures. It has also been shown that ectopic expression of *LEC2* can induce the formation of structures similar to somatic embryos on vegetative tissue (Stone et al., 2001). The use of transcription factors in boosting TAG accumulation in vegetative tissues, however, has focused mainly on the use of LEC2 and WRI1. Santos Mendoza et al. (2005) used a dexamethasone-inducible promoter to demonstrate that induction of *LEC2* expression in Arabidopsis leaves of young seedlings led to the accumulation of seed-specific TAG containing very-long-chain fatty acids when plants were grown on one percent sucrose plates. Seed-specific mRNAs encoding S3 oleosin, a sucrose synthase and a seed storage protein, all exhibited a high level of accumulation, indicating that there was a shift in leaf metabolism to promote storage biosynthesis. Mu et al. (2008) also demonstrated that induction of *LEC1* expression increased fatty acid production in Arabidopsis seedlings. Somewhat later, Andrianov et al. (2010) used an acetaldehyde-inducible promoter to induce Arabidopsis *LEC2* expression in tobacco (*Nicotiana tabacum*). The use of an inducible promoter circumvented the abnormal growth that was associated with conventional constitutive expression *LEC2* (Lotan et al., 1998). *LEC2* expression experiments were conducted with a normal and high-sugar line of *N. tabacum* (Andrianov et al., 2010). Treatment with one percent acetaldehyde increased total fatty acids from 2.9% to 6.8% of the dry weight. This effect was accompanied by a shift toward 18:1 and decrease in the proportion of α-18:3 in the total fatty acids. The investigators suggested that this could have been caused by limited availability of the ER desaturases responsible for PUFA formation. It was apparent that studies on transcription factor production in vegetative tissue benefited from the use of sucrose. Sugars can potentially promote lipid accumulation because they serve as precursors of fatty acid biosynthesis (Durrett et al., 2008). Indeed, it was previously demonstrated directly that increasing sucrose concentration can stimulate TAG production and DGAT activity in microspore-derived embryos of *B. napus* L. cv Jet Neuf (Weselake et al., 1998; Nykiforuk et al., 2002). Slocombe et al. (2009) repeated the *LEC2* induction experiments of Santos Mendoza et al. (2005) in both sucrose plates and under soil conditions. When grown in soil (in the absence of sucrose), leaf TAG only accumulated once the leaves entered senescence. During senescence, cytosolic OB and very large plastidial starch grains were observed in transgenic lines. Seedlike TAG accumulated, which was enriched in 20:1 ($20{:}1\Delta^{11cis}$). The authors suggested that fatty acids from increased turnover of membrane during senescence might have been redirected into TAG due to an induction of the seed program resulting from ectopic expression of *LEC2*. In a more recent study, Nookaraju et al. (2014) demonstrated that xylem-specific expression of Arabidopsis *LEC2* in tobacco (*Nicotiana benthamiana*) stems resulted in an 80% increase in total fatty acid production in mature stems of transgenic plants. The authors indicated that it may not be practical to collect leaves from trees such as poplar which have been engineered to produce leaf oil. Instead, more easily harvested woody stems could serve two purposes: cellulose for bioethanol production and oil for biodiesel production.

Expression of *WRI1* cDNA in the Arabidopsis mutant background *wri1–1* under the control of a cauliflower mosaic virus 35S-promoter was shown to lead to increased TAG accumulation in the vegetative tissue of seedlings grown in sucrose-containing media (Cernac and Benning, 2004). The seedlings exhibited a phenotype similar to the embryonic state of developing seeds during oil accumulation. The ectopic expression of *WRI1*, however, did not result in ectopic embryo formation as shown for *LEC1* expression in vegetative tissue (Lotan et al., 1998). Vanhercke et al. (2013) later used a transient leaf expression system in tobacco (*N. benthamiana*) to demonstrate that *WRI1* expression was capable of increasing TAG levels by 22-fold. In addition, Kelly et al. (2013) demonstrated that ectopic expression of *WRI1* substantially increased the TAG content of Arabidopsis roots. *WRI1* expression appears to be enhanced by the presence of sucrose in Arabidopsis leaves (Masaki et al., 2005; Sanjaya et al., 2011).

INCREASING THE AVAILABILITY OF BUILDING BLOCKS FOR TRIACYLGLYCEROL ASSEMBLY BY MODIFYING ENZYME ACTION

As indicated in the previous section, expression of cDNAs encoding transcription factors in vegetative tissue can influence the availability of precursors for TAG assembly through upregulation of genes encoding enzymes in glycolysis and fatty acid biosynthesis. This has been referred to as the "push" approach (Vanhercke et al., 2013, 2014a). In terms of modifying enzyme-catalyzed reactions directly, upregulation of ACCase (Klaus et al., 2004) and AGPase (Sanjaya et al., 2011) represents two examples of engineering the "push" effect. Klaus et al. (2004) constitutively overexpressed Arabidopsis homomeric ACCase (*ACC1*) in potato (*Solanum tuberosum* L.). In transgenic line ACC-29, the TAG content of tubers increased fivefold (from 0.0116 to 0.0580 mg/g on a fresh weight basis). Total fatty acid content increased from 1.08 to 1.39 mg/g (fresh weight) and was incorporated into various lipid classes, although deposition was preferentially into TAG.

In more recent work, Sanjaya et al. (2011) examined the effect of reducing starch synthesis on TAG accumulation in Arabidopsis leaves in the absence or presence of the transcription factor WRI1. RNA interference was used to reduce the expression of *APS1*, which encodes a major catalytic isoform of the small subunit of AGPase, combined with *WRI1* overexpression to increase TAG content in Arabidopsis vegetative tissue. Downregulation of *APS1* reduced starch content by about twofold in the absence or presence of the ectopic expression of *WRI1*. Downregulation of starch synthesis alone, however, was not sufficient to increase the TAG content of seedlings. In contrast, the combination of the two approaches resulted in higher levels of TAG than for the wild type. Transgenic line 29 exhibited a 5.8-fold increase in TAG content. Microscopic observation of leaf and root tissue following Nile red staining revealed distinct OBs. Downregulation of starch synthesis resulted in higher levels of sucrose which probably stimulated the conversion of sucrose into hexoses. The high sugar levels in this transgenic line probably also resulted in an increase in endogenous *WRI1* expression. The overall increase in *WRI1* expression (in the presence of downregulation of starch synthesis) would result in the upregulation of several enzymatic steps in glycolysis and fatty acid synthesis, which would in turn encourage the conversion of glucose into building blocks (fatty acid and G3P) for enhanced TAG assembly.

INCREASING TRIACYLGLYCEROL ASSEMBLY

Metabolic engineering interventions which target TAG assembly are sometimes referred to as the "pull" approach (Vanhercke et al., 2013, 2014a). Bouvier-Navé et al. (2000) were the first to observe the presence of OBs in leaf epidermis cells of tobacco (*N. tabacum*) transformed with cDNA encoding Arabidopsis *DGAT1* using a 35S promoter to drive expression of the cDNA. The best transgenic line exhibited a sevenfold increase in leaf TAG content over the wild type. Andrianov et al. (2010) expressed Arabidopsis *DGAT1* in both a regular and high-sugar line of *N. tabacum* under the control of a strong promoter which drives expression of the gene encoding the small subunit of ribulose-bisphosphate carboxylase. The leaves of the best transgenic lines produced up to 20-fold more TAG than the wild type. Oleic acid was significantly enriched in TAG from transgenic lines (20–25% 18:1) compared with the wild type (1.5% 18:1). In addition, α-18:3 content decreased from 67% in the wild type to 30–40% in transgenic lines. The investigators indicated that the high-sugar background would provide more precursors for fatty acid biosynthesis (see Durrett et al., 2008), and this was reflected in the increased fatty acid content in the vegetative tissue of the high-sugar tobacco variety. A large proportion of the transgenic lines produced using the high-sugar tobacco variety exhibited consistently high levels of TAG of over 5% of the dry weight, compared with transgenic lines produced using regular tobacco. It is also important to note that sucrose has been shown to upregulate *DGAT1* expression and activity in microspore-derived cell suspension cultures of *B. napus* L. cv Jet Neuf (Davoren et al., 2002; Nykiforuk et al., 2002). Furthermore, Lu et al. (2003) showed that Arabidopsis *DGAT1* gene expression was upregulated by glucose. Therefore, the high-sugar background combined with possible increased *DGAT1* expression caused by the sugar may have synergistically contributed to increased TAG accumulation the leaves. In a more recent study, xylem-specific expression of Arabidopsis *DGAT1* in tobacco (*N. benthamiana*) was shown to increase total fatty acid production in stems by 63% (Nookaraju et al., 2014). Using an *N. benthamiana* transient leaf expression system, Vanhercke et al. (2013) demonstrated that coexpression of Arabidopsis *WRI1* and *DGAT1* increased TAG levels in excess of what would be expected from an additive effect, and the fatty acid profile of leaf tissue shifted from PUFAs to monounsaturated fatty acids. Feeding experiments with radiolabeled acetate confirmed that there was an increase in the flux of carbon through fatty acid biosynthesis and TAG assembly. Sanjaya et al. (2013) constitutively expressed an algal (*Chlamydomonas reinhardtii*) *DGAT2* in Arabidopsis. Fifteen-day-old seedlings of transgenic plants contained 22- to 25-fold more TAG than the wild type, and the TAG was enriched in very-long-chain fatty acids. Feeding studies with transgenic plants and caterpillars (*Spodoptera exigua*) indicated that the nutritional value and/or energy density of the transgenic plants was enhanced by ectopic expression of the *DGAT2*.

Overexpression of *PDAT1* in Arabidopsis leaves was shown to increase the TAG content of 7-week-old leaves by 28-fold compared with the wild type (Fan et al., 2013a). *PDAT1* expression resulted in more TAG accumulation than could be achieved by *DGAT1* overexpression. PUFAs (18:2 ($18{:}2\Delta^{9cis,12cis}$) and α-18:3) accounted for over 75% of the acyl chains in the leaf TAG of *PDAT1* overexpressors, but limited amounts of very-long-chain fatty acids were detected. This is in contrast to Santos Mendoza et al. (2005) and Sanjaya et al. (2013), who reported an increase in very-long-chain fatty acids in leaf TAG from Arabidopsis ectopically expressing Arabidopsis *LEC2* or algal *DGAT2*, respectively. *PDAT1* overexpression also stimulated fatty acid synthesis and increased the flux of fatty acids toward the prokaryotic pathway (Fan et al., 2013a). PDAT1-mediated TAG production was also important in reducing fatty acid-induced cell death in leaves by diverting fatty acids from membrane lipid biosynthesis to TAG (Fan et al., 2013b).

TAG content in vegetative tissue has also been increased through the introduction of a monoacylglycerol (MAG) pathway into a transient expression system in *N. benthamiana* leaves, which involves the utilization of MAG for biosynthesis of DAG (Petrie et al., 2012; Divi et al., 2014). Heterologous expression of two forms of mouse (*Mus musculus*) MONOACYLGLYCEROL ACYLTRANSFERASE resulted in DAG production, which in turn increased TAG content in the leaf. TAG increases of 9.2-fold and 7.3-fold were observed for the expression of *MGAT1* and *MGAT2*, respectively (Petrie et al., 2012).

DECREASING TRIACYLGLYCEROL TURNOVER

Reducing TAG turnover in leaves has been achieved by decreasing TAG mobilization, β-oxidation of fatty acids in the peroxisome or by shielding OBs from degradation. Reducing TAG turnover has been referred to as the "protect" approach (Vanhercke et al., 2014a). Four-week-old seedlings of an Arabidopsis mutant (*sdp1–5*) defective in TAG lipase exhibited significant increases in the TAG content of leaf, stem and root tissue relative to the wild type, and more TAG accumulated when the medium contained sucrose (Kelly et al., 2013). Root tissue produced the highest levels of TAG, which increased as the plant matured to 50-fold more in the transgenic plant over the wild type. Further increases in TAG content were achieved by combining knockdown of the TAG lipase with other strategies to boost TAG content.

The mammalian comparative gene identifier-58 (CGI-58) is known to stimulate lipolysis in mammalian tissues. Disruption of a homolog of CGI-58 in Arabidopsis was also shown to increase leaf TAG content without affecting plant development (James et al., 2010). Leaf TAG content was 10-fold higher in CGI-58 plants than wild-type plants. In plant vegetative tissues, the CGI-58 homolog is thought to have a role in controlling lipid turnover (James et al., 2010).

Interfering with fatty acid catabolism is also known to increase the TAG content of leaves. Leaves of Arabidopsis mutant plants with compromised peroxisomal ABC-transporter 1 (PXA1) activity accumulated significant levels of TAG, especially under conditions of extended darkness where transitory starch reserves were depleted (Kunz et al., 2009). PXA1 mediates the uptake of fatty acids into peroxisomes where β-oxidation occurs. Under darkness, the increased requirement for respiratory carbon comes from release of esterified fatty acids. Inactivation of fatty acid transport into the peroxisome results in accumulation of fatty acids in TAG. Arabidopsis mutants defective in peroxisomal β-oxidation were also shown to accumulate TAG in leaves (Slocombe et al., 2009).

Oleosins are known to coat OBs in seeds of oleaginous plants and prevent the OBs from coalescing (Siloto et al., 2006). *OLEOSIN* expression in leaves represents another approach for protecting OBs once they are formed (Fan et al., 2013a; Winichayakul et al., 2013; Vanhercke et al., 2014b). *OLEOSIN1* (*OLE1*) encodes the most abundant seed lipid droplet-protein in Arabidopsis (Huang, 1996). Overexpression of *OLE1* in the Arabidopsis *PDAT1* overexpressor resulted in a 74-fold increase in TAG content compared with the wild type (Fan et al., 2013a). As mentioned previously, *PDAT1* overexpression alone resulted in a 28-fold increase in TAG content (Fan et al., 2013a). In addition, the transgenic plants had clusters of spherical OBs in leaves, unlike the large and irregular OBs seen in Arabidopsis which only overexpressed *PDAT1*.

Winichayakul et al. (2013) introduced cysteine residues into strategic positions in the 15-KDa sesame (*Sesamum indicum*) seed L-oleosin (accession no. AAD42942) to encourage formation of inter-oleosin disulfide bonds, resulting in "cages" of oleosins around OB. Arabidopsis ectopically coexpressing cDNA encoding the modified oleosin along with Arabidopsis *DGAT1* resulted in a 24% increase in carbon dioxide assimilation rate per unit leaf area and 50% increase in biomass. The fatty acid content of mature leaves, senescing leaves and roots increased by two-, three- and fivefold, respectively. The investigators proposed that their coexpression approach prevented the futile cycle of TAG formation/degradation creating demand for de novo lipid biosynthesis, resulting in increased carbon dioxide recycling in the chloroplast.

MULTIGENE STRATEGIES FOR INCREASING TRIACYLGLYCEROL CONTENT IN VEGETATIVE TISSUE

Some examples of two-gene modifications for increasing oil content in vegetative tissue have already been discussed in the preceding sections. A number other proof-of-concept studies for increasing TAG content in vegetative tissue have been conducted using multigene approaches with two-or-more-gene modifications. Examples of multigene strategies leading to substantial levels of TAG in vegetative tissue are presented in Table 15.1.

Coexpression of *PDAT1* and *OLEOSIN* (*OLEO1*) increased leaf TAG content of Arabidopsis to 6.4% of the dry weight without affecting plant growth and the lipid composition of the membrane (Fan et al., 2013b). Somewhat later, Fan et al. (2014) disrupted TAG lipase (SDP1) or PXA1 in a *trigalactosyldiacylglycerol1–1mutant* background in Arabidopsis with enhanced PDAT1-mediated TAG synthesis to reach 9% (dry weight) TAG content.

Arabidopsis TAG levels of 5–8% (dry weight) in leaf, stem and root tissue were achieved in 4-week-old seedlings when Kelly et al. (2013) coexpressed *DGAT1* and *WRI1* in an *sdp1* background, but there was a penalty on growth. When grown in 3% (w/v) sucrose, the TAG content of the roots reached 17%, which was close to that of soybean seeds.

Recently, Vanhercke et al. (2014b) combined the expression of *WRI1*, *DGAT1* and *OLEOSIN* to engineer *N. tabacum* leaves capable of producing more than 15% (dry weight) TAG content without severe effects on plant development. When extrapolated to a high-biomass crop, this amount of TAG accumulation would exceed the oil yield per hectare for most cultivated oilseed crops. Furthermore, application of this particular multigene strategy to an elite tobacco cultivar germplasm which exhibits high seed oil yields could result in a dual-purpose crop (with high TAG content in both leaves and seeds) that could potentially result in oil yields close to that of oil palm (Vanhercke et al., 2014b). This study also examined the effect of different extraction conditions. Acetone extraction resulted in the highest TAG extraction efficiency from fresh *N. tabacum* leaves, compared with hexane or methanol-based extraction.

ENRICHING LEAF TRIACYLGLYCEROL IN INDUSTRIALLY USEFUL FATTY ACIDS

Methyl esters of plant oils enriched in monounsaturated fatty acids such as 18:1 and palmitoleate (16:1Δ^{9cis}) exhibit high ignition quality, oxidative stability and good cold-temperature flow characteristics for biodiesel use (Durrett et al., 2008). Therefore, metabolic engineering approaches that enhance accumulation of leaf TAG enriched in monounsaturated fatty acids could potentially result in a suitable TAG feedstock for production of biodiesel. As previously indicated, *DGAT1* or *LEC2* expression in *N. tabacum* leaves has been shown to result in increased TAG production, with a shift toward 18:1 and concomitant decrease in the proportion of α-18:3 (Andrianov et al., 2010). In addition, the combined expression of *WRI1*, *DGAT1* and *OLEOSIN* in *N. tabacum* resulted in the production of leaf TAG more enriched in 18:1 at the expense of α-18:3 (Vanhercke et al., 2014b). Further improvement in the cold-flow properties of biodiesel

TABLE 15.1 **Examples of Multigene Strategies Resulting in Substantial Triacylglycerol (TAG) Accumulation in Vegetative Tissue of C3 Plants**

Engineered Plant	Modifications	TAG (% Dry Weight)	References
Arabidopsis (wild type)	Overexpression of Arabidopsis *PDAT1* with *OLEOSIN1*	6.4 (leaves)	Fan et al. (2013a,b)
Arabidopsis *tgd1-1* mutant	Coexpression of Arabidopsis *PDAT1* and *OLEOSIN1*	8.6 (leaves)	Fan et al. (2013a)
Arabidopsis *tgd1-1* mutant	Disruption of SDP1 lipase (*sdp1-4*)	9 (leaves)	Fan et al. (2014)
	Disruption of PXA1 (*pxa1-2*)	7 (leaves)	
Arabidopsis lipase mutant (*sdp1*)	Coexpression of Arabidopsis *WRI1* and *DGAT1*	5–8 (roots, stems and leaves) 17 (roots with 3% [w/v] exogenous sucrose)	Kelly et al. (2013)
Tobacco (*Nicotiana tabacum*) (wild type)	Coexpression of Arabidopsis *WRI1* and *DGAT1* with sesame (*Sesamum indicum*) seed *L-OLEOSIN*	>15 (leaves from plants at the seed-setting stage)	Vanhercke et al. (2014b)

The TAG content of 4-day-old wild-type Arabidopsis roots, stems and leaves grown on agar plates was about 0.02% (dry weight), whereas the *sdp1-5* mutant exhibited about 0.03%, 0.3% and 0.4% TAG, respectively, in leaves, stems and roots (Kelly et al., 2013). The TAG content of fully mature leaves from 7-week-old Arabidopsis *tdg1-1* mutants grown on soil was about 0.2% (Fan et al., 2014). The TAG content of wild-type and null segregant control *N. tabacum* leaves was <0.2% at all stages of plant development (Vanhercke et al., 2014b). Abbreviations: *DGAT1*, diacylglycerol acyltransferase 1; *PDAT1*, phospholipid:diacylglycerol acyltransferase 1; *PXA1*, peroxisomal transporter 1; *SDP1*, sugar-dependent 1; *tgdl-1*, trigalactosyldiacylglycerol1-1; *WRI1*, WRINKLED 1.

could potentially come from the use of medium-chain fatty acids (12–16 carbons long) (Durrett et al., 2008).

Medium-chain fatty acids are also used as a feedstock in the production of coatings and detergents. Recently, Reynolds et al. (2015) investigated whether production of medium-chain fatty acids was compatible with increased TAG accumulation in leaves. The transient leaf expression system in *N. benthamiana* was used to demonstrate that leaf TAG containing significant levels of medium-chain fatty acids could be generated by combined expression of *WRI1*, *DGAT1*, medium-chain thioesterases and coconut (*Cocos nucifera*) lysophosphatidic acid acyltransferase (*LPAAT*). LPAAT catalyzes the acylation of the *sn*-2 position of lysophosphatidic acid in the Kennedy pathway to produce phosphatidic acid (Snyder et al., 2009).

ENGINEERING C4 HIGH-BIOMASS PLANTS TO PRODUCE TRIACYLGLYCEROL IN VEGETATIVE TISSUE

C4 plants are more effective in converting solar energy into biomass than C3 plants. C4 photosynthesis is a modification of C3 photosynthesis in which the oxygenase activity of ribulose-bisphosphate carboxylase is reduced to a carbon dioxide–concentrating mechanism involving phosphoenolpyruvate carboxylase (Ehleringer and Cerling, 2002). Although C4 grasses such as *Miscanthus* spp. and switchgrass have been exploited as sources of lignocellulose for production of bioethanol (van der Weijde et al., 2013), there is potential for engineering these crops to produce TAG in their vegetative tissues to supply a large global feedstock for biofuel applications (Ohlrogge and Chapman, 2011). Sugar cane (*Saccharum* spp.) exhibits high biomass production, with 70% of the biomass in stems, and is the most energy efficient C4 perennial plant (Tew and Cobill, 2008). Recently, Zale et al. (2015) metabolically engineered sugarcane to increase TAG accumulation in leaves and stems. Arabidopsis *WRI1*, maize (*Zea mays*) *DGAT1-2* and Arabidopsis *OLEOSIN1* were coexpressed in combination with genetic downregulation of AGPase and a subunit of the PXA1. The overexpressed cDNAs were codon optimized according to the codon bias of sugarcane. Stem accumulation of TAG in transgenic sugarcane reached 0.9% (dry weight), whereas leaf tissue levels reached 1.9% TAG, which was 95-fold higher than the wild type. Mature wild-type sugarcane contained TAG which was composed of saturated fatty acids. In contrast, the leaves of two transgenic lines exhibited about a 50% decrease in saturated fatty acid content with substantial amounts of unsaturated fatty acids (18:1, 18:2 and α-18:3). A similar trend was observed in both immature leaves and stems of these transgenic lines, with the exception that the wild type contained significant amounts of 18:2 in the TAG of immature leaves and stems. Given the high biomass production for sugarcane, the investigators have estimated that each percentage of TAG produced in this crop corresponds to TAG yield from the same land area as oilseed rape. It should be noted, however, that there could be practical limitations with regard to the mass of vegetative tissue that could be processed by large-scale solvent extraction.

CONCLUSION

Studies with knockout mutants in Arabidopsis and the development of strategies to boost TAG content in vegetative tissue have shed increasing light on the function and regulation of TAG metabolism in vegetative tissue, but there is much to be learned. New or complementary approaches for boosting the TAG content of vegetative through metabolic engineering will surely be explored in the coming years. The use of inducible promoters to drive TAG accumulation in later stages of plant development appears to be a useful strategy for minimizing penalties on plant growth caused by various metabolic engineering approaches (Troncoso-Ponce et al., 2013). The noted absence of enolase activity in chloroplasts represents a possible opportunity to increase TAG production in leaves through the introduction of an enolase enzyme into the chloroplast. Engineering more efficient transport of acetyl-CoA precursors, such as pyruvate, may also be worth exploring. Transcriptome analysis of wild-type versus *wri1* Arabidopsis has shown that only a few glycolytic enzymes are down-regulated in the *wri1* mutant, which suggests that a number of enzymes are regulated at the biochemical level (Ruuska et al., 2002). Additional molecular tools for increasing TAG accumulation in both developing seeds and vegetative tissue could come through identification of allosteric mechanisms influencing primary metabolism leading to TAG. For example, it may be possible to suppress metabolite-induced feedback inhibition in a number of instances. In a study with microspore-derived cell suspension cultures and developing seeds of *B. napus*, Andre et al. (2012) demonstrated that 18:1-ACP directly inhibits plastidial ACCase, resulting in reduced fatty acid synthesis. This represents an opportunity to desensitize ACCase to the effects of 18:1-ACP perhaps resulting in increased fatty acid production.

Fatty acid compositional changes in leaf TAG can be affected by abiotic stress or the type of metabolic engineering approaches used to promote increased leaf TAG accumulation. For example, *LEC2* or *DGAT1* overexpression was associated with a shift to 18:1 in leaf TAG (Andrianov et al., 2010) whereas *PDAT1* overexpression resulted in increased PUFA content in leaf TAG (Fan et al., 2013a). A general observation in the various proof-of-concept studies to boost TAG content in vegetative tissue is that TAG increases were enhanced by the presence of sucrose in the growth medium. These are all factors to take into consideration in future research aimed at producing industrial oils in leaf tissue. Thus, both the metabolic engineering approach used and the environmental conditions under which plants are grown could impact the TAG content of vegetative tissue and the ability to enrich specific industrially useful fatty acids in the TAG.

Finally, production of TAG in high-biomass C4 crops such as switchgrass and sugarcane has the potential to create a vast amount of global feedstock for liquid biofuel applications and other industrial applications. In addition, metabolic engineering of rapidly growing trees such as poplar to produce oil in their woody stems (Nookaraju et al., 2014) could also potentially provide another high-biomass source of TAG.

ACKNOWLEDGMENTS

The author is grateful to Alberta Enterprise and Advanced Education, Alberta Innovates Bio Solutions, AVAC Ltd., the Canada Foundation for Innovation, the Canada Research Chairs Program and the Natural Sciences and Engineering Research Council of Canada for supporting his research on storage lipid accumulation in plants. The author thanks Drs. David Hildebrand and Thomas Mckeon for their critical evaluations of the manuscript.

REFERENCES

Andre, C., Froehlich, J.E., Moll, M.R., Benning, C., 2007. A heteromeric plastidic pyruvate kinase complex involved in seed oil biosynthesis in Arabidopsis. Plant Cell 19, 2006–2022.

Andre, C., Haslam, R.P., Shanklin, J., 2012. Feedback regulation of plastidic acetyl-CoA carboxylase by 18:1-acyl carrier protein in *Brassica napus*. Proc. Natl. Acad. Sci. U.S.A. 109, 10107–10112.

Andrianov, V., Borisjuk, N., Pogrebnyak, N., Brinker, A., Dixon, J., Spitsin, S., Flynn, J., Matyszczuk, P., Andryszak, K., Laurelli, M., Golovkin, M., Koprowski, H., 2010. Tobacco as a production platform for biofuel: overexpression of *Arabidopsis DGAT* and *LEC2* genes increases accumulation and shifts the composition of lipids in green biomass. Plant Biotechnol. J. 8, 277–287.

Bates, P.D., Browse, J., 2012. The significance of different diacylglycerol synthesis pathways on plant oil composition and bioengineering. Front. Plant Sci. 3, 147.

Bates, P.D., Stymne, S., Ohlrogge, J., 2013. Biochemical pathways in seed oil synthesis. Curr. Opin. Plant Biol. 16, 358–364.

Baud, S., Wuillème, S., Dubreucq, B., de Almeida, A., Vuagnat, C., Lepiniec, L., Miquel, M., Rochat, C., 2007a. Function of plastidial pyruvate kinases in seeds of *Arabidopsis thaliana*. Plant J. 52, 405–419.

Baud, S., Santos Mendoza, M., To, A., Harscoët, E., Lepiniec, L., Dubreucq, B., 2007b. WRINKLED1 specifies the regulatory action of LEAFY COTYLEDON2 towards fatty acid metabolism during maturation in *Arabidopsis*. Plant J. 50, 825–838.

Baud, S., Dichow, N.R., Kelemen, Z., d'Andréa, S., To, A., Berger, N., Canonge, M., Kronenberger, J., Viterbo, D., Dubreucq, B., Lepiniec, L., Chardot, T., Miquel, M., 2009. Plant Cell Physiol. 50, 1463–1478.

Baud, S., Lepineic, L., 2010. Physiological and developmental regulation of seed oil production. Prog. Lipid Res. 49, 235–249.

Bouvier-Navé, P., Benveniste, P., Oelkers, P., Sturley, S.L., Schaller, H., 2000. Expression in yeast and tobacco of plant cDNAs encoding acyl-CoA:diacylglycerol acyltransferase. Eur. J. Biochem. 267, 85–96.

Braybrook, S.A., Stone, S.L., Park, S., Bui, A.Q., Le, B.H., Fischer, R.L., Goldberg, R.B., Harada, J.J., 2006. Genes directly regulated by LEAFY COTYLEDON2 provide insight into the control of embryo maturation and somatic embryogenesis. Proc. Natl. Acad. Sci. U.S.A. 103, 3468–3473.

Carlsson, A.S., 2009. Plant oils as feedstock alternatives to petroleum—A short survey of potential oil crop platforms. Biochemie 91, 665–670.

Cernac, A., Benning, C., 2004. *WRINKLED1* encodes an AP2/EREB domain protein involved in the control of storage compound biosynthesis in Arabidopsis. Plant J. 40, 575–585.

Cernac, A., Andre, C., Hoffmann-Benning, S., Benning, C., 2006. WRI1 is required for seed germination and seedling establishment. Plant Physiol. 1421, 745–757.

Chapman, K.D., Ohlrogge, J.B., 2012. Compartmentation of triacylglycerol accumulation in plants. J. Biol. Chem. 287, 2288–2294.

Chapman, K.D., Dyer, J.M., Mullen, R.T., 2013. Commentary: why don't plant leaves get fat? Plant Sci. 207, 128–134.

Davoren, J.D., Nykiforuk, C.L., Laroche, A., Weselake, R.J., 2002. Sucrose-induced changes in the transcriptome of cell suspension cultures of oilseed rape reveal genes associated with lipid biosynthesis. Plant Phyisol. Biochem. 40, 719–725.

Divi, U.K., El Tahchy, A., Vanhercke, T., Petrie, J.R., Robles-Martinez, J.A., Singh, S.P., 2014. Transcriptional and biochemical responses of monoacylglycerol acyltransferase-mediated oil synthesis and associated senescence-like responses in *Nicotiana benthamiana*. Front. Plant Sci. 5 Article 204.

Durrett, T.P., Benning, C., Ohlrogge, J., 2008. Plant triacylglycerols as feedstocks for the production of biofuels. Plant J. 54, 593–607.

Ehleringer, J.R., Cerling, T.E., 2002. C3 and C4 photosynthesis. In: Mooney, A., Canadell, J.G. (Eds.), Encyclopedia of Global Environmental Change. John Wiley and Sons Ltd, Chichester, pp. 186–190.

Li, N., Gügel, I.L., Giavalisco, P., Zeisler, Z., Schreiber, L., Soll, J., Philippar, K., 2015. FAX1, a novel membrane protein mediating plastid fatty acid export. PLoS Biol. 13, e1002053.

Fan, J., Yan, C., Zhang, X., Changcheng, X., 2013a. Dual role for phospholipid:diacylglycerol acyltransferase enhancing fatty acid synthesis and diverting fatty acids from membrane lipids to triacylglycerol in *Arabidopsis* leaves. Plant Cell 25, 3506–3518.

Fan, J., Yan, C., Xu, C., 2013b. Phospholipid:diacylglycerol acyltransferase-mediated triacyglycerol biosynthesis is crucial for protection against fatty acid-induced cell death in growing tissues of Arabidopsis. Plant J. 76, 930–942.

Fan, J., Yan, C., Roston, R., Shanklin, J., Xu, C., 2014. Arabidopsis lipins, PDAT1 acyltransferase, and SDP1 triacylglycerol lipase synergistically direct fatty acids toward β-oxidation, thereby maintaining membrane lipid homeostasis. Plant Cell 26, 4119–4134.

Finkelstein, R.R., Wang, M.L., Lynch, T.J., Rao, S., Goodman, H.M., 1998. The Arabidopsis abscisic acid response locus *AB14* encodes an APETALA2 domain protein. Plant Cell 10, 1043–1054.

Flügge, U.I., Häusler, R.E., Ludewig, F., Gierth, M., 2011. The role of transporters in supplying energy to plant plastids. J. Exp. Bot. 62, 2381–2392.

Focks, N., Benning, C., 1998. *wrinkled*1: a novel, low-seed-oil mutant of Arabidopsis with deficiency in the seed-specific regulation of carbohydrate metabolism. Plant Physiol. 118, 91–101.

Geigenberger, P., Kolbe, A., Tiessen, A., 2005. Redox regulation of carbon storage and partitioning in response to light and sugars. J. Exp. Bot. 56, 1469–1479.

Graham, I.A., 2008. Seed storage oil mobilization. Annu. Rev. Plant Biol. 59, 115–142.

He, X., Chen, G.Q., Lin, J.-T., McKeon, T.A., 2006. Diacylglycerol acyltransferase activity and triacylglycerol synthesis in germinating castor seed cotyledons. Lipids 41, 281–285.

Higashi, Y., Okazaki, Y., Myouga, F., Shinozaki, K., Saito, K., 2015. Landscape of the lipidome and transcriptome under heat stress in *Arabidopsis thaliana*. Sci. Rep. 5, 10533.

Huang, A.H.C., 1996. Oleosins and oil bodies in seeds and other organs. Plant Physiol. 110, 1055–1061.

James, C.N., Horn, P.J., Case, C.R., Gidda, S.K., Zhang, D., Mullen, R.T., Dyer, J.M., Anderson, R.G.W., Chapman, K.D., 2010. Disruption of the Arabidopsis CGI-58 homologue produces Chanarin-Dorfman-like lipid droplet accumulation in plants. Proc. Natl. Acad. Sci. U.S.A. 107, 17833–17838.

Junker, A., Bäumlein, H., 2012. Multifunctionality of the LEC1 transcription factor during plant Development. Plant Signal. Behav. 7, 1718–1720.

Kaup, M.T., Froese, C.D., Thompson, J.E., 2002. A role for diacylglycerol acyltransferase during leaf senescence. Plant Physiol. 129, 1616–1626.

Kelly, A.A., van Erp, H., Quettier, A.-L., Shaw, E., Menard, G., Kurup, S., Eastmond, P.J., 2013. The SUGAR-DEPENDENT1 lipase limits triacylglycerol accumulation in vegetative tissues of Arabidopsis. Plant Physiol. 162, 1282–1289.

Kirkbride, R.C., Fischer, R.L., Harada, J.J., 2013. *LEAFY COTYLEDON1*, a key regulator of seed development, is expressed in vegetative and sexual propagules of *Selaginella moellendorffii*. PLoS One 8, e67971.

Klaus, D., Ohlrogge, J.B., neuhaus, H.E., Dörmann, P., 2004. Increased fatty acid production in potato by engineering of acetyl-CoA carboxylase. Planta 219, 389–396.

Kunz, H.-H., Scharnewski, M., Feussner, K., Feussner, I., Flügge, U.-I., Fulda, M., Gierth, M., 2009. The ABC transporter PXA1 and peroxisomal β-oxidation are vital for metabolism in mature leaves of *Arabidopsis* during extended darkness. Plant Cell 21, 2733–2749.

Lin, W., Oliver, D.J., 2008. Role of triacylglycerols in leaves. Plant Sci. 175, 233–237.

Lippold, F., vom Dorp, K., Abaham, M., Hölzl, G., Wewer, V., Yilmz, J.L., Lager, I., Montandon, C., Besagni, C., Kessler, F., Stymne, S., Dörmann, P., 2012. Fatty acid phytyl ester synthesis in chloroplasts of Arabidopsis. Plant Cell 24, 2001–2014.

Liu, Q., Siloto, R.M.P., Lehner, R., Stone, S.J., Weselake, R.J., 2012. Acyl-CoA:diacylglycerol acyltransferase: molecular biology, biochemistry and biotechnology. Prog. Lipid Res. 51, 350–377.

Lotan, T., Ohto, M.-A., Yee, K.M., West, M.A.I., Lo, R., Kwong, R.W., Yamagishi, K., Fischer, R.L., Golberg, R.B., Harada, J.J., 1998. Arabidopsis LEAFY COTYLEDON1 is sufficient to induce embryo development in vegetative cells. Cell 93, 1195–1205.

Lu, C.L., de Noyer, S.B., Hobbs, D.H., Kang, J., Wen, Y., Krachtus, D., Hills, M.J., 2003. Expression pattern of diacylglycerol acyltransferase-1, an enzyme involved in triacylglycerol biosynthesis in *Arabidopsis thaliana*. Plant Mol. Biol. 52, 31–41.

Martin, B.A., Wilson, R.F., 1984. Subcellular localization of triacylglycerol synthesis in spinach leaves. Lipids 19, 117–121.

Masaki, T., Mitsue, N., Tsukagoshi, H., Nishii, T., Morkkami, A., Nakamura, K., 2005. ACTIVATOR of spomin::LUC1/WRINKLED1 of *Arabidopsis thaliana* transactivates sugar-inducible promoters. Plant Cell Physiol. 46, 547–556.

Mehrshahi, P., Stefano, G., Andaloro, J.M., Brandizzi, F., Froehlich, J.E., DellaPenna, D., 2013. Transorganellar complementation redefines the biochemical continuity of endoplasmic reticulum and chloroplasts. Proc. Natl. Acad. Sci. U.S.A. 110, 12126–12131.

Mu, J., Tan, H., Zheng, Q., Fu, F., Liang, Y., Zhang, J., Yang, X., Wang, T., Chong, K., Wang, X.J., Zuo, J., 2008. *LEAFY COTYLEDON1* is a key regulator of fatty acid biosynthesis in Arabidopsis. Plant Physiol. 148, 1042–1054.

Mueller, S.P., Drause, D.M., Mueller, M.J., Fekete, A., 2015. Accumulation of extra-chloroplastic triacylglycerols in *Arabidopsis* seedlings during heat acclimation. J. Exp. Bot. 66, 4517–4526.

Nookaraju, A., Pandey, S.K., Fujino, T., Kim, J.Y., Suh, M.C., Joshi, C.P., 2014. Enhanced accumulation of fatty acids and triacylglycerols in transgenic tobacco stems for enhanced bioenergy production. Plant Cell Rep. 33, 1041–1052.

Nykiforuk, C.L., Furukawa-Stoffer, T.L., Huff, P.W., Sarna, M., Laroche, A., Moloney, M.M., Weselake, R.J., 2002. Characterization of cDNAs encoding diacylglycerol acyltransferase from cultures of *Brassica napus* and sucrose-mediated induction of enzyme biosynthesis. Biochim. Biophys. Acta 1580, 95–109.

Ohlrogge, J., Chapman, K., 2011. Expanding the contribution of plant oils as biofuels. The seeds of green energy. Biochemist 33, 34–38.

Ohlrogge, J., Browse, J., 1995. Lipid biosynthesis. Plant Cell 7, 957–970.

Pan, X., Siloto, R.M.P., Wickramarathna, A.D., Mietkiewska, E., Weselake, R.J., 2013. Identification of a pair of phospholipid:diacylglycerol acyltransferases from developing flax (*Linum usitatissiumum* L.) seed catalyzing the selective production of trilinolenin. J. Biol. Chem. 288, 24173–24188.

Pan, X., Chen, G., Kazachkov, M., Greer, M.S., Caldo, K.M., Zou, J., Weselake, R.J., 2015. *In vivo* and *in vitro* evidence for biochemical coupling of reactions catalyzed by lysophosphatidylcholine acyltransferase and diacylglycerol acyltransferase. J. Biol. Chem. 290, 18068–18078.

Petrie, J.R., Vanhercke, T., Shrestha, P., El Tahchy, A., White, A., Zhou, X.-R., Liu, Q., Mansour, M.P., Nichols, P.D., Singh, S.P., 2012. Recruiting a new substrate for triacylglycerol synthesis in plants: the monoacylglycerol acyltransferase pathway. PLoS One 7, e35214.

Prabhakar, V., Löttgert, T., Geimer, S., Dörmann, P., Krüger, S., Vijayakumar, V., Schreiber, L., Göbel, C., Feussner, K., Feussner, I., Marin, K., Staehr, P., bell, K., Flügge, U.-I., Häusler, R.E., 2010. Phosphoenolpyruvate provision to plastids is essential for gametophyte and sporophyte development in *Arabidopsis thaliana*. Plant Cell 22, 2594–2617.

Pracharoenwattana, I., Cornah, J.E., Smith, S.M., 2005. Arabidopsis peroxisomal citrate synthase is required for fatty acid respiration and seed germination. Plant Cell 2037–2048.

Quettier, A.-L., Eastmond, P.J., 2009. Storage oil hydrolysis during early seedling growth. Plant Physiol. Biochem. 47, 485–490.

Reynolds, K.B., Taylor, M.C., Zhou, X.-R., Vanhercke, T., Wood, C.C., Blanchard, C.L., Singh, S.P., Petrie, J.R., 2015. Metabolic engineering of medium-chain fatty acid biosynthesis in *Nicotiana benthamiana* plant leaf lipids. Front. Plant Sci. 6 Article 164.

Ruuska, S.A., Girke, T., Benning, C., Ohlrogge, J.B., 2002. Contrapuntal networks of gene expression during *Arabidopsis* seed filling. Plant Cell 14, 1191–1206.

Sakaki, T., Saito, K., Kawaguchi, A., Kondo, N., Yamada, M., 1990a. Conversion of mongalactosyldiacylglycerols to triacylglycerols in ozone-fumigated spinach leaves. Plant Physiol. 94, 766–772.

Sakaki, T., Kondo, N., Yamada, M., 1990b. Pathway for the synthesis of triacylglycerols from monogalactosyldiacylglycerols in ozone-fumigated spinach leaves. Plant Physiol. 94, 773–780.

Sanjaya, Durrett, T.P., Weise, S.E., Benning, C., 2011. Increasing the energy density of vegetative tissue by diverting carbon from starch to oil biosynthesis in transgenic Arabidopsis. Plant Biotechnol. J. 9, 874–883.

Sanjaya, Miller, R., Durrett, T.P., Kosma, D.K., Lydic, T.A., Muthan, B., Koo, A.J.K., Bukhman, Y.V., Reid, G.E., Howe, G.A., Ohlrogge, J., Benning, C., 2013. Altered lipid composition and enhanced nutritional value of *Arabidopsis* leaves following introduction of an algal diacylglycerol acyltransferase 2. Plant Cell 25, 677–693.

Santos Mendoza, M., Dubreucq, B., Miquel, M., Caboche, M., Lepiniec, L., 2005. LEAFY COTYLEDON 2 activation is sufficient to trigger the accumulation of oil and seed specific mRNAs in *Arabidopsis* leaves. FEBS Lett. 579, 4666–4670.

Santos Mendoza, M., Dubreucq, B., Baud, S., Parcy, F., Caboche, M., Lepiniec, L., 2008. Deciphering gene regulatory networks that control seed development and maturation in *Arabidopsis*. Plant J. 54, 608–620.

Siloto, R.M.P., Findlay, K., Lopez-Villalobos, A., Yeung, E.C., Nykiforuk, C.L., Moloney, M.M., 2006. The accumulation of oleosins determines the size of seed oilbodies in *Arabidopsis*. Plant Cell 18, 1961–1974.

Slocombe, S.P., Cornah, J., Pinfield-Wells, H., Soady, K., Zhang, A.Y., Gilday, A., Dyer, J.M., Graham, I.A., 2009. Oil accumulation in leaves directed by modification of fatty acid breakdown and lipid synthesis pathways. Plant Biotechnol. J. 7, 694–703.

Singer, S.D., Greer, M.S., Mietkiewska, E., Pan, X., Weselake, R.J., 2013. Genetic engineering of lipid synthesis in seeds. In: Gupta, S.K. (Ed.), Biotechnology of Crucifers. Springer, New York, pp. 111–149.

Snyder, C.L., Yurchenko, O.P., Siloto, R.M.P., Chen, X., Liu, Q., Mietkiewska, E., Weselake, R.J., 2009. Acyltransferase action in the modification of seed oil biosynthesis. New Biotechnol. 26, 11–16.

Stitt, M., Zeeman, S.C., 2012. Starch turnover: pathways, regulation and role in growth. Curr. Opin. Plant Biol. 5, 282–292.

Stone, S.L., Kwong, L.W., Yee, K.M., Pelletier, J., Lepiniec, L., Fischer, R.L., Goldberg, R.B., Harada, J.J., 2001. *LEAFY COTYLEDON2* encodes a B3 domain transcription factor that induces embryo development. Proc. Natl. Acad. Sci. U.S.A. 98, 11806–11811.

Tew, T., Cobill, R., 2008. Genetic improvement of sugarcane (*Saccharum* spp.) as an energy crop. In: Vermerris, W. (Ed.), Genetic Improvement of Bioenergy Crops. Springer, Heidelberg, New York, pp. 273–294.

Thazar, N., Miquel, M., Fobis-Loisy, I., Gaude, T., 2015. Peroxisome extensions deliver the *Arabidopsis* SDP1 lipase to oil bodies. Proc. Natl. Acad. Sci. U.S.A. 112, 4158–4163.

Tjellström, H., Strawsine, M., Ohlrogge, J.B., 2015. Tracking synthesis and turnover of triacylglycerol in leaves. J. Exp. Bot. 66, 1453–1461.

Troncoso-Ponce, M.A., Cao, X., Yang, Z., Ohlrogge, J.B., 2013. Lipid turnover during senescence. Plant Sci. 205–206, 13–19.

Vanhercke, T., El Tahchy, A., Shrestha, P., Zhou, S.-R., Singh, S.P., Petrie, J.R., 2013. Synergistic effect of WRI1 and DGAT1 coexpression on triacylglycerol biosynthesis in plants. FEBS Lett. 587, 364–369.

Vanhercke, T., Petrie, J.R., Singh, S.P., 2014a. Energy densification in vegetative biomass through metabolic engineering. Biocatal. Agric. Biotechnol. 3, 75–80.

Vanhercke, T., Tahchy, A.E., Liu, Q., Zhou, X.-R., Shrestha, P., Divi, U.K., Ral, J.-P., Mansour, M.P., Nichols, P.D., James, C.N., Horn, P.J., Chapman, K.K., Beaudoin, F., Ruiz-López, N., Larkin, P.J., de Feyter, R.C., Singh, S.P., Petrie, J.R., 2014b. Metabolic engineering of biomass for high energy density: oilseed-like triacylglycerol yields from plant leaves. Plant Biotechnol. J. 12, 231–239.

van der Weijde, T., Kamei, C.L.A., Torres, A.F., Vermerris, W., Dolstra, O., Visser, R.G.F., Trindade, L.M., 2013. The potential of C4 grasses for cellulosic biofuel production. Front. Plant Sci. 4 Article 107.

Wang, G., Lin, Q., Xu, Y., 2007. *Tetraenea mongolica* Maxim can accumulate large amounts of triacylglycerol in phloem cells and xylem parenchyma of stems. Phytochemistry 68, 2112–2117.

Wang, Z., Benning, C., 2012. Chloroplast lipid synthesis and lipid trafficking through ER-plastid membrane contact sites. Biochem. Soc. Trans. 40, 457–463.

Wang, L., Shen, W., Kazachkov, M., Chen, G., Chen, Q., Carlsson, A.S., Stymne, S., Weselake, R.J., Zou, J., 2012. Metabolic interactions between the lands cycle and the Kennedy pathway of glycerolipid synthesis in *Arabidopsis* developing seeds. Plant Cell 24, 4652–4669.

Weselake, R.J., Byers, S.D., Davoren, J.M., Laroche, A., Hodges, D.M., Pomeroy, M.K., Furukawa-Stoffer, T.L., 1998. Triacylglycerol biosynthesis and gene expression in microspore-derived cell suspension cultures of oilseed rape. J. Exp. Bot. 49, 33–39.

Weselake, R.J., Taylor, D.C., Rahman, M.H., Shah, S., Laroche, A., McVetty, P.B.E., Harwood, J.L., 2009. Increasing the flow of carbon into seed oil. Biotechnol. Adv. 27, 866–878.

Winichayakul, S., Scott, R.W., Roldan, M., Hatier, J.-H.B., Livingston, S., Cookson, R., Curran, A.C., Roberts, N.J., 2013. In vivo packaging of triacylglycerols enhances Arabidopsis leaf biomass and energy density. Plant Physiol. 162, 626–639.

Xu, C., Fan, J., Cornish, A.J., Benning, C., 2008. Lipid trafficking between the endoplasmic reticulum and the plastid in *Arabidopsis* requires the extraplastidic TGD4 protein. Plant Cell 20, 2190–2204.

Xue, L., McCune, L.M., Kleppinger-Sparace, K.F., Brown, M.J., Pomeroy, M.K., Sparace, S.A., 1997. Characterization of the glycerolipid composition and biosynthetic capacity of pea root plastids. Plant Physiol. 113, 549–557.

Yang, Y., Yu, X., Song, L., An, C., 2011. AB14 activates *DGAT1* expression in Arabidopsis seedlings during nitrogen deficiency. Plant Physiol. 156, 873–883.

Youssef, A., Laizet, Y., Block, M.A., Maréchal, E., Alcaraz, J.-P., Larson, T.R., Pontier, D., Gaffé, J., Kuntz, M., 2010. Plant lipid-associated fibrillin proteins condition jasmonate production under photosynthetic stress. Plant J. 61 (3), 436–445.

Zale, J., Jung, J.H., Kim, J.Y., Pathak, B., Karan, R., Liu, H., Chen, X., Wu, H., Candreva, J., Zhai, Z., Shanklin, J., Altpeter, F., June 8, 2015. Metabolic engineering of sugarcane to accumulate energy-dense triacylglycerols in vegetative biomass. Plant Biotechnol. J. http://dx.doi.org/10.1111/pbi.12411.

Zhang, M., Fan, J., Taylor, D.C., Ohlrogge, J.B., 2009. DGAT1 and PDAT1 acyltransferases have overlapping functions in *Arabidopsis* triacylglycerol biosynthesis and are essential for normal pollen and seed development. Plant Cell 21, 3885–3901.

Index

'*Note*: Page numbers followed by "f" indicate figures and "t" indicate tables'.

B

D

E

F

G

H

I

J

N

W

X

Y

Z

Kolibri 360 GmbH
Ferdinand-Jühlke-Straße 7
99095 Erfurt, Deutschland
produktsicherheit@kolibri360.de